Advances in Fisheries Biotechnology

D1824643

Pramod Kumar Pandey · Janmejay Parhi
Editors

Advances in Fisheries Biotechnology

 Springer

Editors
Pramod Kumar Pandey
ICAR-Directorate of Coldwater
Fisheries Resources
Bhimtal, Uttarakhand, India

Janmejay Parhi
Fish Genetics and Reproduction, College
of Fisheries, CAU (I)
Lembucherra, Tripura, India

ISBN 978-981-16-3217-4   ISBN 978-981-16-3215-0   (eBook)
https://doi.org/10.1007/978-981-16-3215-0

© The Editor(s) (if applicable) and The Author(s), under exclusive license to Springer Nature Singapore Pte Ltd. 2021, corrected publication 2022

This work is subject to copyright. All rights are solely and exclusively licensed by the Publisher, whether the whole or part of the material is concerned, specifically the rights of translation, reprinting, reuse of illustrations, recitation, broadcasting, reproduction on microfilms or in any other physical way, and transmission or information storage and retrieval, electronic adaptation, computer software, or by similar or dissimilar methodology now known or hereafter developed.

The use of general descriptive names, registered names, trademarks, service marks, etc. in this publication does not imply, even in the absence of a specific statement, that such names are exempt from the relevant protective laws and regulations and therefore free for general use.

The publisher, the authors, and the editors are safe to assume that the advice and information in this book are believed to be true and accurate at the date of publication. Neither the publisher nor the authors or the editors give a warranty, expressed or implied, with respect to the material contained herein or for any errors or omissions that may have been made. The publisher remains neutral with regard to jurisdictional claims in published maps and institutional affiliations.

This Springer imprint is published by the registered company Springer Nature Singapore Pte Ltd.
The registered company address is: 152 Beach Road, #21-01/04 Gateway East, Singapore 189721, Singapore

Foreword

Biotechnology has always served as a patronage in the field of biological, agricultural, and industrial sciences. Various fields that involve biotechnological applications include therapeutics, diagnostics, bioremediation, waste management, and energy production. In the field of fisheries, biotechnology has been successful in altering the scenario by introducing advanced methods of genetic, cellular, and molecular technologies. The sector is growing at a faster pace like never before leaving behind the slower orthodox and economy-consuming conventional methods and embracing biotechnology. The fisheries sector has tremendous potential not only to raise the economical status to new heights apart from generating employment opportunities to the rapidly growing population. Globally, fish provides about 16.6% of animal protein and 6.5% of all protein for human consumption, making it very effective in addressing food and nutritional security. Contribution from aquaculture to total fish production is growing year after year from 14% (1986) to 46% (2018). Presently, out of total global fish production of 178.5 mmt, aquaculture contributes 82.1 mmt. Researchers and academicians sought out novel means to boost fish production and productivity in the culture system. In this pursuit, fisheries biotechnology played a frontier role in technologies bootstrapping for the benefits of fish farmers. Over the period of time, many biotechnological tools such as synthetic hormone for induced breeding, disease diagnostic kits, recombinant and DNA vaccines, immunostimulants enriched aqua feeds, fish cell lines, barcoding, genome sequencing, etc. have been developed.

The current book *Advances in Fisheries Biotechnology* is a treasure of information that brings together major biotechnological advances that are not only pertinent to the present day but also will enrich researchers and academicians in their future endeavor. This book consists of 29 chapters covering all the major aspects of fish biotechnology. Modern technologies like genome editing, stem cell technology, genome sequencing, nutrigenomics, transgenesis, epigenetics, nanobiotechnology, and metabolomics have been dealt with in detail for the benefit of the reader. Some chapters deal with molecular and genomic approaches to improve brood stock and molecular markers for different breeding programs. The next few chapters cover various areas related to environment management, disease diagnosis, and health improvement in aquaculture. Chapters dealing with the beneficial use of aquaculture bio-wastes and enzymes from aquatic resources for the benefit of human beings have also been covered.

It is evident from the chapters that it will be of great help not only to researchers but also to students and planners. I congratulate the editors and authors for such a useful and timely publication and wish the team a great success.

Rani Lakshmi Bai Central Agricultural University   Panjab Singh
Jhansi, Uttar Pradesh, India

Preface

The field of genetics and biotechnology has seen a sharp rise in its applications to advance aquaculture. Application of these methods and techniques along with their pros and cons should be explained to students and researchers in a simple and sophisticated manner. This book focuses on applied biotechnology in the field of aquaculture and fisheries science. Fisheries science is a vast subject that includes the study of evolution, biology and metabolism, population structure, aquatic ecology, management of diseases, bringing out the desirable traits for human benefits through fish production, conservation biology, etc. Biotechnology has been playing a very important role in almost all the mentioned aspects which has been described in this book. These advancements in the field of fish production have been elaborated in this book. During the past decades, there have been several discoveries related to biotechnological applications such as biomarkers, fish vaccines, disease diagnostic kits, stem cell techniques, and molecular markers. Latest information such as the assessment of biodiversity status and conservation strategies using molecular tools has been discussed in some chapters thoroughly. The application of modern techniques like genome editing and transgenics will provide a broader perspective to students and researchers alike in order to approach the emerging issues in the fisheries sector.

Fish nutrition is a very crucial and important subject in fisheries since feed holds the maximum investment in fish culture. Enhancement of various traits such as growth and coloration through feed supplementation has been discussed thoroughly. Special attention has been given to nutritional aspects such as nutritional biotechnology. Fish health management holds equal importance for which latest advancements such as immunoprophylactics, vaccines, and disease diagnostic tools involving omics study to analyze disease and stress-related molecular mechanisms have been described in this book.

Advanced topics like genomics, transcriptomics, proteomics, and metabolomics have been vividly written in the context of its application in fisheries. These subjects would help the researcher to understand the genetic and molecular mechanisms underlying various traits. Other revolutionary topics like NGS and epigenetics will not only enhance the knowledge of the reader but will also train them to combat the related issues in a more precise scientific way. Aquatic environment management is another emerging subject that has to be dealt intellectually. Hence, chapters such as biosafety and bioremediation using biotechnological tools have also been included in the book.

This book is a complete package of applied biotechnology and molecular biology in the context of fisheries and aquaculture science. It will surely boost the knowledge and innovativeness of students and researchers to understand the application of biotechnology in the field of fisheries and aquaculture.

The editors thank all the authors who have contributed. The technical support and secretarial assistance provided by Mr. Abhijit Bhattacharjee, Mr. Apurba Debnath, and Mr. Subrat Rudra Paul is profusely acknowledged.

Bhimtal, Uttarakhand, India

Lembucherra, Tripura, India

Pramod Kumar Pandey

Janmejay Parhi

The original version of this book has been updated: The Editor Dr. Janmejay Parhi's affiliation has been updated. A Correction to this book is available at https://doi.org/10.1007/978-981-16-3215-0_30

Contents

About the Editors

Pramod Kumar Pandey has been serving as Director, ICAR-Directorate of Coldwater Fisheries Resources, Bhimtal, Uttarakhand, India. He has worked as Dean, College of Fisheries, Central agricultural University, Imphal from 2015–2021. He has served as Scientist, Senior Scientist, and Principal Scientist in the Indian Council of Agricultural Research for 27 years before joining CAU, Imphal. He has published more than 125 peer-reviewed international and national research papers, 6 books, and has delivered several invited talks and keynote addresses in numerous national and international seminars and symposia. He has been the Team Leader of the mega research project "Centre of Excellence in Fisheries and Aquaculture Biotechnology" funded by the Department of Biotechnology, Government of India. Since 2015, he is serving as President of the North East Society for Fisheries and Aquaculture, Agartala. Before that, he has served as General Secretary of Indian Fisheries Association for 5 years based at ICAR-CIFE, Mumbai, India. He has been conferred with several fellowships, i.e., Zoological Society of India, Gaya; Society of Biological Sciences and Rural Development, Allahabad; and national awards such as Best Teacher Award by Heeralal Chaudhuri Fisheries Foundation, Mumbai; and Paryayvaran Sanrakshak Award 2020 by Bharat Tarun Sangh, Alwar.

Janmejay Parhi has completed his PG (Gold Medalist) and Ph.D. from the ICAR-Central Institute of Fisheries Education, Mumbai. Presently, he is serving as Assistant professor, Fish Genetics and Breeding Department, College of Fisheries, Central Agricultural University, Imphal. He has more than 15 years of experience in the field of fish biotechnology. He has published two books and more than 40 peer-reviewed international and national research articles. He has completed many research projects and has undergone 6 months research training at IRTA, Spain, as a part of research award by the Government of India. He has been conferred with several awards, i.e., Dr. Hiralal Choudhuri young scientist award, Dr. C. V. Kulkarni scientist award by ICAR-CIFE, Mumbai, Prof. Harswarup Gold Medal by Zoological Society of India, Gaya, and Dr. S. Ayyapan Gold Medal by IFSI, ICAR-CIFRI Barrackpore, India.

Abbreviations

μm	Micrometer
AMR	Antimicrobial resistance
BMPs	Better management practices
CAMP	Cyclic adenosine mono-phosphate
cDNA	Complementary deoxyribonucleic acid
CIBA	Central Institute of Brackishwater Aquaculture
CMV	Cytomegalovirus
CpG	Cytosine-phosphate-guanine
CRISPR	Clustered regularly interspaced short palindromic repeats
DEGs	Differently expressed genes
DMSO	Dimethyl sulfoxide
DNA	Deoxyribonucleic acid
DNMTs	DNA methyl transferases
dNTP	Deoxynucleoside triphosphate
ELISA	Enzyme-linked immunosorbent assay
ENU	N-ethyl-N-nitrosourea
ESCs	Embryonic stem cells
EST	Expressed sequence tags
FAO	Food and Agriculture Organization
FCR	Feed Conversion Ratio
FISH	Fluorescence in situ hybridization
GC	Gas chromatography
GH	Growth hormone
gRNA	Guide ribonucleic acid
GWAS	Genome-wide association studies
hmC	Hydroxyl-methyl cytosine
HPLC	High performance liquid chromatography
HRD	Human resource development
HUFA	Highly unsaturated fatty acid
ICAR	Indian Council of Agricultural Research
IGF	Insulin-like growth factor
IHC	Immunohistochemistry
IUCN	International Union for Conservation of Nature

kb	Kilobytes
kg	Kilogram
Mb	Megabytes
Mbp	Million base pairs
mC	Methylcytosines
mg	Milligram
miRNAs	Micro-ribonucleic acid
MITF	Microphthalmia-associated transcription factor
MS	Mass spectrometry
MYA	Million years ago
NBFGR	National Bureau of Fish Genetic Resources
NCBI	National Center for Biotechnology Information
ncRNA	Non-coding ribonucleic acid
NFDB	National Fisheries Development Board
nm	Nanometer
NMR	Nuclear magnetic resonance spectroscopy
NMS	Nuclear magnetic spectrometry
OIE	Office International des Epizooties
OTUs	Operational taxonomic units
PCR	Polymerase chain reaction
PIC	Polymorphism information content
PUFA	Polyunsaturated fatty acid
qPCR	Quantitative polymerase chain reaction
QTLs	Quantitative trait locus
RNA	Ribonucleic acid
siRNA	Small interfering RNA
SNPs	Single nucleotide polymorphisms
SPF	Specific pathogen free
SSRs	Simple sequence repeats
STP	Sewage treatment plant
STRs	Short tandem repeats
TLC	Thin layer chromatography
VNTRs	Variable number tandem repeats
WHA	World Health Assembly
WHO	World Health Organization
WWTP	Wastewater treatment plant

Aquaculture Productivity Enhancement Through Advanced Technologies

1

Kshitish C. Majumdar and Rajesh Ramachandran

Abstract

To cope with the growing demand of protein requirement for the world population, aquaculture should be given more attention as the growth of capture fisheries from both sea and freshwater has stagnated in the last few decades. Looking at the data collected by the FAO, it is now clear that aquaculture productivity has surpassed the capture fisheries substantially. Such development has been possible due to factors such as advancement in stock improvement, better health care, optimal food and feed development and better water quality management for species used in commercial aquaculture. Though advancement in all these areas has collectively enhanced the productivity, the increased population pressure has forced the aquaculturists to look for methods to enhance productivity further. Several technologies are now available which may be used to fulfil the requirements of increasing productivity. Transgenic (both auto-transgenic and allo-transgenic) technology has been available for several fish species from the early 1980s, where it has shown that higher growth rate, better health care and tolerance to environmental stress can be modulated for ensuring higher productivity. Problem in implementation of these technologies was the regulatory stumbling block which took more than 20 years to resolve (the first approval was received in 2019). The use of gene editing techniques, especially CRISPR, though developed recently, is increasing at a very rapid pace in many areas of aquaculture that may enhance productivity. The regulatory authorities have taken a proactive decision (at least in one case concerning a Tilapia species altered for

K. C. Majumdar
CSIR, CSIR-CCMB, Hyderabad, Telangana, India

R. Ramachandran (✉)
Indian Institute of Science Education and Research (IISER) Mohali, Mohali, Punjab, India
e-mail: rajeshra@iisermohali.ac.in

© The Author(s), under exclusive license to Springer Nature Singapore Pte Ltd. 2021

P. K. Pandey, J. Parhi (eds.), *Advances in Fisheries Biotechnology*,
https://doi.org/10.1007/978-981-16-3215-0_1

1

higher growth rate) where alteration/deletion of a few bases is not considered as a new genetically modified organism and therefore, could be cultivated at commercial scale without regulatory approvals. Micro RNA technology is also developing at a very fast pace as an advance technique of DNA sequencing and computation in fish species to test and assign the role of new miRNAs. In this chapter, all the aforementioned techniques are described briefly with a few relevant examples of their usage with respect to productivity enhancement of aquaculture species.

Keywords

Aquaculture productivity · Transgenesis · Gene editing · Micro RNA

1.1   Introduction

Aquaculture is the farming of aquatic organisms including fishes, molluscs, crustaceans and aquatic plants in freshwater or salt water. Farming implies some form of intervention in the rearing process including regular stocking, feeding and protection from predators to enhance productivity. Farming also implies individual or corporate ownership of the stock being cultivated (FAO 2020). The freshwater aquaculture operation may be categorized into extensive, semi-intensive and intensive on the inputs. In an extensive culture system, the organisms survive and grow on the food available in the environment whereas, in the case of semi-intensive culture system, additional food is supplied to enhance the growth and in the intensive culture system, the organisms survive and grow exclusively on the feed supplied from the external source. Productivity is very high in the intensive culture system and so is the expenditure. A modified type of intensive culture system is now in operation viz., the recirculating aquaculture system where all parameters of water quality, feed supply and removal of excretory substances are monitored in real time to ensure maximum productivity. This type of system is used to grow high value fishes including ornamental fishes and is the most expensive culture system.

Basic requirements of any aquaculture operation are the continuous availability of good quality water and its maintenance, seed of the species to be cultured and a balanced feed/diet for the cultivable species. Additionally, maintenance of the health of the cultivable species is equally important to obtain optimum production from any culture system. About 179 million metric tons of fish were produced in the year 2018, out of which 31 million metric ton was from sea aquaculture and the larger contribution of 51 million metric tons was from freshwater aquaculture. At present, China contributes 35% of the world fish production. Of the 179 million metric ton fish produced worldwide, 156 million metric tons are consumed at a rate of 20.5 kg per capita by humans and the remaining 22 million metric tons are used as fish meal to produce feed for fish and other animals and also for oil extraction. It is known that capture fisheries, both freshwater and seawater has stagnated since 1990, whereas aquaculture production, both fresh and sea water has steadily increased although at different rates. It is known that freshwater comprises only 0.01% of the total water body on earth where 12% of the animals live; therefore, water bodies of freshwater, although rich in diversity, find themselves as more threatened. At present, human population has crossed seven billion and any further increase in this number will add more pressure on land and freshwater resources. As mentioned earlier, the rate of consumption of fish has been increasing steadily with an increase in the number of consumers. To keep pace with the increasing demand of fish, there is a need to innovate in order to produce more fish from less water and space.

Production of fish can be increased through conventional methods of superior stock production using a combination of classical breeding and selection method, genomic ploidy manipulation, hybridization and sex reversal along with good water quality maintenance and an optimum balanced feed. In the present section, the discussion will be restricted to the use of modern technologies; some of them are yet to be used on the commercial level. These include transgenesis, gene editing and micro RNA and their role in higher productivity of aquaculture.

1.2 Transgenesis

It is the process of integration of a foreign gene into a living organism that confers upon the organism a new property (phenotype) that it will transmit unaltered to its descendants. The foreign gene integration can be of different types: wide type—where transfer may occur between two different kingdoms (e.g. bacteria to plant or animal): close transfer—where gene transfer may happen within the same kingdom (e.g. one animal species to another animal): tweaking—where gene is already present but its expression may be altered (Tester 1999). For our purpose close transfer is more important and may be subdivided as follows: heterologous—where gene transfer occurs between different phyla (e.g. fish to amphibians); homologous—where transfer occurs between two genera or species and cisgenic, where the organisms are distinct species but may interbreed in certain specialized conditions. The heterologous transfer may again be subdivided into (1) same group transfer—where transfer occurs within the genus, (2) auto-transgenic—where the transfer occurs within the same species and (3) soma transgenic—where the transgene remains and expresses exclusively in the somatic tissue(s).

Before embarking upon the production of transgenic fish, one needs to decide the purpose of such experimentation. Depending upon the aim of the experiment, whether for understanding the basic biological phenomenon or for commercial purpose, the organism and the downstream processes are chosen accordingly. Once the purpose is decided, the steps to be taken to obtain a transgenic individual are as follows: engineer the transgene construct, deliver the transgene, obtain the founder population and finally produce a transgenic homozygous stock. (For a comprehensive account on transgenesis, please see Chen et al. 1995, Devlin et al. 2009.)

1.2.1   Engineering the Transgene Construct

First, the gene to be introduced into the fish has to be identified (depending on the purpose of transgenesis); the other components required are 5′ and 3′ regulatory sequences which will regulate the function of the introduced gene. Ideally the 5′ regulatory sequence is a promoter which may also be chosen on the basis of its own functionality i.e. constitutive or inducible, which may be either strong or weak. In general, the promoter chosen is not necessarily the same promoter of the selected gene to be used in transgene construct. Selection of strong/weak or constitutive/inducible promoter depends on how the transgene function will be utilized in the transgenic organism. As many as 13 different promoters are used for transgenesis in fishes (Devlin et al. 2009). The 3′ regulatory sequence comprises the poly-adenylation signal and termination signal of the transcript. Once all the components of the transgene are selected, it needs to be isolated and cloned individually in a suitable vector so that it is able to grow in large quantities for future use. The final transgene construct is prepared in the following sequence, 5″ promoter—the gene of choice—3′ regulatory sequence and the whole construct is inserted in a suitable vector for maintenance as a stock. Once such a construct is prepared, the functionality of each component can be checked in an in vitro system using suitable cell lines.

1.2.2   Transgene Delivery

Once transgene construct is engineered it needs to be introduced into the (un)-fertilized eggs of the fish for further propagation. Several methods have been adopted for this purpose with varying success.

1.2.2.1 Microinjection

Microinjection into the (unfertilized) eggs is most popular among these methods. For microinjection, DNA is taken in a fine micro capillary tube and delivered into the nucleus immediately after fertilization. In the mouse, both pronuclei are visible after fertilization under an interference microscope. The male pronucleus is much larger, hence, easier to microinject. About 1–2 nL of DNA solution with equivalent copy number 10^6 to 10^7 is injected into each egg. In most fishes the pronuclei is not visible in the fertilized eggs. The DNA (10^6 to 10^7 copy numbers in 1–2 nL volume) is injected directly into the cytoplasm. Normally, the eggs of fishes are surrounded by chorion which gets hardened after the fertilization resulting in difficulty to micro inject. Removal of chorion either manually, by chemical means or even by drilling can be successfully done before micro injection. The fish eggs have a micropyle that is clearly visible, through which the sperm enters for fertilization. Microinjection into the cytoplasm can be done through the micropyle of the egg where nuclei after fertilization reside. One advantage of microinjection in fish is that a large number of eggs can be obtained by the induced breeding technique. The disadvantages are: (1) need trained personnel to perform microinjection at a reasonably good speed, (2) short time duration for injection—around 30 min from fertilization to the first cell division which is critical if done at the two-cell stage it increases the chances of getting a mosaic, individual, (3) rate of integration is only around 20%, (4) at times the transgene can get circularized, forming a concatamer and remain as such for a few cell divisions as a free molecule without integration and (5) occurrence of a high number of mosaic individuals.

1.2.2.2 Electroporation

This method is successfully used for gene transfer in fish eggs. Initially, it was used for transforming different types of cells under in vitro and in vivo conditions. In this method, the eggs are kept in a DNA solution and high voltage is applied for a short period of time, giving multiple pulses. During this process, a temporary pore formation occurs in the membrane which allows the DNA (and other macromolecules) to enter the cell by simple diffusion. Once it enters, the DNA finally moves into the nuclei and integrates into the genome. The advantage of this method is that a large number of eggs can be treated simultaneously and it does not require highly trained manpower. Due to the presence of a chorion, this method is not very successful in the case of fish eggs but upon removal of chorion, higher success rates have been reported in carps, catfishes and zebra fish. Transfer of DNA by this method is very successful for molluscan eggs which are small and have either a very thin chorion or no chorion at all. The method of electroporation is easy to perform and the rate of integration is around 20% or higher, which is similar or slightly higher than the rate achieved with the microinjection technique (Buono and Linser 1992, Lu et al. 1992, Powers et al. 1992,).

1.2.2.3 Sperm Mediated Gene Transfer

It is another method used successfully in transgenesis. The fish sperm cells can remain dormant in the seminal fluid or in an artificial medium for long period and can be activated when right conditions are provided without reducing the rate of fertilization of the eggs. It is also known that DNA can bind to the sperm when incubated. Such sperm cells, when used for fertilization, can result in transgenic fish. Thus, the sperm is acting as a vehicle to carry the transgene into the egg. The frequency of such transfers is unfortunately low. Interestingly, however, sperm cells when electroporated in the presence of DNA results in a dramatic increase in the frequency of transgenic individuals (Müller et al. 1992; Powers et al. 1992; Symonds et al. 1994; Tsai et al. 1995a, b). This procedure also does not require highly trained manpower and can be used in the field conditions on a mass scale. Electroporation (sperm mediated or otherwise) is, therefore, considered as an efficient and versatile gene transfer technology (Powers et al. 1992; Chen et al. 1995; Rajesh and Majumdar 2005).

1.2.2.4 Several Other Methods

Gene transfer including lipofection, biolistics (gene gun) and retroviral infection have been used in fishes with varying success rates. Spermatogonial stem cell modification and transplantation also have the potential to revolutionize the field of fish transgenesis (Tonelli et al. 2017).

1.2.3   Maintenance of Injected Eggs

After gene transfer, the eggs are allowed to grow under sterile conditions, sometimes in the presence of antibiotics to prevent bacterial infection. Generally, tropical fishes hatch within 24 h whereas temperate species take a much longer time to hatch, therefore, requiring more precaution during this period of their development. Once yolk is absorbed, the larvae are nurtured with natural/artificial feed till they are ready for further experimentation.

1.2.4   Transgene Integration

Once the transgene is transferred into the nucleus/cytoplasm of the fertilized eggs, it needs to be observed whether it has integrated into the chromosome of the recipient or not. In most of the cases, it is seen that the transgene does not get integrated immediately rather it replicates and only during the first cleavage stage it gets integrated. Most of the time, this results in G_0 transgenic individuals that are mosaic in nature i.e. all cells do not possess the transgene (Ozato et al. 1986; Dunham et al. 1987, 1992; Stuart et al. 1988; Zhang et al. 1990; Lu et al. 1992). As the rate of integration of transgene is low, various methods are adopted, including (1) a nuclear localization signal protein with transgene (2) a pseudotyped retrovirus with special integration protein or (3) use of transposases along with transgene to enhance the rate of integration. It is also seen that a low percentage of transgenic individual has the transgene integrated in multiple places. The other problem with transgenesis is that the integration happens randomly in the genome rather than at a specific site. A eukaryotic chromosome has both heterochromatic and euchromatic regions, the heterochromatic regions are found to be transcriptionally silent whereas the euchromatic regions are active. It is known that a transcriptionally active gene from the euchromatic region, when transposed to heterochromatic region, gets silenced whereas a gene from the heterochromatic region when transposed to euchromatic region becomes transcriptionally active. Similarly, the transgene gets silenced if integrated near the heterochromatic region of the chromosome.

1.2.5   Detection of Integrated Transgene

Mainly three techniques are followed: (1) dot blot hybridization, (2) PCR and (3) inverse PCR, to locate the transgene in the genome of the putative genetically engineered fish. For these methods, DNA is first isolated from the fish. In dot blot analysis, isolated DNA is denatured, immobilized on a nitrocellulose or equivalent membrane and hybridized with a portion of the transgene previously labelled with a radioactive or a fluorescent nucleotide. In PCR, primers specific for the transgene are used to amplify it, using the transgenic fish DNA as template. In the case of inverse PCR (Ochman et al. 1988), the DNA from the putative transgenic fish is digested such that the transgene remains intact as a single fragment. The fragment is circularized by ligation. Primers are then selected from the transgene such that the amplification products include the end region of the integrated site. In the first two cases, it is only possible to infer the presence of a transgene in the individuals which show positive hybridization or a specific amplification of the transgene after PCR which may present as a free circular/linear DNA moiety. To confirm the integration of the transgene, the best method would be Southern hybridization. DNA isolated from the putative transgenic fish is digested with restriction enzyme(s) which will not split the transgene. When Southern hybridization is done on such DNA with a labelled transgene probe, hybridization will always be seen at a higher molecular region than the transgene. Further, using inverse PCR, Rajesh and Majumdar (2005) have shown that the transgene is integrated in the SINE sequences of genome of transgenic *Labeo rohita*. The transgenic fish can be further analyzed by dot blot hybridization with serial dilution of the DNA to determine the transgene copy number in the genome.

1.2.6 Homozygous Transgenic Fish

When the positive transgenic fish is identified and the integration of the transgene is known, testing the performance with respect to the transgene can be done. To obtain a homozygous stock of individuals, transgenic fish is crossed with a non-transgenic fish. This is expected to produce 50% transgenic progeny in a typical Mendelian cross, if the F_0 individual is heterozygous with respect to the transgene. Unfortunately, in many cases, F_1 progeny falls short of the desired 50% heterozygote transgenic fish, indicating that the F_0 is mosaic in nature. F_1 heterozygous progenies, when crossed finally, results in a homozygous stock. This stock can be used to determine the expected performance of the transgene in a homozygous condition.

1.2.7 Applications of Transgenic Fish

In 1983, the first report on the successful production of transgenic fish called *guanil* or crown carp, a cyprinid fish, was made by a Chinese scientist (China.org.cn, Wang Yaping, a researcher at the Chinese Academy of Sciences, Institute of Hydrobiology), with an aim to enhance the growth rate, the primary aim of any commercial aquaculture operation. Growth hormone gene from a fast-growing grass eating carp was transferred into an omnivorous carp native to the Yellow river; it was shown to grow to marketable size within a year, which is half the time required to grow a non-transgenic fish. Besides this, transgenic fish are made for several applications such as, making a model system to determine biological/cellular function, for identification of regulatory elements in gene expression, to study the role of gene in development of other organisms or its role in specific organs, in drug development and testing, to human disease models or models for environmental monitoring, in modulation of phenotypes in aquarium fishes for commercial use, production of therapeutics for human use, in the development of disease resistant fishes and to study production enhancement through higher growth rate in several commercially important fishes.

1.2.7.1 Disease Resistance

The cecropins, small cationic peptides found originally in the moth *Hyalophora cecropia* (Steiner et al. 1981), show antibacterial properties against many Gram negative bacteria and is nontoxic to the eukaryotic cell (Jaynes et al. 1989; Kjuul et al. 1999). Transgenic catfish with cecropin B construct has shown significant resistance to bacterial infection (Dunham et al. 2002). Similar results were obtained in transgenic Medaka, having a cecropin transgene. When challenged separately with *Pseudomonas fluorescens* and *Vibrio anguillarium* the transgenic Medaka showed 0–10% and 10–30% mortality, respectively when compared to non-transgenic controls (Sarmasik et al. 2002).

1.2.7.2 Production of Therapeutics for Human Use

The human coagulation factor VII a vitamin K-dependent glycoprotein is used to treat patients with haemophilia A or B and severe thrombocytopenia. Transgenes, having CMV promoter and human factor VII cDNA injected into fertilized eggs of Zebra fish (*Danio rerio*), Tilapia (*Oreochromis niloticus*) and African catfish (*Clarias gariepinus*) have shown that the human factor VII could be detected in the embryo of all the three species, indicating the possibility of using fish as bioreactors (Hwang et al. 2004) similar to other bovine species.

1.2.7.3 Colour Variation in Aquarium Fish Species

Transgene constructs, having *Xenopus* elongation factor I alpha enhancer-promoter with cDNA of GFP (Green fluorescent protein), when injected into a zebra fish embryo, showed expression of GFP (without any inducer) within 4 h of initiation of development. The F_2 transgenic fish possessing this construct showed GFP expression even when only one copy of the transgene was present (Amsterdam et al. 1995). Similar results on the expression of GFP was seen in transgenic zebra fish, when promoters such as type II cytokeratin, muscle creatine kinase and acidic ribosomal phosphoprotein gene of the same fish were used separately, but in conjunction with GFP (Ju et al. 1999).

1.2.7.4 Environmental Monitoring and Biosensor

Zebra fish and Medaka are used extensively in detecting and monitoring toxic substances including teratogens in the environment that affect vertebrate development. With transgenic technology this can be done in vivo and in real time. Transgene constructs are prepared with promoters from genes such as Cytochrome P450, metallothionein, small heat shock protein hsp 11 and oestrogen receptor, individually with the GFP gene. When these transgenic fish come in contact with dioxins, heavy metals, pesticides or oestradiol/xenoestrogen, they express GFP, indicating the presence of such substances in the environment (Lee et al. 2015; Bambino and Chu 2017). Interestingly, Lee et al. (2014) have shown that transgenic zebra fish with human CHOP promoter with GFP as transgene detects heavy metals and endocrine disruptors in the environment. It is also shown that treatment with Cu^{2+} triggers GFP expression in the skin and muscle tissue, whereas Cd^{2+} induces GFP expression in the skin, olfactory epithelium and pronephric ducts. When such treated fishes are returned to normal water, the GFP expression diminishes without any morphological defects.

1.2.7.5 Growth Manipulation

Since 1983, many groups have been working on growth manipulation in several fishes including commercially important food fishes through transgenic technology. In most of the cases, growth hormone gene has been used in combination with different constitutive or inducible promoters/enhancers as a transgene. Earlier, non-fish growth hormone gene was used in the transgene construct because of non-availability of the fish growth hormone gene. Similarly, viral/bacterial promoters were used for driving the growth hormone expression. Later on, as the data on sequence of many fish species became known, all-fish constructs (both promoter and growth hormone gene derived from fish) were used to obtain transgenic fishes with higher growth rate (for details see Beardmore and Porter 2003; Devlin et al. 2009).

1.3   Clearance from Regulatory Authorities

Since mid-1980s, experiments on growth enhancement in fish through transgenesis have successfully been conducted in more than 30 species, including some commercially important fishes. Unfortunately, none of the transgenic GM stocks could be used in commercial aquaculture as regulatory authorities throughout the world refused permission due to an unknown risk to human safety upon consumption of GM food. The other risk factor under consideration was genetic contamination if these GM fish escape and breed with natural stock in the environment. One exception was mentioned by Beardmore and Porter (2003) that at the other extreme the use of "autotransgenics" must be seen as a risk which were in orders of magnitude lower than that for "allotransgenics" and probably negligible (Beardmore 1997; Nam et al. 2008).

In 2015, after more than 20 years of review, the US FDA approved a genetically engineered Atlantic salmon, "AquAdvantage" for commercial production, the first genetically engineered animal to get this approval for human consumption. AquAdvantage is a genetically modified Atlantic salmon that has a transgene constructed from a promoter of anti-freeze protein from ocean pout (*Zoarces americanus*) hooked to the growth hormone gene of Chinook salmon (*Oncorhynchus tshawytscha*), a typical transgenic and triploid female fish, generated by standard gynogenesis to make it sterile. This fish takes half the time (about 18 months) to reach marketable size with reduced FCR, indicating shortened growth time with less food consumption. In 1989, this fish was developed by AquaBounty Technologies Inc., a company located in Massachusetts, USA and is now cultivated in artificial tanks on land in Canada and Panama. Though approval to grow the fish was given by the US FDA, the US Congress later stopped commercial sale of the same. In 2016, Health Canada and Canadian Food Inspection Agency approved the sale of AquAdvantage salmon in their country. This is the first genetically engineered animal for consumption in Canada and also in the entire world. The market acceptance of this transgenic fish can be ascertained from the fact that on 4th August 2017 in an open market 4.5 tons of fish was sold at a price of $11.7/kg (Waltz 2017) and to reach this stage the company had to wait for more than 25 years. Finally in March 2019, the US FDA approved the import of AquAdvantage salmon embryo into their containment facility in Indiana for cultivation and sale in the USA as food, labelled as "Bioengineered" as mentioned by Scott Gottlieb, Commissioner of the US FDA. He also said that "the fish is safe to eat, the genetic construct added to the fish genome is safe for the animal and that the manufacturer's claims that it reaches a growth marker important to aquaculture more rapidly than its non-GM farm-raised Atlantic salmon counterpart, has been confirmed". It was also mentioned in a press release that the FDA has determined that the approval of the AquAdvantage salmon would not have any significant environmental impact because of the multiple and redundant measures being taken to contain the fish and prevent their escape and establishment in the environment (US FDA press release April 08, 2019). With this development, a big hurdle of transgenic fish cultivation for commercialization has been hopefully overcome and other countries can take this as precedence for developing their own transgenic fish for human consumption.

1.4   Gene Editing

Site-specific alterations were contemplated ever since DNA double helical structure was deciphered. Since then, our knowledge has increased substantially with respect to site-specific recognition of the DNA by several compounds, mechanism of recombination and DNA repair which eliminates or reduces lethal effects of DNA damage in different organisms. This knowledge allows us to perform modifications targeted to a specific locus in the genome to understand its function. In the early years, several methods were employed like, (1) base pair recognition by small molecules oligonucleotides, (2) triple helix formation with oligonucleotides and their chemical cleavage, (3) chemical cleavage of oligonucleotides, (4) site-specific modification by cross linkers and cleavage agents like bleomycin or psoralen, (5) recognition by peptide nucleic acid and breakage by a cleaving agent like bleomycin or (6) use of self-splicing introns through base pairing with the DNA/RNA. None of these got wide acceptability because of the low efficiency with which it could do specific targeting of DNA.

The discovery of ZFNs (zinc-finger nuclease) and TALENs (transcription activator-like effector nucleases) overcame such deficiencies. These two systems possess a sequence specific DNA-binding domain, fused to a nonspecific DNA cleavage module. The altered DNA-binding domains derived from zincfinger and transcription activator-like effector proteins provides the versatility for the system. Theses enzymes introduce targeted DNA double stranded breaks (DSBs) inducing cellular DNA repair mechanisms like error-prone non-homologous end joining (NHEJ) and homology-directed repair (HDR). Simplicity and flexibility were brought in by zinc-finger nucleases (ZFNs) and transcription activator-like effector nucleases (TALENs) to the forefront of gene modification (See review Gaj et al. 2013). The only problem with these systems that exist is that for every site to be altered one pair of enzymes had to be generated, therefore, making it a lengthy process. The discovery of RNA programmable DNA endonuclease, CRISPR/Cas9 (Clustered Regularly Interspaced Palindromic Repeats/CRISPR associated 9) revolutionized the gene editing system.

The following discussion is based broadly on the views of Cong et al. (2013), Gaj et al. (2013), Mali et al. (2013), Doudna and Charpentier (2014), Kim and Kim (2014), Barrangou and Doudna (2016), Lander (2016). Therefore, individual references are not provided. Besides, the controversies regarding the heroes of CRISPR/Cas system has now been resolved by the award of the Nobel Prize for Chemistry (2020) to Emmanuelle Charpentier and Jennifer A. Doudna.

CRISPR locus, identified in *E. coli* by Ishino et al. (1987), was later detected in many bacteria and archaea. It has multiple repeats of ~30 bp with a spacer of ~36 bp in between the repeats. The spacer sequences are unique whereas the repeats are the same and palindromic in nature. CRISPR in association with other adjacent loci including Cas9 (CRISPR/Cas9) carries out the gene editing function. This acts as an adaptive immune system (Jinek et al. 2012), protecting the microbes from the invasion of virus and plasmids through Cas9, a DNA endonuclease digesting the invading DNA, similar to the spacer sequence in the CRISPR locus (for convenience *Streptococcus pyogenes* CRISPR/Cas9 is discussed as a model). The Cas9 endonuclease has two DNA cleavage domains, HNH and RuvC responsible for double strand breaks in DNA. For specificity of recognition, DNA requires two RNAs, crRNA (CRISPR RNA) and tracrRNA (trans-activating crRNA). crRNA is transcribed from the CRISPR locus as precursor crRNA whereas the tracrRNA is transcribed from an adjacent locus. When the crRNA and tracrRNA bind to their complementary regions, it is processed by RNAse III and Cas9 to make it a shorter gRNA (guide RNA). Interestingly, the crRNA has NGG sequence at the 3' end, called protospacer adjustment motif (PAM) with a 20 bp upstream sequence. When the gRNA (crRNA + tracrRNA) and Cas9 complex binds to the complementary 20 + NGG sequence in the DNA, the Cas9 nicks at 3 bp before the PAM by HNH domain in the complementary strand, whereas the RuvC domain cleaves the opposite strand in the same position, producing blunt ends. Both domains of the Cas9 nicks at the 3' end of the DNA. CRISPR/Cas9 can be used to generate double strand DNA breaks provided the complementary sequence 20 + NGG is known. Precise breakage points may also be determined, as nicking by the Cas9 is exactly 3 bp upstream of the PAM i.e. NGG. Later, it has been shown that crRNA and tracrRNA can be linked together to form a single guide RNA (sgRNA), possessing two critical structures, one at the 5' end of the DNA target and the second being a duplex structure that binds to the Cas9 at the 3' end. Further, it has also been shown that rather than allowing transcription of the individual RNAs in vivo it is now possible to synthesize the sgRNA of ~100 bp lengths with same function asgRNA. In the late 1980s, it has been shown that DNA segments can be transferred to a desired location by homologous recombination using embryonic stem cell technology (review by Capecchi 2005). However, the frequency of such transformations is found to be very low. In the mid-1990s, results from studies in yeast and mammalian systems have shown that the frequency of homology-directed repair (HRD) and non-homologous end joining (NHEJ) could be increased several folds if the desired region of DNA has a double strand break (Haber 2000; Jasin and Rothstein 2013). The gRNA-Cas9 causes double strand breakage which triggers the DNA repair mechanisms thereby resulting in (1) complete error free joining of the strand or (2) addition or deletion of base pairs equivalent to NHEJ or (3) HDR in the presence of suitable homology containing replacement. Besides, when two gRNA-Cas9 complexes are used, large-scale deletions can be generated. If the complexes are used along with a segment of DNA, an inversion or replacement of the desired segment may be possible. The replacement of a segment of DNA (gene) is the basis of gene repair/correction/therapy which has now completely revolutionized the field of molecular genetics.

The question that now arises is whether such a tool can be used universally for eukaryotic organisms or not. In zebra fish, it has been shown that gRNA and Cas9 encoding RNA when co-injected into fertilized eggs, result in target specific mutations with a frequency that varied between 10% and 53% (Hwang et al. 2013), On the other hand, Gratacap et al. (2020) successfully used CRISPR/Cas9 for integration at two loci, EGFP and RIG-I separately in CHSE-214 cell line derived from Chinook salmon (*Oncorhynchus tshawytscha*) using lentivirus vector delivery system. The frequency of integration varied between 43% and 70% under different conditions. Zhang and his group modified the Cas9 amino acid codons with mice specific codons and showed that gene editing could be done in mice with such modified Cas9 and gRNA (Cong et al. 2013). Similarly, Church and his group used CRISPR technology in human cell lines to demonstrate proof of principle of gene editing (Mali et al. 2013). Adult mice with a mutated liver enzyme were corrected through targeted delivery by hydrodynamic tail vein injection of sgRNA and Cas9 RNA conjugated with a liver specific protein (Yin et al. 2014). Transgenic mice carrying a human DMD (Duchene muscular dystrophy) gene could be corrected by Adeno virus associated CRISPR/Cas9 gene editing system delivered by injecting via the intra-peritoneal (post natal Day 1) intra-muscular (post natal Day 12) or retro-orbital (post natal Day 18) routes (Long et al. 2015).

sgRNA-Cas9 complex can be used for various purposes and only a few are listed below which may be useful for the aquaculture industry.

1. Cas9 is a 160 kDa protein that requires both its HNH and RuvC domains for its function of double strand breakage of DNA. Even if both domains are mutated, the sgRNA-Cas9 complex can still bind and remain bound to the specific DNA without creating DNA breakage. This property can be used to determine gene function, as RNA polymerase will be stalled at that site.
2. Single strand breakage can be generated by mutating any of the domains of Cas9. If HNH domain is mutated, the non-complementary strand of the DNA will be nicked and the opposite result may be obtained if RuvC is mutated. Therefore,

single strand breakages at specific regions can be generated in a known DNA segment.

3. Visualization of a gene function is possible in vivo by the localization of fluorescently labelled sgRNA-Cas9 binding to DNA.
4. Interestingly, by using several gRNA and Cas9 simultaneously i.e. by multiplexing, it is possible to edit several genes in the same cell/organism.
5. Mutagenic chain reactions can be done to produce homozygous mutation from heterozygous conditions (Gantz and Bier 2015).
6. Gene drive in insects (Gantz et al. 2015) and mammals Grunwald et al. (2019) can be studied.

1.5 Genome Editing in Aquaculture Species

CRISPR/Cas9-based genome editing has been successfully applied in several major food fish species and in Pacific oyster for changing certain commercial characteristics viz., growth, and pigmentation, control of reproduction, sterility, immunogenicity enhancement, disease resistance and pigment formation (Gratacap et al. 2019). The species studied were Rainbow trout, Atlantic salmon, Tilapia, Channel catfish, Southern cat fish, Sea bream, Common carp, Indian and Chinese carps, Northern Chinese lamprey, Shrimp species and Oyster following the methods and studies that were initiated in the model fish (zebra fish) to establish proof of principle. Typically, mRNA encoding the Cas9 protein was injected together with the sgRNA at a single cell stage of development, to demonstrate high efficiency editing in various fish species. The Cas9 protein (instead of the mRNA) along with the gRNA was shown to be equally effective in gene editing is fishes.

1.5.1 Growth Enhancement

A mention is made of only a few genes whose functions when disrupted cause promotion of growth, which is one of the major criteria of aquaculture production. Myostatin (MSTN), also known as growth differentiation factor 8 (GDF8), is a member of the transforming growth factor-beta (TGF-beta) family that functions as a negative regulator of skeletal muscle development and thereby the overall growth of the fish (Lee 2004). Because of this role of MSTN, any natural mutations rendering it inactive and disrupting the physiological function of the protein are associated with robust muscle growth as seen in the Belgian "double bull" muscle phenotype in cattle (McPherron and Lee 1997) and in mice (McPherron et al. 1997). The conservation of the regulatory function of MSTN in muscle growth from fish to mammals (Khalil et al. 2017) is very promising to design strategies capable of inhibiting the expression of this gene for enhancing growth as shown in numerous studies on aquaculture production (Proudfoot et al. 2020).

Gao et al. (2016) used TALEN in zebra fish to generate homozygous MSTNb deficiency which resulted in 21% higher weight in comparison with the control group. Histological studies indicated increased number of muscle fibres in the MSTNb deficient fish. Wang et al. (2018) mutated both MSTNa and MSTNb by CRISPR/Cas9 in zebra fish. Mutant fish that were homozygous for MSTNb showed 30% higher body weight when compared to the control, whereas mutations in MSTNa has no effect on the body weight. However, when both genes were mutated, immunity of the fish was affected.

Common carp (*Cyprinus carpio*) is a very widely cultured aquaculture species. It has four MSTN genes as it is a tetraploid species. Of these four genes, only MSTNa is expressed in most of the tissues. sgRNA specific for MSTNa along with Cas9 mRNA when injected into the embryo, the mutation rate was found to be more than 70%. Growth with respect to body weight and length parameters was measured at 1, 2 and 3 months of the age. Both parameters showed increase at all-time points of measurement. Histological analysis indicated both hyperplasia and hypertrophy of the muscle cells which correlated with higher growth (Zhong et al. 2016).

In Channel catfish (*Ictalurus punctatus*), the exon one of the MSTN gene was targeted for CRISPR/Cas9 mutagenesis by Khalil et al. (2017). The hatchability after injection was found to be 42%, survivability at feeding stage was 90% and mutation frequency varied between 88% and 100%. The growth and the length of the fry on 40th day was 30% and 7% higher, respectively in the mutated group when compared to the control group.

Growth hormone (GH) secreted from the anterior pituitary, controls growth in vertebrates including fishes. Secretion of GH is a highly regulated process. Somatostatin (SST), a tetra-decapeptide discovered in sheep hypothalamus, was found to be a physiological inhibitor of GH secretion. Mammals have a single SST gene whereas teleost possess six SST encoding genes which exhibit differential tissue expression pattern. Zebra fish with SST4 mutants were obtained by the CRISPR/Cas9, showed its effects on growth and gonadal development. The SST4 knockout zebrafish grew significantly faster and was heavier at the onset of gonadal maturation when compared to the wild type (Sui et al. 2019).

1.5.2 Colour Modification

In aquaculture species, body colour is an important phenotype both from evolutionary and commercial points of view. For example, the Tilapia a group of commercially important fish is cultivated in many tropical countries. Normally their body colour is dark, though red/pale coloured strains (Majumdar et al. 1997) are also available which fetch a price that is at least three times more than the dark coloured one. Therefore, colour alteration in fishes is a good commercial proposition.

Jao et al. (2013) analyzed pigmentation morphology in zebra fish. They chose three genes, tyrosinase (tyr), golden (gol-the putative cation exchanger Slc24a5) and mitfa (required for neural crest-derived melanocyte formation) for manipulation. The tyr gene produces pigmentation in the skin, whereas the gol gene shows hypopigmentation in both retina and skin. CRISPR/Cas9 gene editing in tyr and gol gene showed more than 90% mutation and mitfa gene had 80% mutation which was confirmed by sequence analysis. The phenotype in both tyr and gol mutated groups showed hypopigmentation although mosaic in its distribution but the effect was retained throughout its life, indicating that it was not an epigenetic effect. No alteration was seen in pigmentation pattern in mitfa-mutated group.

Edvardsen et al. (2014) selected two pigment forming genes for mutation by CRISPR/Cas9 gene targeting method in Atlantic salmon (*Salmo salar*). These were tyrosinase (tyr) and solute carrier family 45, member 2 (slc45a2). Mutation screening with respect to skin pigmentation was done at the 17 somite stage. Embryo injected with slc45a2 and tyr showed 40% and 22% mutations, respectively. Hatched embryos without pigmentation were subjected to PCR and sequencing to confirm the presence of mutation in the respective genes, slc45a2 and tyr.

Fang et al. (2018) mutated tyrosinase gene by CRISPR/Cas9 (sgRNA-specific tyrosinase and mRNA of Cas9 were injected into fertilized eggs) method in orange red medaka (*Oryzias latipes*), an aquarium fish used as a model organism. They found more than 90% mutation in F_0 and F_1 progenies leading to an albino phenotype.

In China, the white variety of Crucian carp (*Carassius auratus cuvieri*) is a highly prized fish. Liu et al. (2019) mutated tyrosinase gene in *Carassius auratus cuvieri* and the hybrid (*Carassius auratus cuvieri* female X *Carassius auratus* male—red variety) by CRISPR/Cas9, resulting in 60–90% mutation. The phenotype varied from complete albino to mosaic pigmentation of melanin. The mutated fish had reduced tyrosinase concentration. It was also noticed that several pivotal genes in the pathway of melanin formation (e.g. *tyrp1*, *mitfa*, *mitfb*, *dct* and *sox10*) were down regulated in the mutated fishes, thus causing reduced pigment formation.

1.5.3 Control of Reproduction

Fish is the most specious group of vertebrates. They possess various forms of sexuality e.g. hermaphroditic, gonochoristic, protandrous, protogynous, alpha male or female. Similarly, the sex determination mechanism also varies considerably e.g. by genotypic (monogenic or polygenic), chromosomal, environmental conditions or by social interactions (mainly dominance). Interestingly, the sex phenotype can be modulated by environmental or chemical means (especially hormones), as the sex is not always determined at the time of fertilization as in the case of mammals, therefore, it has plasticity which can be used for beneficial purposes. Besides, many fish species show differential growth rates and sexual dimorphisms. From a commercial point of view, this can be exploited, provided sex phenotypes can be modulated by external means. A few such examples will be discussed below where gene editing technique has been successfully used to change the sex of individuals in a species.

It is known that gonadal steroids play an important role in sex determination of fish. Aromatase plays an important role in steroidogenesis by converting androgens to estrogens, which plays a critical role in the development of the ovary. By inhibiting aromatase, it was shown that the fish changed its sex from female to male. Treatment of zebra fish with an aromatase inhibitor during gonadal differenti-ation resulted in small-male population (Fenske and Segner 2004). Sex reversal of adult male zebra fish to female was also possible by treating it with an aromatase inhibitor (Takatsu et al. 2013). There are two aromatase genes (CYP19) in fish, gonad-specific cyp19a1a that expresses in the ovary and brain-specific cyp19a1b (Zhang et al. 2014). In zebrafish, Lau et al. (2016) generated two different mutants in the exon 1 of cyp19a1a by CRISPR/Cas9. When a comparison of sex ratios was made between homozygous wild, heterozygous mutant and homozygous mutants, no difference was seen in the sex ratios in the homozygous wild and the heterozy-gous mutant which was about 50% males: 50% females, whereas in both the homozygous mutants, it was all male. Histological analysis confirmed the gonadal morphology. When homozygous mutant males were treated with estradiol they reverted back to the female phenotype.

Chinese half tongue sole (*Cynoglossus semilaevis*), a marine flat fish cultured commercially in China, has a ZZ male and ZW female sex determination mecha-nism. Though this fish has a distinct sex determining chromosome, it was seen that around 14% of ZW females changed to physiological males under normal conditions, the percentage could increase up to 73% when they were grown at higher temperature like 28 °C. It is known that dmrt1 is a male determining gene, located in the Z chromosome. Cui et al. (2017) used TALEN derived gene editing technique to mutate exon I of dmrt1 gene in this fish, resulting in a mutation rate of 55%. The ZZ males turned to females and the ZW females became intersex with respect to their sex phenotype. The sex reversed females showed higher expression of female specific genes foxl2 and cyp 19a1a and a lower expression of male specific genes Sox9a and Amh. In this fish, the females were larger in size and grew faster than the males. The dmrt1 mutated sex changed females grew faster than the males and were almost equivalent to wild type females in their rate of growth. Therefore, it can be seen from these examples that sex change has an added advantage of higher growth rates in a particular sex, resulting in commercial benefits to the cultivators.

Li et al. (2014) generated mutations in *Oreochromis niloticus* by CRISPR/Cas9 in nanos2 (38%), nanos3 (49%), foxl2 (42%) and dmrt1 (22%) genes. The mutated G_0 population of foxl2 and dmrt1 genes transmit the mutations to the F_1 generation. Nanos3 mutation in the XX female Tilapia shows germ cell deficiency and sex reversal. These individuals do not express ovary specific cyp19a1a gene. Nanos2 mutation in XY male shows germ cell deficiency in the testis but does not indicate sex reversal. Dmrt1 mutation causes male to female sex reversal and induces aromatase expression. Similarly, foxl2 mutation induces reduction of aromatase and causes sex reversal in XX females, similar to the mutation generated in the same gene using TALEN by Li et al. (2013).

1.5.4 CRISPR/Cas9 in Shrimp and Oyster

In shrimp species *Exopalaemon carinicauda*, Zhang et al. (2018) isolated moult inhibiting hormone (MIH) possessing three exons and two introns, which is a negative regulator for suppressing the moulting process, thus reducing the growth rate. This gene is expressed mainly in the eyestalks. Mutation at the exon two in the MIH gene is generated by CRISPR/Cas9 with a frequency of 4.8%. The mutated larvae show body length increase and shortening of metamorphosis time from mysis to post larva stage, without any ill effects on health or death rate, indicating the possibility of using this technique in the breeding of this shrimp. In the case of a bivalve mollusc—the Pacific oyster, *Crassostrea gigas* mutations were generated successfully by CRISPR/Cas9 in two genes, myostatin and twist (Yu et al. 2019). This shows the versatility of the CRISPR/Cas9 gene editing technology.

1.5.5 Disease Resistance

Around 40% loss occurs in aquaculture production because of infectious diseases. Disease resistance phenotypes are generally polygeneic in nature and do not show high heritability. To overcome these obstacles, Gratacap et al. (2019) provided a detailed flow chart to combine gene editing and classical selective breeding to obtain disease resistant aquatic species of commercial importance.

1.5.6 Clearance from Regulatory Authority

The time taken to obtain regulatory clearance for transgenic fish is rather long. Fortunately, gene edited plants and fish have been given regulatory clearances much sooner than is usually the case, as these are classified as non-GMO. It is clear that *in-del* mutation, generated by CRISPR/Cas9 in an individual, will be considered as non-GMO. It will be interesting to see whether the replacement of a gene or part of it, obtained through CRISPR technology, may follow the same path or not.

Button mushroom (*Agaricus bisporus*) after harvest gets discoloured from white to brown within a few days, thus, losing market value. This browning is caused by a group of enzymes called polyphenol oxidases (PPO). Yinong Yang, a plant pathologist of Pennsylvania state university, USA mutated one of the six genes of PPO by CRISPR/Cas9 gene editing tool. This reduced the activity of the PPO enzyme by 30% preventing browning and thereby, increasing the shelf life of the mushroom. Waltz (2016) reported in "Nature" that "the USDA will not regulate a mushroom that has been genetically modified with gene editing tool CRISPR/Cas9 and it can be cultivated and sold without passing through the agency's regulatory process-making it the first CRISPR edited organism to receive a green light from the US government". It was also mentioned that Yang's mushroom did not trigger USDA oversight because it did not contain foreign DNA such as viruses or bacteria (Waltz 2016), therefore, the crop does not pose any health risk. In response to this development Dr Joyce Van Eck, Assistant Professor at Boyce Thompson Institute mentioned if USDA decided that this did not require regulation, it would definitely encourage many people already using CRISPR technology.

In December 2018, in a press release from two companies, Intrexon and Aquabounty; it was reported that the Tilapia strain FLT 01 developed by them by a gene editing tool would not be considered as GMO by National Advisory Commission on Agricultural Biotechnology, Argentina (CONABIA). This was the first gene edited animal to get a regulatory clearance after it was designated as non-GMO. The researcher claimed that the fish has 70% more fillets and showed a 16% higher growth rate with a 14% increase in food conversion ratio. Dr Martin Lema (Chairman of CONABIA, Biotechnology Directorate, State Secretariat of Foodstuff and Bioeconomy in Argentina) mentioned in a paper presented in the symposium at the International Society for Biosafety Research held in Terragona, Spain in 2019, that Gene edited Tilapia line FLT 01 trait, improved fillet yield through loss of function mutation; microinjection with nuclease mRNA (no RNA involved); small deletion created an early stop codon; homologous at final product; cero off target site by design; seven low probability off target sites were verified to be non-modified by sequencing; and no risk hypothesis reported.

However, with regard to gene editing in the case of humans, several unanswered questions remain. The organizing committee on First International Summit on Human Gene Editing mentioned that CRISPR and other gene editing techniques were to be used for somatic cell based therapies that were "intended to affect only the individuals who receive them". It would be irresponsible to proceed with any clinical use of germ line editing unless and until (1) the relevant safety and efficacy issues have been resolved, based on appropriate understanding and balancing of risks, potential benefits and alternatives, and (2) there is broad societal consensus about appropriateness of the proposed application (Thrasher et al. 2016).

1.6   Micro RNA

The discussion below on micro RNA is based on the following articles and therefore, individual citations have not been made in several places in the text: Wahid et al. (2010), Bizuayehu and Babiak (2014), Rasal et al. (2016), Schier and Giraldez (2016), Andreassen and Hoyheim (2017), Herkenhoff et al. (2018).

In 1993, a small RNA lin4, isolated from *C. elegans* (Lee et al. 1993) was shown to regulate the function of lin14 gene (Wightman et al. 1993). This initiated the isolation and characterization of a group of small RNAs, called microRNA (miRNA) from various species of plants, animals and viruses whose numbers are increasing exponentially. The database called miRBASE (Released 22 December 2018) has data from 271 species, including nine teleost fish with 38,589 entries of hairpin precursor miRNAs which may give 48,885 mature miRNA. miRNAs are shown to function in the developmental processes, neuronal pathway formation, cell proliferation, cell death, muscle formation and growth, regulation of metabolic pathways, haematopoietic differentiation, immunity, reproduction and response to environmental stimuli in an organism. Thousands of miRNAs have been isolated and characterized and they have shown to be conserved between species. Sometimes the genes for the miRNAs may occur in clusters and such clusters also show conservation in different vertebrates. For example, 26 miRNA gene clusters are conserved in Atlantic salmon, Atlantic Cod, Zebra fish and humans.

Among the teleost fishes, miRNA was first discovered in zebra fish (Lim et al. 2003). The miRNA is about 21–25 nucleotides long single stranded RNA, produced from a precursor RNA. The transcribed products of these miRNA genes have the capacity to form a hairpin like structure. Initially, miRNA genes were identified in the inter-generic regions but later it was also found in the intronic regions of the gene. Generally, these genes are clustered in a region of the genome that is transcribed as a polycistronic message by the same promoter; however, single miRNA genes have also been located in areas that give monocistronic messages. Depending on their location in the genome, miRNAs are classified into the following categories: (a) intronic miRNA in coding transcription unit (TU) e.g. miRNA 101-2 cluster which is found in the intron of a coding RNA gene HSPC338 (b) intronic miRNA in noncoding TU e.g. miRNA 135a-2 cluster in the gene NCRMS (c) exonic miRNA in coding TU e.g. miRNA985 in the gene CACNG8 (d) exonic miRNA in noncoding TU e.g. miRMA 206 in the gene 7H4. Most of the miRNA are transcribed by RNA pol II with the exception of a small group in Alu repeats transcribed by RNA pol III in the nucleus. The transcripts, produced by both the polymerases, may go up to several Kb in length and are called primary miRNA (Pri-miRNA), possessing a stem-loop-stem structure with the following three defining features: terminal loop, internal bulges and a double stranded stem. The pri-miRNA is cleaved by Drosha in the presence of cofactor DiGeorge Syndrome Critical region gene 8 (DGCR8) in human and Pasha in Drosophila and worm, to get a strand of about 60–70 nucleotides long comprising of about 33 base pairs of the stem loop structure, a terminal loop and single strand RNA as a precursor miRNA (Pre-miRNA) with a 2–3 nucleotide overhang in the $3'$ region. Interestingly, the Drosha marks the $5'$ region of the mature miRNA. The pre-miRNA is transported from the nucleus to the cytoplasm through the nuclear pore by RanGTP dependent exportin-5 (EXP5), a nuclear transport receptor. In the cytoplasm, RNase III and Dicer a conserved enzyme cleaves the Pre-miRNA near the terminal loop of the Drosha cutting site giving rise to ~22 nucleotides long double stranded miRNA. This results in two mature miRNA molecules, one from the $5'$ and the other from the $3'$ end of the pre-miRNA. Dicer along with Argonaute and other proteins viz TRBP and PACT are responsible for miRNA processing and finally loading into the RNA induced silencing complex (RISC) by forming miRNA-RISC. A helicase yet to be discovered will be found to be responsible for the separation of the duplex RNA. One of the strands, the mature miRNA strand remains with Argonaute whereas the other strand, the passenger miRNA is degraded. It is also shown that endonucleolytic enzyme activity of Argonaute is responsible for the removal of the passenger miRNA. There are exceptions where both strands are found to be functional. Of the two strands, which one will be functional is to a certain extant determined by the thermodynamic stability of the strand. The miRNA function can only be initiated once the miRNA-RISC is formed which either (1) inhibits the expression or (2) causes degradation of the mRNA transcript in the cytoplasm. To perform these functions, the miRNA-RISC may bind in the $3'$UTR of the mRNA where in most of the cases the miRNA recognition elements (MRE) are predominantly located. It is also shown that it may bind to the exonic region or the 5'UTR of the mRNA. The mRNA regulation depends on the amount of complementarity of the miRNA seed region, a domain spanning nucleotide positions 2–7 at the 5' of the miRNA and the MRE of the mRNA. Therefore, a small number of miRNA can regulate many mRNAs. Based on this property, it is calculated that 60% of the protein coding genes in mammals may be regulated by miRNAs and its final role may be gene function regulation. This has both an advantage and a disadvantage while determining the assignment of a specific miRNA. The easy sequencing strategies available at present for RNA and DNA help in identifying a large number of miRNA in a short span of time. Once identified, one needs to find out the site where these will bind on the mRNAs. Identification of such sites is possible using bioinformatics tools. The whole process depends on the availability of genome sequence and transcriptome data of a particular species. As we are aware that more than 25,000 species of fish are known whereas genome sequence data and transcriptome data are not available, even for all important food fish species, which is one of the drawbacks for studies on miRNA biology in fishes. As mentioned earlier, single miRNA may have complementary sequences in several places in a single mRNA. It is also seen that the same miRNA may have complementary sequences in different mRNAs, originating from different genes/species. To find out the correct miRNA and its true binding site, binding analysis and inhibition of the gene as the regulatory function of a particular miRNA need to be done in an in vitro analysis which is a time consuming exercise.

Only a few examples have been chosen in this chapter where the role of miRNA is evident on a few genes affecting particular phenotypes, although a huge amount of information available in literature.

1.6.1 miRNA Usage, Colour Modulation

Experimental data from mice and Drosophila established the role of miRNA in body colour pattern. Common carp (*Cyprinus carpio*) has several body colours viz, red, white, orange and black which follows Mendelian inheritance. Similarly, Nile tilapia (*Oreochromis niloticus*) also has red and white body colours. It is shown that four miRNAs (miRNA 25, miRNA 137, miRNA 203a and miRNA 429) are strongly expressed in the skin of both the fishes. When miRNA 429 antagomir was injected to red coloured live fishes (5 g body weight) each day for 30 days the miR 429 got silenced and a reduction of the melanin content was seen in the skin. In silico analysis showed that 3'UTR region of Foxd3 has a potential target site for miRNA 429. In vitro study showed that Foxd3 gene activity was reduced in HEK2BT cells when transfected with miRNA 429. In vivo study with miRNA 429 antagomir injected into fishes showed an increase of Foxd3 activity. These studies indicated that Foxd3 gene was regulated by the interaction of miRNA 429 and the 3'UTR of Foxd3 mRNA. It is known that melanocyte inducing transcription factor (MITF) gene regulates TYR, TRP1 and TRP2 genes which are involved in melanin production. Foxd3 product binds to the promoter of MITF gene and represses its function.

Down regulation of miRNA 429 increases Foxd3 gene expression and will down regulate MITF which in turn will down regulate the downstream genes, thereby, regulating melanin formation in the skin. Hence, the miRNA 429 was shown to regulate melanin formation in two different species of the fish (Yan et al. 2013a).

1.6.2 Growth Studies

MyoD, a member of the helix loop helix transcription factors, is known to play a great role in controlling myogenic regulatory pathway. Yan et al. (2013b) isolated several miRNAs (miRNA 23b, miRNA190, miRNA190b, miRNA 203b) from the skeletal muscle of Nile tilapia (*Oreochromis niloticus*). MyoD is highly expressed in juvenile fish, it is hardly detectable in adult fish but expressed again moderately during the senile phase of the fish. However, miRNA 203b expression is lowest in the juvenile fish, highest in the adult fish and moderate in the senile fish. When fishes were treated with miRNA 203b antagomir, it significantly reduced miRNA 203b expression. Besides, the miRNA 203b has a potential target in the 3′UTR of MyoD. These data indicate that miRNA 203b regulates negatively the function of MyoD gene, thereby, the myogenic pathway in this fish (Yan et al. 2013b).

Myostatin (mstn) gene negatively regulates the development of muscle in most vertebrates including fish. In Nile Tilapia (*Oreochromis niloticus*) two variations of mstn are found viz. mstna and mstnb. The mstnb gene is expressed in different tissues including the muscle, whereas mstna is expressed only in the brain. It is shown that miRNA-181b-5p is expressed in different tissues including white muscle. It binds to the 3′UTR of the mstnb, specifically in two regions, 307–313 bp and 1022–1029 bp. Overexpression of miRNA 181b-5p decreases MSTNb protein, whereas knockdown of the same causes upregulation of the protein. In conclusion, it was mentioned by Zhao et al. (2019) that miRNA-181b-5p decreased MSTN b protein production, thereby promoting muscle growth in Tilapia.

1.6.3 Cold Adaptation

Litopenaeus vannamei, a tropical shrimp, when grown in water at lower temperatures, parameters like survivability and growth are affected. He et al. (2018) analyzed miRNAs related to cold adaptation in this species. In the cold treated group, 34 miRNAs (21 up regulated and 13 down regulated) were differentially expressed in comparison to the shrimps grown at a warmer temperature (28 °C). In another group, where the shrimps were kept first at 16 °C (cold adapted) for 6 days and then adapted back to 28 °C showed 31 miRNAs (16 up regulated and 15 down regulated) differentially expressed, when compared to the shrimp grown at 28 °C. Validation of five miRNAs (three known miRNAs, mja-miR-6491, mja-miR-6494, bta-miR-2478 and two newly identified, novel-68 and novel-5) was done that were differentially expressed significantly in the hepatopancreas and muscle tissues by RT-qPCR. Of these three miRNAs, mja-miR6491, mja-miR-6494 and novel-5 were upregulated in both hepatopancreas and muscle, whereas bta-miR-2478 and novel-68 were expressed more in the muscle in the cold treated group. Pathway analysis of these miRNAs showed that target genes were mostly related to fatty acid metabolism which was expected in the cold adapted individual for regulating membrane fluidity.

To identify miRNAs involved in cold adaptation, Ji et al. (2020) grew zebra fish embryonic cells (ZF4) at 18 °C for 30 days. Cells grown in cold had 24 up regulated (19 known and five novel) and 23 down regulated (nine known and 14 novel) miRNAs in comparison to the cells grown at 28 °C. It was noted that the selected four up regulated miRNAs (dre-miR-16b, dre-miR-729, dre-miR-338-5p, dre-miR-193b-5p) and four down regulated miRNAs (dre-miR-100-3p, dre-miR-100-5p, dre-miR-2184, dre-miR-19a-3p), reverted back to their normal expression when cells were shifted back to 28 °C culture condition, indicating the adaptive nature of these miRNAs. miR-100-3p down regulated genes atad5a, cyp2ae1 and lamp1, whereas miR-16b targets genes rilp, atxn7, tnika and btbd9 by reducing their function, indicating their role in cold adaptation (Ji et al. 2020).

1.6.4 Control of Reproduction

In many vertebrate species, miR-202 is expressed predominantly in the gonads of both the sexes. Gay et al. (2018) have shown in medaka that mature miR-202-5p is present in large quantities in granulosa cells of the ovary and in unfertilized eggs. When they knocked out the miR-202 gene by CRISPR-Cas9 method to determine its role in gonadal function, the mutated females showed no egg production or produced dramatically reduced number of eggs that could not be fertilized, thereby, causing sterility in the female medaka.

1.6.5 Clearance from Regulatory Authority

Research activities in miRNA are moving at a very fast pace, especially, in the areas of diagnostics and therapeutics in humans. The first miRNA-based therapeutic drug was discovered by Lindow and Kauppinen (2012); a locked nucleic acid (LNA) modified antisense oligonucleotide "Miravirsen" which targets miRNA 122 expressed in the liver. Clinical trials are currently going on for the treatment of hepatitis C virus infection by Miravirsen in humans (Chakraborty et al. 2017). In 2015, Inter space Biosciences Inc, USA launched for the first time, a microRNA based gene expression classifier, "Thyromir" to diagnose benign and malignant thyroid cancer in humans. Unfortunately, as mentioned in Bonneau et al. (2019) "all the miRNA based drugs are currently in clinical trials and none have yet reached the regulatory approval stage for pharmaceutical breakthrough". In such a scenario, regulatory approval of diagnostic and therapeutic usage of miRNA-based technology in aquaculture may not happen in the near future.

1.7 Conclusion

It is abundantly clear that aquaculture productivity has increased several folds in the last few decades due to advancement in classical technologies like improved water quality management, food/feed development and better stock development through selective breeding. However, the application of modern molecular techniques like transgenesis, gene targeting and microRNA can revolutionize productivity in intensive aquaculture (both fresh and salt water) by reducing the time period for growth, improving feed intake, lowering the feed conversion ratio and enhancement of stock by directly modifying the gene(s) responsible for growth, disease resistance or control of reproduction. All these technologies will reduce the production cost substantially and thereby provide a cheaper protein even for the economically weaker sections of the population. It is also seen that important pharmaceutical products can be made using aquatic organisms as bioreactors. Considering all these potentially beneficial aspects, it is desirable for the aqua culturists to adopt newer technologies that are being developed by the scientists and technologists to enhance aquaculture productivity and production.

Acknowledgement We are indebted to Dr Uma Devi Komath for correcting the manuscript.

References

Amsterdam AS, Lin S, Hopkins N (1995) The *Aequorea victoria* green fluorescent protein can be used as a reporter in live zebra fish embryos. Dev Biol 171:123–129

Andreassen R, Hoyheim B (2017) miRNA associated with immune response in teleost fish. Dev Comp Immunol 75:77–85

Bambino K, Chu J (2017) Zebra fish in toxicology and environmental health. Curr Top Dev Biol 124:331–367. https://doi.org/10.1016/bs.ctdb.2016.10.007

Barrangou R, Doudna JA (2016) Applications of CRISPR technologies in research and beyond. Nat Biotechnol 34:933–941. https://doi.org/10.1038/nbt.3659

Beardmore JA (1997) Transgenics: autotransgenics and allotransgenics. Transgenic Res 6:107–108

Beardmore JA, Porter JS (2003) Genetically modified organisms and aquaculture. In: FAO fisheries circular No. 989. FAO, Rome. 38 p

Bizuayehu TT, Babiak I (2014) MicroRNA in teleost fish. Genome Biol Evol 6:1911–1937

Bonneau E, Neveu B, Kostantin E, Tsongalis GJ, De Guire V (2019) How close are miRNAs from clinical practice? A perspective on the diagnostic and therapeutic market. JIFCC 30:114–127

Buono RJ, Linser PJ (1992) Transient expression of RSVCAT in transgenic zebrafish made by electroporation. Mol Mar Biol Biotechnol 1:271–275. PMID: 1339227

Capecchi MR (2005) Gene targeting in mice: functional analysis of the mammalian genome for the twenty-first century. Nat Rev Genet 6:507–512

Chakraborty C, Sharma AR, Sharma G, Doss CGP, Lee S-S (2017) Therapeutic miRNA and siRNA: moving from bench to clinic as next generation medicine. Mol Ther Nucl Acids 8:132–143

Chen TT, Lu JK, Shamblott MJ, Chench CM, Lin CM, Burns JC, Reimschuessel R, Chatakondi N, Dunham RA (1995) Transgenic fish: ideal models for basic research and biotechnological applications. Zool Stud 34:215–234

Cong L, Ran FA, Cox D, Lin S, Barretto R, Habib N, Hsu PD, Wu X, Jiang W, Marraffini LA, Zhang F (2013) Multiplex genome engineering using CRISPR/Cas systems. Science 339:819–823

Cui Z, Liu Y, Wang W, Wang Q, Zhang N, Lin F, Wang N, Shao C, Dong Z, Li Y, Yang Y, Hu M, Li H, Gao F, Wei Z, Meng L, Liu Y, Wei M, Zhu Y, Guo H, Cheng CHK, Schartl M, Chen S (2017) Genome editing reveals *dmrt1* as an essential male sex-determining gene in Chinese tongue sole (*Cynoglossus semilaevis*). Sci Rep 7:42213. https://doi.org/10.1038/srep42213

Devlin RH, Raven PA, Sundstrdnt LF, Uh M (2009) Issues and methodology for development of transgenic fish for aquaculture with a focus on growth enhancement. In: Overturf K (ed) Molecular research in aquaculture. Wiley-Blackwell, London, pp 217–260

Doudna JA, Charpentier E (2014) Genome editing. The new frontier of genome engineering with CRISPR–Cas9. Science 346:1258096

Dunham RA, Eash J, Askins J, Townes TM (1987) Transfer of the metallothionein–human growth hormone fusion gene into channel catfish. Trans Am Fish Soc 116:87–91

Dunham RA, Ramboux AC, Duncan PL, Hayat M, Chen TT, Lin CM, Kight K, Gonzalez-Villasenor I, Powers DA (1992) Transfer, expression, and inheritance of salmonid growth hormone genes in channel catfish, *Ictalurus punctatus*, and effects on performance traits. Mol Mar Biol Biotechnol 1:380–389

Dunham RA, Warr G, Nichols A, Duncan PL, Argue B, Middleton D, Liu Z (2002) Enhanced bacterial disease resistance of transgenic channel catfish, *Ictalurus punctatus*, possessing cecropin genes. Mar Biotechnol 4:338–344

Edvardsen RB, Leininger S, Kleppe L, Skaftnesmo KO, Wargelius A (2014) Targeted mutagenesis in Atlantic salmon (*Salmo salar* L.) using the CRISPR/Cas9 system induces complete knockout individuals in the F0 generation. PLoS One 9:e108622. https://doi.org/10.1371/journal.pone.0108622

Fang J, Chen T, Pan Q, Wang Q (2018) Generation of albino medaka (*Oryzias latipes*) by CRISPR/Cas9. J Expt Zool B Mol Dev Evol 330:242–246

FAO (2020) The State of World Fisheries and Aquaculture 2020. Sustainability in action. FAO, Rome. https://doi.org/10.4060/ca9229en

Fenske M, Segner H (2004) Aromatase modulation alters gonadal differentiation in developing zebrafish (*Danio rerio*). Aquat Toxicol 67:105–126

Gaj T, Gersbach CA, Barbas CF (2013) ZFN, TALEN and CRISPR/Cas-based methods for genome engineering. Trends Biotechnol 31:397–405. https://doi.org/10.1016/j.tibtech.2013.04.004

Gantz VM, Bier E (2015) The mutagenic chain reaction: a method for converting heterozygous to homozygous mutations. Science 348:442–444

Gantz VM, Jasinskieneb N, Tatarenkovab O, Fazekasb A, Maciasb VM, Biera E, Jamesb AA (2015) Highly efficient Cas9-mediated gene drive for population modification of the malaria vector mosquito *Anopheles stephensi*. Proc Natl Acad Sci U S A 112:E6736–E6743

Gao Y, Dai Z, Shi C, Zhai G, Jin X, He J, Lou Q, Yin Z (2016) Depletion of myostatin b promotes somatic growth and lipid metabolism in zebrafish. Front Endocrinol 7:88., 1–10. https://doi.org/10.3389/fendo.2016.00088

Gay S, Bugeon J, Bouchareb A, Henry L, Delahaye C, Legeai F, Montfort J, Le Cam A, Siegel A, Bobe J, Thermes V (2018) MiR-202 controls female fecundity by regulating medaka oogenesis. PLoS Genet 14:e1007593. https://doi.org/10.1371/journal.pgen.1007593

Gratacap RL, Wargelius A, Edvardson RB, Houston RD (2019) Potential of genome editing to improve aquaculture breeding and production. Trends Genet 35:672–684

Gratacap RL, Regan T, Dehler CE, Martin SAM, Boudinot P, Collet B, Ross D, Houston RD (2020) Efficient CRISPR/Cas9 genome editing in a salmonid fish cell line using a lentivirus delivery system. BMC Biotechnol 20:35–44

Grunwald HA, Gantz VM, Poplawski G, Xu X-r S, Bier E, Cooper KL (2019) Super-Mendelian inheritance mediated by CRISPR/Cas9 in the female mouse germline. Nature 566:105–109

Haber JE (2000) Lucky breaks: analysis of recombination in *Saccharomyces*. Mutat Res 451:53–69

He P, Wei P, Zhang B, Zhao Y, Li Q, Chen X, Zeng D, Peng M, Yang C, Peng J, Chen X (2018) Identification of microRNAs involved in cold adaptation of *Litopenaeus vannamei* by high-throughput sequencing. Gene 677:24–31

Herkenhoff ME, Oliveira AC, Nachtigall PG, Costa JM, Campos VF, Hilsdorf AWS, Pinhal D (2018) Fishing into the microRNA transcriptome. Front Genet 9:1–15

Hwang G, Müller F, Rahman MA, Williams DW, Murdock PJ, Pasi KJ, Goldspink G, Farahmand H, Maclean N (2004) Fish as bioreactors: transgene expression of human coagulation factor VII in fish embryos. Mar Biotechnol 6:485–492. https://doi.org/10.1007/s10126-004-3121-2

Hwang WY, Fu Y, Reyon D, Maeder ML, Tsai SQ, Sander JD, Peterson RT, Yeh JR, Joung JK (2013) Efficient genome editing in zebrafish using a CRISPR-Cas system. Nat Biotechnol 31:227–229

Ishino Y, Shinagawa H, Makino K, Amemura M, Nakata A (1987) Nucleotide sequence of the iap gene, responsible for alkaline phosphatase isozyme conversion in *Escherichia coli*, and identification of the gene product. J Bacteriol 169:5429–5433

Jao L-E, Wente SR, Chen W (2013) Efficient multiplex biallelic zebrafish genome editing using a CRISPR nuclease system. Proc Natl Acad Sci U S A 34:13904–13909

Jasin M, Rothstein R (2013) Repair of strand breaks by homologous recombination. Cold Spring Harb Perspect Biol 5:a012740

Jaynes JM, Julian GR, Jeffers GW, White KL, Enright FM (1989) *In vitro* cytocidal effect of lytic peptides on several transformed mammalian cell lines. Pept Res 2:157–160

Ji X, Jiang P, Luo J, Li M, Bai Y, Zhang J, Han B (2020) Identification and characterization of miRNAs involved in cold acclimation of zebra fish ZF4 cells. PLoS One 15:e0226905. https://doi.org/10.1371/journal.pone.0226905

Jinek M, Chylinski K, Fonfara I, Hauer M, Doudna JA, Charpentier E (2012) A programmable dual-RNA-guided DNA endonuclease in adaptive bacterial immunity. Science 337:816–821

Ju B, Xu Y, He J, Liao J, Yan T, Hew CL, Lam TJ, Gong Z (1999) Faithful expression of green fluorescent protein (GFP) in transgenic zebra fish embryos under control of zebra fish gene promoters. Dev Genet 25:158–167. https://doi.org/10.1002/(SICI)1520-6408(1999)25:2<158::AID-DVG10>3.0.CO;2-6

Khalil K, Elayat M, Khalifa E, Daghash S, Elaswad A, Miller M, Abdelrahman H, Ye Z, Odin R, Drescher D, Vo K, Gosh K, Bugg W, Robinson D, Dunham R (2017) Generation of *Myostatin* gene-edited channel catfish (*Ictalurus punctatus*) via zygote injection of CRISPR/Cas9 System. Sci Rep 7:7301. https://doi.org/10.1038/s41598-017-07223-7

Kim H, Kim JS (2014) A guide to genome engineering with programmable nucleases. Nat Rev Genet 15:321–334

Kjuul AK, Bullesbach EE, Espelid S, Dunham RA, Jorgensen TO, Warr GW, Styrvold OB (1999) Effects of cecropin peptides on bacteria pathogenic to fish. J Fish Dis 22:387–394

Lander ES (2016) The heroes of CRISPR. Cell 164:18–28

Lau ES-W, Zhang Z, Qin M, Ge W (2016) Knockout of zebra fish ovarian aromatase gene (*cyp19a1a*) by TALEN and CRISPR/Cas9 leads to all-male offspring due to failed ovarian differentiation. Sci Rep 6:37357. https://doi.org/10.1038/srep37357

Lee S-J (2004) Regulation of muscle mass by myostatin. Annu Rev Cell Dev Biol 20:61–86

Lee RC, Feinbaum RL, Ambros V (1993) The *C. elegans* heterochronic gene lin-4encodes small RNAs with antisense complementarity to lin-14. Cell 75:843–854

Lee H-C, Lu P-N, Huang H-L, Chu C, Li H-P, Tsai H-J (2014) Zebrafish transgenic line *huORFZ* is an effective living bioindicator for detecting environmental toxicants. PLoS One 9:e90160. https://doi.org/10.1371/journal.pone.0090160

Lee O, Green JM, Tyler CR (2015) Transgenic fish systems and their application in ecotoxicology. Crit Rev Toxicol 45:124–141. PubMed: 25394772

Li MH, Yang HH, Li MR, Sun YL, Jiang XL, Xie Q-P, Wang T-R, Shi H-J, Sun L-N, Zhou L-Y, Wang D-S (2013) Antagonistic roles of Dmrt1 and Foxl2 in sex differentiation via estrogen production in tilapia as demonstrated by TALENS. Endocrinology 154:4814–4825

Li M, Yang H, Zhao J, Fang L, Shi H, Li M, Sun Y, Zhang X, Jiang D, Zhou L, Wang D (2014) Efficient and heritable gene targeting in tilapia by CRISPR/Cas9. Genetics 197:591–599

Lim LP, Glasner ME, Yekta S, Burge CB, Bartel DP (2003) Vertebrate microRNA genes. Science 299:1540

Lindow M, Kauppinen S (2012) Discovering the first microRNA-targeted drug. J Cell Biol 199:407–412. https://doi.org/10.1083/jcb.201208082

Liu Q, Qi Y, Liang Q, Song Q, Liu J, Li W, Shu Y, Tau M, Zhang C, Qin Q, Wang J, Liu S (2019) Targeted disruption of tyrosinase causes melanin reduction in *Carassius auratus cuvieri* and its hybrid progeny. Sci China Life Sci 62:1194–1202. https://doi.org/10.1007/s11427-018-9404-7

Long C, Amoasii L, Mireault AA, McAnally JR, Li H, Sanchez-Ortiz E, Bhattacharyya S, Shelton JM, Bassel-Duby R, Olson EN (2015) Postnatal genome editing partially restores dystrophin expression in a mouse model of muscular dystrophy. Science 351:400–403

Lu JK, Chen TT, Chrisman CL, Andrisani OM, Dixon JE (1992) Integration, expression and germline transmission of foreign growth hormone genes in medaka (*Oryzias latipes*). Mol Mar Biol Biotechnol 1:366–375. PMID: 1285009

Majumdar KC, Nasaruddin K, Ravinder K (1997) Pink body colour in Tilapia shows single gene inheritance. Aquac Res 28:581–589

Mali P, Yang L, Esvelt KM, Aach J, Guell M, DiCarlo JE, Norville JE, Church GM (2013) RNA-guided human genome engineering via Cas9. Science 339:823–826

McPherron AC, Lee S-J (1997) Double muscling in cattle due to mutations in the myostatin gene. Proc Natl Acad Sci U S A 94:12457–12461

McPherron AC, Lawler AM, Lee S-J (1997) Regulation of skeletal muscle mass in mice by a new TGF-beta superfamily member. Nature 387:83–90

Müller F, Ivics Z, Erdélyi F, Papp T, Váradi L, Horváth L, Maclean N, Orbán L (1992) Introducing foreign genes into fish eggs with electroporated sperm as a carrier. Mol Mar Biol Biotechnol 1:276–281

Nam YK, Maclean N, Hwang G Kim DS (2008) Autotransgenic and allotransgenic manipulation of growth traits in fish for aquaculture: a review. J Fish Biol 72:1–26

Ochman H, Gerber AS, Hartl DL (1988) Genetic applications of an inverse polymerase chain reaction. Genetics 120:621–623

Ozato K, Kondoh H, Inohara H, Iwamatsu T, Wakamatsu Y, Okada TS (1986) Production of transgenic fish: introduction and expression of chickenδ-crystallin gene in medaka embryos. Cell Differ Dev 19:237–244

Powers DA, Cole T, Creech K, Chen TT, Lin CM, Kight K, Dunham R (1992) Electroporation: a method for transferring genes into the gametes of zebrafish, *Brachydanio rerio*, channel catfish, *Ictalurus punctatus*, and common carp, *Cyprinus carpio*. Mol Mar Biol Biotechnol 1:301–309

Proudfoot C, McFarlane G, Whitelaw B, Lillico S (2020) Livestock breeding for the 21st century: the promise of the editing revolution. Front Agric Sci Eng 7:129–135

Rajesh R, Majumdar KC (2005) Transgene integration- an analysis in Auto transgenic *Labeo rohita* Hamilton (Pisces: Cyprinidae). Fish Physiol Biochem 31:281–287

Rasal KD, Nandanpawar PC, Swain P, Badhe MR, Sundaray JK, Jayasankar P (2016) MicroRNA in aquaculture fishes: a way forward with high-throughput sequencing and a computational approach. Rev Fish Biol Fish 26:199–212. https://doi.org/10.1007/s11160-016-9421-6

Sarmasik A, Warr G, Chen TT (2002) Production of transgenic medaka with increased resistance to bacterial pathogens. Mar Biotechnol 4:310–322

Schier AF, Giraldez AJ (2016) MicroRNA function and mechanism: insights from zebra fish. Cold Spring Harb Symp Biol 71:195–203

Smith K (2019) Time to start intervening in the human germline? A utilitarian perspective. Bioethics 34:90–104. https://doi.org/10.1111/bioe.12691

Steiner H, Hultmark D, Engstrom A, Bennick H, Boman HG (1981) Sequence and specificity of two antibacterial proteins involved in antibacterial immunity. Nature 292:246–248

Stuart GW, McMurray JV, Westerfield M (1988) Replication, integration and stable germ-line transmission of foreign sequences injected into early zebra fish embryos. Development 103:403–412

Sui C, Chena J, Ma J, Zhaoa W, Canárioa AVM, Martinsd RST (2019) Somatostatin 4 regulates growth and modulates gametogenesis in zebrafish. Aquac Fisher 4:239–246. https://doi.org/10.1016/j.aaf.2019.05.002

Symonds JE, Walker SP, Sin FYT (1994) Electroporation of salmon sperm with plasmid DNA: evidence of enhanced sperm/DNA association. Aquaculture 119:313–327

Takatsu K, Miyaoku K, Roy SR, Murono Y, Sago T, Itagaki H, Nakamura M, Tokumoto T (2013) Induction of female-to-male sex change in adult zebrafish by aromatase inhibitor treatment. Sci Rep 3:3400

Tester M (1999) Seeking clarity in the debate over the safety of GM foods. Nature 402:575

Thrasher A, Baltimore D, Pei D, Lander ES, Winnacker E-L, Baylis F, Daley GQ, Doudna JA, Berg P, Ossorio P, Zhou Q, Lovell-Badge R (2016) On human gene editing: international summit statement by the organizing committee. Issue Sci Technol 32(3)

Tonelli FMP, Lacerda SMSN, Tonelli FCP, Costa GMJ, França LR, Resende RR (2017) Progress and biotechnological prospects in fish transgenesis. Biotechnol Adv 35:832–844

Tsai HJ, Tseng TS, Liao IC (1995a) Electroporation of sperm to introduce foreign DNA into the genome of loach (*Misgurnus anguillicaudatus*). Can J Fish Aquat Sci 52:776–787

Tsai HJ, Wang SH, Inoue K, Takagi S, Kimura M, Wakamatsu Y, Ozato K (1995b) Initiation of the transgenic *lacZ* gene expression in medaka (*Oryzias latipes*) embryos. Mol Mar Biol Biotechnol 4:1–9

Wahid F, Shehzad A, Khan T, Kim YY (2010) MicroRNAs: synthesis, mechanism, function, and recent clinical trials. Biochim Biophys Acta Mol Cell Res 1803:1231–1243

Waltz E (2016) Gene-edited CRISPR mushroom escapes US regulation. Nature 532:293

Waltz E (2017) First genetically engineered salmon sold in Canada. Nature 548:148

Wang C, Chen Y-L, Bian W-P, Xie S-L, Qi G-L, Liu L, Strauss PR, Zou J-X, Pei D-S (2018) Deletion of mstna and mstnb impairs the immune system and affects growth performance in Zebra fish. Fish Shellfish Immunol 72:572–580

Wightman B, Ha I, Ruvkun G (1993) Posttranscriptional regulation of the heterochronic gene lin-14 by lin-4 mediates temporal pattern formation in C. elegans. Cell 75:855–862

Yan B, Liu B, Zhu C-D, Li K-L, Yue L-J, Zhao J-L, Gong X-L, Wang C-H (2013a) microRNA regulation of skin pigmentation in fish. J Cell Sci 126:3401–3408. https://doi.org/10.1242/jcs.125831

Yan B, Guo J-T, Zhu C-D, Zhao L-H, Zhao J-L (2013b) miR-203b: a novel regulator of MyoD expression in tilapia skeletal muscle. J Exp Biol 216:447–451. https://doi.org/10.1242/jeb.076315

Yin H, Xue W, Chen S, Bogorad RL, Benedetti E, Grompe M, Koteliansky V, Sharp PA, Jacks T, Anderson DG (2014) Genome editing with Cas9 in adult mice corrects a disease mutation and phenotype. Nat Biotechnol 32:551–553

Yu H, Li H, Li Q, Xu R, Yue C, Du S (2019) Targeted gene disruption in Pacific oyster based on CRISPR/Cas9 ribonucleoprotein complexes. Mar Biotechnol 21:301–309. https://doi.org/10.1007/s10126-019-09885-y

Zhang PJ, Hayat M, Joyce C, Gonzalez VL, Lin CM, Dunham RA, Chen TT, Powers DA (1990) Gene transfer, expression and inheritance of pRSV-rainbow trout-GH cDNA in the common carp, *Cyprinus carpio* (Linnaeus). Mol Reprod Dev 25:3–13

Zhang Y, Zhang S, Lu H, Zhang L, Zhang W (2014) Genes encoding aromatases in teleosts: evolution and expression regulation. Gen Comp Endocrinol 205:151–158

Zhang J, Song F, Sun Y, Yu K, Xiang J (2018) CRISPR/Cas9-mediated deletion of *EcMIH* shortens metamorphosis time from mysis larva to postlarva of *Exopalaemon carinicauda*. Fish Shellfish Immunol 77:244–251

Zhao Z, Yu X, Jia J, Yang G, Sun C, Li W (2019) miR-181b-5p may regulate muscle growth in Tilapia by targeting Myostatin b. Front Endocrinol 10:812. https://doi.org/10.3389/fendo.2019.00812

Zhong Z, Niu P, Wang M, Huang G, Xu S, Sun Y, Xu X, Hou Y, Sun X, Yan Y, Wang H (2016) Targeted disruption of sp7 and myostatin with CRISPR-Cas9 results in severe bone defects and more muscular cells in common carp. Sci Rep 6(22953):2016. https://doi.org/10.1038/srep22953

Indigenous Germplasm as Valued Genetic Resources

Ananya Khatei, Dibyajyoti Sahoo, Janmejay Parhi, and Pramod Kumar Pandey

Abstract

Indian fisheries is one of the leading sectors of the world. With vast stretch of water bodies, it is the home to many valuable aquatic organisms including fishes. However, exploitation of these indigenous fishes has proven to have deleterious effects on the ecosystem. Human interventions such as building of dams, dividing the river courses, introduction of exotic fishes, pollutions etc. have not only led to habitat destruction but also disturbed the population structure. In order to improve the fish traits, various techniques such as selection, line-development and hybridization has been done. However, these processes have also adversely affected the natural indigenous populations. Loss of some traits, reduction in genetic diversity and extinction of few species are some of the adverse effects that have come to light recently. Proper legislation and conservation strategies need to be implemented strictly so that the indigenous fish germplasm can be protected. Proper information about the suitability of invasive fish species must be collected before introducing them to the existing ecosystem. Advanced biotechnology such as application of molecular markers and cryopreservation can be utilized effectively for identification of the taxonomical status and restoration of indigenous fish populations, respectively.

The original version of this chapter has been updated: The authors Ananya Khatei, Dibyajyoti Sahoo and Janmejay Parhi affiliation has been updated. A Correction to this chapter is available at https://doi.org/10.1007/978-981-16-3215-0_30

A. Khatei · D. Sahoo · J. Parhi (✉)
Fish Genetics and Reproduction, College of Fisheries, CAU (I), Lembucherra, Tripura, India
e-mail: parhi.fgr.cof@cau.ac.in

P. K. Pandey
ICAR-Directorate of Coldwater Fisheries Resources, Bhimtal, Uttarakhand, India

© The Author(s), under exclusive license to Springer Nature Singapore Pte Ltd. 2021, corrected publication 2022
P. K. Pandey, J. Parhi (eds.), *Advances in Fisheries Biotechnology*,
https://doi.org/10.1007/978-981-16-3215-0_2

Keywords

Indigenous · Population · Conservation · Genetic diversity · Cryopreservation ·
Markers

2.1   Introduction

Death of an organism is inevitable and is, hence, acceptable but the disappearance of
lineages is a serious loss to the biosphere that should be combated with potential
human intelligence. Economic profits and biasedness towards the exploitation of the
fisheries resources have been to such an extent that we often ignore the stability of
the ecosystem upon which the continuity of the race depends. Life evolved first in
water and presently we are facing the extinctions and loss of lineages of the
organisms which forms the origin of vertebrates i.e. Pisces (Baillie et al. 2010).

   We have been meddling with the genetic resources since time immemorial for the
benefits of human kind. The arena where it has a greater impact is agriculture. The
benefits of manipulating the population genetics in wild and captivity, purposeful
selection ad hybridization can be found at every corner of the world. These
implications have also revolutionized aquaculture. However, it is until this time,
we have only admired the economic success and have remained largely ignorant
towards the havoc that has been silently occurring to the aquatic landscape. Decades
of domestication will obviously take decades to restore the natural population.
Meanwhile, we can always put effort to understand the importance of the existing
genetic resources.

2.2   Genetic Resources

As defined by the Nagoya protocol which came to force on 12th October 2014, any
biological material or entity that contains genes which are useful in research and
product development are designated as genetic resources (Laird and Wynberg 2018).
According to the convention on biological diversity, Genetic material refers to any
material of plant, animal, microbial or other origin containing functional units of
heredity. "Genetic resources" refer to genetic material of actual or potential value.

   Utilization of genetic resources is crucial in case of agriculture. It is essential to
explore the genetic resources in order to meet the food security without ignoring the
conservation status of varieties simultaneously. Aquaculture forms the backbone of
the blue revolution and fishes are exploited to benefit the growing human population.
Fishes which are the most diverse group of vertebrates serve as rich genetic
resources.

2.3   Indigenous Fish Resources of India

India is an abode of biodiversity and home to many wild plant and animal species. This makes it one of the richest grounds for genetic resources. Out of the 36 biodiversity hot spots, four are present in India and hence it can be said that it is the home to innumerable species of animals which includes fishes. There are 14 major river systems, 44 medium and many small rivers. It has 3.15 million ha reservoirs, 0.2 million ha floodplain wetlands and 0.72 million ha upland lakes, besides 0.3 million ha estuaries and 0.9 million ha backwaters and lagoons. This gives the account of the vast stretch of inland water bodies that holds a large proportion of the world inland fish fauna. The coastal line stretches for 7516.6 km.

This divides the indigenous fish resources broadly into inland resources, marine resources and brackish water resources. Based on the temperature regimes, the resources can be divided into cold water fisheries and warm water fisheries. The major indigenous cold water species are *Tor tor, T. putitora, T. mosal, T. khudree, T. mussullah, T. progeneius, T. malabaricus, Naziritor chelynoides, N. hexagonolepis, Neolissochielus wynaadensis, Schizothoraichthys progastus, S. esocinus, Schizothorax richardsonii, S. plagiostomus, S. curvifrons, S. micropogon, S. kumaonensis, Barilius bendelisis, B. vagra, B. shacra, B. (Raiamas) bola, Bangana dero, Labeo dyocheilus, Crossocheilus periyarensis, Garra lamta, G. gotyla gotyla, Glyptothorax pectinopterus, G. brevipinnis, G. stoliczkae* and *Lepidopygopsis typus.*

There are several economically important indigenous warm water fishes. The most valued of them are the major and minor carps such as *Labeo rohita, Catla catla, Cirrhinus mrigala, L. calbasu, L. bata, L. fimbriatus, L. dussumieri, C. cirrhosa, C. reba, Puntius dubius, P. carnaticus,* etc. Catfishes are also a major group contributing to the fisheries sector since they have high commercial value. Some of the indigenous catfishes are *Clarias batrachus, Heteropneustes fossils, Sperata aor, S. seenghala, Wallago attu, Pangasius pangasius, Silonia silondia, Bagarius bagarius, Rita rita, Eutropiichthys vacha,* etc. Murrels are also important group of fishes with high medicinal and nutritional value. The indigenous murrels in India include *Channa striata, O. marulius, C. punctata, C. diplogramma, Anabas testudineus, Chitala chitala* and *Notopterus notopterus.*

The indigenous brackishwater fishes include *Mugil cephalus, Lates calcarifer, Tenualosa ilisha, Chanos chanos, Etroplus suratensis, Liza macrolepis, L. tade, L. parsia, Megalops cyprinoides, Elops saurus, E. machnata, Valamugil seheli, V. cunnesius, Ephinephelus tauvina, Rhinomugil corsula, Mystus gulio, Nematolosa nasus, Pseudosciaena coibor, Gerres setifer, G. oyena, Sillago sihama, Polynemus tetradactylus, P. paradiseus, Eleutheronema tetradactylum* and *Lutjanus argentimaculatus.* The marine fishes comprises of several species of tuna, sardines, pomfrets, carangids, catfishes, ribbon fishes, barracudas etc.

With such a great availability of genetic resource, India could be considered as a large pool of genetic diversity that forms 11.7% of the world's fish diversity and can promote crop improvement establishing food security utilizing the existing germplasm (Nair and Dinesh Kumar 2018). This introduces us to two different terms the "germplasm" and "genetic diversity".

2.4   Germplasm

The living genetic resources that are preserved or maintained for research purpose
are called germplasm. It could be seed in the seed bank or breeding line of animals
that are maintained or gene banks etc. This germplasm is a necessity for any crop
improvement program. This germplasm could be both wild and captive species.

The objectives behind collecting and maintaining germplasms are:

- Exploration and collection of indigenous and exotic genetic resources.
- To characterize genetic resources and promote their utility.
- To conduct and spread awareness about the genetic resources.
- To organize conservation strategies.
- To carry out researches for human resource development.

In India, there are five organizations that are involved in germplasm collection
namely: National Bureau of Plant Genetic Resources (NBPGR), National Bureau of
Animal Genetic Resources NBAGR, National Bureau of Fish Genetic Resources
(NBFGR), National Bureau of Agriculturally Important Microorganisms (NBAIM)
and National Bureau of Agricultural Insect Resources (NBAIR). National Bureau of
Fish Genetic Resources (NBFGR) aims at fish germplasm collection. It is responsible for conservation of fish genetic resources and its sustainable utilization. It also
deals with the evaluation of both indigenous and exotic germplasm evaluation to
detect and prevent potential health risk. The most important aspect of a germplasm
collection is that it ensures the maintenance of genetic diversity. Genetic diversity is
a property of a population which will determine its sustainability.

2.5   Genetic Diversity

The genetic variability found within a species population is called genetic diversity.
This genetic diversity is mandatory for the population to adapt to the surrounding
environment and its adverse conditions. In aquaculture, a population needs to have
higher genetic diversity in order to make room for improvement. All the variations
within the species are exploited and the desirable traits are selected for further
process like development of lines. However, if the genetic diversity in a population
is low, there is very less chance to exploit.

Researchers have utilized their knowledge from decades for improving fish stock
by exploiting genetic diversity. But on the contrary this has put down the diversity
score. Continual purposeful selection focusing on a certain trait just for commercial
benefit has a detrimental effect on the overall diversity index which is slowly taking
its pace. It can be said that today we may have a genetically improved fish variety but
at a terrible price.

2.6   Status of Indigenous Fishes in India

We have been witnessing all the propagandas in development of fish cultivars but we tend to neglect the harm it causes to the existing indigenous fish species in India. There are several Small Indigenous Fishes (SIFs) which not only form a larger part of the aquatic diversity but also are high valued food fishes. Around 450 species in Indian freshwater are categorized into SIFs. These fishes inhabit the local ponds, rice fields, river tributaries, lowland areas, beels and small reservoirs. These fishes have the potential to meet the nutritional requirements but are neglected to a larger extent. These species are highly threatened due to various anthropological activities such as pollution, overexploitation, fragmentation of water channels due to construction activities and habitat loss.

The maximum record of these SIFs has been obtained from the north-east India, then the Western Ghats and central part. *Amblypharyngdon mola, A. microlepis, N. notopterus, P. sarana, L. bata, P. ticto, C. reba, Salmostoma bacaila, Nandus nandus, A. testudineus, Esomus danricus, P. chola, P. sarana, Glossogobius giuris, Danio devario, Chanda nama,* etc. Other potential species for aquaculture includes *L. gonius, L. bata, Labeo boggut, L. dussumieri, L. fimbriatus, Barbodes carnaticus, P. pulchellus, P. kolus, P. sarana* and *Cirrhinus cirrhosa.* The air-breathing and non-air-breathing species: *O. marulius, O. stewartii, O. gachua, O. striatus, O. punctatus, O. aurantimaculatata.*

Other than these small indigenous fish, some fishes like the Indian major carps, *Etroplus suratensis, Osteobrama belangiri, Tor sps.* etc. contribute hugely to the Indian economy. However, these indigenous populations face potential threats that have been reported from hatcheries as well as from wild. India's Conservation Assessment and Management Plan (CAMP 1998) has identified 327 indigenous freshwater fishes under threatened condition. Since the enforcement of the Biological Diversity Act (2002) the status of fish diversity has come to limelight. It has not gone unnoticed that many indigenous fishes have been facing drastic depletion due to several reasons. This is a quite alarming situation for all of us, as along with these species, the ecological balance and rich genetic resources are being eroded.

2.7   Taxon Status

According to the International Union for Conservation of Nature (IUCN), 2000 red list the species can be categorized into following status:

1. Extinct (EX): species no longer extant
2. Extinct in the wild (EW): species only in captive condition
3. Critically endangered (CR): in a critical state i.e closer to extinction
4. Endangered (EN): high risk of extinction
5. Vulnerable (VU): under risk of extinction caused due to human activities

6. Other categories: Near threatened (NT), Least Concern (LC), Data deficient (DD) and Not evaluated (NE)

Recent data show that there are 19 critically endangered and five endangered fish species in India.

2.8　Potential Threats

2.8.1　Introduction of Invasive Species

During the few past decades, aquaculture has attended a fair pace of growth due to lots of advancements in methods and techniques. One of them were introduction of exotic species for various purposes which included weed control, mosquito control, better growth, sport fishes and aquarium fishes. The term "exotic" has nothing to do with being "invasive" unless they exhibit problems to other existing species. For example three exotic species that have well established in Indian waters, are the common carp, grass carp and silver carp. However, they are potential threats to the existing gene pool. The exotic invasive species can alter the natural aquatic ecosystems through predation, competition for existing resources, prolific breeding and thereby, domination, habitat loss and disease transmission. The introduction of Silver carp in 1959 showed success rate in meeting the commercial demands. However, in the Govind Sagar reservoir it was found that it affected drastically the population of indigenous fishes like catla and *Tor putitora* (Sharma 2018).

One of the most dangerous threats an exotic fish possess is production of unwanted hybrids that contaminate the indigenous fish germplasm and genetic makeup of a population. They also hamper the socioeconomic values of the farmers. More than 300 exotic species have been reported in the natural water bodies of India (Kumar 2000). One of the most dominating fish among them is the Tilapia (*Oreochromis mossambicus*). It is well known by now how this invasive species is found everywhere in the aquatic bodies. This fish was introduced in 1952, in order to meet the nutritional demand of the growing population, owing to its high growth rate and breeding habit. Very soon this fish began to dominate the reservoirs and other natural water bodies in Kerala and Tamil Nadu. Being a prolific breeder, it took very less time for this fish to expand its population over other existing indigenous fish. *E. suratensis*, a fish native to Kerala was under threat (Padmakumar et al. 2012). Other fish populations including the Indian major carps were also affected due to the proliferation of Tilapia at an alarming rate.

Clarias gariepinus is another invasive species which imposes high risk to the indigenous fish diversity. Being a predator, it diminishes the population of other small fishes residing in the territory to which it belongs. This fish has the potential to cause large scale extinction of the existing indigenous fishes. Introduction of Nile perch (*Lates niloticus*), into Lake Victoria has resulted in 50% depletion of as many as 400 species of indigenous fishes in the lake.

The red-bellied Pacu is also a highly invasive species which was introduced in 2004. Initially it was imported from Bangladesh for ornamental purpose. However, it expanded at a fast rate and is now reported in many natural water bodies. It has been found to infest the Vemban and and Periyar Lake. Besides these food fishes, many ornamental fishes also introduced in India; which have now proliferated into natural water bodies. One of the noteworthy examples is the sucker mouth catfish. This competes with other small hill stream fishes and depletes their population.

2.8.2 Overexploitation

Overfishing has a very disastrous effect on the population structure of the aquatic ecosystem. It is even more destructive when the fishing is selective. Such effect is more in case of freshwater than marine waters. The concept of Evolutionary Significant Unit (ESU) is often neglected while fishing. Due to overfishing, the population size decreases and population fragmentation occurs which produces small and confined populations. This leads to genetic differentiation. During this genetic differentiation, the gene flow is restricted. Many of the variations in certain genes are lost. Hence, proper administration and analysis of techniques is essential before setting the fishing goals.

It is a common phenomenon that when the commercially important species approach extinction, the harvest pressure shifts to other species which only increase the risk on biodiversity. There are two types of overfishing: overfishing of brood stock called recruitment overfishing and overfishing of juveniles called growth overfishing. Another aspect of this overfishing is habitat destruction. The impact of habitat destruction is more for migratory fishes. The destruction of breeding grounds and changes in trophic levels in the ecosystem has large impacts on the fish germplasm.

2.8.3 Other Anthropological Interference

This includes disturbances in the aquatic bodies due to constructions and pollution. Constructions of dams and other man-made structures across rivers and streams lead to fragmentation of populations. This possesses a barrier to gene flow and also disturbs the population structure. For example, the construction of Farraka barrage over Ganga River has caused severe depletion in *T. ilisha* stock. Several dams and barrages are built across the Bramhaputra river basin which is the cause of reduction in abundance of species like *Schizothorax richardsonii*, *S. plagiostomus*, mahaseer, etc. Pengba the native fish of Manipur is declared as regionally extinct in the wild due to the construction of Ithai barrage.

Aquatic pollution is another reason for fish stock depletion. With increase in human population and rapid industrialization, the rate of pollution is also increasing day by day. Disposal of sewage, heavy metals and toxic industrial wastes leads to habitat destruction. The most prominent example of sewage pollution is observed in the Ganga waters. The BOD load in Ganga is more than 2500 million kg/day (Sharma et al. 2019). These affect the fish health and habitat. Pollution in the breeding grounds leads to a poor survivability of the hatchlings.

2.8.3.1 Climate Changes

One inevitable reason for loss of species diversity is the progressive climate change. The shift in temperature has resulted in the degradation of hill stream population such as the loaches and catfishes. For example, fishes such as *P. tincto*, *G. giuris* and *M. vittatus* were earlier found in middle stretches of Ganga. But due to temperature rise they have recently shown a geographical shift towards only the upper and middle stretches of this river.

Another aspect of climate change is the disturbances in the breeding cycles of the species. Due to the irregularity in the onset of monsoons the reproductive behaviour of fishes like IMCs and catfishes has been affected.

2.8.4   Impacts on Indigenous Germplasm

To monitor the threatened or endangered indigenous fishes the primary need shall be the assessment of genetic variability. Populations of various fishes are divided into many sub-populations which maintain their own genetic makeup. However, forces such as mutation, migration and selection are responsible for causing differentiation. These population structures need to be maintained in order for the sustenance of the indigenous germplasm. Once lost, it would be almost irreversible to get back the original genetic makeup. Hence, management strategies should include thorough investigation of population genetics of fish stock through molecular data. There is higher effective population size in case of marine fishes as compared to the freshwater fishes. It is found that there is low genetic differentiation and higher genetic diversity in case of marine fishes. This is unlike freshwater species since it has a lower effective population size.

Such assessments of genetic information are essential for planning the rehabilitation of endangered fish species. For this purpose, molecular markers and biotechnology can be used. For example, microsatellite markers have been developed for *L. rohita*, *C. chitala* and *C. catla* by NBFGR (2004). There has been a profound development of genetic markers for the indigenous fishes of India. Such markers will help to transfer information from map-rich species to map-poor species. Over the course of time, microsatellite has become the most widely used marker in fisheries for the assessment of genetic diversity. However, SNPS are also now paid attention due to their abundance in the genome.

These molecular markers have also been used for resolving taxonomic ambiguities. Taxonomic ambiguities are a major issue while identifying and characterizing fish species. There are numerous indigenous fishes in India that are yet not identified and characterized. Use of mtDNA has recently become very popular among the fish taxonomists. The NBFGR has developed DNA barcode for indigenous fishes of India (Chakraborty and Ghosh 2014). There are several species-specific markers that have been developed by NBFGR. These markers are mostly allozymes and mtDNA haplotypes and are used for determining species-specific differences. Such markers have been used to distinguish the stocks of the invasive *C. gariepinus* and native *C. batrachus*. M13 probes have been used to obtain DNA fingerprints for Indian Major Carps.

Techniques such as Fluoroscent In Situ Hybridization (FISH) has been applied to develop species-specific markers and identifying the sex chromosomes. Other cyto-genetic techniques are also widely used for identification and differentiation of species and their sex chromosomes such as C banding, NOR banding etc. such cytogenetic tools have also been utilized for the identification of many native species of western ghats such as *L. dussumieri*, *P. filamentosus* and *Horabagrus brachysoma*.

2.8.4.1 Population Size Changes

Understanding population size is very important in order to assess the fate of fish germplasm in wild and captive. The kind of steps taken as conservation strategies or for breeding purpose must involve prior assessment of existing population size. The transfer of traits from one generation to the next is a demographic process (Lakra et al. 2007). The phenomenon such as genetic drift is affected by population size. Small populations show larger genetic drift. There are chances of diminishing rare alleles. In this case, there is allele fixation for which there is a loss of genetic diversity. Another aspect of population size change is the bottleneck effect. Such bottlenecks can have drastic impact on genetic variability. In the Challakudy and Valapatnam Rivers of Kerala, such bottlenecks were found to affect indigenous fish *P. denisonii* by using microsatellite markers (Kurup et al. 2004).

2.8.5 Strategies

Under the above mentioned circumstances, it has become evident that many of the indigenous fishes are at threat. If the conservation measures are not implemented properly, there are chances that these indigenous populations will be wiped out entirely. This is a menace for future since there will be no genetic diversity left to exploit. This will hamper the sustainability of aquaculture.

2.8.5.1 Conservation in the Natural Habitat (In Situ)

Conservation of fish germplasm in the wild or in their natural habitat for the maintenance and recovery of the threatened and vulnerable species is defined as in situ conservation (Olver et al. 1995). These programmes include exploration of natural habitats and evolutionary history. Protection of fish germplasm in natural wild conditions can be tedious since majority of the fish farmers depend on the indigenous fish solely for their livelihood. However, restricted regulations must be applied to the overexploited areas. One of the exemplary decisions accepted world-wide including India is the Ramsar Convention for the conservation of wetland biodiversity.

For the conservation of marine biodiversity, several measures have been taken. There are six national marine parks in India namely:

1. Marine National Park, Gujarat.
2. Mahatma Gandhi Marine National Park, Andaman and Nicobar Islands.
3. Gahirmatha Marine Sanctuary, Odisha.
4. Gulf of Mannar Marine National Park, Tamil Nadu.
5. Rani Jhansi Marine National Park, Andaman and Nicobar Islands.
6. Malvan Marine Wildlife Sanctuary, Maharashtra.

2.8.5.2 Replenishing the Degraded Populations

This could be achieved through collecting the fish species from the wild and then releasing the stocks again to the natural environment after captive breeding. However, the genetic environmental effects of such kind of practices should be taken into account. Other consequences in this practice can be the inbreeding effect. All such analysis can be done since recently there have been numerous advances in biotechnology and computational biology.

2.8.5.3 Conservation Strategies Outside Natural Habitats (Ex Situ)

The ex situ conservation techniques are more widely used for the preservation of indigenous fishes and also for commercially important species. Gene banks are established for species-specific genetic improvement programmes. Such measures have been taken up by NBFGR, Lucknow.

This method is highly potential for the preservation of the indigenous fish germplasm. Breeding success using the cryopreservation protocol has been achieved for several indigenous and commercially important species such as *L. rohita*, *T. putitora* and *L. calcarifer* (Lakra et al. 2005). Moreover, NBFGR has also been developing several diseases screening procedure for the Specific Pathogen Free (SPF) species introduced. This shall help to prevent the foreign pathogens contaminate the indigenous fishes.

2.9   Conclusion

India has a vast diversity of indigenous fish germplasm. Such a huge resource needs proper management so that it can be sustained to meet our own demands in the long term. Proper knowledge, molecular assessments, conservation and breeding strategies should be implemented so that these existing variations in the indigenous fish population can be utilized to further develop the aquaculture sector as well as keep the ecological balance intact.

References

Baillie JE, Griffiths J, Turvey ST, Loh J, Collen B (2010) Evolution lost: status and trends of the world's vertebrates. Zoological Society of London, London

Chakraborty M, Ghosh SK (2014) An assessment of the DNA barcodes of Indian freshwater fishes. Gene 537(1):20–28

Kumar AB (2000) Exotic fishes and freshwater fish diversity. Zoos' Print J 15(11):363–367

Kurup BM, Radhakrishnan KV, Manojkumar TG (2004) Biodiversity status of fishes inhabiting rivers of Kerala (S. India) with special reference to endemism, threats and conservation measures. In: Proceedings of LARS2. 2nd Large Rivers symposium. Mekong River Commission and Food and Agricultural Organization, pp 163–182

Laird S, Wynberg R (2018) A fact-finding and scoping study on digital sequence information on genetic resources in the context of the Convention on Biological Diversity and the Nagoya Protocol. Convention on Biological Diversity, Rio de Janeiro

Lakra WS, Behera MR, Sivakumar N, Goswami M, Bhonde RR (2005) Development of cell culture from liver and kidney of Indian major carp, Labeo rohita (Hamilton). Indian J Fisher 52 (3):373–376

Lakra WS, Mohindra V, Lal KK (2007) Fish genetics and conservation research in India: status and perspectives. Fish Physiol Biochem 33(4):475–487

Nair RJ, Dinesh Kumar S (2018) Overview of the fish diversity of Indian waters. CMFRI, Kochi

Olver CH, Shuter BJ, Minns CK (1995) Toward a definition of conservation principles for fisheries management. Can J Fish Aquat Sci 52(7):1584–1594

Padmakumar KG, Bindu L, Manu PS (2012) Etroplus suratensis (Bloch), the state fish of Kerala. J Biosci 37(1):925–931

Sharma I (2018) Status of fishery versus exotic fauna in Gobind Sagar Dam Wetland (HP & Punjab), India. Int J Fisher Aqua Stud 6(2):396–398

Sharma BM, Bečanová J, Scheringer M, Sharma A, Bharat GK, Whitehead PG, Klánová J, Nizzetto L (2019) Health and ecological risk assessment of emerging contaminants (pharmaceuticals, personal care products, and artificial sweeteners) in surface and groundwater (drinking water) in the Ganges River Basin, India. Sci Total Environ 646:1459–1467

Applications of Next-Generation Sequencing in Aquaculture and Fisheries

3

Pragyan Paramita Swain, Lakshman Sahoo, Rajesh Kumar, and Jitendra Kumar Sundaray

Abstract

The aquaculture industry supplies a substantial portion to meet the demand for protein rich food. It also serves as a means of livelihood for the farmers by providing employment opportunities to them. Despite increased demand for fish meat, the aquaculture industry faces several challenges that hamper the adequate supply of fishes. Some of those challenges, faced worldwide, are overfishing, anthropogenic disturbances, obstruction of migration, inconsistency in growth rate, weakened immunity, problem of diseases, dysbiosis of the microbiome, and toxicity due to xenobiotics. One of the setbacks in the attempts to tackle these issues has been the lack of genetic information such as whole-genome and transcriptome. With the advancements in the next-generation sequencing technologies (NGS), such genetic information can be obtained rapidly at an affordable cost. This has led to the application of whole genome and whole transcriptome in aquaculture to understand the molecular mechanisms during infections, migration, sexual development, and response to toxicants. It has also led to the application of metagenomics to explore the symbiotic relationship between the fish and its microbiome. Such studies have brought forth understanding of the genes and SNPs that are involved in gonadal development, sex determination, colonization resistance, migration, dissemination of genetic

P. P. Swain · L. Sahoo · J. K. Sundaray (✉)
Division of Fish Genetics and Biotechnology, ICAR-Central Institute of Freshwater Aquaculture, Bhubaneswar, Odisha, India

R. Kumar
Division of Aquaculture Production and Environment, ICAR-Central Institute of Freshwater Aquaculture, Bhubaneswar, Odisha, India

Division of Fish Genetics and Biotechnology, ICAR-Central Institute of Freshwater Aquaculture, Bhubaneswar, Odisha, India

© The Author(s), under exclusive license to Springer Nature Singapore Pte Ltd. 2021
P. K. Pandey, J. Parhi (eds.), *Advances in Fisheries Biotechnology*, https://doi.org/10.1007/978-981-16-3215-0_3

diversity and rapid growth. The cumulative knowledge of these studies can help in selective breeding of the desired line, sustain the supply of fish meat, and also greatly help in the conservation of wildlife in its natural habitat. In this chapter, we report the current challenges and recent advances in the aquaculture industry through the application of next-generation sequencing.

Keywords

Aquaculture · Fisheries · Next-generation sequencing · Whole genome · Transcriptome · GWAS · SNPs · QTLs

3.1 Introduction

Aquaculture plays an important role in the eradication of poverty by providing over 15% of animal protein to 4.5 billion populations (Bernatchez et al. 2017; Pauly and Zeller 2019). Due to rapid increase in human population, there is a growing demand for food fish consumption (Pauly and Zeller 2019). The average annual consumption of food fish increased by 3.2% which was nearly twice the increase in global population growth (1.6%) as well as meat from terrestrial animals (2.8%) (Béné 2006; McIntyre et al. 2016). Despite its increased production, fisheries and aquaculture face severe challenges worldwide due to problems of diseases, habitat degradations, overfishing, and climate change, etc. (Bernatchez 2016). Therefore, to maintain sustainable development of the aquaculture industry, implementation of proper management strategies is the need of the hour (Melnychuk et al. 2017). The development and availability of Next-Generation Sequencing (NGS) platforms commercially has provided the necessary tools (Tables 3.1 and 3.2) to tackle some of these challenges and ensures the sustainability and profitability of aquaculture production.

Briefly, NGS technology begins with library preparation wherein the double-stranded DNA (dsDNA) is fragmented into platform-specific sizes, followed by adaptor ligation and Polymerase Chain Reaction (PCR) or clonal amplification of the fragments. This is followed by sequencing of the amplified fragments and detection of the nucleotides, using fluorescent dyes or enzymes. The sequenced fragments are obtained as reads and are mapped to the reference genome for alignment. In the absence of reference genome, de novo assembly is performed (Buermans and Den Dunnen 2014).

The breeding and production of fishes for consumption and other uses are highly dependent on the process of selective breeding as well as the brood stock population. Desirable traits, related to performance and quality, are of great economic importance to the aquaculture industry with a view to meet consumer demands. Genome-wide association studies (GWAS) can be implemented to identify genes and their complex association with important traits such as growth, disease resistance, and reproduction which are important in ornamental and food fishes equally. Such studies are greatly effective due to the application of NGS such as Illumina,

Table 3.1 List of Next Generation Sequencing platforms

Sr. No.	Platform	Sequencing technology	Accuracy (%)	Advantages
1	Solexa/ Illumina	Sequencing by synthesis	98	High throughput
	Illumina GAIIx			
	Illumina HiSeq1000			
	Illumina HiSeq1500			
	Illumina HiSeq2000			
	Illumina HiSeq2500			
	Illumina MiSeq			
	Illumina NextSeq 500			
	Illumina HiSeq X Ten			
2	ABI/SOLiD	Sequencing by Ligation	99.94	Read length and accuracy
	SOLiD 4			
	SOLiD 5500			
	SOLiD 5500xl			
	SOLiD 5500 W			
	SOLiD 5500xl W			
3	Roche/454	Pyrosequencing	99.9	Read length and speed
	454 FLX Titanium			
	454 FLX+			
	454 GS Junior			
	454 GS Junior+			
4	Pacific Biosciences RS	Single-molecule, real-time DNA sequencing by synthesis	99.8	Long read lengths, high consensus accuracy, a low degree of bias
	Pacific Biosciences Sequel			
5	Ion PGM	Semiconductor-based sequencing by synthesis	99.6	Cost-effective
	Ion Proton			

Table 3.2 Performance of different NGS platforms

Sr. No.	NGS platform	Average read length (bp)	Runtime	Data output
1	Illumina MiSeq	300 (PE)	21–56 h	13.2–15 Gb
	Illumina HiSeq 3000/4000	150 (PE)	1–3.5 days	650–700 Gb
	Illumina HiSeqX	150 (PE)	3 days	800–900 Gb
2	SOLiD 5500	75	6 days	120 Gb
	SOLiD 5500xl	75	10 days	240 Gb
3	454 GS Junior	600	10 h	35 Mb
	454 GS Junior+	1000	18 h	70 Mb
	454 GS FLX Titanium XLR70	600	10 h	450 Mb
	454 GS FLX Titanium XL+	1000	23 h	700 Mb
4	Pacific Biosciences RS-II	20 kb	4 h	500 Mb to 1 Gb
	Pacific Biosciences Sequel	8–12 kb	0.5–6 h	3.5–7 Gb
5	Oxford Nanopore MK1 MinION	200 kb	48 h	20 kb
6	Ion PGM	400	4–7 h	1–2 Gb
	Ion Proton	200	2–4 h	10 Gb

Pyrosequencing, Solid sequencing, Ion torrent, and Nanopore, by sequencing the whole genome (Table 3.3) for scanning molecular markers such as Single Nucleotide Polymorphisms (SNPs) that are associated with the desirable traits. Furthermore, gene expression studies through RNA-Seq can also be implemented to capture expression profiles of the fishes with desirable traits. In addition to determining the changes in gene expression levels, RNA-seq is also being used in sequencing microRNAs (miRNAs) and long non-coding RNAs (lncRNAs) that are involved in gene regulation. This technology has been very helpful in understanding the transcriptome complexity by investigating the coding and non-coding RNAs, alternative splicing events, structure and the function of transcripts.

Despite such wide applicability of the NGS technology in aquaculture, studies such as GWAS, whole-genome sequencing, transcriptomics, and related issues in aquaculture have not received enough attention as compared to the terrestrial livestock (Kumar and Kocour 2017). A low-density SNP array is used in most of the studies, causing a low level of linkage disequilibrium; thus, a high-resolution genetic map should be used in the identification of markers associated with desirable traits for selective breeding. In recent years, the application of genomic technologies in fisheries has been accelerated by the rapid development of NGS technologies and the reduction in the cost per Gb. This has helped in narrowing down the gap in understanding the traits and their genome. In this chapter, we discuss some of the recent advances related to the application of NGS-based studies in aquaculture and fisheries (Fig. 3.1).

Table 3.3 Whole genome sequencing of various fish species using NGS technologies

Sr. No.	Species	NGS platform	Assembled genome size	N50	Longest scaffold	Reference
1	Japanese eel (*Anguilla japonica*)	Illumina GAIIx and HiSeq 2000	1.15 Gbp	52,849 bp	1.14 Mbp	Henkel et al. (2012)
2	Channel catfish (*Ictalurus punctatus*)	Illumina GAIIx and HiSeq 2000	783 Mb	Contig N50:77,200 bp; scaffold N50: 7,726,806 bp	22,613,484	Liu et al. (2016)
3	Common carp (*Cyprinus carpio*)	Roche 454, Illumina and SOLiD	1.83 Gb	Contig N50: 68.4 kb Scaffold N50: 1.0 Mb	~875 Mb	Xu et al. (2014)
4	Half-smooth tongue sole (*Cynoglossus semilaevis*)	Illumina HiSeq2000	477 Mb	scaffold N50: 867 kb	-	Chen et al. (2014)
5	Grass carp (*Ctenopharyngodon idellus*)	Illumina HiSeq 2000	female (0.90-Gb) and male (1.07-Gb)	Female contig N50: 40,781 bp; Scaffold N50: 6,456,983 bp; Male Contig N50: 18,252 bp; Scaffold N50: 2,279,965 bp	Female 19,571,558 bp; Male 16,339,329 bp	Wang et al. (2015)
6	Asian arowana (*Scleropages formosus*)	Illumina HiSeq 2000	708 Mb	N50 scaffold length is 58,849 bp	616,488 bp	Austin et al. (2015)
7.	Chinese mitten crab (*Eriocheir sinensis*)	Illumina HiSeq2000	1.12 Gb	N50 of 224 kb contig N50: 6.02 kb scaffold N50:224 kb.		Song et al. (2016)
8	Pacific bluefin tuna (*Thunnus orientalis*)	Roche 454 FLX Titanium and Illumina GAIIx	454–786.6 Mb; Illumina-740.3 Mb	Contig N50: 7588; Scaffold: N50 136,950	454–786.6 Mb Illumina-1,021,118	Nakamura et al. (2013)
9	Antarctic bullhead (*Notothenia coriiceps*)	Illumina Hiseq 2000; GS FLX titanium	602 Mb	N50 Contig size of 8581 bp; N50 scaffold size of 219 kb	-	Shin et al (2014)
10.	Turbot (*Scophthalmus maximus*)	Illumina GAIIx and HiSeq2000	544 Mb,	Contig N50 is 31.2 kb and the scaffold N50 is 4.3 Mb.	19 Mb	Figueras et al. (2016)

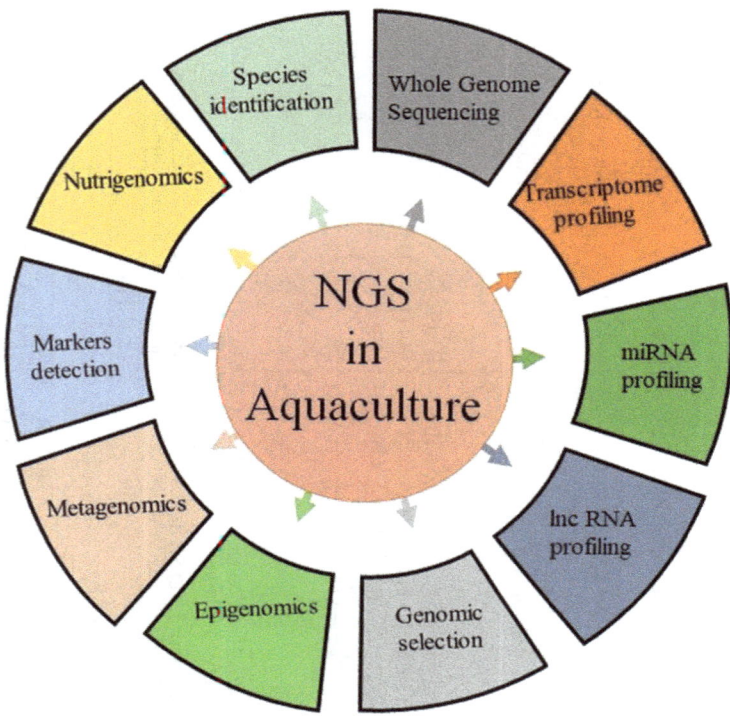

Fig. 3.1 Applications of NGS in aquaculture

3.2 Fish Migration

Migration is a complex phenomenon of population movement from one habitat to another, governed by morphological, physiological, behavioral traits as well as environmental cues (Dingle 2006; Dingle and Drake 2007). Variation in migration patterns is observed not only among different species but also within populations of a species. Such diverse migration, exhibited by the aquatic species, brings about distribution in nutrients and shapes the environment. There has been a rapid worldwide decline of these species due to limited knowledge in genetic factors, governing migration and anthropogenic disturbances such as dam construction, overfishing and pollution (Wilcove and Wikelski 2008). Such limited knowledge about genes, governing migration, has caused difficulty in the prevention of their rapid decline. Certain genes are responsible for the phenotypic variation in migratory traits. However, the candidate genes are obscured due to the lack of extensive genomic data of the migratory fishes (Liedvogel et al. 2011). Traits involved in migration are highly complex and involve interaction among several genes. Many of the traits, associated with migration, are supposedly known to have similar genetic basis due to similar environmental and physiological challenges, experienced by many migratory species. This could be due to the convergent evolution of such migratory traits. Hence, understanding the molecular mechanisms and candidate genes responsible for migration in one species could provide valuable information with regard to other related species (Dingle 2006). This can further help in formulating strategies for the conservation and protection of such species.

The advancement of next-generation sequencing has increased the possibility to genotype thousands of SNPs (Hecht et al. 2013). GWAS based studies on identification of quantitative trait loci (QTLs) have been used to determine the association of their genomes with migratory activities and the probability of migration in the fresh as well as marine fishes such as rainbow trout (*Oncorhynchus mykiss*) in different geographical areas (Hecht et al. 2013; Hale et al. 2013). Resident rainbow trout and anadromous steelhead trout offer an excellent model to delineate the association of SNPs with migratory traits since both populations belong to the same species *O. mykiss*. The anadromous steelhead trout migrates to the sea and returns to freshwater after attaining sexual maturity, whereas the resident rainbow trout has been obstructed from migration for more than 50 years due to hydropower dam. Mapping of SNPs ($n = 15,239$) in steelhead (*O. mykiss*) identified that the neutral divergence was common in individuals that migrated during summer and winter (Hess et al. 2016). Interestingly, the authors also identified three SNPs in the GREB1-like gene those were involved in the migration. NGS-based restriction-site associated DNA (RAD) tags study offers an excellent avenue to investigate the SNP differences among populations. RAD tag studies of two wild populations, resident rainbow trout and anadromous steelhead trout in the United States, identified 504 RAD SNP markers, associated with migration (Hecht et al. 2013). Similarly, RAD tag sequencing identified over 6500 SNPs in migratory and non-migratory *mykiss* ecotypes, out of which the candidate SNPs ($n = 1423$) for selection were associated with migratory-related traits (Hale et al. 2013). Such identification of SNPs could help in migration prediction and conservation of the steelhead species. The RAD tag sequencing has also been implemented to study the anadromous Hilsa shad (*Tenualosa ilisha*) from nine different migratory habitats and identified over 15,453 SNPs out of which more than 49 were associated with the divergent selective process. This led to the identification and categorization of hilsa population into two genetically distinct clusters wherein hilsa from the various freshwater rivers fell into one genetic cluster and the ones from sea and estuaries fell into another genetic cluster (Asaduzzaman et al. 2020).

Overfishing, construction of dams, introduction of exotic species, and pollution pose a major threat to migratory fishes (Carolsfeld 2003; Agostinho et al. 2005). These barriers have created obstacles in the migratory pathways of these fishes which can gradually lead to reproductive isolation by obstructing the gene flow which ultimately affects the diversity, reproduction, and the services these fishes provide (Agostinho et al. 2005). A decrease in migratory fish population has been observed in South American rivers largely due to anthropogenic activities. However, long-distance migratory fish *Prochilodus lineatus* from six different locations in Brazil exhibited high genetic diversity with minimum spatial genetic structuring in the nuclear and mitochondrial DNA (mtDNA) indicating adequate gene flow (Ferreira et al. 2017). The authors emphasized the importance of unimpeded tributaries to avoid obstruction of migratory fishes. To minimize such a negative impact on the migratory routes in the Parana River basin in Brazil, fish passages and stocking programs have been implemented (Agostinho et al. 2005, 2008). Awareness and implementation of such programs should be considered to be made on a global scale for the conservation of wildlife in its natural habitat and to protect the endangered species.

3.3   Growth in Fishes

The profitability of food aquaculture production is majorly influenced by the fast growth rate of the aquaculture species. It is necessary to understand the molecular mechanisms underlying rapid growth in order to meet the consumers' demand and to maximize economic benefits. Furthermore, the selection of individuals showing a rapid growth rate is important for a successful selective breeding program. The pattern of inheritance of genes for a particular trait and its variation among individuals (allelic variation) can be detected by the construction of genetic maps, using low-density markers such as Amplified Fragment Length Polymorphism (AFLP), Random Amplified Polymorphic DNA (RAPD), and Restriction Fragment Length Polymorphism (RFLP). However, such low-density markers lack even distribution across the genome. Therefore, the use of high density and widely distributed SNP markers across the genome are suitable for genome-wide association studies (Wang et al. 2008). The availability of next-generation sequencing platforms has accelerated the genome-wide SNP identification and GWAS studies. The traditional phenotype-based selection is tedious and requires surveillance of several generations to select for desirable traits. The use of high-throughput sequencing offers the detection of markers associated with QTLs related to growth differences. Hence, the selection of fishes with desirable growth traits can be identified rapidly using NGS technologies (Liu et al. 2014; Fu et al. 2016). SNPs detection is the most common technique for evaluation of the genetic differences between individuals. The SNPs, located near or within a promoter region, are considered to be more likely involved in altering gene expression. Hence, SNPs are used to find markers associated with growth traits. Also, SNPs are responsible for more than 90% of the differences at the genetic level between individual fish (Salem et al. 2012).

RNA-seq analysis of slow and fast-growing rainbow trout population identified several synonymous SNPs ($n = 54$) in 778 individuals for growth phenotypes over a span of three generations. This led to the confirmation of 22 SNPs which were significantly involved in growth traits (Salem et al. 2012). Another RNA-seq study of muscle and liver tissues from slow and fast-growing saline tilapia identified differentially expressed genes (DEGs) such as *foxK1, cyp7b1, crot, sparc, smad3, usp38, impa1, fadps, sqlea*, and *gss* located within QTL for growth and omega-3, suggesting their probable association with growth and omega-3 fatty acid content

(Lin et al. 2019). GWAS by double digested restriction-site associated DNA sequencing (ddRADseq) analysis of a commercially important fish brown-marbled grouper, *Epinephelus fuscoguttatus*, identified five candidate genes having significant association with SNPs for growth traits. Three DEGs *bmp2k*, *wasf1*, and *acyp2*, involved in skeletal development, mitochondrial structure maintenance, and metabolism indicating the possible association of these genes to growth traits, were identified (Yang et al. 2020). Such studies show a high variability in SNPs and DEGs among the commercially important fish species. Similarly, the hybrid mandarin fish in China is of great value due to their delicacy and commercial importance. RNA-seq analysis of the genes and pathways which are activated in response to growth performance in hybrid mandarin fish identified several SNPs and 32 DEGs related to fatty acid biosynthesis (e.g., *FASN* and *ACACB*), collecting duct acid secretion (e.g., *ATP6E* and *KCC4*), cell cycle (e.g., *CDC20* and *CCNB*), and the insulin-like growth factor system (*IGFBP1*) (Guan and Qiu 2020). The authors hypothesized that the growth trait could be complex and involved the functions under metabolism together with the SNP alleles. Although several studies, involving the NGS technology, have made substantial advancement in the identification of growth trait molecular markers, the high variability of the SNPs and DEGs between the fish species warrants further studies. In addition, the application of marker-assisted selection (MAS) could be promoted to ensure rapid development of the aquaculture industry.

3.4 Toxicity

The lack of proper plastic waste management has resulted in an exponential increase in plastic production. In 2010, the amount of plastic waste entering the ocean ranged from 4.8 to 12.7 million tonnes which posed a severe threat to aquatic organisms. Microplastics belong to the classification of plastics measuring less than 5 mm in dimension (Lusher et al. 2017). Direct or indirect ingestion and assimilation of microplastics by the aquatic organisms and a long residence time have caused high risk to aquatic as well as terrestrial organisms since they can enter food webs (Anderson et al. 2016). RNA-seq, using Illumina or 454 pyrosequencing technologies, has been utilized in studying the effect of toxicants on different tissues (Zheng et al. 2018; Cai et al. 2019). It is a powerful technology to study the impact of toxicants on the expression of target genes, the molecular pathways, and its impact on animal physiology. RNA-seq studies of zebra fish (*Danio rerio*) embryos, exposed to polystyrene (PS) particles at different developmental stages, identified the up-regulation of genes involved in alternative complement pathway such as *cfb*, *cfh*, *c3a.2*, *c3a.3*, *c3a.6*, and *c9* (Veneman et al. 2017). In addition to microplastics, heavy metals cause severe injuries to multiple organs. Exposure of yellow perch (*Perca flavescens*) to Cadmium (Cd) and Copper (Cu) reduces the expression of genes involved in immune responses, protein biosynthesis, metabolism of lipids as well as chromatin modification or DNA methylation (Pierron et al. 2011). Iron sulfide (FeS) nanoparticles also induce aberrations in the expression of several

Table 3.4 MicroRNA (miRNA) discovery in aquatic species using NGS technologies

Sr. No.	Species	NGS platform used	No of miRNA identified	Year	Reference
1	Atlantic halibut (*Hippoglossus hippoglossus* L.)	SOLiD deep sequencing	199 conserved, one novel antisense, and one miRNA* mature form	2012	Bizuayehu et al. (2012)
2	Rainbow trout (*Oncorhynchus mykiss*)	Illumina GAIIx	266 known identified and 230 putatively novel miRNAs	2012	Ma et al. (2012)
3	Common carp (*Cyprinus Carpio* L.)	Illumina Solexa	188 conserved miRNAs and seven novel miRNAs	2012	Yan et al. (2012)
4	Common carp (*Cyprinus carpio*)	IlluminaGAIIx	113 mature miRNAs, 92 conserved miRNAs and 21 common carp specific miRNAs	2012	Zhu et al. (2012)
5	Blunt snout bream (*Megalobrama amblycephala*)	Illumina/ Solexa GA	347 conserved miRNAs (belonging to 123 families) and 22 novel miRNAs	2013	Yi et al. (2013)
6	Japanese flounder (*Paralichthys olivaceus*)	Illumina Hiseq2000	381 host miRNAs, 130 were discovered for the first time	2014	Zhang et al. (2014)
7	Common Carp (*Cyprinus carpio*)	Illumina Hiseq 2500	698 miRNAs, 8 miRNAs first identified in common carp and 556 predicted mi RNAs	2016	Zhao et al. (2016)
8	Rainbow trout (*Oncorhynchus mykiss*)	Illumina HiSeq-1000	445 conserved putative precursors correspond to 123 mature conserved miRNAs	2016	Juanchich et al. (2016)
9	*Colossoma macropomum*	Illumina HiSeq 2000	279 conserved miRNAs, 257 from liver and 272 from skin	2017	Gomes et al. (2017)
10	*Nibea albiflora* (Yellow drum)	Illumina HiSeq	180 known and 71 novel miRNAs	2019	Xie et al. (2019)

genes in the liver including *stat2*, *sod3a*, *cyp1a*, *krt18*, *pdoa4*, *flot2a*, *cp*, etc. which are involved indetoxification, immunity, inflammatory response, oxidative stress, DNA damage/repair (Zheng et al. 2018). Similarly, van Aerle et al. (2013) carried out a study on zebra fish embryos to understand the effect of silver nanoparticles and found dose-dependent decrease in oxygen consumption. Zebra fish embryos exposed to 2, 3, 7, 8-tetrachlorodibenzo-p-dioxin (TCDD) for a short time showed changes in miRNA expression, resulting to impairment in cardiovascular development and hematopoiesis (Jenny et al. 2012). Such studies show the necessity for proper regulations in the application of nanoparticle-based medicines and fertilizers

Table 3.5 Long noncoding RNA discoveries in fish species using NGS technologies

Sr. No.	Species	NGS platform used	No. of long non-coding RNA identified	Year	Reference
1	Atlantic salmon (*Salmo salar*)	Illumina MiSeq	5636 putative lncRNAs	2016	Boltaña et al. (2016)
2	Tilapia (*Oreochromis niloticus*)	Illumina Hiseq 2500	797 putative lncRNAs	2017	Luo et al. (2017)
3	Tilapia	Illumina HiSeq2500	72,276 high-confidence lncRNAs	2018	Li et al. (2018)
4	*Crassostrea gigas*	Illumina Hiseq. 4000	12,243 lncRNA transcripts	2018	Feng et al. (2018)
5	Common carp (*Cyprinus carpio*)	Illumina HiSeq X	14,199 lncRNA, 124 differentially expressed	2019	Song et al. (2019)
6	Nile tilapia (*Oreochromis niloticus*)	Illumina Hiseq 2000	3401 noncoding lncRNAs	2019	Cai et al. (2019)
7	Koi carp (*Cyprinus carpio* L.)	Illumina HiSeq X	77,159 lncRNAs, 4252 previously annotated and 72,907 novel lncRNAs	2019	Luo et al. (2019)
8	Coho salmon (*Oncorhynchus kisutch*)	Illumina HiSeq 2000	4975 lncRNAs	2020	Leiva et al. (2020)
9	Grasscarp (*Ctenopharyngodon idella*)	Illumina HiSeq 2000	4240 lncRNA transcripts	2020	Gan et al. (2020)
10	White leg shrimp (*Penaeus vannamei*)	PacBio Sequel	441 putative lncRNAs were identified	2020	Ren et al. (2020)

with heavy metals in agriculture. MicroRNA (Table 3.4) and long non-coding RNA (Table 3.5) studies have been carried out in several fish species to study its role in developmental process. Interestingly, increased expression of transcripts encoding peroxidases, heat shock protein, and superoxide dismutase was identified in the biodegradation and metabolism of xenobiotics in *Tigriopus japonicas*, indicating its resilience against such xenobiotics (Kim et al. 2015).

Ammonia-based fertilizers are common in agriculture. Excess usage of ammonia is often accumulated in the surrounding water bodies. According to a transcriptomic analysis, toxicity of ammonia in white spot syndrome virus (WSSV) infected shrimp (*Litopenaeus vannamei*) increases expression of genes related to pathogenicity and disease formation (Lu et al. 2019). The ammonia toxicity also significantly reduced the expression of several genes related to inhibition of viral pathogenesis, multiplication and proteolysis, scavenging of antioxidants, transcription regulation and DNA repair mechanisms. Further, ammonia is metabolized to nitrite by ammonia-

Table 3.6 Transcriptomics studies of various fish species using next-generation sequencing technologies

Sr. No	Species	NGS platform used	Key findings	Year	Reference
1	Chinese mitten crab (*Eriocheir sinensis*)	Illumina HiSeq™ 2000	Candidate genes involved in reproduction	2012	He et al. (2012)
2	Nile tilapia (*Oreochromis niloticus*)	Illumina HiSeq™ 2000	Immune-related DEGs upon *Streptococcus agalactiae* infection	2013	Zhang et al. (2013)
3	Pacific white shrimp (*Litopenaeus vannamei*)	454 GS FLX	DEGs upon WSSV infection	2013	Chen et al. (2013)
4	Pacific white shrimp (*Litopenaeus vannamei*)	Illumina HiSeq 2000	Characterization of transcriptome during five different early development stages	2014	Wei et al. (2014)
5	Bluefin tunas (*Thunnus* spp.)	Illumina HiSeq2500	Key genes and molecular pathways involved in	2016	Bar et al. (2016)
6	Grass carp (*Ctenopharyngodon idella*)	Illumina Hiseq 2500™	Immune-related DEGs upon *A. hydrophila* infection	2016	Yang et al. (2016)
7	Grass carp (*Ctenopharyngodon idella*)	Illumina HiSeq4000	Immune-relevant genes and pathways in grass crap between 1-year and 3-years of age	2017	Li et al. (2017)
8	*Channa punctatus*	Illumina HiSeq 2000	Transcripts associated with testicular activities from preparatory to postspawning	2017	Roy et al. (2017)
9	Zebrafish (*Danio rerio*)	Illumina Hiseq. 4000	Altered gene expression on exposure to CMC-FeSnanoparticle	2018	Zheng et al. (2018)
10	Oriental river prawn (*Macrobrachium nipponense*)	Illumina HiSeq2005™	Differentially expressed immune-related genes and signaling pathways	2018	Zhao et al. (2018)

oxidizing microbes. Exposure of *Megalobrama amblycephala* to nitrite caused alteration in the expression of genes in the liver involved in proteins and fats metabolism, oxygen transport, oxidative stress, immune responses, and apoptotic pathways indicating the seriousness of nitrite toxicity (Sun et al. 2014).

A myriad of studies has shown the high-throughput nature of the transcriptome (Table 3.6) and its suitability in obtaining a complete snapshot of molecular mechanisms in response to various external toxicants and their genotoxic effects. Contamination of aquatic ecosystems with heavy metals can be hazardous, causing serious health issues to the aquatic organisms. This can further affect the human health through the food chain. The response of aquatic organisms to microplastics and such toxicants has been poorly understood. Hence, further research should be undertaken focussing on the insights of the molecular mechanisms to understand the response. Also, proper management of waste disposal should be carried out to avoid entry of the same into the water bodies.

3.5   Microbiome

The gut microbiota is of high importance due to its role in digestion, reproduction, growth, immunity and overall health of the fish. Advances in DNA sequencing technologies and bioinformatics have led to a deeper understanding of the complex microbial communities' profiles in the fish gut and dynamics of this profile in response to variation in feeding strategy, developmental stages, water temperature, salinity, etc. (Ghanbari et al. 2015). The microbial community structure comprises of different microbial species that coexist in a particular environment and influence each other's function. The resident microbiota in the gut is constantly exposed to the environmental microbiome through food and water. These microorganisms, from the surroundings, affect the resident microbiota and thus, influence fish physiology. An understanding of the phylogenetic composition and metabolites secreted by these microorganisms can help to deduce the relationship between gut microbiota and physiology (Austin 2006). The culture-dependent methods detect only a limited fraction of the microbial community because some microorganisms cannot grow in laboratory conditions but can play an essential role in the microbiome of the fish gut. Hence, culture-dependent methods can limit the detection of these beneficial microorganisms present in the microbiome. Illumina and 454/Roche pyrosequencing have been frequently used for fish microbiota studies (Table 3.7).

The heath of the host is partly governed by the microbiome it harbors. Therefore, aberrations in the microbiome can greatly disturb the normal physiology of the host. Hence, understanding the influence of the microbiome by the feeding habits, environment, and other biotic and abiotic factors is essential. A study was carried out by Eichmiller et al. (2016) to delineate the effects of environment in the gut microbiota of carps. The authors discovered that the alpha diversity of the fecal microbiome, obtained from captive and wild carps, exhibited Shannon index ranging from 2.3 to 4.5, indicating a wide difference in the bacterial diversity. The similarity in the microbiome was observed among the captive fishes rather than in the wild fishes. The differences in the microbiome of wild fishes were not associated with feeding habits because the captive fishes exhibited similarity despite differences in their feeding habits. This indicates that the environment plays a major role in shaping the microbiome. In spite of such differences, more than 75% of the reads were assigned to Fusobacteria, Proteobacteria, and Firmicutes. Acore microbiome consisting of five Operational Taxonomic Units (OTUs) was also identified with an abundance of ~40% between the lab and wild carps (Eichmiller et al. 2016). This shows that the surrounding in addition to the feeding habits is a major determining factor in the gut microbiome of fishes. Hence, fish management strategies should also take the environmental surroundings into consideration for the proper development of the gut microbiome.

Table 3.7 Metagenomic studies in fish species using NGS technologies

Sr. No.	Species	NGS platform used	Purpose	Year	Reference
1	Grass carp (*Ctenopharyngodon idellus*)	454/Roche GS FLX Titanium	Characterize intestinal bacterial community and reveal the association between gut microbiota and microbiota from the associated environments	2012	Wu et al. (2012)
2	Wood-eating catfishes (*Panaque* sp.)	454 Life Sciences Genome Sequencer System	Microbial population dynamics in the feces of wood-eating catfishes (*Panaque* sp.)	2013	Di Maiuta et al. (2013)
3	Turbot (*Scophthalmus maximus*)	Illumina Hiseq 2000	Characterize gastrointestinal (GI) microbiome and reveal the relationship between the GI microbiome and its surrounding environment	2013	Xing et al. (2013)
4	Invasive Asian carp (*Hypophthalmichthys molitrix*); Indigenous American fish (*Hypophthalmichthys molitrix*)	454 GS FLX Titanium	Intestinal microbial flora in invasive and indigenous species to understand how the gut microbiota influences the physiology and behavior	2014	Ye et al. (2014)
5	Channel catfish (*Ictalurus punctatus*); Largemouth bass (*Micropterus salmoides*); Bluegill (*Lepomis macrochirus*)	Roche titanium 454	Compare the bacterial diversity associated with the gut of three commercially valuable warm water fish species	2014	Larsen et al. (2014)
6	Rainbow trout (*Oncorhynchus mykiss*)	Illumina HiSeq™ 2000	Effect of the administered diet (marine-based versus plant-based) to *Y. ruckeri* infection	2014	Ingerslev et al. (2014)
7	Prussian carp (*Carassius gibelio*)	ABI 3130XL	Correlations between microbial community and environment including soil, water, and fish feed.	2015	Kashinskaya et al. (2015)
8	Zebrafish (*Danio rerio*)		Change in microbial communities with key	2016	Stephens et al. (2016)

(continued)

Table 3.7 (continued)

Sr. No.	Species	NGS platform used	Purpose	Year	Reference
		Illumina HiSeq 2000	developmental, environmental and dietary transitions of the host		
9	Common, Silver, and Bighead carps	Illumina HiSeq 2000	To determine the effect of environment on the microbiome of invasive carps	2016	Eichmiller et al. (2016)
10	Zebrafish (*Danio rerio*)	Illumina MiSeq	Effect of Triclosan exposure on the structure and ecological dynamics of the gut microbiome	2016	Gaulke et al. (2016)
11	Atlantic salmon (*Salmo salar* L.).	Illumina MiSeq	Alteration in intestinal microbiota when transferred from freshwater to seawater	2017	Dehler et al. (2017)

Microbial infection in the fish is one of the major causes of economic loss. To avoid such bacterial infections, selective breeding has been implemented to find the resistant lines. However, with the advent of metagenomics, it has been understood that the microbiome could also play a significant role to ward off the bacterial infections. This is because the fish microbiome is involved in the resistance to colonization of pathogenic microbes through competition of food and space. Similarly, commensal microbiota in mucosal surfaces serves to protect the host against bacterial infection. *Flavobacterium psychrophilum is a well*-known causative agent of bacterial disease *in salmonids. Selective breeding of F. psychrophilum resistant salmonids changed the microbiome in the gut and gills compared to the susceptible salmonids* (Brown et al. 2019). *The changes were reflected by an increase in alpha diversity and Brevinema andersonii, an opportunistic pathogen, in the mid-gut of the susceptible salmonids. However, depletion in the abundance of Mycoplasma sp. was observed in the susceptible salmonids. Such differences in the microbiome indicate the roles of microbiome in disease resistance. It should also be noted that the* microbiome composition and abundance are dependent on the developmental stages of the host. Since the microbiome is involved in the host's health, it is essential to delineate the changes in microbiome during the development. A study on zebrafish ($n = 135$), during its various developmental stages, identified that the gut microbiome varies during the various developmental stages (Stephens et al. 2016). The difference in micro biome was also prominent among the individuals even at the same developmental stages, especially during the change in diet and environment. Such studies show the importance of extensive microbiome analysis, not only at the species level but also during the developmental stages as well. With the rapid progress in the NGS technologies, future work could be expected to deliver necessary pre- and pro-biotic aimed specifically to address the issues related to the various developmental stages.

3.6 Sexual Development

Genes involved in sex identification and sexual maturation provides valuable information to understand the molecular mechanisms in sexual development and take necessary steps for maximum yield of the fish. Identification of genes and methylation patterns (Table 3.8) involved in sex differentiation is important particularly in fishes that lack sexual dimorphism such as sturgeons. The identification of sex at an early developmental stage could help to separate commercially important females from males (Burcea et al. 2018). Investigation of expression profiles during molecular sex differentiation provides a clear understanding of sex differentiation. Introduction of massive parallel sequencing has greatly enhanced the identification of genes, involved in sex differentiation in several aquatic species through RNA-seq/ transcriptome profiling. RNA-seq study of undifferentiated gonads from juvenile sturgeon *Acipenser gueldenstaedtii* identified *gsdf, foxl2, hsd17b1,* and *cyp19a1a* genes as the sexing markers in sturgeon at early developmental stage (Hagihara et al. 2014). Gene *foxl2, hsd17b1,* and *cyp19a1a* were specific to female sex differentiation and *gsdf* was specific to male sex differentiation. Similarly, RNA-seq of differentially expressed miRNAs in the ovary and testis of *Trachinotus ovatus* identified increased expression of microRNAs dre-miR-26, dre-miR-143, dre-miR-101, and dre-miR-202-5p in testis, suggesting its involvement in testicular development and also increased expression of dre-miR-727-5p and dre-let-7c-5p, suggesting its involvement in ovarian development (He et al. 2019). Another major obstacle in fisheries is the determination of sex for selection of parents for selective breeding. An attempt to identify the genes, involved in sex determination in Nile tilapia (*Oreochromis niloticus*), revealed *amh* and *amhy* to be in full concordance with male sex determination while *cr/20β-hsd, gpa33, rtn4ipl,* and *zp3* genes were over expressed in females (Eshel et al. 2014). Nine microRNAs were further identified to exhibit sexually dimorphic expression patterns in the tilapia embryos.

Growth, reproduction, and meat quality of fishes are strongly influenced by the stage at which the fish attains its sexual maturity. Some of the commercially important fishes exhibit sexual precocity in which the individuals develop sexual characteristics before the onset of minimum age. This leads to huge economic loss to the farmers. There is a need for understanding the molecular mechanisms in the reproduction of such species. In a GWAS analysis of about 480 individuals of Atlantic salmon, several markers such as Ssa2 Ssa10, Ssa12, Ssa13, and Ssa25 were found to be associated to early sexual maturation (grilsing) (Gutierrez et al. 2015). The authors also identified that the late sexual maturation was associated with Ssa01, Ssa21, and Ssa28 markers. Besides the role of early sexual maturation, the development of gonadal tissues also plays a crucial role in the sexual development of fishes. Transcriptomic analysis of gonadal tissues viz. testicular and ovarian tissues,

Table 3.8 Studies on epigenetic modifications (DNA methylation) in fishes using NGS

Sr. No	Species	NGS platform used	Purpose	Year	Reference
1	Half-smooth tongue sole (*Cynoglossus semilaevis*)	Illumina HiSeq 2000	Role of DNA methylation in transition from Genetic sex determination (GSD) to environmental sex determination (ESD)	2014	Shao et al. (2014)
2	*Oncorhynchus mykiss*	Illumina HiSeq 2000	Role of epigenetic regulation (DNA methylation) in smoltification	2016	Baerwald et al. (2016)
3	Rainbow trout (*Oncorhynchus mykiss*)	Ion Torrent PGM™	Study effect of nutritional status on global epigenome modifications	2016	Marandel et al. (2016)
4	Nile tilapia (*Oreochromis niloticus*)	Illumina Hiseq 2500	Compare global DNA methylation changes in female and male gonads between control and high temperature-induced groups.	2016	Sun et al. (2016)
5	Three-spined stickleback (*Gasterosteus aculeatus*)	Illumina 2500	Role of DNA methylation in the short-term and long-term adaptation to changed salinity	2017	Artemov et al. (2017)
6	Atlantic salmon (*Salmo salar*)	Pyrosequencing	Impact of genetic background and embryonic temperature on the epigenetic regulation of body growth and white muscle phenotype	2017	Burgerhout et al. (2017)
7	Pacific salmon (*Oncorhynchus kisutch*)	Illumina HiSeq. 2000	Compare methylation patterns in white muscle tissue of hatchery-reared and wild coho salmon	2017	Le Luyer et al. (2017)
8	Nile tilapia (*Oreochromis niloticus* Linnaeus).	Illumina HiSeq 2000	DNA methylation role in sex differentiation or maintenance	2017	Chen et al. (2017)
9	Zebrafish (*Danio rerio*)	Illumina HiSeq 2000	Effect of a parental high ARA diet on hepatic DNA methylation patterns	2019	Adam et al. (2019)
10	Atlantic salmon	Illumina HiSeq 2500	Role of DNA methylation in inducing stress response	2019	Robinson et al. (2019)

of common carp (*Cyprinus carpio*) identified over 1.2 million transcripts consisting of more than 6000 Clusters of Orthologous Groups (COGs), 26,000 Simple Sequence Repeats (SSRs), and 298 DEGs (Anitha et al. 2019). The authors further verified the expression of the DEGs such as *nanos*, *ad4bp/sf-1*, and *gdf9* through RT-qPCR. Such studies can provide a foundation for future research on the identification of gender at the early developmental stage, sexual maturation, and discovering a master sex-determining gene. Despite the availability of such huge transcriptomic datasets and whole-genome sequences, there is a lack of extensive in vivo studies to confirm the role of such SNPs, DEGs, and related molecular markers. Also, some of the fishes such as Chinese mitten crab (*Eriocheir sinensis*) need a complex environment for spawning (He et al. 2012). Further studies on the effect of the environment and its impact in sexual development could pave the way for new insights into the reproductivity and health of the fishes to develop.

3.7 Conclusion

The current aquaculture industry is challenged with a myriad of difficulties that hampers the production of fishes which leads to a great deal of economic loss. Implementation of the NGS technologies has paved the path to understand the SNPs, microRNAs and the genes, involved in rapid growth rate, sexual determination and development, response to diseases and xenobiotics, and the impact of anthropogenic actions on the fish migration and genetic diversity. Despite the advancement in the genomics and transcriptomics, extensive studies are the need of the hour owing to the high diversity of economically and ecologically significant fishes. Advancements in the field of genomics and big data analysis are expected to pinpoint the complex network of genes and the pathways in the selection of superior traits.

Acknowledgment The authors are thankful to Director, ICAR—Central Institute of Freshwater Aquaculture, Bhubaneswar, Odisha for the facilitation of the work. The authors are grateful to CABIN project on Centre of Agricultural Bioinformatics (CABin) of ICAR-IASRI, New Delhi for the financial support as a fellowship to first author.

References

Adam AC, Lie KK, Whatmore P, Jakt LM, Moren M, Skjærven KH (2019) Profiling DNA methylation patterns of zebrafish liver associated with parental high dietary arachidonic acid. PLoS One 14(8):e0220934

van Aerle R, Lange A, Moorhouse A, Paszkiewicz K, Ball K, Johnston BD, De-Bastos E, Booth T, Tyler CR, Santos EM (2013) Molecular mechanisms of toxicity of silver nanoparticles in zebrafish embryos. Environ Sci Technol 47(14):8005–8014

Agostinho AA, Thomaz SM, Gomes LC (2005) Conservation of the biodiversity of Brazil's inland waters. Conserv Biol 19(3):646–652

Agostinho AA, Pelicice FM, Gomes LC (2008) Dams and the fish fauna of the Neotropical region: impacts and management related to diversity and fisheries. Braz J Biol 68(4):1119–1132

Anderson JC, Park BJ, Palace VP (2016) Microplastics in aquatic environments: implications for Canadian ecosystems. Environ Pollut 218:269–280

Anitha A, Gupta YR, Deepa S, Ningappa M, Rajanna KB, Senthilkumaran B (2019) Gonadal transcriptome analysis of the common carp, Cyprinus carpio: identification of differentially expressed genes and SSRs. Gen Comp Endocrinol 279:67–77

Artemov AV, Mugue NS, Rastorguev SM, Zhenilo S, Mazur AM, Tsygankova SV, Boulygina ES, Kaplun D, Nedoluzhko AV, Medvedeva YA, Prokhortchouk EB (2017) Genome-wide DNA methylation profiling reveals epigenetic adaptation of stickleback to marine and freshwater conditions. Mol Biol Evol 34(9):2203–2213

Asaduzzaman M, Igarashi Y, Wahab MA, Nahiduzzaman M, Rahman MJ, Phillips MJ, Huang S, Asakawa S, Rahman MM, Wong LL (2020) Population genomics of an Anadromous Hilsa Shad Tenualosailisha Species across its diverse migratory habitats: discrimination by fine-scale local adaptation. Genes 11(1):46

Austin B (2006) The bacterial microflora of fish, revised. Sci World J 6:931–945

Austin CM, Tan MH, Croft LJ, Hammer MP, Gan HM (2015) Whole genome sequencing of the Asian arowana (Scleropages formosus) provides insights into the evolution of ray-finned fishes. Genome Biol Evol 7(10):2885–2895

Baerwald MR, Meek MH, Stephens MR, Nagarajan RP, Goodbla AM, Tomalty KM, Thorgaard GH, May B, Nichols KM (2016) Migration-related phenotypic divergence is associated with epigenetic modifications in rainbow trout. Mol Ecol 25(8):1785–1800

Bar I, Cummins S, Elizur A (2016) Transcriptome analysis reveals differentially expressed genes associated with germ cell and gonad development in the Southern bluefin tuna (Thunnus maccoyii). BMC Genomics 17(1):217

Béné C (2006) Small-scale fisheries: assessing their contribution to rural livelihoods in developing countries. FAO Fish Circ 1008:46

Bernatchez L (2016) On the maintenance of genetic variation and adaptation to environmental change: considerations from population genomics in fishes. J Fish Biol 89(6):2519–2556

Bernatchez L, Wellenreuther M, Araneda C, Ashton DT, Barth JM, Beacham TD, Maes GE, Martinsohn JT, Miller KM, Naish KA, Ovenden JR (2017) Harnessing the power of genomics to secure the future of seafood. Trends Ecol Evol 32(9):665–680

Bizuayehu TT, Lanes CF, Furmanek T, Karlsen BO, Fernandes JM, Johansen SD, Babiak I (2012) Differential expression patterns of conserved miRNAs and isomiRs during Atlantic halibut development. BMC Genomics 13(1):11

Boltaña S, Valenzuela-Miranda D, Aguilar A, Mackenzie S, Gallardo-Escárate C (2016) Long noncoding RNAs (lncRNAs) dynamics evidence immunomodulation during ISAV-Infected Atlantic salmon (Salmo salar). Sci Rep 6(1):1–13

Brown RM, Wiens GD, Salinas I (2019) Analysis of the gut and gill microbiome of resistant and susceptible lines of rainbow trout (Oncorhynchus mykiss). Fish Shellfish Immunol 86:497–506

Buermans HPJ, Den Dunnen JT (2014) Next generation sequencing technology: advances and applications. Biochim Biophys Acta Mol Basis Dis 1842(10):1932–1941

Burcea A, Popa GO, Maereanu M, Dudu A, Georgescu SE, Costache M (2018) Expression characterization of six genes possibly involved in gonad development for stellate sturgeon individuals (Acipenser stellatus, Pallas 1771). Int J Genomics 2018:7835637

Burgerhout E, Mommens M, Johnsen H, Aunsmo A, Santi N, Andersen Ø (2017) Genetic background and embryonic temperature affect DNA methylation and expression of myogenin and muscle development in Atlantic salmon (Salmo salar). PLoS One 12(6):e0179918

Cai J, Li L, Song L, Xie L, Luo F, Sun S, Chakraborty T, Zhou L, Wang D (2019) Effects of long term antiprogestine mifepristone (RU486) exposure on sexually dimorphic lncRNA expression and gonadal masculinization in Nile tilapia (Oreochromis niloticus). Aquat Toxicol 215:105289

Carolsfeld J (2003) Migratory fishes of South America: biology, fisheries and conservation status. IDRC, Ottawa, ON

Chen X, Zeng D, Chen X, Xie D, Zhao Y, Yang C, Li Y, Ma N, Li M, Yang Q, Liao Z (2013) Transcriptome analysis of Litopenaeus vannamei in response to white spot syndrome virus infection. PLoS One 8(8):e73218

Chen S, Zhang G, Shao C, Huang Q, Liu G, Zhang P, Song W, An N, Chalopin D, Volff JN, Hong Y (2014) Whole-genome sequence of a flatfish provides insights into ZW sex chromosome evolution and adaptation to a benthic lifestyle. Nat Genet 46(3):253–260

Chen X, Wang Z, Tang S, Zhao Y, Zhao J (2017) Genome-wide mapping of DNA methylation in Nile Tilapia. Hydrobiologia 791(1):247–257

Dehler CE, Secombes CJ, Martin SA (2017) Seawater transfer alters the intestinal microbiota profiles of Atlantic salmon (Salmo salar L.). Sci Rep 7(1):1–11

Di Maiuta N, Schwarzentruber P, Schenker M, Schoelkopf J (2013) Microbial population dynamics in the faeces of wood-eating loricariid catfishes. Lett Appl Microbiol 56(6):401–407

Dingle H (2006) Animal migration: is there a common migratory syndrome? J Ornithol 147 (2):212–220

Dingle H, Drake VA (2007) What is migration? Bioscience 57(2):113–121

Eichmiller JJ, Hamilton MJ, Staley C, Sadowsky MJ, Sorensen PW (2016) Environment shapes the fecal microbiome of invasive carp species. Microbiome 4(1):44

Eshel O, Shirak A, Dor L, Band M, Zak T, Markovich-Gordon M, Chalifa-Caspi V, Feldmesser E, Weller JI, Seroussi E, Hulata G (2014) Identification of male-specific amh duplication, sexually differentially expressed genes and microRNAs at early embryonic development of Nile tilapia (Oreochromis niloticus). BMC Genomics 15(1):774

Feng D, Li Q, Yu H, Kong L, Du S (2018) Transcriptional profiling of long non-coding RNAs in mantle of Crassostrea gigas and their association with shell pigmentation. Sci Rep 8(1):1–10

Ferreira DG, Souza-Shibatta L, Shibatta OA, Sofia SH, Carlsson J, Dias JHP, Makrakis S, Makrakis MC (2017) Genetic structure and diversity of migratory freshwater fish in a fragmented Neotropical river system. Rev Fish Biol Fish 27(1):209–231

Figueras A, Robledo D, Corvelo A, Hermida M, Pereiro P, Rubiolo JA, Gómez-Garrido J, Carreté L, Bello X, Gut M, Gut IG (2016) Whole genome sequencing of turbot (Scophthalmus maximus; Pleuronectiformes): a fish adapted to demersal life. DNA Res 23(3):181–192

Fu B, Liu H, Yu X, Tong J (2016) A high-density genetic map and growth related QTL mapping in bighead carp (Hypophthalmichthys nobilis). Sci Rep 6:28679

Gan L, Wang YZ, Chen SJ, Lin ZH, Sun JJ, He YH, Tang HJ, Peng J, Guo HH (2020) Identification and characterization of long non-coding RNAs in muscle sclerosis of grass carp, Ctenopharyngodon idellus fed with faba bean meal. Aquaculture 516:734521

Gaulke CA, Barton CL, Proffitt S, Tanguay RL, Sharpton TJ (2016) Triclosan exposure is associated with rapid restructuring of the microbiome in adult zebrafish. PLoS One 11(5): e0154632

Ghanbari M, Kneifel W, Domig KJ (2015) A new view of the fish gut microbiome: advances from next-generation sequencing. Aquaculture 448:464–475

Gomes F, Watanabe L, Nozawa S, Oliveira L, Cardoso J, Vianez J, Nunes M, Schneider H, Sampaio I (2017) Identification and characterization of the expression profile of the microRNAs in the Amazon species Colossoma macropomum by next generation sequencing. Genomics 109 (2):67–74

Guan WZ, Qiu GF (2020) Transcriptome analysis of the growth performance of hybrid mandarin fish after food conversion. PLoS One 15(10):e0240308

Gutierrez AP, Yáñez JM, Fukui S, Swift B, Davidson WS (2015) Genome-wide association study (GWAS) for growth rate and age at sexual maturation in Atlantic salmon (Salmo salar). PLoS One 10(3):e0119730

Hagihara S, Yamashita R, Yamamoto S, Ishihara M, Abe T, Ijiri S, Adachi S (2014) Identification of genes involved in gonadal sex differentiation and the dimorphic expression pattern in undifferentiated gonads of Russian sturgeon Acipenser gueldenstaedtii Brandt & Ratzeburg, 1833. J Appl Ichthyol 30(6):1557–1564

Hale MC, Thrower FP, Berntson EA, Miller MR, Nichols KM (2013) Evaluating adaptive divergence between migratory and nonmigratory ecotypes of a salmonid fish, Oncorhynchus mykiss. G3 3(8):1273–1285

He L, Wang Q, Jin X, Wang Y, Chen L, Liu L, Wang Y (2012) Transcriptome profiling of testis during sexual maturation stages in Eriocheir sinensis using Illumina sequencing. PLoS One 7 (3):e33735

He P, Wei P, Chen X, Lin Y, Peng J (2019) Identification and characterization of microRNAs in the gonad of Trachinotus ovatus using Solexa sequencing. Compar Biochem Physiol D Genom Proteom 30:312–320

Hecht BC, Campbell NR, Holecek DE, Narum SR (2013) Genome-wide association reveals genetic basis for the propensity to migrate in wild populations of rainbow and steelhead trout. Mol Ecol 22(11):3061–3076

Henkel CV, Dirks RP, de Wijze DL, Minegishi Y, Aoyama J, Jansen HJ, Turner B, Knudsen B, Bundgaard M, Hvam KL, Boetzer M (2012) First draft genome sequence of the Japanese eel, Anguilla japonica. Gene 511(2):195–201

Hess JE, Zendt JS, Matala AR, Narum SR (2016) Genetic basis of adult migration timing in anadromous steelhead discovered through multivariate association testing. Proc R Soc B Biol Sci 283(1830):20153064

Ingerslev HC, Strube ML, von Gersdorff Jørgensen L, Dalsgaard I, Boye M, Madsen L (2014) Diet type dictates the gut microbiota and the immune response against Yersinia ruckeri in rainbow trout (Oncorhynchus mykiss). Fish Shellfish Immunol 40(2):624–633

Jenny MJ, Aluru N, Hahn ME (2012) Effects of short-term exposure to 2, 3, 7, 8-tetrachlorodibenzo-p-dioxin on microRNA expression in zebrafish embryos. Toxicol Appl Pharmacol 264(2):262–273

Juanchich A, Bardou P, Rué O, Gabillard JC, Gaspin C, Bobe J, Guiguen Y (2016) Characterization of an extensive rainbow trout miRNA transcriptome by next generation sequencing. BMC Genomics 17(1):1–12

Kashinskaya EN, Belkova NL, Izvekova GI, Simonov EP, Andree KB, Glupov VV, Baturina OA, Kabilov MR, Solovyev MM (2015) A comparative study on microbiota from the intestine of Prussian carp (Carassius gibelio) and their aquatic environmental compartments, using different molecular methods. J Appl Microbiol 119(4):948–961

Kim HS, Lee BY, Won EJ, Han J, Hwang DS, Park HG, Lee JS (2015) Identification of xenobiotic biodegradation and metabolism-related genes in the copepod Tigriopus japonicus whole transcriptome analysis. Mar Genomics 24:207–208

Kumar G, Kocour M (2017) Applications of next-generation sequencing in fisheries research: a review. Fish Res 186:11–22

Larsen AM, Mohammed HH, Arias CR (2014) Characterization of the gut microbiota of three commercially valuable warmwater fish species. J Appl Microbiol 116(6):1396–1404

Le Luyer J, Laporte M, Beacham TD, Kaukinen KH, Withler RE, Leong JS, Rondeau EB, Koop BF, Bernatchez L (2017) Parallel epigenetic modifications induced by hatchery rearing in a Pacific salmon. Proc Natl Acad Sci 114(49):12964–12969

Leiva F, Rojas-Herrera M, Reyes D, Bravo S, Garcia KK, Moya J, Vidal R (2020) Identification and characterization of miRNAs and lncRNAs of coho salmon (Oncorhynchus kisutch) in normal immune organs. Genomics 112(1):45–54

Li G, Zhao Y, Wang J, Liu B, Sun X, Guo S, Feng J (2017) Transcriptome profiling of developing spleen tissue and discovery of immune-related genes in grass carp (Ctenopharyngodon idella). Fish Shellfish Immunol 60:400–410

Li BJ, Jiang DL, Meng ZN, Zhang Y, Zhu ZX, Lin HR, Xia JH (2018) Genome-wide identification and differentially expression analysis of lncRNAs in tilapia. BMC Genomics 19(1):729

Liedvogel M, Åkesson S, Bensch S (2011) The genetics of migration on the move. Trends Ecol Evol 26(11):561–569

Lin G, Thevasagayam NM, Wan ZY, Ye BQ, Yue GH (2019) Transcriptome analysis identified genes for growth and omega-3/-6 ratio in saline tilapia. Front Genet 10:244

Liu F, Sun F, Xia JH, Li J, Fu GH, Lin G, Tu RJ, Wan ZY, Quek D, Yue GH (2014) A genome scan revealed significant associations of growth traits with a major QTL and GHR2 in tilapia. Sci Rep 4:7256

Liu Z, Liu S, Yao J, Bao L, Zhang J, Li Y, Jiang C, Sun L, Wang R, Zhang Y, Zhou T (2016) The channel catfish genome sequence provides insights into the evolution of scale formation in teleosts. Nat Commun 7(1):1–13

Lu X, Luan S, Dai P, Luo K, Chen B, Cao B, Sun L, Yan Y, Kong J (2019) Insights into the molecular basis of immunosuppression and increasing pathogen infection severity of ammonia toxicity by transcriptome analysis in pacific white shrimp Litopenaeus vannamei. Fish Shellfish Immunol 88:528–539

Luo H, Yang H, Lin Y, Zhang Y, Pan C, Feng P, Yu Y, Chen X (2017) LncRNA and mRNA profiling during activation of tilapia macrophages by HSP70 and Streptococcus agalactiae antigen. Oncotarget 8(58):98455

Luo M, Wang L, Yin H, Zhu W, Fu J, Dong Z (2019) Integrated analysis of long non-coding RNA and mRNA expression in different colored skin of koi carp. BMC Genomics 20(1):515

Lusher A, Hollman P, Mendoza-Hill J (2017) Microplastics in fisheries and aquaculture: status of knowledge on their occurrence and implications for aquatic organisms and food safety. FAO, Rome

Ma H, Hostuttler M, Wei H, Rexroad CE III, Yao J (2012) Characterization of the rainbow trout egg microRNA transcriptome. PLoS One 7(6):e39649

Marandel L, Lepais O, Arbenoits E, Véron V, Dias K, Zion M, Panserat S (2016) Remodelling of the hepatic epigenetic landscape of glucose-intolerant rainbow trout (Oncorhynchus mykiss) by nutritional status and dietary carbohydrates. Sci Rep 6(1):1–12

McIntyre PB, Liermann CAR, Revenga C (2016) Linking freshwater fishery management to global food security and biodiversity conservation. Proc Natl Acad Sci 113(45):12880–12885

Melnychuk MC, Peterson E, Elliott M, Hilborn R (2017) Fisheries management impacts on target species status. Proc Natl Acad Sci 114(1):178–183

Nakamura Y, Mori K, Saitoh K, Oshima K, Mekuchi M, Sugaya T, Shigenobu Y, Ojima N, Muta S, Fujiwara A, Yasuike M (2013) Evolutionary changes of multiple visual pigment genes in the complete genome of Pacific bluefin tuna. Proc Natl Acad Sci 110(27):11061–11066

Pauly D, Zeller D (2019) Agreeing with FAO: comments on SOFIA 2018. Mar Policy 100:332–333

Pierron F, Normandeau E, Defo MA, Campbell PG, Bernatchez L, Couture P (2011) Effects of chronic metal exposure on wild fish populations revealed by high-throughput cDNA sequencing. Ecotoxicology 20(6):1388–1399

Ren Y, Li J, Guo L, Liu JN, Wan H, Meng Q, Wang H, Wang Z, Lv L, Dong X, Zhao W (2020) Full-length transcriptome and long non-coding RNA profiling of whiteleg shrimp Penaeus vannamei hemocytes in response to Spiroplasma eriocheiris infection. Fish Shellfish Immunol 106:876–886

Robinson NA, Johnsen H, Moghadam H, Andersen Ø, Tveiten H (2019) Early developmental stress affects subsequent gene expression response to an acute stress in Atlantic salmon: an approach for creating robust fish for aquaculture? G3 9(5):1597–1611

Roy A, Basak R, Rai U (2017) De novo sequencing and comparative analysis of testicular transcriptome from different reproductive phases in freshwater spotted snakehead Channa punctatus. PLoS One 12(3):e0173178

Salem M, Vallejo RL, Leeds TD, Palti Y, Liu S, Sabbagh A, Rexroad CE III, Yao J (2012) RNA-Seq identifies SNP markers for growth traits in rainbow trout. PLoS One 7(5):e36264

Shao C, Li Q, Chen S, Zhang P, Lian J, Hu Q, Sun B, Jin L, Liu S, Wang Z, Zhao H (2014) Epigenetic modification and inheritance in sexual reversal of fish. Genome Res 24(4):604–615

Shin SC, Ahn DH, Kim SJ, Pyo CW, Lee H, Kim MK, Lee J, Lee JE, Detrich HW, Postlethwait JH, Edwards D (2014) The genome sequence of the Antarctic bullhead notothen reveals evolutionary adaptations to a cold environment. Genome Biol 15(9):1–14

Song L, Bian C, Luo Y, Wang L, You X, Li J, Qiu Y, Ma X, Zhu Z, Ma L, Wang Z (2016) Draft genome of the Chinese mitten crab, Eriocheir sinensis. GigaScience 5(1):s13742–s13016

Song F, Wang L, Zhu W, Dong Z (2019) Long noncoding RNA and mRNA expression profiles following igf3 knockdown in common carp, Cyprinus carpio. Sci Data 6:190024

Stephens WZ, Burns AR, Stagaman K, Wong S, Rawls JF, Guillemin K, Bohannan BJ (2016) The composition of the zebrafish intestinal microbial community varies across development. ISME J 10(3):644–654

Sun S, Ge X, Xuan F, Zhu J, Yu N (2014) Nitrite-induced hepatotoxicity in blunt snout bream (Megalobrama amblycephala): the mechanistic insight from transcriptome to physiology analysis. Environ Toxicol Pharmacol 37(1):55–65

Sun LX, Wang YY, Zhao Y, Wang H, Li N, Ji XS (2016) Global DNA methylation changes in Nile tilapia gonads during high temperature-induced masculinization. PLoS One 11(8):e0158483

Veneman WJ, Spaink HP, Brun NR, Bosker T, Vijver MG (2017) Pathway analysis of systemic transcriptome responses to injected polystyrene particles in zebrafish larvae. Aquat Toxicol 190:112–120

Wang S, Sha Z, Sonstegard TS, Liu H, Xu P, Somridhivej B, Peatman E, Kucuktas H, Liu Z (2008) Quality assessment parameters for EST-derived SNPs from catfish. BMC Genomics 9(1):450

Wang Y, Lu Y, Zhang Y, Ning Z, Li Y, Zhao Q, Lu H, Huang R, Xia X, Feng Q, Liang X (2015) The draft genome of the grass carp (Ctenopharyngodon idellus) provides insights into its evolution and vegetarian adaptation. Nat Genet 47(6):625–631

Wei J, Zhang X, Yu Y, Huang H, Li F, Xiang J (2014) Comparative transcriptomic characterization of the early development in Pacific white shrimp Litopenaeus vannamei. PLoS One 9(9): e106201

Wilcove DS, Wikelski M (2008) Going, going, gone: is animal migration disappearing. PLoS Biol 6(7):e188

Wu S, Wang G, Angert ER, Wang W, Li W, Zou H (2012) Composition, diversity, and origin of the bacterial community in grass carp intestine. PLoS One 7(2):e30440

Xie X, Ma R, Qian D, Yu Y, Liu X, Lei Y, Lin Y, Yin F (2019) MicroRNA regulation during Nibea albiflora immuno-resistant against Cryptocaryon irritans challenge in fish skin. Aquaculture 507:211–221

Xing M, Hou Z, Yuan J, Liu Y, Qu Y, Liu B (2013) Taxonomic and functional metagenomic profiling of gastrointestinal tract microbiome of the farmed adult turbot (Scophthalmus maximus). FEMS Microbiol Ecol 86(3):432–443

Xu P, Zhang X, Wang X, Li J, Liu G, Kuang Y, Xu J, Zheng X, Ren L, Wang G, Zhang Y (2014) Genome sequence and genetic diversity of the common carp, Cyprinus carpio. Nat Genet 46 (11):1212–1219

Yan X, Ding L, Li Y, Zhang X, Liang Y, Sun X, Teng CB (2012) Identification and profiling of microRNAs from skeletal muscle of the common carp. PLoS One 7(1):e30925

Yang Y, Yu H, Li H, Wang A (2016) Transcriptome profiling of grass carp (Ctenopharyngodon idellus) infected with Aeromonas hydrophila. Fish Shellfish Immunol 51:329–336

Yang Y, Wu L, Wu X, Li B, Huang W, Weng Z, Lin Z, Song L, Guo Y, Meng Z, Liu X (2020) Identification of candidate growth-related SNPs and genes using GWAS in brown-marbled grouper (Epinephelus fuscoguttatus). Mar Biotechnol 22:153–166

Ye L, Amberg J, Chapman D, Gaikowski M, Liu WT (2014) Fish gut microbiota analysis differentiates physiology and behavior of invasive Asian carp and indigenous American fish. ISME J 8(3):541–551

Yi S, Gao ZX, Zhao H, Zeng C, Luo W, Chen B, Wang WM (2013) Identification and characterization of microRNAs involved in growth of blunt snout bream (Megalobrama amblycephala) by Solexa sequencing. BMC Genomics 14(1):754

Zhang R, Zhang LL, Ye X, Tian YY, Sun CF, Lu MX, Bai JJ (2013) Transcriptome profiling and digital gene expression analysis of Nile tilapia (Oreochromis niloticus) infected by Streptococcus agalactiae. Mol Biol Rep 40(10):5657–5668

Zhang BC, Zhang J, Sun L (2014) In-depth profiling and analysis of host and viral microRNAs in Japanese flounder (Paralichthys olivaceus) infected with megalocytivirus reveal involvement of microRNAs in host-virus interaction in teleost fish. BMC Genomics 15(1):878

Zhao L, Lu H, Meng Q, Wang J, Wang W, Yang L, Lin L (2016) Profilings of microRNAs in the liver of Common Carp (Cyprinus carpio) infected with flavobacterium columnare. Int J Mol Sci 17(4):566

Zhao C, Fu H, Sun S, Qiao H, Zhang W, Jin S, Jiang S, Xiong Y, Gong Y (2018) A transcriptome study on Macrobrachium nipponense hepatopancreas experimentally challenged with white spot syndrome virus (WSSV). PLoS One 13(7):e0200222

Zheng M, Lu J, Zhao D (2018) Toxicity and transcriptome sequencing (RNA-seq) analyses of adult zebrafish in response to exposure carboxymethyl cellulose stabilized iron sulfide nanoparticles. Sci Rep 8(1):1–11

Zhu YP, Xue W, Wang JT, Wan YM, Wang SL, Xu P, Zhang Y, Li JT, Sun XW (2012) Identification of common carp (Cyprinus carpio) microRNAs and microRNA-related SNPs. BMC Genomics 13(1):413

Genome Sequencing in Fishes

4

Ravindra Kumar, Basdeo Kushwaha, and Mahender Singh

Abstract

In the recent years, the genome sequencing of organisms has been adapted as a tool for understanding genetic variations affecting body functions, developing markers for tagging these variations useful in genome-wide association studies (GWAS), studying G × E and host–pathogen interactions, evolutionary analysis and other resource generation. It has been facilitated by a number of cutting edge technologies, such as high throughput parallel DNA sequencing, high performance computing and various computer algorithms. These genomic tools are being extensively used in various research programs on commercially important fish and shellfish. Large number of genome information and huge amount of expressed sequence tags (EST) are now available in public database for many aquatic species. The availability of DNA sequencing data mobilized genetic research to an unprecedented level to provide insight into diseases and complex traits. The GWAS has reached to a level where dataset of a large number of individuals have been included, leading to the establishment of large scale of bio-bank. Such developments led to the new era of consumer genetics and genetic testing companies, not only for humans, plants and commercial animals, but also for commercial fishes. Moreover, there are chances of improper use of genetic resources and many of the findings, which may be inappropriate such an analysis of polygenic risks. The developments in genomic technologies have been so fast that several issues, viz. legal and ethical consequences of these new techniques have not been addressed properly. Therefore, the researchers should always keep in mind that how the availability of whole genome sequence information is liable to impact other sectors, such as aquaculture, conservation, climate change, carbon

R. Kumar (✉) · B. Kushwaha · M. Singh
ICAR - National Bureau of Fish Genetic Resources, Lucknow, Uttar Pradesh, India
e-mail: ravindra.kumar1@icar.gov.in

© The Author(s), under exclusive license to Springer Nature Singapore Pte Ltd. 2021
P. K. Pandey, J. Parhi (eds.), *Advances in Fisheries Biotechnology*,
https://doi.org/10.1007/978-981-16-3215-0_4

65

sequestration, habitat health etc. In this chapter, the current development in genomic technologies, their potential and beneficial applications for fisheries management and aquaculture, have been discussed.

Keywords

Genome · Genomics · Sequencing platforms

4.1 Introduction

Both aquaculture and fisheries sectors significantly expanded in the past. Total production, trade and consumption reached an all-time record high in 2018. As per an estimate (FAO 2020), the global capture fisheries, aquaculture and total fish production increased to the extent of 14%, 527% and 122%, respectively during 1990–2018. Total global capture fisheries production reached to 96.4 million tonnes in 2018. Similarly, total aquatic animals' production reached to 82.3 million tonnes in 2018. This increase in production has been the result of the technological development in skillful breeding, seed production and grow-out husbandry practices of a number of diversified cultivable finfish and shellfish species all over the world and their large-scale adoption, both at small-scale farming levels and also in commercial production systems. Further, genetic improvement programmes, nutrition and feed technologies, disease management, appropriate soil-water management measures etc. also contributed significantly to the growth pace of aquaculture sector during these years.

The volume of genetic variation present in fishes has not been utilized to its full potential in attaining genetic gain through conventional genetic improvement programme, since these are usually attained based on the genetic parameters of the traits under improvement that are estimated based on individual's own and on relatives' records. With proper scientific management practices and genetic improvement programmes, one could realize around 6–7% genetic gain per generation for the harvested body weight (Luan et al. 2012). These conventional selective breeding tools are effective for easy-to-measure traits, but are less successful in difficult-to-measure traits. Further, the marker-based selection programmes require time for measuring the traits and space for maintaining the families. In such cases, faster improvement in important economic traits, i.e. growth, fertility, disease resistance, feed efficiency etc., with less time and space requirement can be attained through application of genomic strategies and tools.

With the advancements in the area of genomics in recent past, significant progress has been made from the analysis of a single or few genes at a time to whole genome of an organism. With the success of human genome sequencing project, the area has promptly been expanded towards functional level. Development of new strategies has added powerful tools in genomics for enhancing production performances. These advances in genomics have already shown results in certain crops and animals, but the benefits have not been realised in other fields, including fish.

Such inconsistency, thus, need to be explained and addressed through generation of genomic information for understanding and manipulating the biological systems.

4.2   Genomics

Genome is the total genetic content present in a haploid set of chromosomes of the individual/species, whereas the genomics is the branch of molecular biology that concerns with structure, function, evolution, mapping of the genome and related techniques (WHO 2002; WHA 2004). The main difference between the genomics and genetics is that: the genetics scrutinizes the functioning and composition of the single or few genes, while genomics address inter-relationships among the large number of genes in order to identify their combined influence on the development of the character/organism (http://www.who.int/genomics/geneticsVSgenomics/en/).

Genomics has been classified mainly into following categories: (a) structural genomics, (b) functional genomics, and (c) comparative genomics. Structural genomics dealt with the studies at the initial phase of genome analysis and mapping (the construction of high resolution genetic, physical and transcription maps of an organism) (Baker and Sail 2001). Human whole genome sequencing project is a good example of structural genomics. Other examples of model organisms for which genome sequences are either completed or in progress are: plants (*Arabidopsis thaliana*, *Oryza sativa*, *Zea mays* etc.), mammals (*Homo sapiens*, *Mus musculus*, *Rattus norvegicus* etc.), non-mammals (*Anopheles gambiae*, *Caenorhabditis elegans*, *Clarias batrachus*, *Danio rerio*, *Drosophila melanogaster*, *Escherichia coli*, *Catla catla*, *Labeo rohita*, *Oryzias latipes*, *Saccharomyces cerevisiae*, *Tenualosa ilisha*, *Xenopus laevis* etc.). Once the genome structure is known, it is essential to identify the functions of the genes and the proteins encoded by these genes. Functional genomics refers to deciphering the functions of genes present in the genome, for which many high-throughput technologies (Schoulink 2002) and bioinformatics tools (Niazi and Riaz-ud-Din 2006) have been developed for predictions about the biological function using DNA sequence data. The comparative genomics is the field that explores the resemblance and differences among the genomes (Clark 1999). It may be useful to identify genes and their regulatory elements in closely related species, since sequences only with an evolutionary conserved function would be found in both the genomes, while others would diverge significantly. Although comparative genomics can be a first useful resource for functional annotation, but the information must carefully be utilized due to homolog/ortholog/paralog gene natures and for several other reasons.

4.3   Genome Sequencing Projects

The works of genome sequencing began with the Fredric Sanger's sequencing technique developed in 1977. The time consuming, labour intensive gel preparation and running, as well as the cost of such conventional sequencing were replaced by the 'shotgun' sequencing. Deciphering the entire genome and their characterization between organisms was unconceivable until the rapid development in microchips and processors in last two decades. Thus, a synergistic coordination between the sequencing technology and the powerful bioinformatics skills initiated a plethora of ambitious genome sequencing proposals.

The genome sequencing era started with the initiation of Human Genome Project in 1990 with the aim to identify and map approximately all the 20,000–25,000 genes, 3 billion chemical base pairs, *store* this information in database, improve the bioinformatics tools for data analysis, transfer related technologies to various sectors, and address the ethical, legal and social issues that may arise from the project (http://web.ornl.gov/sci/techresources/Human_Genome/redirect.shtml). The successful completion of human genome project and drop in sequencing costs as well as increase in high throughput sequencing opened the gateway for sequencing of various model organisms, which included some fish species too. The research community has taken advantage of the infrastructure built by the human genome project and now 19,566 eukaryotic and 8757 animal genome-sequencing projects has been completed and operational worldwide (https://www.ncbi.nlm.nih.gov/Traces/wgs/; as on September 11, 2020). In fishes, 1406 genome sequencing projects are being generated, assembled and annotated (https://www.ncbi.nlm.nih.gov/genome/; as on September 11, 2020). With the success in genome sequencing works, the international consortium of scientists conceded a proposal, known as Genome 10K project (2009), for assembling a 'genomic zoo' of vertebrate species by performing whole-genome sequencing of 10,000 vertebrate species, approximately one for every vertebrate genus (Genome 10K 2009; Bernardi et al. 2012).

In India, the ICAR-NBFGR, in collaboration with ICAR-CIFA, Bhubaneshwar; ICAR-IASRI, New Delhi and AAU, Anand has initiated whole genome sequencing of *Labeo rohita* and *Clarias magur (batrachus)* in 2013 with the financial assistance from DBT, New Delhi. The genome sequence of *L. rohita* (estimated ~1.4 Gb genome size) and magur (~1 Gb genome size) were generated on multi-platform NGS. The assembly of these genomes is completed with about 93–95% coverage. The annotation of draft genome is complete with 26,400 protein coding genes in *L. rohita* and ~23,748 protein coding genes in *C. magur*. The whole genome sequencing of the two commercially important species, i.e. *Tenualosa ilisha* and *Fenneropenaeus indicus*, has also been completed under the ICAR Consortium Research Project (CRP) on Genomics. The draft genome of *T. ilisha* is about 762.5 Mb in size, with a total of 2864 contigs, N50 of 2.65 Mbp (largest contig length of 17.4 Mbp) and 33,042 predicted genes. ICAR-CIFA has also accomplished genome sequencing of *Labeo catla* on Illumina and Oxford Nanopore platforms. The genome size of catla is about 1.01 Gb, with 5345 scaffolds, N50 value of 0.7 Mb and 25,812 predicted genes. *Ompok bimaculatus* has been sequenced by Illumina short reads and PacBio long reads platforms. The draft genome size is about 718 Mb with N50 of 81 kb and 21,371 predicted genes.

4.4   Next Generation Sequencing Platforms

Technical advancement in the field of molecular biology during the last few years has opened up new ways to generate large-scale sequencing data in time and cost effective manner. The year 1977 gave two DNA sequencing technologies, one based on chain-termination method by Frederick Sanger and the other based on chemical modification method by Allan Maxam and Walter Gilbert in 1977. After a decade, Applied Biosystems used capillary electrophoresis, instead of gel electrophoresis, to automate the sequencing process for faster and accurate results. Sanger's method is considered as a 'first-generation' technology, whereas various newer sequencing technologies, with their own set of characteristics, constitute next generation sequencing (NGS) that rapidly generate huge sequence data in a very reasonable way. The NGS technologies are different from the Sanger method in aspects of parallel, massive and high throughput sequence generation at reduced cost. The real growth in the field of genomics, can be attributed to NGS technology and main technologies are, Pyrosequencing Technology (Machines: Roche 454 series), Reversible Terminator Technology (Illumina Series), and Sequencing by Ligation Technology (Machine: ABI SOLiD). The 454 was launched in 2005 by 454 Life Sciences (later purchased by Roche in 2007) which generates 500 MB bases per run of 8–10 h. Roche 454 was the first NGS, based on pyrophosphates released from dNTPs incorporation during DNA synthesis (Nyren et al. 1993). The DNA fragments on beads are compartmentalized into water-in-oil emulsion microvesicles, often called micro-reactors, where clonal multiplication occurs. Long reads of 700–1000 bp made the downstream bioinformatics analysis easy, but the data output was low and cost per base sequenced was high in comparison to other concurrent NGS technologies.

The reversible terminator technology of Illumina sequencing started with concept by Shankar Balasubramanian and David Klenerman and culminated in platforms like Solexa, MiSeq, HiSeq, NovaSeq. The DNA templates with adapters are attached to the anchors and amplified with primers complementary adapters to generate clonal clusters. In this way, millions of clusters of different DNA molecules are made in one flow cell. The labels on nucleotides also serves as chain terminators during polymerization, so after each dNTP incorporation, the fluorescent dye is imaged and enzymatically removed to allow incorporation of the next nucleotide. The error rate is less, repeat regions are sequenced nicely and the huge amount of data is generated in the form of paired ends (100/150/300 bp) in different platforms.

The Small Oligonucleotide Ligation and Detection System (SOLiD) was developed by George Church in 2005. In the sequencing step of this method, the DNA ligase is used instead of DNA polymerase. DNA fragments flanked with adaptors are attached to beads and amplified clonally by emulsion PCR. The whole process of library preparation, emulsion PCR, bead deposition, sequencing and primer reset take 6–7 days in a SOLiD 5500 system and generate sequence data 120–240 GB with a read length of 75 bases for paired-end with claimed accuracy of 99.99%. Comparatively less data output and shorter read length than Illumina limited the use of technology.

Based on Ion Semiconductor Sequencing, ABI developed Ion Torrent and Ion Proton platforms. When a nucleotide is incorporated into a strand of DNA by a polymerase, a hydrogen ion released, can be detected by semiconductor devices. These machines have high-density array of micro-machined wells, each with different DNA template, to perform parallel sequencing. Bottom of the wells have ion-sensor to detect the changes in the pH of the solution, resulting from incorporation of nucleotides during DNA synthesis. Ion Torrent generates read lengths of around 200 bp, which are used to fill gaps in the assembly produced by other technologies. The short run time of this technique also facilitates multiple runs for generation of more data in a given time.

In case of Oxford Nanopore Sequencing, a pore of 1 nm in internal diameter, made by transmembrane proteins like haemolysis, porins, etc. is used to pass the DNA through. When a DNA strand passes through nanopore, passage of different nucleotides makes specific changes in electric current across the nanopore. DNA may pass through the whole one base at a time, so that the change in the current can be read, and the sequence of the target DNA can be recorded. Reads are much longer than sequenced by other technologies and very helpful in genome assembly, particularly along with short read data. Once the issue of error rate is addressed, the low cost and very fast sequencing process will make it widely used technology in future. The other NGS systems include Polonator G.007, HeliScope from Helicos may have a striking impact on genomic research and the biological field.

4.5   Fish Genomics

The numbers of documented fish species globally are around 33,600 (www.fishbase. org, ver. 06/2017) that is distributed in different ecosystems. It is economic to complete genomic sequences of comparatively larger number of fish species, because the fish genome is comparatively smaller in size and very compact with high gene density. The genome sizes of some of the fish species of the actinopterygii and sarcopterygii lineages are available at Animal Genome Size Database (http:// www.genomesize.com/statistics.php?stats=fish).

Fish genome is being used to determine the part of eukaryote genomes that do not code for proteins, but are functional as non-coding RNA genes and regulatory control regions. Such regions can be detected by aligning genomic sequences of distantly related organisms and searching for regions that have remained similar during evolution, thus, suggesting that these regions might have undergone mutations, but functionality of the genes have remained preserved. The advantage of fish in this context is; firstly, the genome of fish is saturated with such neutral mutations and secondly, the fish genomes have experienced long evolutionary distance, i.e. ~450 million years. A comparative genome-wide discovery of conserved and ultra-conserved regions of unknown function is of fundamental importance across vertebrates and mammalian genome. With increased number of newly annotated fish and mammalian genome sequences, such comparative studies are likely to play an important role in the identification of functional non-coding elements and in underpinning the basis of genomic homology.

The genome information needs to be encouraged for use in molecular characterization of all the genes and gene products of a species, the introduction of single genes conferring potentially useful traits in fish, identification and evaluation of useful traits in breeding programs by the use of marker assisted selection, accurate and quick identification of pathogens by using new diagnostics based molecular characterization of the pathogens and use of modern immunology to develop recombinant DNA vaccines for improved disease control against the lethal diseases.

4.6 Major Fish Genome Sequencing Programs

The genomes of actinoterygiians serve as excellent models for studying comparative genomics, evolutionary processes and fates of duplicated genes. The genome size is small having high gene density and intense number of duplicated genes. The consolidated knowledge of sequenced genomes of *Danio rerio, Takifugu rubripes, Tetraodon nigroviridis, Oryzias latipes, Gasterosteus aculeatus, Cyprinus carpio,* etc. and many stronger frontiers of fish genomics as well as fish biology give light on evolutionary processes. Genomic sequencing of other fish species, as cartilaginous fish (elephant shark), has also generated enormous information for understanding fish genomics.

4.6.1 Zebrafish (*Danio rerio*)

The first model fish organism chosen for understanding vertebrate developmental biology, vertebrate evolution and human diseases was the zebra fish (zf), which is a common and useful model fish. The project was initiated in collaboration between the Sanger Institute and the zf community and was announced during the Sanger Institute Zebra fish Workshop in 2000. The zf genome (~1427.29 Mb size) is about half of the human genome and hence, good candidate for sequencing. The zf genome is being sequenced and analysed at the Wellcome Trust Sanger Institute and z11 is the latest assembly, submitted by Genome Reference Consortium. In recent NCBI zf Annotation Release 106, the total number of protein coding, non-coding and pseudogenes were 26,522, 13,137 and 330, respectively, with 98.59% genome coverage. The number of transcripts per gene and exons per transcript were 1.85 and 12.2, respectively. The zv9 assembly showed an overall repeat content of 52.2%, the highest reported so far in a vertebrate (Howe et al. 2013; https://www.nature.com/nature/journal/v496/n7446/pdf/nature12111.pdf). In India, whole genome re-sequencing of a Wildtype strain of Zebra fish (2009) was completed by Institute of Genomics and Integrative Biology, New Delhi, using Solexa/Illumina sequencing technology and a database 'FishMap' has been created for the display of annotated genomic data.

4.6.2 Fugu (*Takifugu rubripes*)

The Fugu genome sequencing project was initiated in 1989 by S. Brenner and colleagues which was the second vertebrate sequenced genome, after human. The International Fugu Genome Consortium, one of the largest international genome-sequencing projects since Human Genome Project formed in November 2000, was led by the US Department of Energy's Joint Genome Institute (JGI) in Walnut Creek, California and the Institute for Molecular and Cell Biology (IMCB), an institute of Biomedical Research Council of Singapore. The Cambridge based MRC, UK Human Genome Mapping Resource Centre (HGMP-RC), the Cambridge University Department of Oncology and the Institute for Systems Biology in Seattle, Washington were also part of the consortium. This genome is highly purposeful because of its compactness and due to lack of enormous amount of junk DNA, which *make fugu genome sequencing very cost effective and rapid* (http://www.fugu-sg.org/).

The Fugu genome Annotation Release 103 indicates that the genome length is 384.127 Mb, which is among the smallest (about 1/8th the size of human genome) vertebrate genomes and has proved to be a useful 'reference' genome for identifying genes and other functional elements and understanding the structure and evolution of vertebrate genomes. The current assembly fTakRub1.2 contains 128 scaffolds with N50 of 16,705,553, 530 contigs with 3,136,617 N50 and 45.70% GC content. Annotation of the recent assembly resulted in 22,076 protein coding, 4718 noncoding and 546 pseudo genes (https://www.ncbi.nlm.nih.gov/genome/annotation_euk/Takifugu_rubripes/103/). The mean number of transcripts per gene and exons per transcript were 2.05 and 13.22, respectively.

4.6.3 Medaka (*Oryzias latipes*)

The medaka genome project commenced in late 2002 as a collaborative effort of three core laboratories, led by Hiroyuki Takeda (University of Tokyo, Japan), Shinichi Morishita (University of Tokyo, Japan) and Yuji Kohara (National Institute of Genetics, Japan) with the funding support from Grants-in-Aid for Scientific Research in Priority Area "Genome Science" from the Ministry of Education, Culture, Sports, Science and Technology of Japan. The sequencing was carried out at Academia Sequencing Centre of NIG, using whole genome shotgun sequencing approach (Kobayashi and Takeda 2008). The successful completion of a high-quality draft genome sequence of the inbred Hd-rR strain of medaka was reported in 2007.

Medaka is an excellent model system for a wide range of biology, including ecotoxicology, carcinogenesis, sex determination and developmental genetics. It establishes as a first successful sex-reversal candidate in vertebrates and its genome demonstrates identification of male-determining gene, DMY, the first non-mammalian equivalent to SRY. Small genome size and the ability to grow at permissive temperature range (6–40 °C) during its embryonic development increase the chance of medaka for being a model organism.

In genome assembly ASM223467v1 and annotation release 103, the median total length of genome covered is 746.737 Mb, assembled in 25 scaffolds ($=1$ N + MT) with N50 of 31,218,526 and 516 contigs with N50 of 2,530,934, 40.8727% GC content, 22,071 protein coding, 4481 noncoding and 188 pseudo genes. The number of transcripts per gene and exons per transcript were 1.96 and 13.57, respectively (https://www.ncbi.nlm.nih.gov/genome/annotation_euk/Oryzias_latipes/103).

4.6.4 Puffer Fish (*Tetraodon nigroviridis*)

The spotted green puffer fish genome sequencing project was started in 1997 at GenoScope, the French National Sequencing Center, Paris, France. The project was supported by the Consortium National de Recherche en Genomique and the National Human Genome Research Institute. The gene assemblage of the puffer fish genome is very similar to that of other vertebrates, including mammals like humans and mice. For geneticists interested in studying genes, fishes of the Tetraodontiform family have a huge advantage over mammals; their gene assemblage is contained within approximately 8 times less DNA because of the very low content of repetitive DNA (https://www.ncbi.nlm.nih.gov).

The sequence assembly was performed, using Arachne program and the gene predictions by GenoScope using the GAZE computational framework. The assembly has been improved (v8) by using a new fosmid library. The genome data are available on Tetraodon Genome Browser (http://www.genoscope.cns.fr/externe/tetranew/). The genome assembly ASM18073v1 submitted by GenoScope indicate 342.403 Mb genome coverage with 25,773 scaffolds, 41,566 contigs, 29,054 N50, 46.6% GC content and 27,918 genes (https://www.ncbi.nlm.nih.gov). The small and compact genome has high gene density, rapid molecular evolution tendency and high chromosome stability.

4.6.5 Elephant Shark (*Callorhinchus milii*)

Cartilaginous fishes (Chondrichthyes) are phylogenetically the most basal living jawed vertebrates that serve as an important group for understanding the origins of the complex developmental and physiological systems of jawed vertebrates. The elephant shark is a cartilaginous fish having smallest genome among the known cartilaginous genomes (Venkatesh et al. 2005). The genome is inhabited by large number of ultra-conserved elements and the protein sequences that evolved at a slower rate than in other vertebrates. Due to the slow evolution rate, there might have been little change in its genome during the last 450 million years of evolution. Thus, this species has retained more features of the ancestral genome than other vertebrates and hence, stands to be useful model for gaining insight into the ancestral genome.

The degree of synteny between the human and elephant shark genomes are higher than that between human and teleost fish genomes.

The genome assembly Callorhinchus_milii-6.1.3, submitted by IMCB, Singapore, resulted in 974.439 Mb assembled genome length with 21,204 scaffolds, 67,421 contigs, 46,577 N50 42.5998% GC content. In genome Annotation Release 100, there are 18,044 protein coding, 1840 non-coding and 206 pseudo genes in the genome of *C. milli*. The number of transcripts per gene and exons per transcript were 1.56 and 11.15, respectively (https://www.ncbi.nlm.nih.gov).

4.6.6 Common Carp (*Cyprinus carpio*)

The common carp is one of the most important cyprinid that globally accounts for 10% of freshwater aquaculture production and is cultured in over 100 countries worldwide. The genome information provides a valuable resource for the molecular-guided breeding and genetic improvement of the species. The Centre for Applied Aquatic Genomics, Chinese Academy of Fishery Sciences (CAFS), Beijing; Heilongjiang Fisheries Research Institute, CAFS, Harbin; and Beijing Genomics Institute, China did the whole genome sequencing of common carp (strain Songpu) with the grant support from the National High-Technology Research and Development Program of China, National Department Public Benefit Research Foundation of China, National Basic Research Program of China, National Natural Science Foundation of China and Special Scientific Research Funds for Central Non-Profit Institutes of the CAFS.

The genome assembly GCF_000951615.1, submitted by CAFS, resulted in 1546.88 Mb genome lengths with 37.12% GC content. In genome Annotation Release 100, 49,579 protein coding, 10,547 non-coding and 8926 pseudo genes are present in the genome of *C. carpio*. The number of transcripts per gene and exons per transcript were 1.29 and 7.55, respectively (https://www.ncbi.nlm.nih.gov). The assembly contains 52,610 protein-coding genes and ~92.3% coverage of its paleo-tetraploidized genome ($2n = 100$) (Xu et al. 2014).

4.6.7 Coelacanth (*Latimeria chalumnae*)

The sarcopterygian fishes, that occupy a unique phylogenetic position between ray-finned fishes and tetrapods, consist of very stable genome evolved neutrally with few major rearrangements. The two distinguished taxa of sarcopterygii, coelacanths and lung-fishes are of utmost importance in evolutionary studies. Large genome sizes of lung-fish (>100 Gb) make them poor candidates for genomic sequencing. Coelacanths are lobe-finned fish, more closely related to tetra pods than to ray-finned fish. The slow rate of evolution of the coelacanth genome might have contributed to the retention of ancestral genes that might have been lost in teleost and tetra pod lineages. Two modern coelacanth species, viz. *Latimeria chalumnae* and *L. menadoensis*, are accountable to whole genome sequencing due to their smaller genome size. The genome provides access to the phenotypic and genomic transitions leading to the emergence of tetrapods.

The genome assembly LatCha1 of *L. chalumnae*, submitted by Broad Institute, resulted in 2798.46 Mb genome lengths with 42.55% GC content. In genome Annotation Release 101, there are 21,021 protein coding, 5218 noncoding and 400 pseudo genes in the genome of *L. chalumnae*. The number of transcripts per gene and exons per transcript were 1.56 and 9.68, respectively (https://www.ncbi.nlm.nih.gov).

4.6.8   Other Genomes

Cichlid genome is an important source for understanding vertebrate evolution as well as for sex determination, behavioural, immunological and toxicological aspects. The genome research in tilapia can add steps to study social dominance, territoriality, sexual selection, feeding behaviours and also contribute to the maintenance and improvement of the tilapia as an important food source across developing countries. The Broad Institute, Cambridge generated high quality draft genomes of *Oreochromis niloticus* and four additional cichlid species (*Astatotilapia burtoni*, *Maylandia zebra*, *Pundamilia nyererei* and *Neolamprologus brichardi*) from East African lake lineages (http://www.broadinstitute.org/models/tilapia).

The genome assembly O_niloticus_UMD_NMBU of *O. niloticus*, submitted by University of Maryland, resulted in 993.469 Mb genome lengths with 40.67% GC content. In genome Annotation Release 104, there are 29,550 protein coding, 12,030 noncoding and 689 pseudo genes present in the genome. The number of transcripts per gene and exons per transcript were 1.95 and 11.27, respectively (https://www.ncbi.nlm.nih.gov/genome/annotation_euk/Oreochromis_niloticus/104/).

Sticklebacks are freshwater fish species that has undergone a dramatic evolutionary radiation since the last ice age. Ancestral marine sticklebacks populated the newly created lakes and subsequently adapted to different environments. A number of subspecies have recently evolved multiple changes in their anatomical and physiological traits. Stickleback species are, therefore, a good model to study adaptive evolution (https://www.ncbi.nlm.nih.gov). The genome assembly NID of freshwater three-spine stickleback (*Gasterosteus aculeatus*), submitted by Institute of Ecology and Evolution, is of size 445.644 Mb, which includes 10,242 scaffolds with N50 3,715,221, 32,646 contigs with 38,090 N50 and 43.4% GC content (https://www.ncbi.nlm.nih.gov/assembly/GCA_006229165.1/).

Besides these, Atlantic salmon genome (3.27 pg) has also been sequenced, which is fairly similar to those of warm-blooded vertebrates with respect to size and overall base composition. An International Collaboration to Sequence the Atlantic Salmon Genome (ICSASG), representing researchers, funding agencies and industry from Canada, Chile and Norway, was formed to undertake this genome sequencing project. The genome information of Atlantic salmon will facilitate exploitation of genomic information in a wide range of ecological, evolutionary, conservation and production biology settings within salmonids. The genome information would also provide novel insights into vertebrate post-WGD evolution that may contribute to a more thorough understanding of the underlying mechanisms as well as the long-term importance of WGD for adaptation (Lien et al. 2016). European subspecies (*Salmo salar europensis*), with 29 chromosome pairs, was sequenced. The genome assembly ICSASG_v2 of *S. salar* submitted by ICSASG resulted in 2966.89 Mb genome lengths with 241,573 scaffolds, 368,060 contigs, N50 value 57,618 and 43.89% GC content (https://www.ncbi.rlm.nih.gov). In genome Annotation Release 100, there are 48,775 protein coding, 6452 non-coding and 2556 pseudo genes, present in the genome of *S. salar*. The number of transcripts per gene and exons per transcript were 1.99 and 11.62, respectively.

4.7 Other Fish Genomic Resources

As the availability of genomic resources is increasing constantly for aquatic species, more research is being undertaken worldwide to improve brood stocks for many key aquaculture species. There are numerous genomic resources, other than WGS, and tools that can provide extensive insight into a genome and can be used for applications in molecular breeding and other therapeutics, both for the species of interest and other closely related species. These resources include genetic markers (cytogenetic, microsatellite, SNP, etc.), expressed sequence tags, linkage maps, BAC libraries, QTL, expression profiling (microarrays and qPCR), bioinformatics tools, genomic databases, etc.

4.8 Applications of Genomics

The considerable benefit from the use of genomic information in aquaculture development is either achieved and/or expected in the following areas in future:

4.8.1 Molecular Markers in Selection Programs

Identification of genetic markers, surrounding quantitative trait loci (QTL), forms the basis for MAS. These molecular markers can help in genetic management of the species to optimize the production. Quantitative variation, controlled by polygenes and environmental factors, characterizes economically important traits, such as growth, meat quality, disease resistance etc., in the organisms (Liu 2007).

Gene associated markers, like EST-SSRs and SNPs, etc., have become the marker of choice for genome-wide association studies. In order to select best genomes for performance and production traits, a large number of relatively evenly distributed Gene-associated SNPs are needed. Globally, SNP technology is now being developed for several species, including catfish, rainbow trout, Atlantic salmon, cod, oysters, shrimp, etc. (Browdy et al. 2012). These SNP markers can also facilitate the development of a variety of DNA-based genetic markers, used for the management of wild and cultured populations. The advantages of genotyping polymorphic SNPs with high-throughput assays have created much interest. However, the scarcity of available DNA sequence data for non-model species can limit marker development. Further, because of comparatively low mutation rates, cross-species primers amplification for SNP analyses did not yield the same results as for microsatellites, therefore, the SNP assays or probes developed for one species were not likely to be useful in others, even though primers may cross-amplify. This information can be utilized in aquaculture to maximise the rate of genetic gain through selective breeding programmes.

4.8.2 Selection for Growth Trait

Fast growth is one of the most desired traits, affecting the profitability of farm production systems. Traditional phenotype-based selection is typically used to select animals for growth traits; however, it does not allow for optimal control over all phenotypic characteristics of the growth parameters (Salem et al. 2012). As an alternative, MAS aims to expedite genetic improvement in breeding programmes. SNPs are the most suitable tool for MAS, since they explain 90% of the phenotypic differences between individuals. SNPs found within or flanking a coding sequence is of particular interest because they are more likely to alter the biological function of a protein. For example, in rainbow trout, putative type-I SNP markers, observed through whole transcriptome shotgun sequence analysis, were found to be associated with growth rate and 104 markers were validated for association with the fast growth (Salem et al. 2012). Similarly, prolactin SNP c.264+269T>C in Asian sea bass was found to be significantly associated with body weight, total length, standard length and Fulton's condition factor (He et al. 2012).

4.8.3 Genetic Diversity and Resource Analysis

Many strains/lines are used in most of the genetic improvement programs, but the precise genetic relationship among strains is often unknown. For analysing the genetic diversity in aquaculture brood stocks, tools are required to reveal genetic variations among the stocks. The gene-wide molecular markers can very well be utilized for this purpose. In addition, these markers can be used for determining parentage/pedigrees, estimating genetic parameters (heritability and genetic correlations) and breeding values of the traits, while optimising selection and mating plan, in order to avoid inbreeding and physical tagging of the individual. DNA based pedigrees analysis is useful in conservation programmes, where individual tagging is difficult. This is especially useful for the species where no extensive resources are available or for new species whose reproduction cannot be fully controlled. Furthermore, there are studies showing these molecular markers useful for determining the effective number of parents and their individual reproductive success as well as level of inbreeding in mass selection programmes. Similarly, the impact of hatchery practices on the genetic variability of progenies can be monitored as well as genotyping can be also useful in assessment of introgression to/from a natural population (Chittenden et al. 2010).

4.8.4 Fish Health Management in Aquaculture

Fish possess an immune system that is highly developed and the basic mechanisms of immunity in fish and mammals are quite similar. Fish do not possess bone marrow and lymph nodes, like mammals, and the main lymphoid organs in teleosts are thymus, head kidney (pronephros) and spleen. Leukocytes are also present in the blood, intestine and epithelia of skin and gills. Genomics can help to overcome problems related to infectious diseases by better understanding of host defence systems and identifying QTL or candidate genes. The genome-wide expression profiling would provide a better understanding of the molecular mechanisms that control or contribute to the anti-infectious response of the fish. A group of genes and cellular immune functions could be identified as potential targets for mechanisms of pathogen resistance and escape to the immune response. This will open the way to in-depth characterization of pathogen-specific gene expression signatures and the identification of those physiological traits controlling survival capacity and subsequently for a better understanding of the phenomenon of mortality due to diseases (de Lorgeril et al. 2011).

Systematic research on fish immuno-genetics is indispensable in understanding the origin and evolution of immune systems. This has long been a challenging task because of the limited availability of deep sequencing technologies and genome backgrounds of non-model fish. The newly developed Illumina RNA-Seq and Digital gene expression (DGE) are high-throughput sequencing approaches for genomic studies at the transcriptome level. Globally, identification of genes involved in immune response by EST approach has been carried out in several aquaculture species, like Asian seabass, Atlantic halibut, Atlantic salmon, channel catfish, common carp, grass carp, Japanese flounder, rainbow trout, etc. (Duan et al. 2013; Esteban 2012). The sequences of immune-related genes have also been studied in lymphoid tissues, following experimental infection with viruses, bacteria and parasite, and also following vaccination and immuno-modulation (Uribe et al. 2011).

Further, the use of SNPs, associated with disease/immunity related traits in aquaculture species, can facilitate the selection of fish with superior genetic material as well as a better understanding of host-pathogens interaction and disease resistance. A project on improved disease resistance of rohu and tiger shrimp was jointly implemented by ICAR-Central Institute of Freshwater Aquaculture (CIFA), Bhubaneshwar and ICAR-Central Institute of Brackish water Aquaculture (CIBA), Chennai, India. Putative SNPs, showing the highest allele frequency differences between susceptible and resistant lines, identified fish demonstrated utility for identifying genes, associated with disease resistance. Statistically significant presence of DNA marker in disease resistant populations and their absence in disease susceptible populations were shown in *Penaeus monodon* (http://www.google.com/patents/WO2007057915A1?cl=en).

4.8.5 Development of Alternative Feed

Fish meal is a major source of high-quality protein in fish feeds and the increasing demand for the aquaculture production have compelled the need for developing alternate fish feeds, which may comprise of economical and highly digestible protein sources of plant and/or animal origin and has no adverse effect on fish performance and the environment. Nutritional genomics or nutrigenomics studies the interactions between the diet and the genome to reveal the effects of nutrient intake on genetic responses (Müller and Kersten 2003). It can contribute to production of new kinds of feed for cultured fishes, which can facilitate the selection of genotypes that produce a good performance when fed with a low fish oil or protein diet.

4.8.6 Phylogeny Analysis

Fish genome has played important role in providing penetrating insight into vertebrate genome evolution as whole genome duplication (WGD), a major evolutionary event that occurred 320–400 MYA in teleost ancestor, which is considered the sole reason for shaping the present genome organization in vertebrates. WGD had structured the genome of entire vertebrates by affecting them twice, and the timing has been estimated to be around 350 MYA, using molecular clock hypothesis and phylogenetic analysis. The sequencing of genes and gene families from teleost fishes had unexpectedly revealed the presence of duplicate genes for several human genes which in turn led to the hypothesis that WGD occurred in the ray-finned fish lineage before the diversification of teleost fishes.

   Fishes like ghost/elephant shark, sturgeons etc. had undergone two rounds of WGD. A third round of WGD bifurcated Actinoterygii into teleost and non-teleost lineages. Most of the tetrapods, on the other hand, have not experienced additional WGD. However, they have experienced repeated chromosomal rearrangements throughout the whole genome. Further, the sequencing and comparative analysis of whole-genome sequences of teleost fishes, such as fugu, tetraodon and medaka, have provided compelling evidence for WGD event in fish lineage.

   Following WGD, the teleost genome had undergone eight major chromosomal rearrangements. A comparison of medaka genome with the zebra fish, tetraodon and human genomes has revealed that eight major inter-chromosomal rearrangements occurred within a relatively short period of approximately 50 million years after the WGD in the fish lineage (Kasahara et al. 2007). Subsequently, three major rearrangements occurred in the tetraodon lineage, while medaka lineage experienced no major inter-chromosomal rearrangements. In contrast, the zebra fish lineage has experienced many inter-chromosomal rearrangements after it diverged from the medaka lineage. A comparison of gene order across large regions supports a higher rate of chromosomal rearrangements in teleost fishes as compared to other vertebrates (Venkatesh 2003). At present, the teleost lineage has flourished by half of all vertebrate species (including *Danio rerio, Oryzias latipes, Takifugu rubripes, Tetraodon nigroviridis*) which have adapted to a variety of marine and freshwater habitats and their genome evolution and diversification are important subjects for the understanding of vertebrate evolution.

4.9   Fish Genomes Sequencing in India

The fish genome sequencing work in India was initiated and the whole genome re-sequencing of wild zebra fish was done by CSIR-IGIB, New Delhi, while sequencing of *L. rohita, C. magur), T. ilisha, F. indicus,* etc. genomes have been completed at annotation level. In India, a nationwide project has been started to generate genome information of all known species of plants and animals in the country, with National Institute of Plant Genome Research, New Delhi as the coordinating centre involving a total of 24 institutes from the country.

The genome size of *L. rohita*, an important food fish in India and adjoining countries, is about 1.4–1.5 Gb. De novo whole genome sequencing, assembly and scaffolding of *L. rohita* resulted in 13,477 scaffolds, with maximum scaffold size 15.2 Mb, covering around 95% genome. Gene prediction revealed 26,400 genes, where more than 85% genes were annotated (Das et al. 2020). *C. magur* has approximately 0.9–1.01 Gb genome size. Genome sequencing, assembly and scaffolding of this species has resulted 3484 scaffolds covering 98% of the total genome and 1,316,675 bp N50 (Accession:). The largest scaffold was 13.16 Mb and 245 scaffolds were of >1 Mb size. The genome has predicted to contain 23,748 proteins encoding genes of which 82.71% genes were supported by EST or RNA-Seq evidence (Kushwaha et al. 2021). The gene annotation, using Blast2GO, revealed 99.7% of the annotated genes showed homology to protein present in NR database. The draft genome assembly of *T. ilisha* has been reported (Mohindra et al. 2019) with a total of 2864 contigs, 96.4% completeness, 2.65 Mb N50 value and 17.4 Mb size of the largest contig. A total of 33,042 protein coding genes were predicted, where 512 genes classified under 61 Gene Ontology (GO) terms were associated with various homeostasis processes.

4.10   Conclusions

All the major fish genome projects have generated a vast array of structural and functional genomics data and this baseline data are being utilised as a platform for further research. Elucidation of the nucleotide sequence information, genes, markers and their association to various linkage groups in the genome will provide the tools for genetic improvement of traits through whole genome selection. The genome information of commercially important fish may uncover a veritable mine of genetic information in the form of gene and regulatory sequences, SNPs, STRs, etc., which can further be used to develop a HAPMAP for documenting genetic variability and identifying DNA regions for early and efficient selection. Association of the SNPs with particular traits has been outlined on genomic scale and utilised for selection of individual/strain with superior quality.

However, the fish genome sequencing work in India is at infant stage. The Indian fisheries sector still needs to aggravate its potential by generating a lot of raw genomic data from genome sequence of Indian fishes. Completely assembled genomic sequences and their annotations would provide a clear view of genome organisation of Indian fishes as well as generate fish specific information. Thus, sequencing of these genomes would help to identify and even manipulate specific genes for useful traits such as for growth rate, colour, disease resistance, cold and salinity tolerance. The data generated at genomic level will extensively be utilised in designing of SNP chips as well as expression arrays for detection of traits of economic importance.

Acknowledgements   The authors are grateful to Dr. Trilochan Mohapatra, Secretary, DARE, and Director General, ICAR, Ministry of Agriculture and Farmers' Welfare, New Delhi; Dr. J K Jena, DDG (Fy.), ICAR, New Delhi, India, and Director, ICAR-NBFGR, Lucknow, for their support, encouragement, and guidance. The authors are also thankful to Dr. N S Nagpure, Principal Scientist, ICAR-CIFE, Mumbai (earlier worked at ICAR-NBFGR), and Dr. Vindhya Mohindra, Principal Scientist, ICAR-NBFGR, Lucknow, for their help and inputs in preparing this chapter.

References

Baker D, Sail A (2001) Protein structure prediction and structural genomics. Science 29:93–96

Bernardi G, Wiley EO, Mansour H, Miller MR, Orti G, Haussler D, O'Brien SJ, Ryder OA, Venkatesh B (2012) The fishes of Genome 10K. Mar Genomics 7:3–6

Browdy CL, Hulata G, Liu Z, Allan GL et al (2012) Novel and emerging technologies: can they contribute to improving aquaculture sustainability? In: Subasinghe RP, Arthur JR, Bartley DM, De Silva SS, Halwart M, Hishamunda N, Mohan CV, Sorgeloos P (eds) Farming the waters for people and food. Proceedings of the Global Conference on Aquaculture 2010, Phuket, Thailand. 22–25 September 2010. FAO/NACA, Rome/Bangkok, pp 149–191

Chittenden CM, Biagi CA, Davidsen JG, Davidsen AG, Kondo H et al (2010) Genetic versus rearing-environment effects on phenotype: hatchery and natural rearing effects on hatchery- and wild-born coho salmon. PLoS One 5(8):e12261

Clark MS (1999) Comparative genomics: key to understanding the Human Genome Project. Bioassay 21:21–30

Das P, Sahoo L, Das SP, Bit A, Joshi CG, Kushwaha B, Kumar D et al (2020) *De novo* assembly and genome-wide SNP discovery in rohu carp, *Labeo rohita*. Front Genet Livest Genomics 11:386. https://doi.org/10.3389/fgene.2020

Duan Y, Liu P, Li J, Li J, Chen P (2013) Immune gene discovery by expressed sequence tag (EST) analysis of hemocytes in the ridgetail white prawn *Exopalaemon carinicauda*. Fish Shellfish Immunol 34(1):173–182

Esteban MA (2012) An overview of the immunological defences in fish skin. SRN Immunol 2012:853470. 29 p

FAO (2020) The State of World Fisheries and Aquaculture 2020: Interactive Story, SOFIA 2020. FAO, Rome. 200 p. http://www.fao.org/state-of-fisheries-aquaculture

Genome 10K (2009) A proposal to obtain whole-genome sequence for 10000 vertebrate species. Genome 10K Community of Scientists. J Hered 100(6):659–674

He XP, Xia JH et al (2012) Significant associations of polymorphisms in the prolactin gene with growth traits in Asian seabass (*Lates calcarifer*). Anim Genet 43(2):233–236

Howe K, Clark M, Torroja C et al (2013) The zebrafish reference genome sequence and its relationship to the human genome. Nature 496:498–503

Kasahara M, Naruse K, Sasaki S, Nakatani Y, Qu W, Ahsan B, Yamada T, Nagayasu Y, Doi K, Kasai Y et al (2007) The medaka draft genome and insights into vertebrate genome evolution. Nature 447:714–719

Kobayashi D, Takeda H (2008) Medaka genome project. Brief Funct Genomics 7(6):415–426

Kushwaha B, Pandey M, Das P, Joshi CG, Nagpure NS, Kumar R, Kumar D et al (2021) The genome of walking catfish *Clarias magur* (Hamilton, 1822) unveils the genetic basis that may have facilitated the development of environmental and terrestrial adaptation systems in air-breathing catfishes. DNA Res 28:dsaa031

Lien S, Koop BF, Sandve SR et al (2016) The Atlantic salmon genome provides insights into rediploidization. Nature 533:200–205

Liu ZJ (2007) Aquaculture genome technologies. Blackwell Publishing, Ames, IA. ISBN 13: 978-0-8138-0203-9

de Lorgeril J, Zenagui R, Rosa RD, Piquemal D, Bachére E (2011) Whole transcriptome profiling of successful immune response to *Vibrio* infections in the oyster *Crassostrea gigas* by digital gene expression analysis. PLoS One 6(8):e23142

Luan S, Yang B, Wang J, Luo K, Zhang Y, Gao Q, Hu H, Kong J (2012) Genetic parameters and response to selection for harvest body weight of the giant freshwater prawn *Macrobrachium rosenbergii*. Aquaculture 362–363:88–96

Mohindra V, Dangi T, Tripathi RK et al (2019) Draft genome assembly of *Tenualosa ilisha*, Hilsa shad, provides resource for osmoregulation studies. Sci Rep 9:16511

Müller M, Kersten S (2003) Nutrigenomics: goals and strategies. Nat Rev Genet 4:315

Niazi GA, Riaz-ud-Din S (2006) Biotechnology and genomics in medicine - a review. World J Med Sci 1(2):72–81

Nyren P, Petersson B, Uhlen M (1993) Solid phase DNA minisequencing by an enzymatic luminometric inorganic pyrophosphate detection assay. Anal Biochem 208(1):171–175. https://doi.org/10.1006/abio.1993.1024

Salem M, Vallejo RL, Leeds TD, Palti Y, Liu S et al (2012) RNA-Seq identifies SNP markers for growth traits in rainbow trout. PLoS One 7(5):e36264

Schoulink KG (2002) Functional and comparative genomics of pathogenic bacteria. Curr Opin Microbiol 5:20–26

Uribe C, Folch H, Enriquez R, Moran G (2011) Innate and adaptive immunity in teleost fish: a review. Vet Med 56(10):486–503

Venkatesh B (2003) Evolution and diversity of fish genomes. Curr Opin Genet Dev 13(6):588–592

Venkatesh B, Tay A, Dandona N, Patil JG, Brenner S (2005) A compact cartilaginous fish model genome. Curr Biol 15(3):R82–R83

WHA (2004) 57.13: Genomics and World Health, Fifty Seventh World Health Assembly Resolution. WHA, Geneva. http://apps.who.int/gb/ebwha/pdf_files/WHA57/A57_R13-en.pdf

WHO (2002) Genomics and World Health: Report of the Advisory Committee on Health Research. WHO, Geneva. http://whqlibdoc.who.int/hq/2002/a74580.pdf

Xu P, Zhang X, Wang X et al (2014) Genome sequence and genetic diversity of the common carp, *Cyprinus carpio*. Nat Genet 46:1212–1219

Omics in Aquaculture

5

Partha Sarathi Tripathy, Ananya Khatei, and Janmejay Parhi

Abstract

The omics is playing a major role in the modern aquaculture sector and its associated research works. The omics studies like genomics, proteomics, transcriptomics, epigenomics, lipidomics, glycomics, metabolomics, etc. are opening new ways for fisheries research as well as new developments to the sector. Genomics includes the study of structure, function, mapping, and editing of genomes. Many new omics studies are at present is studied under genomics like cognitive genomics, comparative genomics, functional genomics, metagenomics, pangenomics, and neurogenomics. Many researches based on genome are shaping new ideas for selective breeding, growth, adaptations, etc. for different fish species. Change of gene expression for different economically important traits i.e., independent of change of DNA, which is said to be epigenetics are studied under epigenomics. New study areas of omics like glycomics i.e., study of complex sugar molecules; lipidomics i.e., study of lipid profile of cell; metabolomics i.e., study of intermediate products of metabolism are under research in fisheries. Discovery of unknown genes and studying the expression pattern through fish RNA-sequencing or next-generation sequencing i.e., transcriptomics is helpful in various genomics studies as well as in genome comparisons. Proteomics also helps in studying the protein profile of fishes as well as determining meat quality. The omics studies in

The original version of this chapter has been updated: The authors Ananya Khatei and Janmejay Parhi affiliation has been updated. A Correction to this chapter is available at https://doi.org/10.1007/978-981-16-3215-0_30

P. S. Tripathy
Faculty of Biosciences and Aquaculture, Nord University, Bodø, Norway

A. Khatei · J. Parhi (✉)
Fish Genetics and Reproduction, College of Fisheries, CAU (I), Lembucherra, Tripura, India
e-mail: parhi.fgr.cof@cau.ac.in

© The Author(s), under exclusive license to Springer Nature Singapore Pte Ltd. 2021, corrected publication 2022
P. K. Pandey, J. Parhi (eds.), *Advances in Fisheries Biotechnology*, https://doi.org/10.1007/978-981-16-3215-0_5

combinations like in foodomics are also important for future researchers. The power of omics studies is important for achieving future blue revolution and fighting world food problems.

Keywords

Aquaculture · Genomics · Epigenomics · Transcriptomics · Metabolomics

5.1   Introduction

Culture of aquatic organisms and plants, more precisely called as aquaculture, is a rapidly developing branch for the purpose of food and livelihood. Development of any sector that deals with living beings requires the analysis of stock, genetic makeup, identification of unknown species, functional aspects of genes, etc. Genetics and biotechnology is helping in this regard.

"Omics" is a terminology that literally means "study." The branches of science in biology whose names with suffix -omics are together said to be omics study. Important among them are, genomics, proteomics, metabolomics, and glycomics. All these omics study are aimed at understanding the mechanism behind structure, function, and dynamics of an organism or organisms.

5.2   Genomics

Genomics is the field of biology that includes the understanding of structure, function, evolution, mapping and editing of genomes. An organism's complete set of DNA is said to be genome. Genomics involves the sequencing, assembling and analyzing the function and structure of entire genome. This is also a subdiscipline of genetics that combine classical and molecular biology with the goal of sequencing and understanding genes, gene interaction, genetic elements, structure and evolution of genomes. Under genomics, different resources like whole genome reference sequences, high-density SNP genotyping arrays and genotyping-by-sequencing are in action for several aquaculture species. These resources in combination can be powerful tools for many researches as well as in determination of the genetic factors involved in the regulation of complex traits in aquaculture sector. Analysis of fish genome through genomics can help in various aspects from establishing breeding strategies to analyzing reasons behind failure of captive breeding of any species and growth of fish. Tools of genomics can help in determining breeding plans for certain commercially important traits such as disease resistance. Under genomics research Zebra fish (*Danio rerio*) has been identified as a potential model organism and has been used in various researches for improvement of breeding strategies in aquaculture. But due to the vast evolutionary distance between the model organism and many other aquaculture species, it is clear that direct research on the species of interest can often be the most feasible and informative. That's why genomics is emerging as a vital science for aquaculture.

An important aspect understands the role of brain in regulation of social behavior and social processes. These areas are an emerging sector in future challenges of contemporary neuroscience. This area is categorized as **cognitive genomics**. Zebra fish has now been selected as a model for studying, the information processing mechanisms in brain those are inbuilt for social behavior and neuroscience. This will certainly help in the development of aquaculture sector development and as well as in fighting mammalian diseases like Down's syndrome, autism, many depressions and Alzheimer's disease. According to Krogh's principle, "for many problems there is an animal on which it can be most conveniently studied" (Krebs 1975). Cognitive genomics with the power of aquaculture is the future solution for researches in neural disorders. Social values, social preferences, social recognition, cognitive appraisal, social learning, inter temporal choice and traffic rules can be studied in Zebra fish and most importantly in a small aquaculture system.

An important part of genomics is **comparative genomics**, in which the genomic features of different organisms are compared with the help of Next-Generation Sequencing (NGS) technology. At present, the major focus in aquaculture sector is growth enhancement, disease resistance, and improved meat quality in fish. Comparative genomics helps in comparing the genes between biological groups and even from different environment to screen out the reasons behind slow growth, low meat quality of cultured fish, and stress conditions. The most fascinating thing regarding comparative genomics is that it can help in studying different phenotypes and related genes at the same time. In the near future this will help in development of reproductive biotechnology and selective breeding programs in aquaculture. At present, comparative genomics has been used in studying pigmentation (Mandal et al. 2020), evolution (Yuan et al. 2018), muscle (Sun et al. 2016), etc.

Functional genomics is another branch of genomics that studies the gene and protein functions and their interactions. This branch uses the modern NGS and transcriptomics approaches to find the gene functions. Gene transcription, translation, gene expression and protein–protein interactions which are very important to understand the basic mechanisms and pathways responsible for growth, metabolism, stress, etc. These can be easily studied with the help of functional genomics. The most fascinating feature of functional genomics is to use genome-wide approach to address the questions related to aquaculture genomics through high-throughput methods rather than a more traditional "gene-by-gene" approach. Modern goals of functional genomics are to gain better understanding in the field development, growth, metabolism, immunity, behavior, reproduction, and various other processes in organisms (Liu et al. 2012) and these will certainly help in future aquaculture sector. According to Liu et al. (2012), the functional genomics generally includes, functional correlation of gene expression pattern and later on its influence on gene functions, functional correlation of gene positions and later on their influence on candidate gene functions, gene pathways analysis and gene functions testing. At present, the most commonly used approaches in aquaculture sector for gene expression profiling are, Expressed Sequence Tags (EST), microarrays, and RNA-Seq. After RNA isolation and cDNA library preparation, the functional genomics approach gives list of genes or proteins, which are later taken for pathway analysis.

5.3 Transcriptomics

The study of transcriptome is called transcriptomics i.e., the sum of all of the RNA transcripts of an organism. From the past decade, the study of complete set of transcripts of a specific organ or tissue through new next-generation sequencing technologies has revolutionized the molecular biology techniques for aquaculture development. Modern RNA-sequencing technology through transcriptomics approach has helped the studies regarding control and a treated specimen in the same experiment so as to compare different physiological or environmental conditions. In transcriptomics, after sample collection and RNA isolation, library construction and sequencing are done through Next-Generation Sequencing (NGS) like Illumina, Pyrosequencing, PacBio, etc. The raw reads generated from NGS are converted to high-quality reads by FastQC or Trimmomatic software. Then the high-quality reads are taken for de novo pooled transcriptome assembly preparation in a single file using Trinity software. Different gene fragments i.e., Unigenes are predicted and validated using CD Hit test, BWA, and SAMtools. These unigenes are used for prediction of Coding Sequences (CDS) by TransDecoder, BWA, and SAMtools. Then three types of analysis are done later on using the predicted CDS. First of all, the functions of different genes i.e., functional annotation are predicted by BLASTX and then genes are classified to their respective functional categories i.e., gene ontology by Blast2GO. Quantification of CDS is done for differential gene expression analysis using DESeq. This determines the expression of different genes from the samples. The gene pathway analysis is done in another part by Kyoto Encyclopedia of Genes and Genome (KEGG). The transcriptomics approach at present is helping in finding unique genes and their expression for aquaculture species development.

5.4 Metabolomics

Under metabolomics, metabolites i.e., the small molecule substrates, intermediates, and products of metabolism, are studied. In other simple terms, it can be explained as the systematic study of the unique chemical fingerprints of specific cellular processes. The metabolome is the term associated with metabolomics that represents the complete set of metabolites in a biological cell, tissue, organ, or organism that are the end products of different cellular processes. At present two common metabolomic approaches are used in aquaculture i.e., Nuclear Magnetic Resonance (NMR) and Mass Spectrometry (MS). Fingerprinting is an approach of metabolomics i.e., based on the high-throughput analysis of a large set of fish species to build a model based on all analytical variables generated from metabolite detection. This approach can be used for fish farming as well as fish and feed authentication. Similarly, profiling is another approach i.e., based on the identification and/or quantification of analytical variables generated from the nonselective analysis of the biological system (Roques et al. 2020). Identification of the metabolites is essential for qualitative analysis of metabolic pathways due to internal or external factors. In aquaculture, metabolomics has helped in researches related to environmental impact on fish, fish health and welfare and fish nutrition. The role of physical, chemical, and biological stressors, impacting fish metabolism are studied under environmental metabolomics. Through metabolomics, farmed fish disease due to pathogen infection is now being studied. The most important approach of metabolomics on aquaculture system is the researches related to fish nutrition. The basics of metabolomics workflow include, sampling and analytical sample preparation, data acquisition, data processing, statistical analysis, and biological interpretation. Muscle and liver are the two main tissues that are studied for metabolism and metabolic pathways. Some important environmental and biological bias needs to be taken care of during sampling for metabolomic approaches i.e., season, tank effect, fish gender, sexual maturation, and environmental or water quality parameters. Other samples such as plasma, serum, gut content, feces, mucus, skin, etc. are also studied for various metabolic pathway studies in aquaculture.

5.5   Proteomics

Proteomics is the large-scale study of proteins or simply can be said proteome. Proteome is the entire set of proteins that is produced by an organism. The main workflow of proteomics starts from the separation of proteins from a given tissue, cell line, or even the whole organism (embryo). The proteins can be separated and later digested or vice versa (Forné et al. 2010). The digested proteins are then analyzed by Peptide Mass Fingerprinting (PMF) or MS for Two-Dimensional Electrophoresis (2-DE), or by Liquid Chromatography–Mass Spectrometry (LC–MS/MS) for One-Dimensional Electrophoresis (1-DE). These MS data are then processed by search engines to identify proteins of interest. The downstream analysis to address the biological question can be done by Western Blotting, Flow Cytometry, Immunohistochemistry (IHC), qPCR, microarrays or by metabolomics. The role of proteomics in aquaculture sector starts with estimation of seafood and fish quality along with safety. Proteomics is being used in aquaculture for studying proteins and protein networks, protein expression, protein biomarkers for therapeutic intervention, parallel analysis of genome and proteome for posttranslational modifications and proteolytic events, large-scale protein structures, etc. Proteomics can be divided into (1) immunoproteomics: the study of proteins involved in immune response, (2) nutriproteomics: identifying molecular basis of nutritive and nonnutritive components of the diet through protein analysis, (3) proteogenomics: proteome analysis for studying gene annotations, and (4) structural genomics: the study of three-dimensional structure of every protein encoded by a given genome.

5.6   Metagenomics

Metagenomics is that part of genomics that studies genetic material recovered directly from environmental samples. Some synonyms related to metagenomics is being used by other researchers like, environmental genomics, ecogenomics, and community genomics. The major problem that aquaculture industry is facing is the captive stress conditions leading to various diseases. The gut of fish carries microbial communities similar to other vertebrates, to increase the overall size, immune response and ultimately defence against other pathogens. These are called gut microbiome, whereas environmentally friendly microbial communities are said to be microbiome. In the recent years, there has been an increase in the antimicrobial-resistant (AMR) bacteria leading to fish diseases and this can be overcome with the help of probiotics (Yukgehnaish et al. 2020). The studies based on the metagenomics of gut microbiome is important to use them as probiotics for aquaculture. Use of probiotics are much better than traditional antibiotics in aquaculture, as the antibiotics develop AMR bacteria with due course of time. There are many remarkable achievements with regard to the use of probiotics in aquaculture. *Aeromonas hydrophila* has been used as a probiotic in *Oncorhynchus mykiss* (Rainbow trout) to fight against *Aeromonas salmonicida* infection (Irianto and Austin 2002). *Agarivorans albus* F1-UMA in *Haliotis rufescens* (Abalone) (Silva-Aciares et al. 2011) and *Alteromonas* CA2 in Pacific oyster has been used as probiotic for better survivability (Douillet and Langdon 1994). Similarly, *Burkholderi acepacia* Y021 used in *Crassostrea corteziensis* (Lions-pay scallop) for better growth and survivability (Granados-Amores et al. 2012). Metagenomics is not limited to the studies on anti-pathogenic effects of different gut microbiota, but it can also help in detecting beneficial bacteria from soil and water. This will help in monitoring future water quality parameters in aquaculture. The steps of metagenomics analysis initially start with collection of samples from natural sources (soil or water) or fish gut, depending on the type of experiment. DNA extraction has to be done from these bacterial niches, followed by metagenomic DNA library construction. The libraries are used for the mining of clones and DNA sequences. These DNA sequences are later classified as two categories i.e., long insert (Fosmids/Bacterial Artificial Chromosomes (BACs)/cosmids) and short insert (plasmids). The long inserts are subjected to functional analysis, whereas short inserts are taken for sequence-based analysis. Later on, good clones and desired sequences are taken for unique bacteriological applications or other biotechnological applications. Similarly, environmental DNA (eDNA) i.e., collected water or soil samples, are now-a-days used for metagenomic analysis of biodiversity, detecting fish species from a natural source and even calculation of pond biomass.

5.7 Neurogenomics

Neurogenomics[1] includes the studies of the genome of an organism in relation to the nervous system. This field unites functional genomics and neurobiology. Parental care in some ornamental fish species and commercial fishes are very important for the survival of young ones, perpetuation of race, defending young ones against predators, etc. Darter fish, Stickleback fish, African lung fish, Fighter fish, Tilapia, Cichlids, etc. shows parental care. Studying these parental cares and the related neural behavior is important to derive benefits through neurogenomics. Neurogenomics can help in studying parental care at the molecular level and analyzing related life events i.e., stages of territory establishment, stages of pair-bonding, stages of dispersal, etc. (Bukhari et al. 2019). The neurogenomic dynamics in fish can be analyzed by measuring gene expression (RNA-Seq) on male and female brain. This will give more insight into the genes and their expression during parental care and to reveal mysteries behind the mating, caring for eggs, hatching and caring for fry. The neurogenomics methodology includes RNA-Seq followed by RNA-Seq informatics and later defining different DEGs (Differentially Expressed Genes). After identifying unique genes, Transcriptional Regulatory Network (TRN) analysis is done followed by functional analysis. Neurogenomics will also help in studying neural behavior related to photoperiodism, reproduction, hormonal secretion from brain, etc.

5.8 Pangenomics

Pangenomics is another emerging branch of genomics that studies pan-genome. Pan-genome is the entire gene set of all strains of a species. Disease outbreaks in aquaculture industry are a major problem, especially in prawn or shrimp farming. It is important to detect the entire set of strains and their virulence of causative organism to cope with the disease. Through pangenomics the core-genome containing genes present in all strains within the clade and genes that are major causes for strain-specific virulence can be detected. The genome sequences of causative bacteria and virus can be obtained through RNA-Seq or from earlier submitted sequences in National Center for Biotechnology Information (NCBI) database and the predicted genes can be annotated against the NCBI non-redundant database, Gene Ontology (GO) and Swiss-Prot databases using DIAMOND software (Buchfink et al. 2014; Nourdin-Galindo et al. 2017). After this the metabolic pathways are recovered from the Kyoto Encyclopedia of Genes and Genomes (KEGG) database. The phylogenetic and phylogenomic analysis are done later on based on the gene sequences and taking an out-group. The

---

[1]Under neurogenomics, the zebrafish model system and mutants at present is already been used to study neurodegeneration and severe polygenic human diseases like cancer and heart disease.

core-genome[2] and pan-genome are analyzed and virulence factors in pan-genome are detected with the help of NCBI BLASTp (Altschul et al. 1990) program and Virulence Factor Database. After this RNAmmer software (Lagesen et al. 2007) is used to predict 5S, 23S, and 16S rRNAs gene sequences of all the strains for genomic comparisons among the different strains and genogroups using NCBI BLASTn (Altschul et al. 1990).

5.9   Epigenomics

Epigenomics is the omics study that deals with the complete set of epigenetic modifications on the genetic material of a cell, which is said to be epigenome. Epigenetics is the study of heritable changes to the gene expression that are independent of changes to the DNA sequence. Epigenetic mechanisms in fish are an emerging area of research. This can reveal the response of aquaculture species with respect to their environment for phenotypic plasticity i.e., ability of one genotype to show many phenotypes under different environmental conditions. Moreover, the features like high diversity, high fecundity, remarkable plasticity and commercial economic values of aquaculture species possess excellent research opportunities in the field of fisheries epigenomics for future blue revolution. DNA methylation is the widely studied technique in aquaculture epigenomics, which is defined as the addition of methyl group (CH_3) to position 5 of the pyrimidine ring of a cytosine (5mC). In case of eukaryotic cells, these 5mC mostly occur on cytosine-phosphate-guanine (CpG) island i.e., transcription start site of a gene, which causes changes to the gene expression. Moreover, the DNA methylation process can also regulate the transcriptional activity of neighboring genes by change in chromatic structure. These epigenetic mechanisms can vary among fish species, can be maintained at individual level and eventually is transmitted to subsequent generations, providing an added source of heritable variation that is independent of genetic variation (Metzger and Schulte 2016). The methods of studying epigenomics through DNA methylation is by low resolution assessment of global DNA methylation, methods with intermediate resolution and methods with single base-pair resolution.

Whole genome DNA methylation levels at low resolution rate can be assessed by one of the most quantitative method i.e., reverse-phase high performance liquid chromatography (RP-HPLC). In this method, DNA is digested initially to single nucleotides. The molecular mass of cytosines and methylated cytosines is different, that's why they can be easily quantified by HPLC or liquid chromatography coupled with tandem mass spectrometry (LC–MS/MS). ELISA-based methods can also help in DNA methylation detection by low resolution assessment. By ELISA-based methods, purified genomic DNA is hybridized to assay wells initially. After this, an antibody specific to methylated cytosines is used to detect methylated DNA. A

---

[2]The core-genome is the set of homologous genes that are present in all genomes of an analyzed dataset.

secondary enzyme-linked antibody is used for colorimetric quantification later on. At present commercial kits are available from different companies for these methods. The major disadvantages of these methods are low resolution rate and very large differences can only be detected.

The methylation-sensitive amplified polymorphism analysis is another method to assess DNA methylation patterns at intermediate resolution (Reyna-López et al. 1997). This method is also called MSAP or MS-AFLP. First of all, the genomic DNA is digested by methylation-sensitive and non-methylation-sensitive isoschizomeric restriction enzymes i.e., HpaII and MspI, respectively. The DNA fragments generated are then ligated to adapters and PCR amplified using fluorescent-tagged primers. The PCR products are analyzed later on by capillary electrophoresis or gel electrophoresis. The band pattern of digested DNA with MspI and HpaII restriction enzyme determine the methylation state of a particular locus. These differentially methylated bands as seen on gel electrophoresis or capillary electrophoresis can then be isolated and sequenced to identify differentially methylated genes. Again, the limitations to this method are, low resolution rates, anonymous loci screening and poor reproducibility.

DNA methylation with single base-pair resolution can be done by bisulfite conversion method. In this method, the DNA is first treated with sodium bisulfite or metabisulfite. As a result of this bisulfite treatment, the unmethylated cytosine bases (C) are converted into uracil (U). But, methylated cytosines (5mC) are not converted to uracil. By PCR amplification, the original sequence is changed from a C to a T in unmethylated positions, whereas methylated positions remain as C's. This helps in accurate estimation of DNA methylation. New methods are present to reduce the cost of Whole Genome Bisulfite Sequencing (WGBS) like, methylated DNA immunoprecipitation and sequencing (MeDIPseq) and methyl CpG binding domain immunoprecipitation and sequencing (MDBseq). There are several restriction-enzyme-based methods of DNA methylation are done now-a-days like, reduced representation bisulfite sequencing (RRBS) and EPIRADSeq. In RRBS method, a methylation-insensitive restriction enzyme like MspI is used to digest the genomic DNA to specifically target CpG motifs only. Then these fragments are then bisulfite treated and sequenced. This allows sequencing efforts to be concentrated primarily on regions of the genome that are likely to be methylated (Gu et al. 2010) EpiRADseq is an alternative to the RRBS technique. In EpiRADseq, two restriction enzymes that is methylation-insensitive (i.e., PstI) and methylation-sensitive (i.e., HpaII) are used to double digest the genomic DNA. The methylation-sensitive restriction enzyme will only cut unmethylated DNA and thus the methylation can be assessed by comparing the frequencies at which loci are sampled. So, this technique only provides information about methylation state at the enzyme cut-sites.

Epigenomics is already in action in the field of aquaculture in respect of fish muscle quality development, selective breeding, pigmentation, location based commercially important phenotypic changes, etc.

5.10   Lipidomics

Lipidomics is the broad study area related to pathways and networks of cellular lipids in biological systems. The "lipidome" is the complete lipid profile within a cell, tissue, organism, or ecosystem. At present, on-going research works are targeting mostly fish muscle quality enhancement. For this purpose, it is important to study initially the lipid profile of muscle and different environmental conditions which are affecting muscle quality. Initially, fish muscle is sampled and kept inside liquid nitrogen vials. The muscle samples are homogenized and freeze-dried and taken for lipids extraction. Then these lipid extracts are injected in an ultrahigh performance liquid chromatography (UHPLC). Then the separated samples are subjected to mass spectrometry (MS). The mass spectrometry is an analytical technique that helps in the ionization of samples and measuring the ratio of their mass-to-charge (m/z). The MS raw data are then plotted for further statistical analysis and interpretation. Lipidomics in aquaculture can help in the analysis of Fatty Acids, Fatty Acids Derivatives, Fatty Acids Metabolism, Glycerolipids, Glycerophospholipids, Sphingolipids, Isoprenoids, Sterols, Lipids CLASS, Phospholipids, Wax Esters, Olive Oil Phenols, etc.

5.11   Glycomics

Glycomics is the study of entire set of free or more complex molecules of sugars of an organism. The glycome is the entire carbohydrate content of cell and glycans are chain-like structures that are composed of single sugar molecules (monosaccharides) and found attached to proteins and lipids in living organisms. Many glycan-based diseases related to central nervous system development, cartilage formation, muscular dystrophies, skeletal disorders, neuromuscular transmission defects, neurocutaneous disorder and inflammatory bowel disease are present in human. These can be studied based on the zebrafish model glycomics and can be treated in the near future. It is believed that, around 1–2% of the vertebrate's genome is associated with the synthesis and processing of glycans and is a vital process for many biological activities (Yamakawa et al. 2018). Initially for fish glycomics, tissue samples are dissected, washed, homogenized with a mechanical disperser and later freeze-dried. Then the tissues are taken for purification of glycans i.e., N-linked glycans (NGs), O-linked glycans (OGs), and glycosphingolipids (GSLs). Following their purification, the glycans are permethylated using the NaOH/dimethyl sulfoxide, extracted in chloroform and repeatedly washed with water and then taken for MS analysis. Analysis of monosaccharides, derivatizations of lipids, IHC, WB, qPCR are done for further analysis. Glycomics can help in recognizing bacterial and viral detection, as glycoproteins and glycolipids are the cell surface detectors for these infections. It can help in studying cellular signaling pathways, innate immunity, cancer development, stability and folding of proteins, pathway and fate of glycoproteins, etc. In the near future, many novel drugs and glycoconjugate vaccines can be developed based on glycomics.

5.12   Conclusion

The omics in aquaculture with the advancement of RNA-Seq and NGS have impacted a lot of new strategies for unique gene identification, marker-assisted selection, selective breeding, reproductive biotechnology, disease diagnosis and nutrition. These emerging studies will help in aquaculture development leading to blue revolution. In near future many new omics studies that have been studied on mammals i.e., foodomics, pharmacogenomics, toxicogenomics, psychogenomics, connectomics, tomomics, cellomics, etc. can be applied in fisheries and aquaculture biotechnology sector.

References

Altschul SF, Gish W, Miller W, Myers EW, Lipman DJ (1990) Basic local alignment search tool. J Mol Biol 215:403–410. https://doi.org/10.1016/S0022-2836(05)80360-2

Buchfink B, Xie C, Huson DH (2014) Fast and sensitive protein alignment using diamond. Nat Methods 12:59–60. https://doi.org/10.1038/nmeth.3176

Bukhari SA, Saul MC, James N et al (2019) Neurogenomic insights into paternal care and its relation to territorial aggression. Nat Commun 10:4437. https://doi.org/10.1038/s41467-019-12212-7

Douillet PA, Langdon CJ (1994) Use of probiotic for the culture of larvae of the Pacific oyster (Crassostrea gigas Thurnberg). Aquaculture 119:25–40

Forné I, Abián J, Cerdà J (2010) Fish proteome analysis: model organisms and non-sequenced species. Proteomics 10(4):858–872. https://doi.org/10.1002/pmic.200900609

Granados-Amores A, Campa-Cordova AI, Araya R, Mazon Suastegui JM, Saucedo PE (2012) Growth, survival and enzyme activity of lions-paw scallop (Nodipecten subnodosus) spat treated with probiotics at the hatchery. Aquac Res 43:1335–1343

Gu H, Bock C, Mikkelsen TS, Jäger N, Smith ZD, Tomazou E, Gnirke A, Lander ES, Meissner A (2010) Genome-scale DNA methylation mapping of clinical samples at single-nucleotide resolution. Nat Methods 7:133–136. https://doi.org/10.1038/nmeth.1414

Irianto A, Austin B (2002) Use of probiotics to control furunculosis in rainbow trout, Oncorhynchus mykiss (Walbaum). J Fish Dis 25:333–342

Krebs HA (1975) The august Krogh principle: "for many problems there is an animal on which it can be most conveniently studied". J Exp Zool 194:221–226. https://doi.org/10.1002/jez.1401940115

Lagesen K, Hallin P, Rødland EA, Staerfeldt H-H, Rognes T, Ussery DW (2007) RNAmmer: consistent and rapid annotation of ribosomal RNA genes. Nucleic Acids Res 35:3100–3108. https://doi.org/10.1093/nar/gkm160

Liu S, Zhang Y, Sun F, Jiang Y, Wang R, Li C, Zhang J, Liu Z (2012) Functional genomics research in aquaculture: principles and general approaches. In: Functional genomics in aquaculture. Wiley Blackwell, Ames, IA, pp 1–40. https://doi.org/10.1002/9781118350041.ch1

Mandal SC, Tripathy PS, Khatei A, Behera DU, Ghosh A, Pandey PK, Parhi J (2020) Genetics of colour variation in wild versus cultured queen loach, Botiadario (Hamilton, 1822). Genomics 112(5):3256–3267. https://doi.org/10.1016/j.ygeno.2020.06.012

Metzger DCH, Schulte PM (2016) Epigenomics in marine fishes. Mar Genomics 30:43–54. https://doi.org/10.1016/j.margen.2016.01.004

Nourdin-Galindo G, Sánchez P, Molina CF, Espinoza-Rojas DA, Oliver C, Ruiz P et al (2017) Comparative pan-genome analysis of Piscirickettsia salmonis reveals genomic divergences within Genogroups. Front Cell Infect Microbiol 7:459. https://doi.org/10.3389/fcimb.2017.00459

Reyna-López GE, Simpson J, Ruiz-Herrera J (1997) Differences in DNA methylation patterns are detectable during the dimorphic transition of fungi by amplification of restriction polymorphisms. Mol Gen Genet 253:703–710. https://doi.org/10.1007/s004380050374

Roques S, Deborde C, Richard N, Skiba-Cassy S, Moing A, Fauconneau B (2020) Metabolomics and fish nutrition: a review in the context of sustainable feed development. Rev Aquac 12 (1):261–282

Silva-Aciares FR, Carvajal PO, Mejias CA, Riquelme CE (2011) Use of macroalgae supplemented with probiotics in the Haliotis rufescens (Swainson, 1822) culture in Northern Chile. Aquac Res 42:953–961

Sun Y, Huang Y, Hu G, Zhang X, Ruan Z, Zhao X, Guo C, Tang Z, Li X, You X, Lin H, Zhang Y, Shi Q (2016) Comparative transcriptomic study of muscle provides new insights into the growth superiority of a novel grouper hybrid. PLoS One 11(12):e0168802. https://doi.org/10.1371/journal.pone.0168802

Yamakawa N, Vanbeselaere J, Chang LY et al (2018) Systems glycomics of adult zebrafish identifies organ-specific sialylation and glycosylation patterns. Nat Commun 9:4647. https://doi.org/10.1038/s41467-018-06950-3

Yuan Z, Liu S, Zhou T et al (2018) Comparative genome analysis of 52 fish species suggests differential associations of repetitive elements with their living aquatic environments. BMC Genomics 19:141. https://doi.org/10.1186/s12864-018-4516-1

Yukgehnaish K, Kumar P, Sivachandran P, Marimuthu K, Arshad A, Paray BA, Arockiaraj J (2020) Gut microbiota metagenomics in aquaculture: factors influencing gut microbiome and its physiological role in fish. Rev Aquac 12:1903. https://doi.org/10.1111/raq.12416

Growth Hormone Transgenesis in Aquaculture

6

Kirankumar Santhakumar

Abstract

For more than three decades, extensive research work has been done on growth hormone transgenesis in various teleost species. Transgene constructs of piscine origin have been observed to be successful in consistently generating transgenic fish lines with enhanced somatic growth. Yet, several concerns remain, both public and scientific, for growth hormone transgenesis in teleosts. In this chapter, a brief overview about the beginning of growth hormone transgenic technology in fish, role of IGF in growth hormone signalling, muscle development in transgenic fish overexpressing growth hormone are given. This chapter also discusses the role of selected environmental factors on survival and adaptation of the transgenic fish.

Keywords

Transgenic fish · Growth hormone · Microinjection · IGF · Somatic growth · Transgene

6.1 Introduction

Transgenesis refers to the insertion of foreign DNA molecule into an organism by artificial procedures. Richard Palmiter's generation of transgenic mice in 1982, overexpressing growth hormone, exhibited accelerated growth and doubled body size compared to non-transgenic littermates (Palmiter et al. 1982). This study

K. Santhakumar (✉)
Zebrafish Genetics Laboratory, Department of Genetic Engineering, SRM Institute of Science and Technology, Kattankulathur, Tamil Nadu, India
e-mail: kirankus@srmist.edu.in

© The Author(s), under exclusive license to Springer Nature Singapore Pte Ltd. 2021
P. K. Pandey, J. Parhi (eds.), *Advances in Fisheries Biotechnology*,
https://doi.org/10.1007/978-981-16-3215-0_6

generated enormous interest in employing genetic engineering techniques to improve phenotypic traits, like somatic growth, in fish. Global fisheries harvest from natural waters is fast approaching the extreme sustainable yield and this scenario has emphasized huge importance on aquaculture production and the genetic improvement of domesticated fish stocks and strains. In the last 35 years, in several teleost species, successful introduction and expression of *growth hormone* transgene has been accomplished (Zhu et al. 1985, 1986; Rokkones et al. 1989; Zhang et al. 1990) and in several cases enhanced somatic growth have been recorded. USA has developed a hybrid Atlantic salmon, AquAdvantage salmon that expresses *growth hormone* gene from Chinook salmon, which can grow to marketable size in almost half the time of a wild-type salmon. Thus, growth hormone transgenesis presents a promising application in aquaculture. In this chapter, a brief summary of the historical progression, current status of growth hormone transgenesis technology and the potential applications and challenges of this biotechnological strategy are discussed.

6.2   Origin of Growth Hormone Biology in Fish

The field of growth hormone transgenic biology in teleost started with the use of gene constructs, comprised of mammalian or viral promoters and mammalian GH genes. One of the initial success in generating faster growing transgenic fish was accomplished by introducing a recombinant human *growth hormone* (hGH) transgene, driven by a mouse *metallothionein1* (*MT*) gene promoter into the eggs of goldfish 1984 (Zhu et al. 1985). The chorion of the goldfish eggs was removed by treatment with 0.25% trypsin solution and about 1–2 nL of *MThGH* transgene, containing DNA solution was dispensed by microinjection. All the microinjected eggs were incubated in Holtfreter's balanced salt solution and the embryonic development was monitored. Several of the generated transgenic fish exhibited accelerated somatic growth. In another earlier study, Zhang et al. (1990) utilized a transgene, containing rainbow trout *GH* cDNA with the long terminal repeat (LTR) element of Avian Rous sarcoma virus (RSV) and generated transgenic carp embryos (Zhang et al. 1990). The generated transgenic carps were observed to be 22% larger than their siblings (Zhang et al. 1990). Penman et al. (1991) studied the inheritance of the transgene *MTrGH* in transgenic trout to the offspring by crossing them with the wild-type fish and demonstrated the germ line transmission of the *growth hormone* transgene (Penman et al. 1991). Subsequently, various research groups started using transgene constructs of piscine origin. Du et al. (1992) made a *GH* transgene construct with promoter sequence of *antifreeze protein* gene from ocean pout (*Macrozoarces americanus*) and chinook salmon (*Oncorhynchus tshawytscha*) *GH* cDNA and injected the transgene into Atlantic salmon zygotes and observed an upsurge in transgenic salmon size of about fivefold over the respective age-matched controls (Du et al. 1992).

6.3   Fish Growth Hormone Signalling

Proper understanding of the molecular basis of regulation of somatic growth and its interface with feed intake and energy metabolism are vital for efficient aquaculture management. The somatotropic axis, the endocrine system that controls somatic growth, involves hypothalamus, pituitary gland and liver and the key molecule orchestrating the process of growth promotion is growth hormone (GH), released from the pituitary gland. The growth hormone acts by stimulating IGF-I production in hepatic tissues, which in turn induces cell proliferation in the somatic tissue. In addition, growth hormone has been observed to act directly on the skeletal muscle and adipose tissue to trigger lipolysis and protein production (Björnsson 1997). Two growth hormone receptors, GHR-I and GHR-II, are present in teleosts that can bind to the growth hormone and both GH receptors (ghr) are observed to be expressed in multiple tissues (Di Prinzio et al. 2010). In specific stages of development, like osmotic stress and selective nutritional condition the GHR expression pattern differs accordingly (Pierce et al. 2007; Malkuch et al. 2008). Interestingly, both GH receptors have been observed to elicit overlapping functions and the different GHR isoforms play an important role in maintaining the balance between somatic growth and the lipolysis functions of GH (Bergan-Roller and Sheridan 2018). GH synthesis and secretion in the pituitary gland is controlled by multiple factors secreted by the hypothalamus like growth hormone releasing hormone (GHRH), neuropeptide Y (NPY), somatostatin (SS), etc. IGF-I is the major negative feedback regulator of growth hormone secretion (Butler and LeRoith 2001).

6.4   Techniques for Transgenesis in Teleosts

Because of external fertilization observed in most of the teleost species, microinjection of transgene into just fertilized eggs is the very common method for introducing transgenes in the teleosts. Injection of DNA solution from pulled glass capillaries into newly fertilized eggs have proven to give efficient transgene expression and germ line inheritance. Using microinjection method, successful transgenic fish have been generated in several species e.g., tilapia (*Oreochromis niloticus*: Brem et al. 1988), common carp (*Cyprinus carpio*: Zhang et al. 1990) etc.

There has been lot of interest in electroporation, a technique that can facilitate mass gene transfer and has been proven useful for species like trout, carp and salmon which possess cyclical availability of gametes. Electroporation comprises the use of brief electrical pulses to the cell membrane for permeabilization, facilitating entry of transgenes into the egg or sperm cell and it has been successfully used in various species (Buono and Linser 1992; Powers et al. 1992). Venugopal et al. (1998) used rainbow trout *growth hormone* gene to perform sperm electroporation-mediated transgenesis in three species of Indian major carps namely mrigal (*Cirrhinus mrigala*), rohu (*Labeo rohita*), and catla (*Catla catla*).

6.5   Impact of GH Transgenes on Teleost Growth

Growth acceleration effects of GH have been documented in various review articles
(Zbikowska 2003; Devlin 2011). Briefly, effects of growth hormone over-expression
have been observed to vary widely depending on species and/or strain, ranging from
very mild increase in few species to 37-fold increase in coho salmon body weight
(Devlin 2011). In the case of fast-growing fish species like zebra fish, tilapia, carps
and catfishes, the metabolic apparatus is observed to function at full capacity and
further stimulation of growth potential is limited whereas in the slow-growing coldwater species like salmon and rainbow trout significant growth acceleration is
possible by triggering growth hormone over-expression. In coho salmon, GH
transgenesis leads to swift embryogenesis and robust stimulation of growth in the
subsequent life stages. Bovine *growth hormone* (bGH) transgene microinjected into
just fertilized northern pike eggs, led to increase in somatic growth compared to
age-matched controls (Gross et al. 1992). Interestingly, in few experiments only
transgenic male northern pikes exhibited growth enhancement. Nam et al. (2001)
generated transgenic mud loaches (*Misgurnus mizolepis*) through microinjection of
the GH transgene, wherein all components of the transgene were from the same
species, mud loach *β-actin* gene promoter and mud loach *growth hormone* (mlGH)
cDNA sequences. Interestingly, 7.5% of transgenic fish exhibited increased growth,
up to a maximum of 35-fold faster growth than their age-matched controls (Nam
et al. 2001). Several fast-growing GH-transgenic mud loaches exhibited gigantic
growth. Nam et al. (2001) also analysed the growth performance of the F1 generation transgenic individuals and observed dramatic acceleration of up to 35-fold,
albeit with lots of variability among transgenic lines. Nam et al. (2001) established
three transgenic mud loach families up to the F4 generation and observed the
inheritance of the transgene in typical Mendelian fashion.

6.6   Muscle Development in GH Over-Expressing
　　　　Transgenic Fish

Growth hormone (GH) functions via the insulin-like growth factor I (IGF-I) that gets
generated due to the activation of the growth hormone receptor by the growth
hormone. Kuradomi et al. (2011) analysed the morphology of white skeletal muscle
tissue in the GH-transgenic zebra fish (*Danio rerio*) and observed that transgenic
zebra fish showed enhanced muscle hypertrophy compared to non-transgenic
siblings. Interestingly, they noted that the transgenic zebra fish females to be more
hypertrophic. Importantly, the transgenic hypertrophy occurred independent of the
local activation of *insulin-like growth factor I* gene (*igfI*). In the case of transgenic
males exhibiting muscle hypertrophy, increased expression of *myogenin* (*myo*) was
observed, suggesting a role for myogenin in the hypertrophic growth. In another
study, Figueiredo et al. (2007) compared the hemizygous and homozygous
genotypes of GH-transgenic zebra fish and observed that the homozygous transgenic
zebra fish possess lower condition factor, a potential catabolic state. This

physiological condition of the homozygous GH-transgenic zebra fish was due to decreased *igf1* and *ghr* gene expression levels due to GH resistance. Figueiredo et al. (2007) concluded that homozygous GH-transgenic fish resembled starvation-induced fish on the basis of various functional parameters.

Silva et al. (2015) generated double transgenic zebra fish overexpressing *gh* and its receptor, *ghr*, and analysed the muscle growth in these transgenic fish. GH/GHR double transgenic zebra fish exhibited identical lengths to that of single GH-transgenic fish. But the double transgenic fish displayed significantly less body weight and low condition factor, suggesting an unbalanced growth. GH/GHR double transgenic zebra fish is found to possess reduced thick muscle fibres compared to GH-transgenic fish. Expectedly, GH/GHR genotype led to reduced expression of muscle growth genes. Silva et al. (2015) proposed that overexpression of *gh* and *ghr* genes in double transgenic zebra fish led to decreased somatotrophic axis intracellular signalling mediated by STAT5.1 transcription factor.

6.7   IGF Expression Levels in GH Over-Expressing Transgenic Fish

The insulin-like growth factors, IGF-I and IGF-II, have structural similarities to insulin and IGF-I is known to promote growth and differentiation in various target tissues. Most of the effects of GH are mediated through IGF-I and hence, it is imperative to know the expression status of *igf* genes in various tissues of growth hormone overexpressing transgenic fish. Expression levels of *IGF-I* and *IGF-II* in liver and other organs of adult transgenic tilapia expressing growth hormone and corresponding control samples were determined by quantitative PCR (Eppler et al. 2010). Both *IGF-I* and *IGF-II* mRNA were observed to have increased in kidney, gills, heart, intestine, testes and brain of the transgenic fish (Eppler et al. 2010). But, changes were not observed in *IGF-I* or *IGF-II* mRNA levels in the pituitary gland. In contrast in the spleen tissue, *IGF-I* and *IGF-II* mRNA were observed to be reduced in the transgenic fish (Eppler et al. 2010).

Eppler et al. (2007) analysed 17 months old transgenic tilapia, *Oreochromis niloticus*, overexpressing growth hormone and observed them to exhibit 1.5-fold increase in length and 2.3-fold increase in weight compared to non-transgenic controls. Subsequently, when they performed radioimmunoassay for the serum IGF-I levels, the transgenic fish showed lower levels of IGF-I (6.22 ± 0.75 ng/mL) than the serum from controls (15.01 ± 1.49 ng/mL) (Eppler et al. 2007). But IGF-I levels in the liver tissue of transgenic fish was 4.2-times higher than in the corresponding control liver samples. Hepatocytes in the control samples showed negligible IGF-I immunoreactivity whereas hepatocytes in transgenic samples exhibited strong IGF-I immunoreactivity (Eppler et al. 2007). Expectedly, in situ hybridization displayed more numbers of hepatocytes with strong *IGF-I* mRNA staining in the transgenic liver samples. Twofold elevation in *IGF-I* mRNA levels was observed in the skeletal muscle tissue of transgenic fish samples compared to the control samples. Based on their analysis, Eppler et al. (2007) proposed that the increased growth of the transgenic tilapia was due to enriched autocrine and paracrine functions of IGF-I in extra-hepatic tissues.

6.8   Impact of Environmental Factors on Performance of GH-Transgenic Fish

Environmental risk assessment of transgenic fish involves analysis of their fitness and invasiveness in relation to conspecific members and various ecological factors. Devlin et al. (2004) performed an interesting study on the growth hormone transgenic coho salmon (*Oncorhynchus kisutch*), exhibiting robust feeding capacity and increased body size (>7-fold) relative to non-transgenic salmon. When they cohabitated the transgenic and non-transgenic salmons and allowed them to compete for different quantities of food, transgenic salmon always outgrew the wild types. This scenario could disturb the growth of non-transgenic fish except when there was high food availability. In an experiment, when food availability was reduced, all the groups containing transgenic salmon were observed to undergo population crashes, while the groups containing only non-transgenic salmon had good survival (Devlin et al. 2004). When food availability was low, emergence of dominant individuals happened, usually the transgenic fish, which led to strong agonistic and cannibalistic behaviour and controlled possession of the reduced food resources (Devlin et al. 2004). Thus, Devlin et al. (2004) proposed that in experimental populations mixed with growth hormone transgenic salmon the effects were mediated by factors like food availability and population structure.

One of the important concerns of the growth hormone transgenic fish is the potential risks if the transgenic fish manage to escape rearing facilities. In an interesting study, Leggatt et al. (2017) analysed the growth and survival of fast-growing transgenic coho salmon (*Oncorhynchus kisutch*), wherein the growth hormone transgene is driven by *metallothionein* (MT) gene promoter in hatchery conditions and semi-natural stream tanks with different levels of food and predators (Leggatt et al. 2017). In spite of enhanced growth in hatchery conditions, the transgenic coho salmon fry showed little or no growth increase in stream condition. Leggatt et al. (2017) also observed that the transgenic coho salmon exhibited better survival under limited food conditions.

In another interesting study, experiments performed in semi-natural stream mesocosms demonstrated greater risk of predation for growth hormone transgenic salmons compared to their non-transgenic siblings, possibly due to the risky foraging behaviours associated with greater metabolic demands of transgenic fish. Crossin and Devlin (2017) performed a study analysing the effects of genotype on burst-swimming capacity against a sustained water flow in transgenic trout. The transgenic rainbow trout exhibited sustained burst-performance for longer durations than their wild-type siblings, both in predator as well as predator-free conditions (Crossin and Devlin 2017). Crucially, this burst-performance effect happened before variations in growth rate were noticeable. Crossin and Devlin (2017) suggested a possible interaction between predation and metabolic grooming, which led to greater burst capacity in the transgenic fish.

6.9   Conclusion

As the global population is growing at a rapid pace, the aquaculture industry needs to play a critical role in food security. Though, over the last century, classical selective breeding strategies have been successfully used to increase the fish yield, the advent of genetic engineering technology in the last few decades have ushered in a new era of opportunities for genetic improvement of fish stocks. This chapter, in a concise summary, revealed the progress made in the field of growth hormone mediated growth enhancement in teleosts.

Acknowledgement The author thanks SRM Institute of Science and Technology, Kattankulathur for providing support and the necessary facilities.

References

Bergan-Roller HE, Sheridan MA (2018) The growth hormone signaling system: insights into coordinating the anabolic and catabolic actions of growth hormone. Gen Comp Endocrinol 258:119–133

Björnsson BT (1997) The biology of salmon growth hormone: from daylight to dominance. Fish Physiol Biochem 17:9–24

Brem G, Brenig B, Holstgen-Schwark G et al (1988) Gene transfer in tilapia (*Oreochromis niloticus*). Aquaculture 68:209–219

Buono RJ, Linser PJ (1992) Transient expression of RSVCAT in transgenic zebrafish made by electroporation. Mol Mar Biol Biotechnol 1:271–275

Butler AA, LeRoith D (2001) Control of growth by the somatotropic axis: growth hormone and the insulin-like growth factors have related and independent roles. Annu Rev Physiol 63:141–164

Crossin GT, Devlin RH (2017) Predation, metabolic priming and early life-history rearing environment affect the swimming capabilities of growth hormone transgenic rainbow trout. Biol Lett 13:20170279

Devlin RH (2011) Growth hormone overexpression in transgenic fish. In: Encyclopedia of fish physiology: cellular, molecular, genomics and biomedical approaches. Elsevier, Amsterdam, pp 2016–2024

Devlin RH, D'Andrade M, Uh M et al (2004) Population effects of growth hormone transgenic coho salmon depend on food availability and genotype by environment interactions. Proc Natl Acad Sci U S A 101:9303–9308

Di Prinzio CM, Botta PE, Barriga EH et al (2010) Growth hormone receptors in zebrafish (*Daniorerio*): adult and embryonic expression patterns. Gene Expr Patterns 10:214–225

Du SJ, Gong ZY, Fletcher GL et al (1992) Growth enhancement in transgenic Atlantic salmon by the use of an 'all fish' chimeric growth hormone gene construct. Biotechnology (NY) 10:176–181

Eppler E, Caelers A, Shved N et al (2007) Insulin-like growth factor I (IGF-I) in a growth-enhanced transgenic (GH-overexpressing) bony fish, the tilapia(*Oreochromis niloticus*): indication for a higher impact of autocrine/paracrine than of endocrine IGF-I. Transgenic Res 16:479–489

Eppler E, Berishvili G, Mazel P et al (2010) Distinct organ-specific up- and down-regulation of IGF-I and IGF-II mRNA in various organs of a GH-overexpressing transgenic Nile tilapia. Transgenic Res 19:231–240

Figueiredo MA, Lanes CFC, Almeida DV et al (2007) The effect of GH overexpression on GHR and IGF-I gene regulation in different genotypes of GH transgenic zebrafish. Comp Biochem Phys D 2:228–233

Gross ML, Schneuder JF, Moav N et al (1992) Molecular analysis and growth evaluation of northern pike (*Esox Iucius*) microinjected with growth hormone genes. Aquaculture 103:253–273

Kuradomi RY, Figueiredo MA, Lanes CFC et al (2011) GH overexpression causes muscle hypertrophy independent from local IGF-I in a zebrafish transgenic model. Transgenic Res 20:513–521

Leggatt RA, Sundstrom LF, Woodward K et al (2017) Growth-enhanced transgenic Coho Salmon (*Oncorhynchus kisutch*) strains have varied success in simulated streams: implications for risk assessment. PLoS One 12:e0169991

Malkuch H, Waloch C, Kittilson JD et al (2008) Differential expression of preprosomatostatin- and somatostatin receptor-encoding mRNAs in association with the growth hormone-insulin-like growth factor system during embryonic development of rainbow trout (*Oncorhynchus mykiss*). Gen Comp Endocrinol 159:136–142

Nam YK, Noh JK, Cho YS et al (2001) Dramatically accelerated growth and extraordinary gigantism of transgenic mud loach *Misgurnus mizolepis*. Transgenic Res 10:353–362

Palmiter RD, Brinster RL, Hammer RE et al (1982) Dramatic growth of mice that develop from eggs microinjected with metallothionein-growth hormone fusion genes. Nature 300:611–615

Penman DJ, Iyengar A, Beeching AJ et al (1991) Patterns of transgene inheritance in rainbow trout (*Oncorhynchus mykiss*). Mol Reprod Dev 30:201–206

Pierce AL, Fox BK, Davis LK et al (2007) Prolactin receptor, growth hormone receptor, and putative somatolactin receptor in Mozambique tilapia: tissue specific expression and differential regulation by salinity and fasting. Gen Comp Endocrinol 154:31–40

Powers DA, Hereford L, Cole T et al (1992) Electroporation: a method for transferring genes into the gametes of zebrafish (*Brachydanio rerio*), channel catfish (*Ictalurus punctatus*), and common carp (*Cyprinus carpio*). Mol Mar Biol Biotechnol 1:301–308

Rokkones E, Alestrom P, Skjervold H et al (1989) Microinjection and expression of a mouse metallothionein human growth hormone fusion gene in fertilized salmonid eggs. J Comp Physiol 158:751–758

Silva ACG, Almeida DV, Nornberg BF et al (2015) Effects of double transgenesis of somatotrophic axis (GH/GHR) on skeletal muscle growth of zebrafish (*Danio rerio*). Zebrafish 12:408–413

Venugopal T, Pandian TJ, Mathavan S et al (1998) Gene transfer in Indian major carps by electroporation. Curr Sci 74:636–638

Zbikowska HM (2003) Fish can be first – advances in fish transgenesis for commercial applications. Transgenic Res 12:379–389

Zhang P, Hayat M, Joyce C et al (1990) Gene transfer, expression and inheritance of pRSV rainbow trout GH cDNA in common carp, Cyprinus carpio. Mol Reprod Dev 25:13–25

Zhu Z, Li G, He L et al (1985) Novel gene transfer into the fertilized eggs of goldfish (Carassius auratus L. 1758). Z Angew Ichthyol 1:31–34

Zhu Z, Xu K, Li G et al (1986) Biological effects of human growth hormone gene microinjected into the fertilized eggs of loach. Kexue Tongbao, Acad Sinica 31:988–990

Genome Editing in Fish Reproduction

7

Sipra Mohapatra and Tapas Chakraborty

Abstract

Fish eggs are an excellent system for studying how cells organize dynamically into exceedingly complex structures. It is accessible to experimental and genetic manipulations and is easier to scrutinize the development from fertilization to hatching, ex vivo. Genome editing in fish eggs, though challenging, is a powerful technique to develop genetically edited animals with minimal changes to the whole genome and is a huge leap in achieving reproductive stability and fertility enhancement. Usually by applying different techniques like CRISPR/Cas9, which is a faster and cost-effective gene targeting technique, it is convenient to conduct reproductive manipulation study and also produce knockout (KO) and knockin models, which would give us an opportunity to obtain better understanding about fish reproduction or the reproductive behavior of several fish.

Keywords

Reproduction · Genome editing · CRISPR/Cas9 · Fish egg · Future aquaculture

7.1 Introduction

Since ages, agriculture, animal (cattle, pigs, etc.) and fish farming have been traditionally used to fulfill the nutritional requirements of humans. Although agriculture and livestock farming dates back to 12,000 years ago, fish aquaculture has only started about 200 years ago. Aquaculture, defined as the farming of fish and

S. Mohapatra · T. Chakraborty (✉)
Laboratory of Marine Biology, Department of Bioresources, Faculty of Agriculture, Fukuoka, Japan
e-mail: sipra_mo@agr.kyushu-u.ac.jp; tapas_ch@agr.kyushu-u.ac.jp

© The Author(s), under exclusive license to Springer Nature Singapore Pte Ltd. 2021

P. K. Pandey, J. Parhi (eds.), *Advances in Fisheries Biotechnology*,
https://doi.org/10.1007/978-981-16-3215-0_7

103

other aquatic plants and animals, is the fastest growing food industry in the world. Aquaculture sector has increased dramatically since 2000–2012 by 6.2%, and by 2030, it is expected that nearly 60% of the total fish supply will be supplied for direct human consumption. Advanced technologies like Aquaponics and IMTA (Integrated Multitrophic Aquaculture) have also raised the probability of improved sustainable growth of aquaculture.

Aquaculture, despite being more diverse and global than other farming methods, has enormous geographical and ecological restrictions. Different regions of the world have varied preference in their consumption of seafood, wherein, America and Europe share a likeness for clams, oysters and salmon, Japan cultures seaweed, Tuna and yellowtail extensively, and India prefers the carps. Although several fish species have made way into the international consumer market, it is only the Norwegian salmon and GIFT tilapia that has obtained the commercial popularity in the recent years. This has been possible owing to meticulous research, selective breeding and better strain development, steady supply of high-quality seed and one-stop consultancy programs.

With the popularity of DNA since 1950, the journey of molecular biology has seen several breakthroughs such as sequencing, recombinant DNA technology, whole-genome sequence analysis and genome editing techniques. In the last few decades, genome editing in fish and mostly in reproductive biology of the fish, has been intensively investigated. Gonadal development is a complex process, controlled by precise coordination of multiple factors and successful application of gene editing technique in mammals and fish, and has significantly enhanced our understanding about the molecular mechanism of reproduction. In this chapter, we will discuss about the issues of reproduction in aquaculture and the potential of genome editing in fish reproduction.

7.2 Reproduction

Reproduction is the process of attainment of next generation necessary for species survival and ecosystem sustenance. In reality, reproduction is a compromise between balancing the individual growth and species maintenance. Broadly, reproduction can be subdivided into two categories—asexual and sexual. In the former, one parent divides the genetic material and produces identical progenies. While in the latter, genetic materials from two genetically identically parents hereafter regarded as male and female, combines and unique sets of progenies are produced. Majority of the fish follows sexual mode of reproduction, though there are few fish species who often choose between asexual (clonal) and sexual. For instance, dojo loach *Misgurnus anguillicaudatus*, an east African native, harbors 50 chromosomes in somatic cell and 25 in germ cells. DNA barcoding analysis shows that only a small population of this fish, living in Hokkaido areas of Japan, possesses 50 chromosomes in germ cells which were maintained by specifically doubling the chromosomal materials during cell division and hence, maintain clonal reproduction (Kuroda et al. 2018). Successful reproduction depends on several factors, some of them are intrinsic (genetic, physiological, age of first reproduction) and others are environmentally driven (temperature, feeding, reproductive effort). While the fish is sexually immature, assimilated energy is fully allocated for survival and on-growth, but once maturation starts, part of the energy is required for gamete production and reproductive behavior. Being one of the most diverse group of vertebrates, fish exhibit great variability in reproductive traits and strategies; namely:

7.2.1 Gender Identity Systems

Most of the species are gonochoristic, i.e., sex is determined and fixed prior to maturation. Hermaphroditism occurs basically in fish among vertebrates, and although is not common (only in 2% of the teleosts), it is present in many commercial species as groupers, wrasses or porgies. Hermaphroditism in fish can be simultaneous (acting as male and female at the same time, e.g., *Serranus cabrilla*) or more commonly sequential, either protandrous (male as initial gender, as in clownfish), or protogynous (female as initial gender, as parrotfish). In a few species, the same individual can change sex in both directions (bidirectional sex change, e.g., *Centropyge fisheri*).

7.2.2 Breeding Performance Systems

Most of the species are iteroparous, i.e., they breed more than once, expanding from two (some anchovies) to more than 50 years (species of genus *Sebastes*). In some species, however, the individuals breed only once after the maturation and then die; they are semelparous species like Pacific salmons (*Oncorhynchus* spp.), capelin (*Mallotus villosus*), and eels (*Anguilla* spp.). Although reproductive cycles are commonly annual, several species exhibit shorter cycles, i.e., more than one per year or even spawning nearly continuously (e.g., Medaka, Zebrafish), especially in tropical and subtropical waters, where the reduced environmental seasonality allows for the production of several cycles during the year (e.g., *Oreochromis niloticus*). Furthermore, fish produce either pelagic (Blue fin tuna) or demersal (thread sail file fish) eggs of adhesive or nonadhesive types.

7.2.3 Fertilization and Embryonic Development Systems

External fertilization is, by far, the most common means in which oviparous females release haploid eggs that are fertilized outside the reproductive system. Internal fertilization occurs in several fish species (in few teleosts and all elasmobranchs) and it implies that female release embryos. However, embryos can be released at an incredibly early stage (i.e., zygoparity, e.g., *Helicolenus dactylopterus*); at the larval stage as it occurs in *Sebastes* or at the juvenile stage (e.g., viviparous sharks as *Prionace glauca*). Viviparity occurs in more than 500 species (mostly in chondrichthyans), where the embryos develop within the uterus or ovary. Viviparity can be either lecithotrophic (embryos develop without any specialized vascular exchange organ and rely solely on the yolk sac for nutrition) or matrotrophic (where the nutrients are provided by the mother during the gestation in a variety of ways, including through a placental structure or not).

7.2.4 Parental Care Systems

Approximately, 5000 fish species demonstrate various forms of parental care including matrotrophic viviparity. Guarding of eggs by hiding (e.g., salmonids), nest construction (e.g., *Labrus bergylta*), mouthbrooding (both paternal and maternal, as in many cichlids, e.g., Tilapia) or carrying embryos in specialized body structures (e.g., seahorses) are most common forms of parental cares in fish. Several species show biparental care, but the most common is uniparental care, and more often provided by the male.

7.3 General Development of Reproductive Organ in Fish

Sexually reproducing organisms are bestowed with increased amount of genetic diversity, which is achieved through the mixing of genomes by the union of gametes, sperm and ovum from different individuals. The biological process by which the sexual fate of an individual is decided is known as sex determination (SD). A diverse array of SD mechanisms, either environmental SD (ESD) or genetic SD (GSD), has been found to exist among vertebrates (ranging from fish to mammals). In GSD vertebrates, SD occurs at the time of fertilization when the two gametes fuse. pH, salinity, and other social factors are found to have a role in fish SD (Devlin and Nagahama 2002; Baroiller et al. 2009; Nagahama et al. 2020). Most recently, green light irradiation during sex differentiation was reported to induce female-to-male sex reversal in some XX medaka (15.9%) (Hayasaka et al. 2019). Social factors are found to be more prevalent among hermaphroditic fish. Vertebrate gonads arise in both sexes from a bipotential gonad, an organ that has the potential to develop either as a testis or an ovary. Sex differentiation is the process of gonad development after the sex has been determined and is one of the many manifestations of sexual differentiation. All these mechanisms primarily aim to orchestrate the sexual development of the bipotential gonad in embryos in a way to create a suitable microenvironment for further gonadal development. Gamete development involves the processes of generation, migration, proliferation, differentiation and ovulation/spermiation, which are generally similar among fish. Most of these processes occur before meiosis starts, but once the oocyte and spermatocyte are formed the meiosis is arrested in prophase I (also known as primary growth period). Later during gonadotropin dependent secondary growth period, germ cell starts developing further and become mature. In female, they pass through cortical alveoli (CA), vitellogenic (VIT), and oocyte maturation (OM) stages. In matrotrophic viviparous species, vitellogenesis does not take place. During OM, depending on the availability of gonadal steroids and hormones, oocytes undergo germinal vesicle migration and breakdown and ovulation. OM may also include the formation of large oil droplets or lipid coalescence, yolk coalescence and hydration to increase the gamete volume and bouncy (e.g., Chub mackerel, Japanese anchovy). At ovulation, the follicle (the nurturing layers enclosing the oocyte) ruptures and collapses for the release of mature oocytes into the ovarian lumen producing a haploid egg to be fertilized. Once the oocyte reaches the final maturation stage, the energetic investment becomes irreversible. However, before that point oocytes can be reabsorbed via follicular and germ cell atresia. The incidence of atresia is often related with poor fish condition associated with nutritional deprivation or other stress factors. Though sexual maturation process is remarkably similar among fish, but the timings of puberty, onset of maturation, triggering factors are genetically and environmentally controlled. With the intervention of genome editing, several unknowns of fish reproduction have been deciphered and thereby has become pivotal for aquaculture improvement.

7.4 Genome Editing

Reverse genetics approaches are critical for exploring gene function, genome editing and genetic engineering (Zhu and Ge 2018). Traditionally, generating a gene editing model is largely dependent on the availability of embryonic stem cells (ESCs) that has been developed in only a few model organisms, such as mouse and rat. Homologous recombination (HR) based gene editing in mice opened the door to genome editing. Due to the lack of methods for HR and especially the development of ESCs, there has been no effective technology for generating gene editing in non-rodent organisms, especially culturable fish. Now, this situation has thoroughly changed with the recent development of engineered nucleases, which have been used to develop genome editing models precisely in a wide variety of organisms from invertebrates to vertebrates. In contrast to transgenesis, which involves the transfer of a gene from one organism to another, genome editing allows specific, targeted and often minor changes to the genome of the species of interest without foreign gene sequence and thus becoming a legal choice for food grade "Gene modified organism."

7.4.1 Practical Advantages of Genome Editing in Aquaculture

Some practical reasons why genome editing has such potential for research and applications in aquaculture species are:

1. The ease of access to many thousands of externally fertilized embryos and relatively established gene transfer methodologies. This helps to obtain large nuclear families, enables a degree of control of background genetic effects, and

increases individual variation for downstream application. This not only allows retaining genetic variability of a strain but also helps to create a significantly different strain from their unedited full-sibling counterparts.

2. The ability to perform extensive "phenotyping" is often also feasible, for example using either invasive capillary sampling or noninvasive ultrasound procedure; we can easily judge the reproductive status of an individual. Finally, should favorable alleles for a target trait (e.g., disease resistance) be created or discovered, and then there is potential for widespread dissemination of the improved germplasm for rapid impact via selective breeding programs.

3. Parallely, high-quality, well-annotated reference genomes are available for most of the key species. A high-quality species-specific reference genome is essential for the effective design of target guide with high specificity and minimum change of off-target editing, in particular given the relatively recent whole-genome duplication events that are features of several finfish lineages, including commercially important fish like, Salmon, Trout, Herring, Barramundi, European sea bass, tilapia, channel catfish, common carp, eel, Fugu, sea bream, sole, Turbot, and Yellowtail.

7.4.2   Types of Genome Editing

Several genome editing techniques have been developed over the past few decades and used to study reproduction in fish.

7.4.2.1 Chemical Mutagenesis

Targeting Induced Local Lesions IN Genomes (TILLING) TILLING is a reverse genetics strategy that identifies induced mutations in specific genes of interest in chemically mutagenized populations with high-throughput discovery technologies. TILLING, first described in 2000 for mutation detection in Arabidopsis, is now used in a wide range of plants and animals including fish like medaka, zebrafish, and Fugu (Kuroyanagi et al. 2013). The first step in TILLING is chemical mutagenesis. In plants, mutations have been induced by using ethylmethane sulfonate (EMS) and methylnitrosourea (MNU), while in animals, EMS is used mainly for invertebrate species and *N*-ethyl-*N*-nitrosourea (ENU) is used for vertebrate males. ENU acts as an alkylating agent and transfers its ethyl group to nucleophilic nitrogen or oxygen sites on deoxyribonucleotides, leading to base mismatch during DNA replication. Single-base substitutions that resemble natural spontaneous mutations are mainly induced by ENU treatment and all genes are mutated at random. This protocol induces mutations in germ line stem cells, somutated gametes can be produced least several months after the treatment. These mutations are induced with premeiotic ENU treatment, so that the initial mutation, usually an ethylation of a base on one DNA strand, is fixed by DNA replication prior to production of differentiated sperm, leading to non-mosaic offspring in the next generation (Fig. 7.1).

Fig. 7.1 General schematics for TILLING mutant fish breeding and maintenance for functional analysis. ENU-treated males are crossed with wild females to produce F1 hetero zygotes carrying various mutations. Each male sperm and its respective genomic DNA of F1 progeny is cryopreserved or made into DNA library for further analysis. Whenever required, after screening the DNA, required mutant fish is selected and the respective cryopreserved sperm is used for fertilizing wild female eggs through *in vitro fertilization* (IVF) to produce the F2 generation of mutants

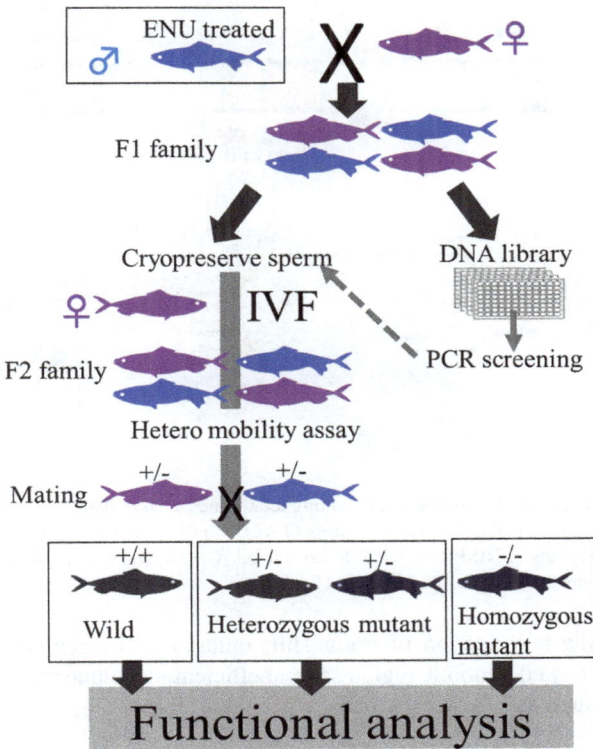

In zebra fish and medaka, the mutations are introduced in spermatogonia by soaking founder fish in ENU solution but many aquaculturable species are of large body size which creates limitation for TILLING-based mutagenesis. After the ENU treatment, a large F1 population is generated to choose the mutations for the purpose from many random heterozygous mutations in their genomes. Sanger sequencing, CEL I nuclease-mediated screening of heteroduplexes and high-resolution melting (HRM) curve analysis are widely used to detect mutations in TILLING.

7.4.2.2 Enzymatic Mutagenesis

Targeted genome engineering or editing helps the researchers to modify genomic loci of the interest in a precise manner, having enormous applications in food industry. In most organisms, efficient genome editing relies on a site-specific DNA double strand break (DSB). The subsequent DNA repair by homology directed repair (HDR) or nonhomologous end joining (NHEJ) generates desired genetic modifications. HDR aids to mediate nucleotide exchange between an endogenous genomic region and an exogenous DNA fragment through flanking homologous sequences, resulting in DNA insertion, deletion, or replacement. The DNA repaired by NHEJ usually contains small deletions and insertions at the break sites. Therefore, the imprecise DSB repair through NHEJ can be applied for gene disruption by

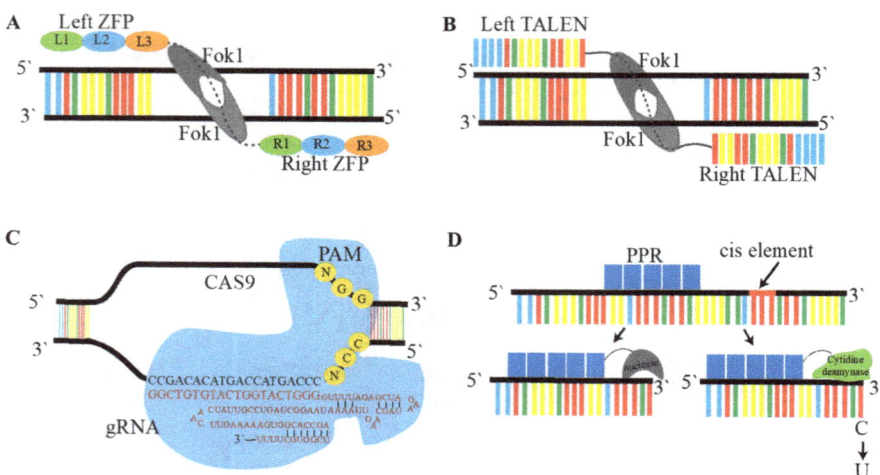

Fig. 7.2 Various genome editing techniques used in fish (**a**) Zinc finger nuclease (**b**) Transcription activator-like effector nucleases (TALENs) (**c**) Clustered, regularly interspaced, short palindromic repeats (CRISPR)–CRISP R-associated (Cas) systems (CRISPR/Cas9), and (**d**) Pentatricopeptide repeat (PPR)

the introduction of frame-shift mutations. To generate a site-specific DSB at the desired genomic region several efficient techniques have been discovered (Fig. 7.2), such as:

Zinc Finger Nuclease (ZFN)

These are artificial proteins, constructed by fusing several ZF domains to a sequence independent cleavage domain of the type IIS restriction endonuclease FokI. Each ZF domain comprises 30 amino acids (aa) and coordinates one zinc atom using two His and two Cys residues. An α-helix in each ZF domain recognizes a specific DNA triplet, so linkage of several ZF domains enables ZFNs to recognize long DNA sequences. Because the FokI nuclease domain functions as a dimer, functional ZFNs are formed when two sets of ZF domains are positioned in proximity and in appropriate orientations. The DNA-binding and cleavage domains are functionally independent and can be optimized in isolation for efficient DNA cleavage against custom-designed target sites. Although ZFNs are used for targeted genome editing in various organisms, two major limitations prevent their wider applications, such as ZF domains have limited modularity due to the context-dependent DNA-binding effects, making it difficult for ZFNs to target any desired DNA sequence (Ramirez et al. 2008), and the lack of specificity of some ZF domains can generate off-target cleavage, leading to undesired mutations and chromosomal aberrations (Pattanayak et al. 2011; Radecke et al. 2010).

Transcription Activator-Like Effector Nucleases (TALENs)

TALENs have recently acquired the genome editing monopoly from ZFNs. TALENs exhibit significantly reduced off-target effects and cytotoxicities compared with ZFNs, making them an efficient genome editing tool (Sun and Zhao 2013). TALENs are fusions of transcription activator-like (TAL) proteins and a FokI nuclease and almost follow the same principle as ZFNs. However, the DNA-binding domains of TALENs are composed of a series of tandem repeats as in TALEs of the plant pathogenic bacteria from the genus *Xanthomonas* (Wei et al. 2013). TAL proteins are composed of 33–34 amino acid repeating motifs with two variable positions that have a strong recognition for specific nucleotides.

Clustered, Regularly Interspaced, Short Palindromic Repeats (CRISPR)–CRISP R-Associated (Cas) Systems (CRISPR/Cas9)

Several initial progresses using other technologies have been largely superseded by the advent of the repurposed CRISPR/Cas9 system which allows easy, cheap and efficient targeted editing of the genome. The system enables imperfect or targeted repair to create alterations to the sequence of the genomic DNA. It is based on the *Streptococcus pyogenes* adaptive immune system which requires three components, nuclease Cas9, a targeting crRNA and a transactivating crRNA (tracrRNA). Further, improvement of this system was performed by engineering dual crRNA: tracrRNA to form a single guide RNA (gRNA), which contains a designed hairpin that mimics the crRNA–tracrRNA complex, as it is sufficient to direct Cas9 protein to cleave DNA sequence. Different from ZFN and TALEN using DNA-binding proteins to recognize DNA sequence, CRISPR/Cas9 utilizes 20 bp long gRNA to bind its target via Watson–Crick base pairing. Additionally, the protospacer adjacent motif (PAM, NGG) immediately downstream of the target sequence together with gRNA determine the specificity of this system. Importantly, CRISPR/Cas9 has the capacity to edit multiple genes simultaneously, producing conditional alleles, and generating tagged protein with high efficiency. Compared with ZFN and TALEN, CRISPR/Cas9 is found to be much easier, less expensive, much simpler and more efficient technology to precisely modify the genome of any organisms during development.

Pentatricopeptide Repeat (PPR)

Most of the genes editing nucleases are based on the microbial physiology. Recently, a group of researchers at Kyushu University (Japan) have found PPR proteins in plants which have the capability of RNA editing and other RNA processing events in plant mitochondria and chloroplasts (Yagi et al. 2013). Briefly, PPR protein is a sequence-specific RNA binding protein that identifies specific "C" residues for editing. Discovery of the RNA recognition code for PPR motifs includes the verification and prediction of the individual RNA editing site and its corresponding PPR protein. Although the technology is yet to be used for aquaculture, however given the fact that it can effectively edit the genomes of various organisms, it has a huge potential for future aquaculture.

7.4.3   Methods of Genome Editing

In most bony fishes, the eggs undergo meroblastic cleavage and consequently, a thin region of the cytoplasm at the animal pole forms the embryo. After fertilization, the widely distributed cytoplasm moves toward the animal pole to form the blastodisc. Fish eggs vary widely among species in size, shape, and physical characteristics and can be categorized into two types using features of the yolk. Some fish have a single-yolk mass in their eggs, others have multiple yolk granules. In the former group, the cytoplasm moves toward the animal pole on the surface of the unitary yolk to form the blastodisc. In the latter group, the cytoplasm moves between the yolk granules to the animal pole. In eggs with a single-yolk mass, an injectant needs to be directly introduced into the blastodisc since molecules cannot pass through the yolk-cytoplasm boundary. By contrast, in eggs with multiple yolk granules, such as the Japanese anchovy, an injectant can be introduced either directly into the blastodisc or into the yolk region. In both the cases, the injectant is held in the cytoplasm and is able to move to the blastoderm with the cytoplasm during embryonic development. Therefore, the nature of the egg determines the position of injection and is of primary importance for the success of gene editing (Goto et al. 2019). Moreover, the nature of eggs (pelagic or demersal; adhesive or demersal), the peri-vitelline space (Fig. 7.3), and hardness of chorion also affects the genome editing success.

7.4.3.1 Microinjection

It is the most laborious and skill intensive but most preferred method of choice for gene editing. The requirements for microinjection are: (a) bright field stereomicroscope, (b) micro injector, (c) three-dimensional coarse manual manipulator, (d) glass micropipette puller grinder, and (e) Precision glass capillary tube (Fig. 7.4). The microinjection method was initially developed in the model fish species (zebra fish and medaka) and was modified further to meet the specific demands of various species. For instance, due to less peri-vitelline space and higher interior pressure in pelagic marine eggs, some researchers have made a constricted microinjection needle for efficient injection and better results.

7.4.3.2 Electroporation

Compared to microinjection, it is relatively easy and can be used to edit several thousands of eggs in a shorter duration of time. It is a preferred method for gene transfer in cell lines. Electroporation was experimentally used in a model fish to transfer DNA construct into the fertilized eggs. The requirements are: (a) electroporator, (b) electroporating buffer, and (c) electroporation cuvette. As electroporation involves electric pulses, adjusting the right pulse is the major hurdle. In this regard, it has been found that fish eggs require multiple square wave pulses of lower voltage instead of single high voltage pulse. Moreover, though it is possible to transfect large number of eggs at a time, but the amount of gene dose cannot be controlled in electroporation and thus remains as a non-popular method.

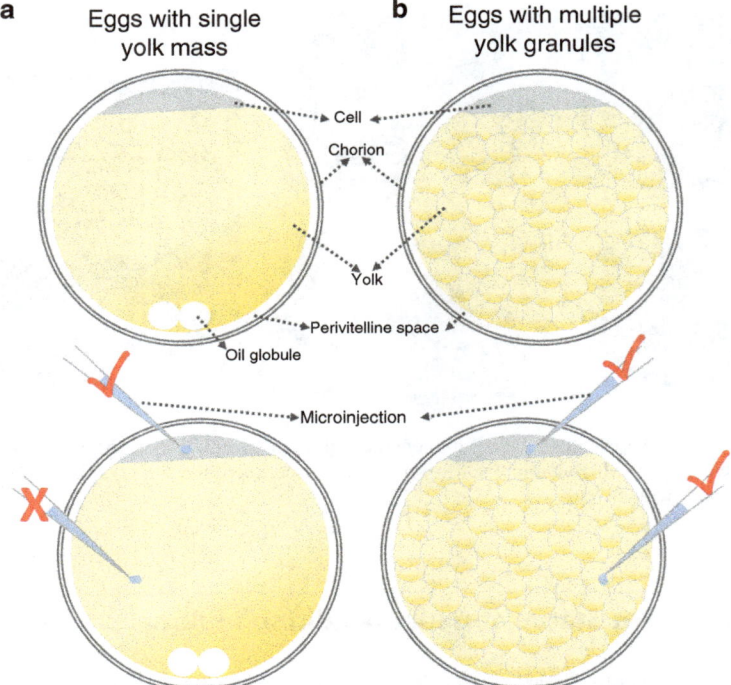

Fig. 7.3 Methods of microinjection used in various types of fish eggs (**a**) Microinjection into the blastodisc of an egg with a single-yolk mass. The injectant must be injected directly into the blastodisc to achieve best results. (**b**) The injectant must be introduced into the blastula or the space between the yolk granules to achieve good result

7.4.3.3 In Vivo Transfection/Transduction

Lipid conjugated gene delivery was extensively used in in vitro studies. Recently, several chemicals (e.g., in vivo-jetRNA®) are identified to transfect an adult organ without much difficultly. Additionally, nanoparticle-based gene delivery and chitosan conjugated gene delivery technologies are gaining momentum in experimental gene editing. However, these techniques are applied in large animals. In future, with the help of improved recombinant RNA-protein conjugate we might be able to extensively use these techniques to overcome the difficulties of microinjection and electroporation and thereby, form a relatively common procedure for commercial grade genome editing methodology.

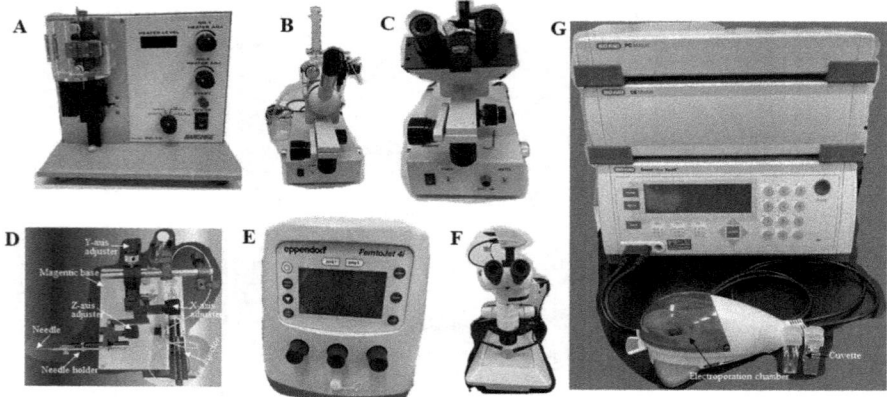

Fig. 7.4 Minimal apparatus required for genome editing (**a**) Needle puller, (**b**) Needle grinder, (**c**) Needle selector, (**d**) Micromanipulator, (**e**) Air compressed microinjector, (**f**) Microscope, and (**g**) Electroporator with cuvette

7.5   Current Status of Genome Editing in Aquaculture Species

Genome editing relies heavily on fully sequenced and annotated genomes. The lack of complete genomes is a current bottleneck for utilizing the technology efficiently in some fish species. However, the recent developments of high-throughput sequencing have, in recent years, led to the full characterization of many fish genomes for several key species in aquaculture, such as Atlantic salmon (Lien et al. 2016). This has allowed the establishment of gene editing in Tilapia, Atlantic salmon, carp, and catfish species. Genome editing, using CRISPR/Cas9, was recently successfully applied in vivo and/or in cell lines of several major aquaculture species of Cyprinidae (Indian major carp, grass carp and common carp), Salmonidae (Atlantic salmon and rainbow trout), Scombridae (Chub mackerel, eastern little tuna), Siluridae (channel and southern catfish), as well as gilthead sea bream (*Sparus aurata*), Nile tilapia (*Oreochromis niloticus*), and Pacific oyster (*Crassostrea gigas*). Most studies have a proof-of-principle focus, typically followed CRISPR/Cas9 protocols developed in a model organisms, such as zebra fish (*Danio rerio*), and have often targeted genes with a clearly observable phenotype to test editing success (e.g., pigmentation). The standard methodology to induce in vivo mutations in aquaculture species is injection of the CRISPR/Cas9 complex into newly fertilized eggs as close as possible to the one-cell stage of development. Typically, mRNA encoding the Cas9 protein is injected together with the guide (g) RNA, leading to the high efficiency of editing demonstrated in various species to date; using the Cas9 protein in place of mRNA is also effective. While most studies have used nonhomologous end joining (NHEJ) to induce mutations, homology directed repair (HDR) has been successfully used to insert a template DNA in rohu carp. Furthermore, successful germline transmission of edits has been reported in several of the studies to date. Mosaicism is common in edited animals, implying that the Cas9-induced cutting and editing continues past the one-cell stage; this is an issue that needs to be tackled with future research. To overcome that, genome editing has been tried directly in germ cells of zebra fish and medaka. This improvement though does not produce any expected phenotype in the F0 generation but has the potential to produce heterozygous knockout individual at F1 generation and particularly might be useful for lethal genes. Target production traits for genome-editing studies in aquaculture species to date have included sterility, growth, and disease resistance (Gratacap et al. 2019). Creating sterile animals for aquaculture is desirable to prevent introgression with wild stock and to avoid the negative production consequences of early maturation. In this context, CRISPR/Cas9 has also been used to induce sterility in Atlantic salmon and catfish. For growth-associated traits, several groups have edited the myostatin gene, resulting in larger fish. To date, this has been performed in channel catfish and common carp. Immunity and disease resistance have already been investigated using genome editing in rohu and grass carp, respectively, and it is expected that this area of research will flourish as a route to improving and understanding disease resistance as a key target trait for aquaculture. In addition to improving reproductive traits, genome editing can also be applied to develop models for studying fundamental immunology, such as the targeted disruption of the TLR22 gene in carp. Such models can improve our fundamental understanding of host response to infection in fish and may lead to more effective treatment protocols. Along the similar lines, it is plausible to use genome editing technology to generate improved cell lines for fish species, for example by enabling more efficient production of viruses for future vaccine development by knocking out key components of the interferon pathway.

7.6   Genome Editing and Reproductive Manipulation

A hallmark for use of gene-edited animals in the aquaculture field is sterility, as this phenotype prevents genetic introgression of farmed animals into wild populations upon escape. There are two significant additional benefits of using sterile fish. First, early maturation is prevented, which avoids the associated negative phenotypes, such as reduced growth, lower flesh quality and higher susceptibility to disease. Second, sterility in producted fish may safeguard Intellectual Property for the breeding companies. Currently, triploidisation is a way to induce sterility in fish species. However, this cause welfare problems in fish and has a major impact on the genome of the fish, not comparable to the side effect of gene editing such as knockout or medication of a single gene. As such, germ cell-free animals by gene editing (knockout) have been developed in Atlantic salmon (Wargelius et al. 2016; Kleppe et al. 2017). The gene encoding dead end (DND) has been targeted to induce sterility in salmon, preventing the formation of germ cells. This was done using targeted mutagenesis against DND with CRISPR/Cas9, thereby creating a gene-edited sterile fish. Germ cell-free salmon are 100% sterile and do not enter maturity.

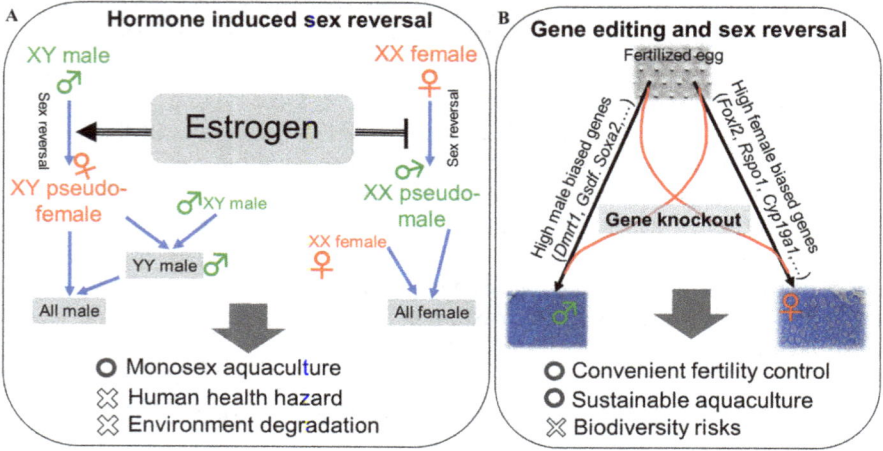

Fig. 7.5 Advantageous of genome editing in sex reversal of fish (**a**) Hormone induced sex reversal is carried out by using several estrogen or anti-estrogen chemicals to change the hormone composition of the fish and induce sex change. (**b**) Gene editing, especially by knockout or knockin techniques, is applied to target specific female oriented or male specific genes to obtain sex reversal

In addition to DND, knockout of primordial germ cells (PGCs) related *nanos2* and *nanos3* in tilapia using CRISPR/Cas9 produce infertile individuals in F0 generation (Li and Wang 2017). In Channel catfish, a reproductive hormone has been abolished and then added again to make the fish mature (Qin et al. 2016). It is anticipated that gene editing technology, in future, will contribute toward stable production of sterile fish. Practical application of such sterile fish in breeding programs will require developments in genome editing, including knock-in, which could lead to the production of an inducible on-off system for sterility. Such mechanisms have been developed for the model fish species medaka and zebra fish.

Sex reversal is another aspect of reproduction which is addressed by genome editing (Fig. 7.5). Several genes, i.e., *foxl2*, *dmrt1*, *cyp19a1a*, *dmrt6*, *amhy*, *amhrII*, *gsdf*, *cyp26a1*, *aldh1a2*, *sf-1,* and *igf3*, have been mutated with extreme efficiencies as high as 95% in the F0 generation and found to be effective for producing high percentage of female-to-male or male-to-female sex reversal fish without external hormonal manipulation. However, technologies for hormone free production of 100% monosex population are still a far reality. For instance, only partial sex reversal (ovotestis) was observed in *Cyp19a1a* deficiency F0 XX fish. However, complete female-to-male functional sex reversal occurs in F2 individuals.

Brain pituitary gonadal (BPG) axis and their components i.e., sex steroids (androgen, estrogen, and progestogen) and gonadotropins (Follicle stimulating hormone (FSH) and luteinizing hormone (LH)) are the core of fertility. Sex steroids play important roles at all stages of the reproductive cycle in fish (Fig. 7.6). Using gene knockout based functional analysis it was found that both estrogen and estrogen receptors (ER) are pivotal for gonadal sexuality and fertility in medaka (Nakamoto

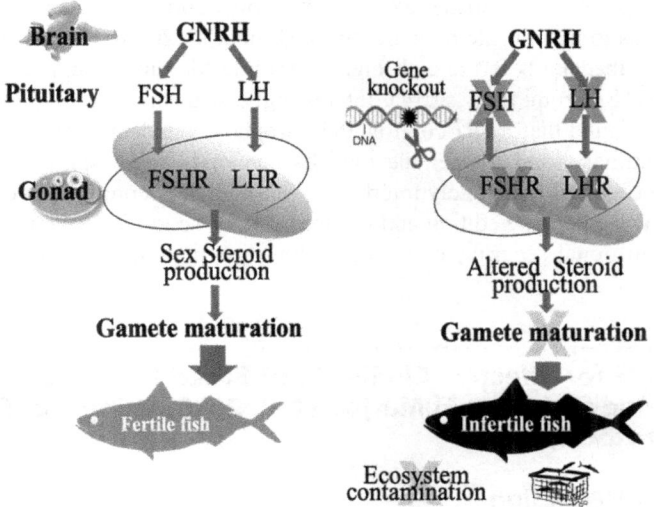

Fig. 7.6 Proposed application of genome editing in biosecurity of future aquaculture. Gene knockout of gonadotropin signaling pathway genes expectedly disrupts fertility by restricting the gonadal maturation and thereby producing infertile fish. The genome edited infertile fish when cultured in open environment are unable to breed with wild population and hence reduces the risk of ecosystem contamination and increases biosecurity of aquaculture

et al. 2018; Chakraborty et al. 2019). It was determined that, alternation of germ cell physiology could induce sex reversal and thereby, affect fertility. Further, remodulation of somatic cell physiology also affects fertility status. In fish, knockout of ovarian aromatase (enzyme that converts testosterone to estrogen) causes granulosa to Sertoli cell transdifferentiating and further results in germ cell sex change and functional female-to-male sex reversal (Nakamoto et al. 2018). Furthermore, ablation of most germ cells in adult zebrafish causes female fish to develop as phenotypic males (Dranow et al. 2013). These results suggest that germ-soma interactions are critical for proper sexual development and fertility management.

Notably, gonadotropins (FSH and LH) bind to their receptors, viz FSH receptor (FSHR) and LH receptor (LHR), located in the gonads to orchestrate the sex steroid production, gonadal growth, and development (Fig. 7.6). In fish, it has been suggested that FSH signaling is mainly responsible for the early stage of gonad development and LH signaling mainly regulates the processes at the later stage of gonad development. *lhb* and *lhr* mutant zebra fish lines demonstrated that LH signaling is crucial for oocyte maturation and ovulation but is dispensable for testis development. Addition to that, *fshb* and *fshr* single mutant lines as well as *lhb; fshb*, *lhr; fshr*, and *fshb; lhr* double mutant lines show that FSH signaling is mainly responsible for promoting follicular growth, whereas LH signaling is mainly responsible for stimulating oocyte maturation and ovulation (Chu et al. 2014). Importantly, *fshb* mutant zebra fish females are sub-fertile (produce significantly low number of offspring), whereas *fshr* mutants are infertile (produce no offspring at all, when mated with wild fish of opposite sex). However, only double knockout of *fshb*, *lhb* or *fshr*; *lhr* leads to all infertile male offspring (Chu et al. 2015). Similarly, the FSHR mutant male medaka is fertile and females are infertile due to suppressed estrogen production (Murozumi et al. 2014). Interestingly, with the help of $ERB1^{-/-}$ Kayo et al. (2019) found that *ERB1* controls the estrogen dependent gonadotropin release from the pituitary, and the female medaka cannot undergo spontaneous gamete release and becomes completely infertility. Although fish reproduction can be easily manipulated via genome editing and has tremendous scope for sustainable aquaculture and nutritional security, it still has a long journey ahead of it to become an actuality.

7.7   TIPS for Effective CRISPR/Cas9 Based Knockout in Fish (Modified from El Marjou et al. 2020, Using Our Own Protocol)

7.7.1   sgRNA Design

Obtain the target DNA sequence using a genome browser (e.g., Ensembl website). The next step is to choose a specific gRNA against this specific genomic location like the CRISPOR web tool (http://crispor.tefor.net/). Choose the sgRNA with the highest specificity score (>80%). sgRNA can be order-made from companies or prepared in lab, using PCR based cloning and in vitro transcription. CRISPOR website also provides the sequences required for cloning and specific vector ligation. To generate KO models, it is recommended to select at leasttwo different sgRNAs to delete the critical exon common to all transcript variants. Alternatively, KO models can also be created by the deletion of a functional domain of the protein using two sgRNAs.

7.7.2   Preparation of CRISPR/Cas9 Solution

sgRNAs can either be produced by in vitro transcription or can be commercially obtained to prepare the sgRNA/Cas9 solution. Fish researchers prefer sgRNA over CRISPR RNA (crRNA) and transactivating RNA (tracrRNA) complex, so if anybody is using the later, crRNA and tracrRNA can be resuspended in nuclease-free water at 1μg/μL. To obtain a chimeric sgRNA, 5μg of crRNA and 10μg of tracrRNA (5μL of 1μg/μL crRNA and 10μL of 1μg/μL tracrRNA) are mixed. This solution is then put into a thermocycler with the following conditions: 95 °C for 5 min and then immediately transferred to ice water. This solution can be aliquoted and stored at −80 °C. From this step, CRISPR/Cas9 delivery can be performed by either microinjection or electroporation. Electroporation is bit tricky with fish egg and each species requiring different protocol and buffer for successful electroporation, hence here after we will discuss about microinjection. For microinjection, a final volume of 6μL is prepared in nuclease-free water by adding phenol red, sgRNA and Cas9 protein at the following final concentration: 200 ng/μL sgRNA and 600 ng/μL Cas9 protein. Cas9 protein and sgRNA are mixed by pipetting. The solution is incubated for 15 min at room temperature (RT) to allow for the formation for ribonucleoprotein (RNP) complex. The solution is centrifuged at 14,000 RPM for 5 min and the supernatant is kept on ice for injection.

7.7.3   Collection of One-Cell Stage Embryo

The fertile males and females are allowed to mate, and the fertilized eggs are collected in a petri dish. Generally, within 30 min to 1 h of fertilization, the first cell division occurs. So, it is advisable to keep the eggs in at least 4–5 °C lower temperature for relatively slower development.

7.7.4   Cytoplasmic Microinjection

To prepare the injection dish for embryos, the eggs are arranged on an agarose (1.5%) plate, containing grooves (diameter almost same as egg diameter, but depth slightly smaller than egg diameter) in such a way that the oil globule stays at the bottom and the cell on the top. The plate is then put under the microscope. The microinjection capillaries are made using a two-step glass capillary puller and a needle grinder. 1–2μL of Cas9/sgRNA solution, kept at 4 °C during the injection session, is loaded into the microinjection capillary with a thin pipette tip. The capillary is then mounted onto the micromanipulator under an angle of approximately 25–30 °C which will lead to a successful zygote cytoplasm penetration. The tip of the microinjection capillary needs to be broken to allow for Cas9/sgRNA mixture release. To ensure that the pipette was broken correctly, the microinjection pipette is positioned into solution of the agarose plate; pressing the "clean function" button on Femtojet apparatus will induce a high pressure in the pipette leading to small drops exiting from the hole, validating that it is open. After this step, one can proceed to injections. A single embryo is kept immobilized in the groove under the microscope so that one of its nuclei is in focus. Using the micromanipulator, the microinjection pipette is carefully moved to enter the egg and reach the cytoplasm. Then, the Cas9/sgRNA solution is injected using the specific Femtojet parameters. Injected embryos are then transferred to antibiotic-containing sterilized water or Ringer solution and incubated till hatching with daily solution exchange.

7.7.5   Genotyping

Genomic DNA preparation—(a) caudal fin samples are collected in 100% methanol from 1 to 2 month old juveniles. (b) incubated at 98 °C for 5–6 min to evaporate the methanol (c) 50μL of Tris-EDTA (pH 8.0) Proteinase K (20 mg/mL) lysis buffer is added (d) incubated at 56 °C for 3 h to overnight (e) Proteinase K is inactivated by heating samples at 85 °C for 30 min.

PCR analysis (a) Primers are chosen within the flanking regions of the targeted DNA region and inside the deleted region. Several couples of primers should be designed to fully characterize the targeted region. Keep in mind that the deleted fragment can be reinserted in inverted sense or as a tandem repeat head to head or head to tail. Primers should be designed to verify the absence of this type of DNA rearrangements. (b) Genotyping PCR using a high-fidelity PCR kit is performed. Typically, 1µL representing 50–100 ng genomic DNA and 0.5µM of each primer is added in 20µL final volume of the PCR mix. Touchdown PCR program is recommended because it provides high specificity and efficiency. The PCR product is separated using 2.5–3% agarose gel prepared in Tris Burette EDTA (TBE) buffer.

Sequencing Analysis needs to be run on genotyping PCR product to further analyze the putative founders. (a) PCR products are purified using a PCR purification kit. (b) Sent for sequencing with appropriate primers to a specialized facility or a sequencing company. The exact sgRNA cut location can be checked using Sanger sequencing approaches.

7.7.6 Germ Line Transmission

The F0 founder (confirmed by PCR and sequencing), raised in a P1A fish facility, is crossed with an inbred control fish to achieve germ line transmission. Genotype and sequence the F1 progenies to confirm whether the genomic modification is inherited or not.

7.8 Future Direction

The United Nations Food and Agriculture Organization (FAO) have reported that aquaculture continues to grow faster than other major food production sectors. The world production of aquaculture has been increasing rapidly and expected to contribute "zero hunger" (SDG2) and "life below water" (SDG14) aims of Sustainable Development Goals (SDGs) adopted by all member of United Nations (UN) in 2015. Sciences for creating improved conditions for sustainable development of the Ocean has also been encouraged by a Decade of Ocean Science for Sustainable Development (2021–2030) proclaimed by UN in 2017. If the public and government regulatory bodies permit, genome editing technologies are likely to be used in commercial aquaculture breeding in the coming years. However, for widespread adoption with maximal benefit and minimal risk, it is necessary that these technologies are seamlessly integrated with well-managed Quantitative trait loci (QTL) based breeding programs. Achieving this will not only ensure careful management of genetic diversity and avoidance of potential inbreeding depression but also help to control the biodiversity deterioration through aquaculture escapees and reduce intellectual property right issues.

Effective integration of multiple edits simultaneously into brood stock animals to target multiple traits, or multiple causative alleles for the same trait are a must for the future. Thorough testing of edited animals is required to assess and exclude possibilities of unintended and potential detrimental pleiotropic effects of edits before any application in production. Additionally, technologies for inducible and reversible editing targets, mass delivery of CRISPR/Cas9 to edit production or multiplier animals and editing entire brood stock populations to carry the desirable alleles in the germplasm (living genetic resources like seeds or tissues that are maintained for the purpose of breeding, preservation, and other research uses) will be more practical choice. Once these issues have been addressed, widespread and rapid positive impacts could be achieved, because the high fecundity of most aquaculture species may enable dissemination to production systems without the need for pyramid breeding schemes typical to terrestrial livestock species.

Acknowledgments This work was in part supported by Grants from Ministry of Education, Culture, Sports, Science and Technology (MEXT), Japan; Japanese Society for the Promotion of Science (JSPS) Kakenhi, Grant Nos. JP16H04981, JP18K14520, JP19H03049, and Sumitomo Grant no. 180959.

References

Baroiller JF, D'Cotta H, Bezault E, Wessels S, Hoerstgen-Schwark G (2009) Tilapia sex determination: where temperature and genetics meet. Comp Biochem Physiol A Mol Integr Physiol 153:30–38

Chakraborty T, Mohapatra S, Zhou LY, Ohta K, Matsubara T, Iguchi T, Nagahama Y (2019) Estrogen receptor β2 oversees germ cell maintenance and gonadal sex differentiation in medaka, *Oryzias latipes*. Stem Cell Rep 13(2):419–433

Chu L, Li J, Liu Y, Hu W, Cheng CHK (2014) Targeted gene disruption in zebrafish reveals non-canonical functions of LH signaling in reproduction. Mol Endocrinol 28:1785–1795

Chu L, Li Y, Liu Y, Cheng CHK (2015) Gonadotropin signaling in zebrafish ovary and testis development: insights from gene knockout study. Mol Endocrinol 29(12):1743–1758

Devlin RH, Nagahama Y (2002) Sex determination and sex differentiation in fish: an overview of genetic, physiological, and environmental influences. Aquaculture 208:191–364

Dranow DB, Tucker RP, Draper BW (2013) Germ cells are required to maintain a stable sexual phenotype in adult zebrafish. Dev Biol 376(1):43–50

El Marjou F, Jouhanneau C, Krndija D (2020) Targeted transgenic mice using CRISPR/Cas9 technology. In: Epigenetic reprogramming during mouse embryogenesis: methods and protocols. Springer protocols. Humana Press, New York, NY, pp 125–141

Goto R, Saito T, Matsubara T, Yamaha E (2019) Microinjection of marine fish eggs. In: Microinjection: methods and protocols. Springer protocols. Humana Press, New York, NY, pp 475–487

Gratacap RL, Wargelius A, Edvardsen RB, Houston RD (2019) Potential of genome editing to improve aquaculture breeding and production. Trends Genet 35(9):672–684

Hayasaka O, Takeuchi Y, Shiozaki K, Anraku K, Kotani T (2019) Green light irradiation during sex differentiation induces female-to-male sex reversal in the medaka *Oryzias latipes*. Sci Rep 9:2383

Kayo D, Zempo B, Tomihara S, Oka Y, Kanda S (2019) Gene knockout analysis reveals essentiality of estrogen receptor β1 (Esr2a) for female reproduction in medaka. Sci Rep 9:8868

Kleppe L, Andersson E, Skaftnesmo KO, Edvardsen RB, Fjelldal PG, Norberg B, Bogerd J, Schulz RW, Wargelius A (2017) Sex steroid production associated with puberty is absent in germ cell-free salmon. Sci Rep 7:12584

Kuroda M, Fujimoto T, Murakami M, Yamaha E, Arai K (2018) Clonal reproduction assured by sister chromosome pairing in dojo loach, a teleost fish. Chromosom Res 26:243–253

Kuroyanagi M, Katayama T, Imai T, Yamamoto Y, Chisada S-I, Yoshiura Y, Ushijima T, Matsushita T, Fujita M, Nozawa A, Suzuki Y, Kikuchi K, Okamoto H (2013) New approaches for fish breeding by chemical mutagenesis: establishment of TILLING method in fugu (*Takifugu rubripes*) with ENU mutagenesis. BMC Genomics 14:786

Li M, Wang D (2017) Gene editing nuclease and its application in tilapia. Sci Bull 62:165–173

Lien S, Koop B, Sandve S et al (2016) The Atlantic salmon genome provides insights into rediploidization. Nature 533:200–205

Murozumi N, Nakashima R, Hirai T, Kamei Y, Ishikawa-Fujiwara T, Todo T, Kitano T (2014) Loss of follicle-stimulating hormone receptor function causes masculinization and suppression of ovarian development in genetically female medaka. Endocrinology 155(8):3136–3145

Nagahama Y, Chakraborty T, Bindhu P-P, Ohta K, Nakamura M (2020) Sex determination, gonadal sex differentiation and plasticity in vertebrate species. Physiol Rev

Nakamoto M, Shibata Y, Ohno K, Usami T, Kamei Y, Taniguchi Y, Todo T, Sakamoto T, Young G, Swanson P, Naruse K, Nagahama Y (2018) Ovarian aromatase loss-of-function mutant medaka undergo ovary degeneration and partial female-to-male sex reversal after puberty. Mol Cell Endocrinol 460:104–122

Pattanayak V, Ramirez CL, Joung JK, Liu DR (2011) Revealing off-target cleavage specificities of zinc-finger nucleases by *in vitro* selection. Nat Methods 8(9):765–770

Qin Z, Li Y, Su B, Cheng Q, Ye Z, Perera D, Fobes M, Shang M, Dunham RA (2016) Editing of the luteinizing hormone gene to sterilize channel catfish, *Ictalurus punctatus*, using a modified zinc finger nuclease technology with electroporation. Mar Biotechnol (NY) 18(2):255–263

Radecke S, Radecke F, Cathomen T, Schwarz K (2010) Zinc-finger nuclease-induced gene repair with oligodeoxynucleotides: wanted and unwanted target locus modifications. Mol Ther 18:743–753

Ramirez CL, Foley JE, Wright DA, Muller-Lerch F, Rahman SH, Cornu TI, Winfrey RJ, Sander JD, Fu F, Townsend JA et al (2008) Unexpected failure rates for modular assembly of engineered zinc fingers. Nat Methods 5:374–375

Sun N, Zhao H (2013) Transcription activator-like effector nucleases (TALENs): a highly efficient and versatile tool for genome editing. Biotechnol Bioeng 110(7):1811–1821

Wargelius A, Leininger S, Skaftnesmo KO, Kleppe L, Andersson E, Taranger GL, Schulz RW, Edvardsen RB (2016) Dnd knockout ablates germ cells and demonstrates germ cell independent sex differentiation in Atlantic salmon. Sci Rep 6:21284

Wei C, Liu J, Yu Z, Zhang B, Gao G, Jiao R (2013) TALEN or Cas9 – rapid, efficient and specific choices for genome modifications. J Genet Genomics 40(6):281–289

Yagi Y, Tachikawa M, Noguchi H, Satoh S, Obokata J, Nakamura T (2013) Pentatricopeptide repeat proteins involved in plant organellar RNA editing. RNA Biol 10(9):1419–1425

Zhu B, Ge W (2018) Genome editing in fishes and their applications. Gen Comp Endocrinol 257:3–12

Metabolomics: A Novel Technology for Health Management in Aquaculture

8

Lopamudra Sahoo, Chandan Debnath, Hurien Bharti, and Basant K. Kandpal

Abstract

Metabolomics is a unique tool in which metabolites are analyzed through various analytical tools like GC-MS, HPLC, NMS, etc. It is one of the omics approaches which provide proper understanding of the functioning of the organism at cellular level. Metabolites are the end products of a metabolic pathway and it can be used to design species wise and pathogen wise customized and personalized treatment. Metabolomics can be used as novel tool for the health management in aquaculture.

Keywords

Metabolomics · Metabolites · Health management · Aquaculture

8.1   Introduction

Systems biology includes the 'OMICS' Sciences i.e. Genomics (DNA), Transcriptomics (mRNA), Proteomics (Proteins) and metabolomics (metabolites). Metabolomics is the systematic study of metabolites within cells, tissues and bio-fluids in a biological system. In other words, it can be said that metabolomics is the characterization of metabolic profile of a living organism. Usually, metabolites have a size range of less than 1000 Da. Metabolites belong to different categories like carbohydrates, fatty acids, nucleotides, pigments, steroids, etc.

Metabolites in plants are usually referred to as primary and secondary metabolites where as in animals, they are referred to as exogenous and endogenous metabolites. Endogenous metabolites are the end products of metabolism of a given organism

L. Sahoo (✉) · C. Debnath · H. Bharti · B. K. Kandpal
ICAR RC for NEHR, Tripura, Lembucherra, Tripura, India

© The Author(s), under exclusive license to Springer Nature Singapore Pte
Ltd. 2021
P. K. Pandey, J. Parhi (eds.), *Advances in Fisheries Biotechnology*,
https://doi.org/10.1007/978-981-16-3215-0_8

123

where as exogenous metabolites are derived from extraneous sources like medicines, drugs, etc. In general, metabolomes are the end products of various molecular and cellular processes. Metabolomic profiling provides us the overall physiological process that has taken place inside a cell. Along with genomics transcriptomics and proteomics, the study of metabolomics should also also be integrated to get a thorough interpretation of cellular processes.

There are two terms that we come across like metabolomics and metabonomics. There are certain disagreement on the usage of the two terms among the researchers. But majority of the researchers have come to a common consensus that metabolomics is the characterization of metabolites at cellular level and metabonomics is the analysis of metabolites due to enviromental alterations in the form of drugs, diets or any genomic influences.

8.2   Why Metabolomics?

As metabolites are the end products of any metabolic process in the cell, metabolic profiling can be considered as penultimate result of an organism to any internal or external perturbations. Although genomics, transcriptomics and proteomics hold their own advantages and given the fact that they are either used alone or in combination with metabolomics, still metabolomics offer quite extensive advantages in terms of not only accurate profiling of the status of an organism but also ease in terms of sample collection, sample preparation and data analysis. Compared to genes and proteins, there are few classes of metabolites that are available for analysis. Hence analysis of metabolites is less cumbersome. It can also be used in plasma and faecal material which is particularly useful when we are studying animal model like fish and other aquatic organisms. It is unique because it depicts the cellular state of an organism and it also takes into consideration gene expression, enzyme regulation and alteration in metabolic reactions.

Genomics and transcriptomics are almost always species-specific, but metabolomics is not confined to any species. Also, there is ease in access of its application across non model organisms without any prior knowledge of its gene, protein or metabolites. Each and every organism in order to achieve a biological balance has a tendency to produce metabolites at a constant rate. It is a known fact that genes and proteins are subjected to epigenetic and post translational modifications. Hence, as metabolites represent the downstream end product of an organism; it showcases the most accurate state of the biological state of the said organism.

Another aspect of metabolomics is that these are the footprints of several ecological processes taking place. Metabolomics forms the bridge for communication between the organisms. The behaviour of organism such as breeding or predation is often guided by the abundance or reduction in release of certain metabolites. These analyses can be done with the metabolomics study of the environment that includes both biotic and abiotic factors.

8.3   Factors Affecting the Metabolic Profiling

There are various factors which affect the metabolic profiling of an organism. They are age, diseases, drugs, environment, genetic factors, lifestyle and nutrition.

8.4   Metabolomics Workflow

This involves the study of biomolecules and metabolites through separation techniques that include chromatography analysis and use of spectrophotometry for detection. Further, statistical analysis is carried out to validate the outcomes which may include the concentration of metabolites using various statistical-software.

8.4.1   Separation Methods

The method of separation depends on the type of metabolites which generate massive data sets in order for proper detection and analysis. Two major techniques that are involved in identification of metabolites are Mass spectrometry and nuclear magnetic spectrometry (NMR). However, there are many important endogenous metabolites in the biological systems that need a broader analysis through separation. Hence several other techniques are utilized such as chromatography. Chromatography is the separation of various phase components in a mixture. Depending on the phases i.e. mobile and stationary, the chromatography technique is divided into liquid chromatography and gas chromatography.

8.4.1.1 Gas Chromatography (GC)
GC coupled with mass spectrometry is a widely used and popular method for separation of the metabolites. The advantage with GC is that it offers high chromatographic resolution, but analysis of bio molecules will require chemical derivatization. Few modern analytical instruments allow 2D chromatography for better and enhanced resolution. Volatile chemicals do not require derivatization. One of the limitations with GC is in its inability to analyze large and polar metabolites.

8.4.1.2 High Performance Liquid Chromatography (HPLC)
In comparison to GC, HPLC has lower chromatographic resolution, but has the advantage of measuring and analyzing wide range of metabolites. HPLC is considered as a versatile method as is hence used widely in pharmaceutical industries. HPLC is more advanced than the conventional liquid chromatography since the latter depends on only gravity but the former is subjected to high pressure and is hence also termed as 'High Pressure Liquid Chromatography'.

8.4.1.3 Capillary Electrophoresis (CE)
CE is mostly used for charged analytes. It has higher efficiency than HPLC and is suitable for a wide range of metabolites. While carrying out capillary electrophoresis

various conditions such as the capillary length, buffer strength and pH should be carefully maintained since it affects the electro osmotic flow thereby affecting the separation process. Hydrophilic interactions can be well studied through this technique. Micellar Electrokinetic Chromatography is a kind of technique that can be used for separating of neutral compounds.

8.4.2 Detection Methods

8.4.2.1 Mass Spectrometry (MS)

Mass spectrometry is a detection method used to quantify and identify metabolites after the initial separation in any of the techniques discussed above. GC-MS is the preferred combination for separation and detection of metabolites by many researchers and clinicians. This technique has the unique advantage of being both sensitive and specific at the same time. In quite a few studies, MS has been used to both separate and detect metabolites with good results.

8.4.2.2 Nuclear Magnetic Resonance (NMR) Spectroscopy

NMR technique is the single technique that does not depend on the pre-separation of analytes. The sample preparation for NMR Spectroscopy is also less cumbersome. The analytical reproducibility is quite high. But the main drawback of MNR in comparison to MS is in its insensitivity in detection of metabolites.

Although NMR and MS are the most widely used techniques, other methods of detection that have been used include electrochemical detection (coupled to HPLC) and radiolabel (when combined with thin layer chromatography).

NMR and MS are two main analytical techniques for metabolic analysis. Both these techniques can identify large number of small molecules in a complex sample. It is pertinent to mention here that although NMR Spectroscopy can identify and quantitatively analyze numerous analytes, but the sensitivity is lower and it requires large number of samples in comparison to MS. So in cases where sample size is less, MS based technology is used particularly in clinical research.

8.4.3 Statistical Methods

After the separation and detection of metabolites we get measurements of metabolites which may be in the form of digitized spectra or the list of the concentration of various metabolites. The data needs to be analyzed statistically. Most metabolomics data analyzed by principal components analysis and partial least squares regression.

8.5   Application of Metabolomics in Health Management in Aquaculture

Species in an aquaculture system are always prone to diseases that may be due to pathogens or environmental alterations. Disease in aquaculture is a huge concern resulting in economic losses particularly in periods of stresses. Many types of antibiotics are used (overused) to control the disease leading to antibiotic resistant bacterial strains and hence these treatments become ineffective. An LD50 represents the individual dose required to kill 50% of a population of test animals. However, why the dose leads to half death and others survive in the population is largely unknown. This may probably be related to differential metabolomes between the survivors and the dead groups. So there is definitely some involvement of metabolic networks in the initiation and proliferation of diseases. In such circumstances, evaluation of metabolites of a healthy and diseased stock can help in understanding the reason behind susceptibility of certain stocks to diseases. In addition to that metabolites released due to treatment with drugs/antibiotics/probiotics etc. can be used as biomarkers and can be used to design customized disease treatment at the species level. Metabolomics as a tool can be used to investigate preliminary stage of infection and biomarkers can be identified so that disease can be identified and necessary steps can be taken to contain the disease at the initial level. Analytical approaches like NMR Spectroscopy and mass spectrometry can be employed to provide metabolic fingerprinting of a diseased organism. Metabolomics can also be employed to identify resistant phenotype of aquatic species. Hence metabolomics can provide unique insights into the effects of pathogen exposure and the mechanisms of resistance. This will aid in developing metabolic biomarker to pathogens which can aid in managing several devastating diseases in aquaculture.

Peng et al. (2015) has identified simple therapeutic strategy to increase disease resistance of Tilapia to bacterial pathogens. They suggested that Glucose was the key biomarker which was upregulated and downregulated in the survival and the dying groups. Exogenous addition of glucose by injection or by oral route enhanced the ability of Tilapia to resist *E. tarda* infections. Their study highlights that living organism use metabolic strategy to combat bacterial infections.

In another study Du et al. (2017) has studied on inducing L-Leucine exogenously to eliminate *Streptococcus iniae* in Tilapia. They found that Serine and Proline are the key metabolites that helped in L-Leucine mediated control of *S. iniae*. From this study it is clearly evident that antibiotics and chemical compounds are not the only solution to control bacterial diseases. Exogenous modulators can be used to induce metabolome in a desired host organism not only to enhance immunity but also to eliminate pathogens.

Cheng et al. (2014) studied on GC/MS based metabolic profiling of Tilapia liver and the result of the profiling was compared between the survivors and the dying fish post challenge with *S. iniae*. There they identified *N*-acetyl glucosamine is being produced in the survivor fish which is acting as a crucial metabolite in aid in resistance to *S. iniae*. So they concluded that exogenous feeding of Tilapia with

N-acetyl glucosamine can significantly elevate the survival of Tilapia post challenge with *S. iniae*.

Guo et al. (2014) established the metabolic profile of *Carassius auratus* after challenging with *Edwardsiella tarda*. After comparing the metabolome of the survivor and the dead fish, they found that increasing palmitic acid and decreasing D-mannose as the most crucial metabolite that is differentiating between the two groups. In their results they have cited that metabolomics can be an effective approach to identify early signs of pathogens and infections that could be mitigated at the onset of a crisis. In a different study by the same researcher, after analyzing the metabolic profile in zebra fish to live *E. tarda* vaccine, oleate was found to be a significant metabolite that induces effective protection to *E. tarda* vaccines. This finding suggests that novel therapeutic strategies can be devised based on metabolomics to bacterial infections. In addition, the improved understanding of biochemical pathways involved in immunological responses may be extremely valuable when selecting optimal conditions for growth of cultured species. Also they have reiterated that the strength of the metabolomics approach is that it provides a wide scan of physiological activities and identifies areas where problems may be encountered. Thus, this is a powerful approach to inform more detailed studies of production issues such as health threats. So these compounds can be used to control *Edwardsiella tarda* infections in other fishes as productions of metabolites are global and universal across living organisms.

To date there are still dearth of vaccines to control diseases in aquaculture. Moreover, the exact mechanism by which vaccines are rendering protection is largely unknown. In this context, a very interesting study has been conducted by Xu et al. (2019) where they have explored the metabolic profile of Zebra fish with inactivated and live *E. tarda* vaccine. It is interesting to note here that both the vaccines promoted biosynthesis of unsaturated fatty acids and the TCA Cycle. Live vaccine promoted higher production of palmitate in the host organism. They concluded that live and inactivated *E. tarda* produce palmitic acid differentially in Zebra fish thus contributing to differential immunities against infection. Thus metabolites are key factors in deciding the success of any vaccination as they direct the host innate immunity to outside pathogens.

Shrimp Industry is a lucrative industry having tremendous market value. Schock et al. (2013) assessed the health of shrimp in a super intensive production system with NMR-based metabolomics. In this study, the authors have beautifully recommended management practices for enhancing shrimp production sans stress. Potential biomarkers that were identified in this study are Inosine and Trehalose. So they have recommended that Inosine and Trehalose test kits can be developed and farmers could use these kits to ensure that his crops is growing with minimal stress and also he can detect early stressful situation and can take up necessary mitigation strategies before any further damage is done.

In another study by the same author, NMR-based metabolomics was used to assess the health of Cobia fish (*Rachycentron canadum*) after decreasing dietary fish meal (Schock et al. 2012). Here they found that two metabolite; Tyrosine and Betaine increased while glucose decreased in Cobia with reduced fish meal diet.

The fishes fed with formulated fish meal with inclusion of fish meal showed elevated lactate level which suggest enhanced gut microflora metabolism in response to dietary components. This work underlines the important role; metabolomics can play in aquaculture for assessing health status of desired species.

8.6 Other Applications

Metabolism is a powerful tool for gaining insights into the cell. It has multifarious applications across several fields. Some of the areas where metabolomics is being extensively used are listed below:

- Biomarker discovery
- Diagnostic test development
- Tracking drug and nutrient effects
- Identifying targets for drug development
- Bioprocess optimization
- Extracting value from genomics research
- Precision Medicine
- Pharmaceutical and Biotechnology
- Population Health
- Academic Research
- Nutrition
- Consumer Goods and Personal Care
- Agriculture and Aquaculture

8.7 Application of Metabolomics in Environmental Studies

There are several metabolites present in the environment either endogenously or exogenously secreted by other organisms. Different individuals reciprocate differently to these compounds. Such interactions are largely unexploited. Metabolomics can be beneficial in untying the knots for understanding the ecological interactions. Through such study the environmental stressors could be analyzed effectively. There are many distinct biological responses which involves the environmental microbes that also are able to alter physiological responses of other higher organisms present in that environment. Hence microorganisms are considered the best organisms to study environmental metabolomics. The effects of toxins in environment are also evaluated through metabolomics study of these microbes.

8.8   Limitations of Metabolomics

Although Metabolomics based System Biology has made remarkable progress in the present time, still there are few limitations which needs to be worked upon. The speed, accuracy and sensitivity of the analytical tools like MS and NMR needs to be more precise and accurate. Hence research needs to be directed towards further modifying the analytical tools. NMR and MS based procedure can help us identify large number of metabolites, but only those which already exist in the database. Hence more and more work should be done on identification of the unknown repertoire of metabolites. Mirror form of isomers like Cis and Trans form can also lead to inaccurate identification of metabolites.

Also to put it in a true sense, Metabolomics is not a solo science, in order to understand the biological system in totality it should be coupled with the other 'OMICS' science. In order to simplify the OMICS and to gain a proper understanding of the system biology, bioinformatics area needs to be strengthened.

8.9   Database

The first metabolite database (called METLIN) for searching m/z values from mass spectrometry data was developed by scientists at The Scripps Research Institute in 2005. In January 2007, scientists at the University of Alberta and the University of Calgary completed the first draft of the human metabolome. They catalogued approximately 2500 metabolites, 1200 drugs and 3500 food components that can be found in the human body, as reported in the literature. Some of the available databases are listed in Table 8.1.

8.10   Conclusion

Metabolomics is undoubtedly the 'OMICS' of the current era. Through this science biomarkers can be developed which can be added to the already existing repertoire of biomarkers to further strengthen disease management strategies. Precision and customized drugs can be designed for each individual species targeting individual pathogens. With the rapidly developing tools like big data, artificial intelligence, undoubtedly metabolomics will play a bigger role in the aquaculture health sector.

Table 8.1 Database used in metabolomics

Comprehensive metabolome database	Metabolic pathway databases	Compound or compound-specific databases	Drug databases	Spectral databases	Disease and physiology databases
HMDB	KEGG	PubChem	DrugBank	HMDB	OMIM
BiGG	MetaCyc	ChEBI	Therapeutic Target Database	BMRB	METAGENE
SetupX	HumanCyc	ChemSpider	PharmGKB	MMCD	OMMBID
SYSTOMONAS	BioCyc	KEGG Glycan	STITCH	MassBank	
MetaboLights database	WikiPathways			Metlin	
				Fiehn GC-MS Database	
				BML-NMR	
				MetaboLights database	
				mzCloud	
				Fiehn GC-MS Database	

References

Cheng ZX, Ma YM, Li H, Peng XX (2014) N-acetylglucosamine enhances survival ability of tilapias infected by Streptococcus iniae. Fish Shellfish Immunol 40(2):524–530

Du CC, Yang MJ, Li MY, Yang J, Peng B, Li H, Peng XX (2017) Metabolic mechanism for L-leucine-induced metabolome to eliminate Streptococcus iniae. J Proteome Res 16 (5):1880–1889

Guo C, Huang XY, Yang MJ, Wang S, Ren ST, Li H, Peng XX (2014) GC/MS-based metabolomics approach to identify biomarkers differentiating survivals from death in crucian carps infected by Edwardsiella tarda. Fish Shellfish Immunol 39(2):215–222

Peng B, Ma YM, Zhang JY, Li H (2015) Metabolome strategy against Edwardsiella tarda infection through glucose-enhanced metabolic modulation in tilapias. Fish Shellfish Immunol 45 (2):869–876

Schock TB, Newton S, Brenkert K, Leffler J, Bearden DW (2012) An NMR-based metabolomic assessment of cultured cobia health in response to dietary manipulation. Food Chem 133 (1):90–101

Schock TB, Duke J, Goodson A, Weldon D, Brunson J, Leffler JW, Bearden DW (2013) Evaluation of Pacific white shrimp (Litopenaeus vannamei) health during a superintensive aquaculture growout using NMR-based metabolomics. PLoS One 8(3):e59521

Xu D, Wang J, Guo C, Peng XX, Li H (2019) Elevated biosynthesis of palmitic acid is required for zebrafish against Edwardsiella tarda infection. Fish Shellfish Immunol 92:508–518

Epigenetics: Perspectives and Potential in Aquaculture

9

Suvra Roy, Vikash Kumar, B. K. Behera, and B. K. Das

Abstract

Epigenetics has attracted considerable attention recently and presents great opportunities in the aquaculture sector. Epigenetics literally means 'above' or 'on top of' genetics and epigenetic trait, referring to the stable heritable phenotype resulting from changes in the chromosomes without any changes in the DNA sequence information. Epigenetic regulation is mainly mediated through histone modifications, DNA methylation and noncoding RNAs; e.g. miRNA. Environmental factors can exert influence on epigenetic changes to produce the phenotype and this effect can be passed on to the next subsequent generations/offspring. This creates a huge possibility of epigenetic programming in animal husbandry/ aquaculture sector for selection of the most favourable phenotypic traits and production enhancement. In this chapter, we have discussed the concept of epigenetics and inheritance, main epigenetic mechanisms and current understanding of epigenetics in aquaculture species in fish/shellfish and highlighted the key areas of aquaculture where epigenetics could be applied. Though epigenetics has a huge potential, epigenetic study of economically important aquaculture species are still in its infancy and there remain many unanswered questions. Understanding of epigenetic mechanisms and epigenetic markers for important aquaculture species will contribute to the selection of the most favourable phenotypic traits and help in the expansion of economically viable commercial aquaculture industry.

S. Roy (✉) · V. Kumar · B. K. Behera · B. K. Das
Aquatic Environmental Biotechnology and Nanotechnology (AEBN) Division, ICAR-Central Inland Fisheries Research Institute (CIFRI), Barrackpore, India

© The Author(s), under exclusive license to Springer Nature Singapore Pte Ltd. 2021

P. K. Pandey, J. Parhi (eds.), *Advances in Fisheries Biotechnology*,
https://doi.org/10.1007/978-981-16-3215-0_9

133

Keywords

Epigenetics · Aquaculture · Epigenetic mechanisms · DNA methylation · Histone modifications · Noncoding RNA · Epigenetic inheritance · Aquaculture applications

9.1   Introduction

Over the last decade, interest in the field of epigenetics has rapidly increased and a paradigm shift has been noticed in the rules of inheritance. The word 'epigenetics' consists of the word 'genetics' and the Greek root epi ('over, upon, outside of') used as a prefix, which in epigenetics implies features that are 'on top of' or 'in addition to' the traditional genetic basis of inheritance. Although epigenetics caught the attention of researchers in recent times in many biological fields of research, the term epigenetics and its definition have a long history. The term 'epigenetics' was first coined by the embryologist and developmental biologist Conrad H. Waddington in the year 1942 (Waddington 1942; Jamniczky et al. 2010) to describe the influence of environmental cues to develop the specific phenotypes through genotype–environment interactions (Fig. 9.1). In 1958, David Nanney broadened the view of epigenetics to distinguish between different types of cellular control systems and defined epigenetics as the causes of heritable differences that does not involve DNA sequence changes (Nanney 1958). In general, Conrad H. Waddington was more concerned with gene regulation and genotype–phenotype interactions, while David Nanney was more interested in the stability of expression states and cellular

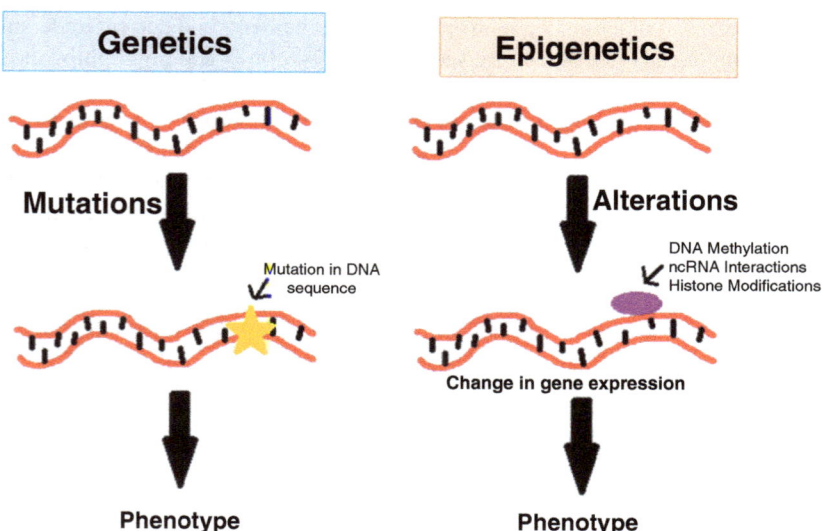

Fig. 9.1 Comparison of genetics and epigenetics mechanism for emerging new phenotype

inheritance. The findings go a long way to explain the changes in gene expression that Waddington termed 'epigenetics'. Many researchers over the last few decades have been discussing about the ambiguity surrounding the field of epigenetics and attempted to redefine it with more specificity, taking a critical look at the terms dependence, DNA sequence, and heritability (Holliday 1994, 2006; Jablonka and Lamb 2002; Wu and Morris 2001; Bird 2007; Berger et al. 2009; Mann 2014). The addition of heritability by Holliday to Waddington's original definition was a significant change; a definition that was formulated in 2008 at a Cold Spring Harbor meeting is as follows: "an epigenetic trait has been defined as a stably heritable phenotype, resulting from the changes in a chromosome without any alterations in the DNA sequence" (Berger et al. 2009). In the modern aspects, epigenetics refers to the "study of phenomena and mechanisms that cause chromosome-bound, mitotically and/or meiotically heritable changes to gene expression or phenotype that are not dependent on any changes to DNA sequence" (Deans and Maggert 2015).

We will introduce key concepts of main epigenetic mechanisms and the current understanding of epigenetics in aquatic animals in both fish and shellfish and the bay areas of aquaculture in which epigenetics could be applied.

9.2   Epigenetic Inheritance

There are two distinct pathways by which epigenetic marks could be transmitted or inherited (a) from one generation of cells to another; and the evidence for epigenetic inheritance through mitosis is strong and (b) from one generation of organisms to another; where the epigenetic modification of certain genes, produced by the certain factor or environmental trigger which could lead to significant changes in an individual's body. This could persist over time and in turn signal the epigenetic reorganization of the subsequent generations, not being explained either by alterations in the DNA sequence or Mendelian genetics (Daxinger and Whitelaw 2012). But the extent of epigenetic inheritance through different generations of organisms varies. Epigenetic inheritance is an important mechanism by which adaption of one generation of organisms to a changed environment can be carried to many subsequent generations. There are three main types of epigenetic inheritance from one generation of organisms to another (Wang et al. 2017) and these are (a) intergenerational epigenetic inheritance representing the transmission of epigenetic marks or impact of the paternal (F0) exposure to particular environmental cues only to the next (F1) generation (Radford et al. 2014); (b) multigenerational epigenetic inheritance (from the F1 to the F2 generation) (Dias and Ressler 2014); and (c) transgenerational epigenetic inheritance that are found in more than three generations, so the transmission of information from grandparents to the grandchildren (Skinner 2007; Greer et al. 2014; Norouzitallab et al. 2019; Roy et al. 2019). In fact, transgenerational inheritance was first described by Skinner (2007) and according to him, transgenerational epigenetic inheritance only can be called so when two criteria are met: exposure to an event in generation F0 and an effect of the event must be observed in the third or fourth generation—i.e. F2 or

F3—depending on whether the mother or father was first exposed or affected (F0). Since, female (mother) exposure to a certain factor or environmental cues during pregnancy might even affect the offspring's germ cells directly, which will become F2 generation eventually, so for which reason only from the third (F3) or beyond generations can be considered event-free and unsullied. In case of male (father), exposure to a certain factor or environmental cues produces an epigenetic change, from F2 generation can be considered event-free (Daxinger and Whitelaw 2012).

9.3   Main Epigenetic Mechanisms

All the genes in cells are not active at all the time and in all organisms, although they carry the same genetic information. So, epigenetics was introduced as an additional mechanism that controls the gene expression in an inheritable manner. To introduce the different epigenetic mechanisms and understand how these can interact and complement each other, it is necessary to understand the basic structure and function of chromatin. In eukaryotic cells, DNA is compacted into the chromatin, in which ~146 bp of DNA is wrapped around histone octamers (two copies of histones H2A, H2B, H3, and H4), forming the nucleosomes which is linked by the exterior linker histone H1, are folded into chromatin fibres, allowing the genetic information to compact inside the nucleus (Conaway 2012). This chromatin structure can be altered and remodelled by different epigenetic mechanisms, which will ultimately define the distinct patterns of gene expression (Siggens and Ekwall 2014). Many mechanisms for underlying epigenetic modifications have been identified in vertebrates and invertebrates and new mechanisms are likely to emerge in future. The key epigenetic mechanisms (or 'marks') include DNA methylation, histone modifications and noncoding/small RNA (ncRNA/sRNA) activity which influence the gene expression primarily through the local modification of chromatin. These epigenetic mechanisms individually or collectively play a key role in turning the gene expression on or off, thus facilitating or inhibiting the production of specific proteins. Epigenetic marks also can be directly influenced by the environment, and therefore, have been shown to be important mediators of phenotypic responses to environmental cues. This section presents a description of key epigenetic mechanisms and marks (Fig. 9.2).

9.3.1   DNA Methylation

One of the most important and best studied epigenetic mechanisms is DNA methylation (Jablonka and Raz 2009). It occurs by addition of methyl group (CH_3), mostly to the fifth carbon $5'$ of the cytosine residues to form 5-methyl cytosines (5mC) predominantly at cytosines, followed by guanines called CpG sites/islands catalyzed by a family of DNA methyl transferases (DNMTs) (Zhu 2009; Moore et al. 2013). DNA methylation affects the coiling of DNA around histone proteins and changes the potential binding affinity of transcriptional factors (TFs). The gene expression regulation of DNA methylation is dependent on the position of methylation in the

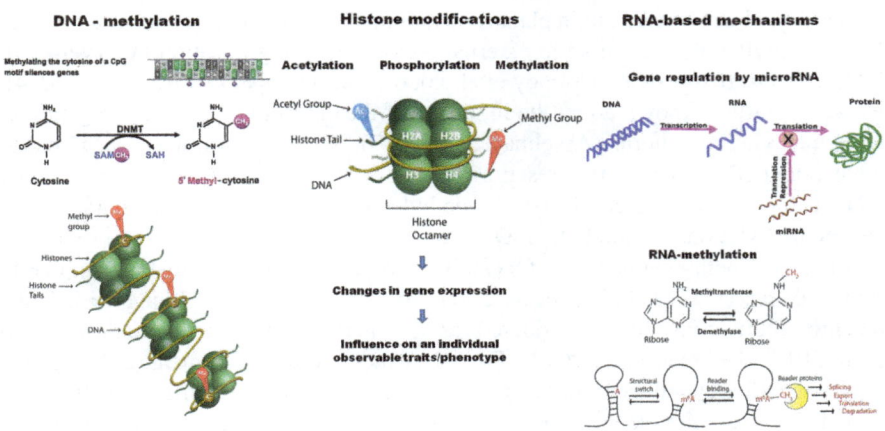

Fig. 9.2 Main epigenetic mechanisms for changing the gene expression and phenotype

gene such as promoter, exon, intron or transcriptional start site. In general, mostly DNA methylation in CpG islands near the promoters of genes suppresses gene transcription or maintains it at a silenced state (Jones 2012). However, DNA methylation doesn't occur just in CpG Island. The studies have also shown that DNA methylation is found throughout genes, but the contribution of gene body DNA methylation to gene regulation is still not fully clear (Feng et al. 2010; Sarda et al. 2012; Zemach et al. 2010). DNA methylation in gene bodies and intragenic regions is reported to be associated with high levels of gene expression (Jones 1999). Although DNA methylation is typically associated with gene silencing, its exact regulatory role is specific to the genomic context. In addition, DNA methylation can also regulate the transcriptional activity of neighbouring genes by altering the structure and function of chromatin. DNA methylation units are relatively conserved within the organism (Hernando-Herraez et al. 2015), however, the absolute levels and patterns of the DNA methylation are highly variable between vertebrates and invertebrates (Sarda et al. 2012). It may also vary substantially between the different tissues, developmental stages and species and can be directly modulated by the environmental cues (Bocklandt et al. 2011; Fritsche et al. 2013). Vertebrate genomes are highly methylated at CpG sites primarily associated with promoter silencing. DNA methylation occurs widely in invertebrate genome. However methylation often occurs in units involved in active transcription (Feng et al. 2010; Zemach et al. 2010). DNA methylation regulates gene expression which is reversible, dynamic, and can change throughout the life span. The potential of DNA methylation to act as a heritable epigenetic mark has been shown in few studies (Lämke and Bäurle 2017; Stadlbauer 2017). Initially, it was thought that passing the DNA methylation marks to the next generation was impossible due to embryonic demethylation; however, recent evidences suggest that DNA methylation marks can escape demethylation and is transferred during meiosis which may be heritable across generations (Roth et al. 2018; Roy et al. 2019). Most of the DNA methylation studies have been carried out in plants and mammals, in which it has been shown that DNA methylation is sensitive to external factors including nutrition (Weaver et al. 2004), exposure to toxins (Dolinoy et al. 2006), and photoperiod (Azzi et al. 2014). However, little is known about the influence of the environmental cues on genome-wide methylation patterns (Asselman et al. 2017). Also, it is still not clear if and to what extent DNA methylation resetting occurs in fish and shellfish. It is known that DNA can be demethylated; 5-hydroxymethylcytosine (5hmC) is well characterized as a demethylation intermediate in DNA demethylation pathways. It is an oxidative product of 5-methylcytosine (5mC) catalyzed by the ten eleven translocation (TET) family of enzymes (Tahiliani et al. 2009; Branco et al. 2012). Along with 5mC, recently 5hmC has been considered as a functional epigenetic mark (Branco et al. 2012; Bachman et al. 2014). For DNA methylome assessment, either global or site-specific several methods are available such as bisulphite sequencing (BS-seq), oxidative bisulphite sequencing (oxBS-seq), reduced representation bisulphite sequencing (RRBS), methylated DNA immune precipitation (MeDIP) and antibody-based ELISA method (Granada et al. 2018; Roy et al. 2019).

9.3.1.1 DNA Methylation in Fish and Shellfish

Among fish and shellfish, DNA methylation is the most studied epigenetic mark. To understand the patterns and functions of DNA methylation, significant work has been done in model fish species such as zebrafish and medaka and also in non-model species (Metzger and Schulte 2016), but the extent of DNA methylation reprogramming in fish is still not clear (Jiang et al. 2016; Potok et al. 2013). However, a recent study showed a clear evidence of transgenerational inheritance of environmentally induced DNA methylation patterns in a fish, suggesting that at least some of the genome escaped putative resetting between the generations (Shao et al. 2014). Plant-based phloroglucinol compound induced transgenerational inheritance of DNA methylation which was observed in an invertebrate species i.e. brine shrimp (Roy et al. 2019). More detailed studies on the extent of DNA methylation in fish or shellfish is needed, particularly for important aquaculture species. DNA methylation patterns and content in invertebrates are also different from the vertebrates; vertebrates exhibit a global pattern of DNA methylation whereas invertebrates show a 'mosaic' pattern and moreover the methylated fraction tends to consist of gene bodies in invertebrates (Gavery and Roberts 2017; Feng et al. 2010).

9.3.2 Histone Modifications

Nucleosomes is composed of core histones proteins (two H3 and H4 homodimers and two H2A/H2B heterodimers) which is encircled by ~147 bp of DNA to coil the DNA into a smaller volume for fitting in the nucleus of the cell. Moreover, apart from the globular structure, histones contain a more flexible and charged NH2-terminus (containing 25–30 basic amino acids rich residues) which protrudes from the nucleosome called 'histone tail' (Jenuwein and Allis 2001). Through the posttranslational modification of histones, chromatin structure can be modified to either enhance or repress transcription (Berger 2007); such as altering the degree of chromatin compaction/organization resulting in either euchromatin (it is active form, referring to open chromatin that support gene expression) or heterochromatin (referring to tightly packed DNA associated with transcriptional silencing) (Araki and Mimura 2017). These modified chromatin states can be inherited both mitotically and meiotically, thus histones are key elements in epigenetic reprogramming (Henikoff and Smith 2015). Histone tail undergoes different type of covalent chemical modifications including acetylation, methylation, phosphorylation, ubiquitylation, and chromatin states dependant on the types of modification and location (e.g. various lysine or arginine residues). Most modifications occur predominantly on the tails of H3 and H4, different modifications are correlated with different functions at chromatin organization and transcription regulation levels (Daneshfozouna et al. 2015). Although there are various types of histone modifications, histone acetylation (lysine residues) and methylation (arginine and lysine residues) are mostly studied. These modifications are controlled by various families of highly specific enzymes, including histone acetylases (HATs), histone deacetylases (HDACs), histone methyltransferases (HMT) and histone demethylases (Greer et al. 2014). Histone acetylation is usually correlated with an open chromatin structure, thus, more accessible for transcription and often associated with transcriptional activation. In histone deacetylation, DNA is tightly bound leading to a repressive chromatin structure and transcription is silenced. On the other hand, histone methylation can result either in transcriptional silencing or promote transcriptional activation, depending on which residue is methylated (Annunziato and Hansen 2001; Verdone 2006). Histone modifications can play an important role in mediating both short- and long-term responses to environmental cues (Talbert and Henikoff 2010; Roy et al. 2019). Some studies suggest that histone modifications are heritable, epigenetic memory and signal can pass on from one generation to the next. Histone modifications also have potential for a role in trained immunity or innate immune memory in invertebrates (Roy et al. 2020; Roth et al. 2018; Norouzitallab et al. 2016). Although histone modifications and DNA methylation require different sets of enzymes and biochemical reactions; increasing evidences suggest that there may be cross-talk between the two processes and these mechanisms act together in modulating gene expressions (Lan et al. 2017). Histone posttranslational modifications are usually detected by using sequence-specific antibodies (e.g. by Western blot analysis) and mass spectrometry methods (including Bottom-Up, Middle-Down and Top-Down proteomics) and for genome-wide approach chromatin immune precipitation (ChIP) method was developed (Granada et al. 2018; Huang et al. 2015).

9.3.2.1 Histone Posttranslational Modifications in Fish and Shellfish

Histone posttranslational modifications and the dynamics have been studied in zebrafish model and histone acetylation was found to be playing a role in embryogenesis and in tissue regeneration (Vastenhouw and Schier 2012). Histone modifications in non-model fish are rare. Recent studies in rainbow trout and

European sea bass showed that diet can influence histone modification levels and can regulate the expression of associated enzymes (Marandel et al. 2016; Panserat et al. 2017). Although, histone modifications are not extensively studied in shellfish, it was found to play a role in mediating the physiological responses of oysters to the environmental cues e.g. temperature (Fellous et al. 2015). Moreover, histone post-translational modifications such as acetylation and methylation play a role in the transgenerational epigenetic inheritance in brine shrimp species (Norouzitallab et al. 2014; Roy et al. 2019).

9.3.3   Noncoding RNA and RNA Modifications

Although, most of the genome is transcribed, only a small part of the transcripts actually code for proteins. The remaining noncoding transcripts, that do not directly code for a protein (originally known as 'junk'), are now recognized to play a role in modulating gene expression and are broadly categorized as non-coding RNA (ncRNA). There are two major classes of ncRNA: long ncRNA (>200 nucleotides) and small ncRNA (<200 nucleotides). Small ncRNAs is generally RNA molecules of about 20–30 nucleotides, such as miRNAs (microRNAs), siRNAs (small interfering RNAs), and piRNAs (Piwi-interacting RNAs). There are evidences that ncRNAs serve as important epigenetic regulators of gene expression at the posttranscriptional level, genome stability, environmental plasticity and embryonic development (Moazed 2009; Bizuayehu and Babiak 2014; Goldstein et al. 2017). Generally, ncRNAs are considered to be 'epigenetic' as they interact with other epigenetic mechanisms such as DNA methylation and histone modifications to silence or activate various parts of the genome (Peschansky and Wahlestedt 2014). There is also evidence that ncRNA, particularly miRNA, have emerged as posttranscriptional regulators of wide-ranging of immune-related biological processes, host defense and transgenerational inheritance of environmental cues to offspring (Klosin et al. 2017; Zhi and Fei 2017). Noncoding RNAs can be identified by several methods, such as microarrays, direct sequencing (RNA-seq), RNA immune precipitation (RIP) and ChIP (Granada et al. 2018).

Very recently along with DNA methylation, RNA methylation has also been reported a potential mechanism in epigenetic control of fine-tuning of gene expression (Yue et al. 2015; Roy et al. 2019). RNA methylation is a reversible, posttranscriptional RNA modification, which affect several biological processes of mRNA and ncRNA such as RNA stability and mRNA translation (Romano et al. 2018).

9.3.3.1 Noncoding RNA in Fish and Shellfish

Most studies on noncoding RNAs in fish and shellfish, including important aquaculture species such as Atlantic salmon and rainbow trout have focused on miRNAs in a physiological context for instance, immune function and embryonic development (Andreassen et al. 2017; Bekaert et al. 2013; Ma et al. 2012; Juanchich et al. 2016). Very less information is available on the other types of small ncRNAs, except such as piRNAs which have been shown to silence transposable elements in zebrafish and several associations have been observed between lncRNA expression and disease in salmonids (Atlantic salmon and rainbow trout) (Paneru et al. 2016). An association between lncRNAs and larval development have been reported in the Pacific oyster (Yu et al. 2016).

9.4   Potential Aquaculture Applications

Aquaculture production is rapidly growing worldwide and to address the increasing demand of the fish there is an urgent need of further technological innovations in different sectors like disease management, breeding, feeds and nutrition among others. Epigenetics has attracted considerable attention with respect to its potential value in many areas of agricultural production and it holds great opportunities for aquaculture sectors, particularly where the environment can be manipulated (Fig. 9.3). The transcriptional impact of epigenetic modifications, triggered by environmental stimuli, has been shown to influence the organism's phenotype. Therefore, the study of epigenetic mechanisms and epigenetic markers, related to disease resistance or other economically important traits which leads to the favourable phenotype in economically important species, present great opportunities and can contribute to the expansion of commercial aquaculture industry with increased economical revenue (Granada et al. 2018; Gavery and Roberts 2017). Moreover, several studies have shown epigenetic effects in various species induced by different conditions for the favourable acquired adaptive phenotypic traits sometimes also evidenced across generations, offering even more relevant in the production context. The fact that the epigenetic markers and corresponding phenotype can be inherited by the subsequent generations further highlights the importance of epigenetics application in aquaculture for a more sustainable practice at environmental and economic levels.

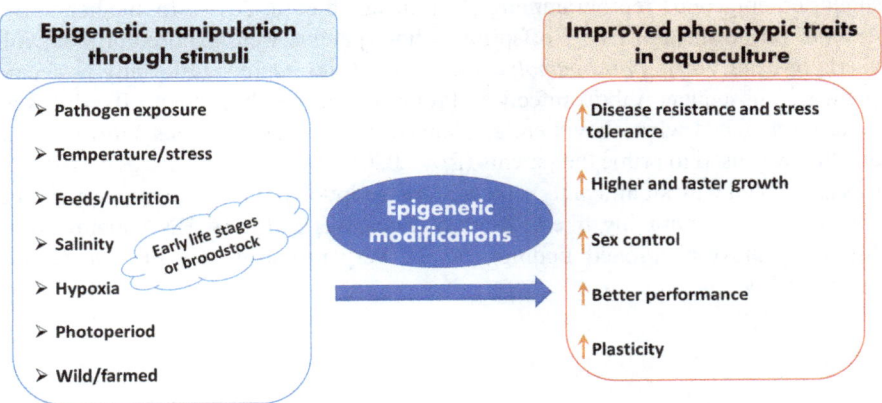

Fig. 9.3 Schematic overview of the potential applications of epigenetic in aquaculture

9.4.1 Disease Resistance and Stress Tolerance

There are a few studies which have linked the exposure to certain abiotic conditions to the altered epigenetic states or changes in epigenetic markers that result in improved disease resistance and increased tolerance to stressors, which can have clear economic advantages (Gonzalez-Recio 2012). Stress experienced during early life is likely to affect the ability of aquatic animals to cope with pathogens and respond to subsequent stressors later in the life. Early life stress (thermal stress) is reported to have effect on the disease resistance, immune response and impacts on the methylome and transcriptome profiles of Atlantic salmon (Moghadam et al. 2017; Uren Webster et al. 2018). In brine shrimp (an invertebrate crustacean model) heat shock-induced thermotolerance and resistance against pathogen have been shown (Norouzitallab et al. 2014). Moreover, in brine shrimp model a plant derived heat shock protein (hsp) inducing compound (phloroglucinol) treatment to the parental generation at early life stages induced transgenerational epigenetic inherited resistance in offspring against *Vibrio* infections and thermal stress (Roy et al. 2019). In the mentioned study, increased resistance was recorded in three subsequent generations and enhanced expression of innate immune genes and epigenetic mechanisms such as histone modifications (acetylation and methylation), DNA methylation, RNA m6A methylation played the underlying role in the observed phenotype (Roy et al. 2019). There is also increasing evidence that prior exposure to an immune challenge can increase the immune response later in life and the memory can be transmitted to the offspring. A study demonstrated that the offspring of Pacific oyster parents treated with poly (I: C) possess enhanced protection against Ostreid herpes virus type I infection (Green et al. 2016). The improved protective phenotype and higher expression level of antiviral-related gene in the oyster larvae originated from poly (I: C)-treated parents suggested heritable epigenetic reprogramming. In *Artemia (an* invertebrate) model study, F1–F3 progenies whose parents were exposed to pathogenic *Vibrio campbellii* at early life stages exhibited resistant phenotype compared to the respective progeny of control and suggested epigenetic reprogramming (Norouzitallab et al. 2016). In another study, *Artemia* (brine shrimp) F1-F3 offspring, whose parents were exposed/primed with viable or dead *Vibrio parahaemolyticus/V. campbellii,* were significantly protected against a subsequent Vibrio infection. Protection in the challenged offspring was more pronounced when they were challenged with the homologous *Vibrio,* so the one that was used to prime the parents (Roy 2020). Moreover, the disease resistance was associated with significantly increased transcription of innate immune genes and epigenetic reprogramming like histone modifications and m6A RNA methylation. All these above-mentioned findings can be very promising for the aquaculture sustainability.

9.4.2 Feeds/Nutritional Programming and Growth

In the field of nutritional epigenetics, the possibility of early nutritional programming and understanding how early life nutritional conditions influence the key phenotypic traits later in life to improve the performance is important to consider. There are few evidences which showed a link between early life nutrition and epigenetics can be extremely useful for the aquaculture. In rainbow trout fish study, fry fed a plant-based diet showed higher growth rates, feed intakes and feed efficiencies (Geurden et al. 2013). Transcriptomic analyses in the follow-up study suggested epigenetic mechanisms involvement in the observed response (Balasubramanian et al. 2016). In another study in rainbow trout fish, vitamin supplementation at first feeding reported changes in global methylation and histone modification 7 months after the supplementation was discontinued, despite the apparent lack of phenotypic responses (Panserat et al. 2017). Brown trout fish (*Salmo trutta*) fed with salt-enriched diet showed higher survival rates when transferred to seawater as compared to trout feeding commercial trout pellets and alteration in the global DNA methylation patterns was observed induced by diets, which can have important economic consequences for trout farming (Moran et al. 2013). The study in Senegalese sole (*Solea senegalensis*) showed temperature-induced phenotypic plasticity of muscle growth, gene expression and DNA methylation in the larval phase (Campos et al. 2013). Similarly, in another study in Atlantic cod juvenile (*Gadus morhua*) an increased growth was observed that was induced by photoperiod and it is accompanied by the differential expression of some DNMTs in fast muscle tissue (Giannetto et al. 2013). These two, above-mentioned studies suggested that DNA methylation may have a role in the regulation of muscle growth rate by temperature and photoperiod.

9.4.3 Sex Reversal and Better Sex Ratios

There are also a few examples of developmental programming in aquaculture such as sex determination in fish where the epigenetic mechanisms were described. In European sea bass (*Dicentrarchus labrax*), exposure to high temperature in early stages was associated with higher proportion of phenotypic males (Navarro-Martín et al. 2009). This early exposure to high temperature was associated with increased DNA methylation in the promoter of the gonadal aromatase gene in adults (Navarro-Martín et al. 2011). In another study in half smooth tongue sole (*Cynoglossus semilaevis*), pseudo males was generated by exposing genetic females to high temperature during the sensitive developmental period (Chen et al. 2014). The pseudo males DNA methylation patterns exhibited in testes consistent with genetic males and it differs from the ovarian methylome of normal females. Moreover, interestingly this pseudo male phenotype and testes-specific methylation patterns are inherited in F1 generation, which suggested the transgenerational epigenetic inheritance of environmentally induced sex reversal phenomena in this fish species (Shao et al. 2014).

9.4.4 Brood-Stock Conditioning and Transgenerational Plasticity/ Epigenetic Inheritance

Brood-stock conditioning is an important area to consider for the potential transmission of environmentally induced epigenetic information between parents and their offspring. These types of environmentally driven nongenetic transmission of phenotypes in the offspring mediated through epigenetic mechanisms are frequently referred to as 'transgenerational plasticity' or 'transgenerational epigenetic inheritance' (Salinas et al. 2013; Roy et al. 2020). Recently there has been growing interest in the brood-stock conditioning and transgenerational epigenetic inheritance of aquatic animals (fish and shellfish), particularly relating to the environmentally induced fitness and also climate change (Roy et al. 2019, 2020; Norouzitallab et al. 2014; Munday 2014). Evidence of transgenerational epigenetic inheritance and higher survival of brine shrimp (*Artemia franciscana*) offspring in three subsequent generations (F1-F3) were recorded against biotic stress (*Vibrio* challenge) and abiotic stress (thermotolerance) whose parental generation was exposed to heat shock or phloroglucinol, a plant-based hsp inducing compound at their early life (Norouzitallab et al. 2014; Roy et al. 2019). In addition, higher survival of offspring (F1 larvae) was also recorded in freshwater giant prawn (*Macrobrachium rosenbergii*) whose parents were treated with phloroglucinol at early life stages (Roy 2020). Another example of environmental manipulation for brood-stock conditioning study was on adult Manila clams and Sydney rock oyster that were exposed to low pH produced faster growing and higher fitness offspring (Zhao et al. 2017; Parker et al. 2012).

9.4.5 Epigenetics in Selection and Domestication

Epigenetic selection could be used in aquaculture, alone or in combination with genetic selection to identify the individuals with desired traits which may increase the reliability of producing animals with desired phenotypes (Gavery and Roberts 2017). Genomic information does not fully account for all the phenotypic variations in animals; phenotype may result from the interplay between the genome and the epigenome. Thus, environmentally induced epigenetic variations could play a critical role in defining individual variations and phenotypic outcomes within generation and also across the generations (Deans and Maggert 2015; Gavery and Roberts 2017; Roy et al. 2019). The concept of epigenetic selection has been applied practically from its theoretical formulation in one agriculture commodity i.e. oil palms, where a critical trait oil content was epigenetically selected (Ong-Abdullah et al. 2015). It is also possible that epigenetic markers could be integrated into brood-stock selection in aquaculture. In fact, a study in Pacific oyster (*Crassostrea gigas*) investigated the genetic and epigenetic variation in the mass selection populations of selective breeding. Results showed some correlation and epigenetic variation might be partly dependent on the genetic context in *C. gigas* (Jiang et al. 2013). Epigenetics could also make genetic selection more challenging, two life stages which may be important particularly to generating within or between generation 'epigenetic memories' are larvae and brood-stock (Moghadam et al. 2015).

Domestication process often involves captivity rearing and selective breeding which contribute to the expression of phenotypes. However, how domestication affects the animal is still not clear, both genetic and epigenetic changes can alter the phenotype. Interestingly, a new role of DNA hydroxyl methylation (5hmC) unveiled in epigenetic regulation for domestication in Nile tilapia (*Oreochromis niloticus*) fish which reared in captivity just for one generation (Konstantinidis et al. 2020). In another study, sex-specific epigenetic regulation of growth in Nile tilapia was observed during early stages of domestication (Podgorniak et al. 2019). It is also known that captive/hatchery rearing impacts the fitness of individuals. Genome-wide methylation profiles were performed to compare the epigenetic differences between Pacific salmon originating from a hatchery and their wild counterparts from two geographically distant rivers. A significant epigenetic variation in hatchery-reared salmon provides the evidence that epigenetic modifications induced by hatchery rearing acts as a potential explanatory mechanism for reduced fitness in hatchery-reared salmon once released in the wild (Le Luyer et al. 2017).

9.5   Conclusions and Future Perspectives

Epigenetics has great potential and application of its knowledge could significantly affect the productivity and sustainability of aquaculture sectors. However, epigenetics study and its aquaculture application are in the infant stage and very scarce. There are many unanswered questions regarding how epigenetic mechanisms may support the aquaculture in future. Understanding the epigenetic mechanisms involved in the regulation of important traits in aquaculture species such as stress and disease resistance, possibility for nutritional programming and environmental adaptation will contribute to the economically viable commercial aquaculture industries. Future research should focus on how the genome sequence impacts the epigenome, how epigenetic events interact with genotype to influence the induction of phenotypic traits under different conditions and use this fundamental information for aquaculture applications. The major constraint to the development of epigenetic solutions in aquaculture includes the unavailability/unknown genome information of several commercially important species and data on the epigenome are even more limited. In addition, these studies often require significant time and costs, mainly for larger fish and crustacean species, due to their long and complex life cycle. So, epigenetic studies with suitable model species which have shorter life cycles may be a better option for the first approach to provide elucidative outcomes in a shorter period.

Acknowledgements We acknowledge the support received from the Director, ICAR-Central Inland Fisheries Research Institute, Barrackpore, India and Dean, College of Fisheries, CAU, Lembucherra, Tripura, India.

References

Andreassen R, Woldemariam NT, Egeland IØ, Agafonov O, Sindre H, Høyheim B (2017) Identification of differentially expressed Atlantic salmon miRNAs responding to salmonid alphavirus (SAV) infection. BMC Genomics 18:349

Annunziato AT, Hansen JC (2001) Role of histone acetylation in the assembly and modulation of chromatin structures. Gene Expr 9(1):37–61

Araki Y, Mimura T (2017) The histone modification code in the pathogenesis of autoimmune diseases. Mediat Inflamm 2017:2608605

Asselman J, Coninck DIM, De Beert E, Janssen CR, Orsini L, Pfrender ME et al (2017) Bisulfite sequencing with *Daphnia* highlights a role for epigenetics in regulating stress response to Microcystis through preferential differential methylation of serine and threonine amino acids. Environ Sci Technol 51:924–931

Azzi A, Dallmann R, Casserly A, Rehrauer H, Patrignani A, Maier B, Kramer A, Brown SA (2014) Circadian behavior is light-reprogrammed by plastic DNA methylation. Nat Neurosci 17:377–382

Bachman M, Uribe-Lewis S, Yang X, Williams M, Murrell A, Balasubramanian S (2014) 5-Hydroxymethylcytosine is a predominantly stable DNA modification. Nat Chem 6:1049–1055

Balasubramanian MN, Panserat S, Dupont-Nivet M, Quillet E, Montfort J, Le Cam A, Medale F, Kaushik SJ, Geurden I (2016) Molecular pathways associated with the nutritional programming of plant-based diet acceptance in rainbow trout following an early feeding exposure. BMC Genomics 17:449

Bekaert M, Lowe NR, Bishop SC, Bron JE, Taggart JB, Houston RD (2013) Sequencing and characterisation of an extensive Atlantic salmon (*Salmo salar* L.) microRNA repertoire. PLoS One 8(7):e70136

Berger SL (2007) The complex language of chromatin regulation during transcription. Nature 447:407–412

Berger SL, Kouzarides T, Shiekhattar R, Shilatifard A (2009) An operational definition of epigenetics. Genes Dev 23:781–783

Bird A (2007) Perceptions of epigenetics. Nature 447:396–398

Bizuayehu TT, Babiak I (2014) MicroRNA in teleost fish. Genome Biol Evol 6:1911–1937

Bocklandt S, Lin W, Sehl ME, Sánchez FJ, Sinsheimer JS, Horvath S et al (2011) Epigenetic predictor of age. PLoS One 6:e14821

Branco MR, Ficz G, Reik W (2012) Uncovering the role of 5-hydroxymethylcytosine in the epigenome. Nat Rev Genet 13:7–13

Campos C, Valente LMP, Conceição LEC, Engrola S, Fernandes JMO (2013) Temperature affects methylation of the myogenin putative promoter, its expression and muscle cellularity in Senegalese sole larvae. Epigenetics 8:389–397

Chen S, Zhang G, Shao C, Huang Q, Liu G, Zhang P et al (2014) Whole-genome sequence of a flatfish provides insights into ZW sex chromosome evolution and adaptation to a benthic lifestyle. Nat Genet 46:253–260, 90

Conaway JW (2012) Introduction to theme "chromatin, epigenetics, and transcription". Annu Rev Biochem 81:61–64

Daneshfozouna H, Panjvinib F, Ghorbanic F, Farahmandd H (2015) A review of epigenetic imprints in aquatic animals. Fish Aquacult J 6:119

Daxinger L, Whitelaw E (2012) Understanding transgenerational epigenetic inheritance via the gametes in mammals. Nat Rev Genet 13(3):153–162

Deans C, Maggert KA (2015) What do you mean, "epigenetic"? Genet Perspect 199:887–896

Dias BG, Ressler KJ (2014) Parental olfactory experience influences behavior and neural structure in subsequent generations. Nat Neurosci 17:89–96

Dolinoy DC, Weidman JR, Waterland RA, Jirtle RL (2006) Maternal genistein alters coat color and protects mouse offspring from obesity by modifying the fetal epigenome. Environ Health Perspect 114(4):567–572

Fellous A, Favrel P, Riviere G (2015) Temperature influences histone methylation and mRNA expression of the Jmj-C histone-demethylase orthologues during the early development of the oyster *Crassostrea gigas*. Mar Genomics 19:23–30

Feng S, Cokus SJ, Zhang X, Chen P-Y, Bostick M, Goll MG, Hetzel J, Jain J, Strauss SH, Halpern ME, Ukomadu C, Sadler KC, Pradhan S, Pellegrini M, Jacobsen SE (2010) Conservation and divergence of methylation patterning in plants and animals. Proc Natl Acad Sci U S A 107:86898694

Fritsche LG, Chen W, Schu M, Yaspan BL, Yu Y, Thorleifsson G et al (2013) Seven new loci associated with age-related macular degeneration. Nat Genet 45:433–439

Gavery MR, Roberts SB (2017) Epigenetic considerations in aquaculture. Peer J 5:e4147

Geurden I, Borchert P, Balasubramanian MN, Schrama JW, Dupont-Nivet M, Quillet E, Kaushik SJ, Panserat S, Médale F (2013) The positive impact of the early-feeding of a plant-based diet on its future acceptance and utilisation in rainbow trout. PLoS One 8:e83162

Giannetto A, Nagasawa K, Fasulo S, Fernandes JMO (2013) Influence of photoperiod on expression of DNA (cytosine 5) methyltransferases in Atlantic cod. Gene 519:222–230

Goldstein B, Agranat-tamir L, Light D, Zgayer OB, Fishman A, Lamm AT (2017) A-to-I RNA editing promotes developmental stage—specific gene and lncRNA expression. Genome Res 27:462–470

González-Recio O (2012) Epigenetics: a new challenge in the post-genomic era of livestock. Front Genet 2:106

Granada L, Lemos MF, Cabral HN, Bossier P, Novais SC (2018) Epigenetics in aquaculture—the last frontier. Rev Aquac 10(4):994–1013

Green TJ, Helbig K, Speck P, Raftos DA (2016) Primed for success: oyster parents treated with poly (I:C) produce offspring with enhanced protection against Ostreid herpesvirus type I infection. Mol Immunol 78:113–120

Greer EL, Beese-sims SE, Brookes E, Spadafora R, Zhu Y, Rothbart SB et al (2014) A histone methylation network regulates transgenerational epigenetic memory in C. elegans. Cell Rep 7:113–126

Henikoff S, Smith MM (2015) Histone variants and epigenetics. Cold Spring Harbor Perspect Biol 7:a019364

Hernando-Herraez I, Heyn H, Fernandez-Callejo M, Vidal E, Fernandez-Bellon H, Prado-Martinez J et al (2015) The interplay between DNA methylation and sequence divergence in recent human evolution. Nucleic Acids Res 43:8204–8214

Holliday R (1994) Epigenetics: an overview. Dev Genet 15:453–457

Holliday R (2006) Epigenetics: a historical overview. Epigenetics 1:76–90

Huang H, Lin S, Garcia BA, Zhao Y (2015) Quantitative proteomic analysis of histone modifications. Chem Rev 115:2376–2418

Jablonka E, Lamb MJ (2002) The changing concept of epigenetics. Ann N Y Acad Sci 981:82–96

Jablonka E, Raz G (2009) Transgenerational epigenetic inheritance: prevalence, mechanisms, and implications for the study of heredity and evolution. Q Rev Biol 84:131–176

Jamniczky HA, Boughner JC, Rolian CP, Gonzalez N, Powell CD et al (2010) Rediscovering Waddington in the postgenomic age. BioEssays 32:553–558

Jenuwein T, Allis CD (2001) Translating the histone code. Science 293:1074–1081

Jiang Q, Li Q, Yu H, Kong LF (2013) Genetic and epigenetic variation in mass selection populations of Pacific oyster *Crassostrea gigas*. Genes Genomics 35:641–647

Jiang Q, Li Q, Yu H, Kong LF (2016) Inheritance and variation of genomic DNA methylation in diploid and triploid pacific oyster (*Crassostrea gigas*). Mar Biotechnol 18:124–132

Jones PA. (1999) The DNA methylation paradox. Trends Genet 15:3437

Jones P (2012) Functions of DNA methylation: islands, start sites, gene bodies and beyond. Nat Rev Genet 13:484–492

Juanchich A, Bardou P, Rué O, Gabillard J-C, Gaspin C, Bobe J, Guiguen Y (2016) Characterization of an extensive rainbow trout miRNA transcriptome by next generation sequencing. BMC Genomics 17:164

Klosin A, Casas E, Hidalgo-Carcedo C, Vavouri T, Lehner B (2017) Transgenerational transmission of environmental information in *C. elegans*. Science 356(6335):320–323

Konstantinidis I, Sætrom P, Mjelle R, Nedoluzhko AV, Robledo D, Fernandes JM (2020) Major gene expression changes and epigenetic remodelling in Nile tilapia muscle after just one generation of domestication. Epigenetics 15(10):1052–1067

Lämke J, Bäurle I (2017) Epigenetic and chromatin-based mechanisms in environmental stress adaptation and stress memory in plants. Genome Biol 18:124

Lan J, Lepikhov K, Giehr P, Walter J (2017) Histone and DNA methylation control by H3 serine 10/threonine 11 phosphorylation in the mouse zygote. Epigenetics Chromatin 10:5

Le Luyer J, Laporte M, Beacham TD, Kaukinen KH, Withler RE, Leong JS, Rondeau EB, Koop BF, Bernatchez L (2017) Parallel epigenetic modifications induced by hatchery rearing in a Pacific salmon. Proc Natl Acad Sci 114(49):12964–12969

Ma H, Hostuttler M, Wei H, Rexroad CE III, Yao J (2012) Characterization of the rainbow trout egg microRNA transcriptome. PLoS One 7:e39649

Mackenzie R. Gavery, Steven B. Roberts, (2017) Epigenetic considerations in aquaculture. PeerJ 5: e4147

Mann JR (2014) Epigenetics and memigenetics. Cell Mol Sci 71:1117–1122

Marandel L, Lepais O, Arbenoits E, Véron V, Dias K, Zion M, Panserat S (2016) Remodelling of the hepatic epigenetic landscape of glucose-intolerant rainbow trout (*Oncorhynchus mykiss*) by nutritional status and dietary carbohydrates. Sci Rep 6:32187

Metzger DCH, Schulte PM (2016) Epigenomics in marine fishes. Mar Genomics 30:43–54

Moazed D (2009) Small RNAs in transcriptional gene silencing and genome defence. Nature 457:413–420

Moghadam H, Mørkøre T, Robinson N (2015) Epigenetics potential for programming fish for aquaculture? J Marine Sci Eng 3:175–192

Moghadam HK, Johnsen H, Robinson N, Andersen Ø, Jørgensen EH, Johnsen HK, Bæhr VJ, Tveiten H (2017) Impacts of early life stress on the methylome and transcriptome of Atlantic salmon. Sci Rep 7(1):1–11

Moore LD, Le T, Fan G (2013) DNA methylation and its basic function. Neuropsychopharmacol Rev 38:23–38

Moran P, Marco-Rius F, Megıas M, Covelo-Soto L, Perez-Figueroa A (2013) Environmental induced methylation changes associated with seawater adaptation in brown trout. Aquaculture 392–395:77–83

Munday PL (2014) Transgenerational acclimation of fishes to climate change and ocean acidification. F1000 Prime Rep 6:99

Nanney DL (1958) Epigenetic control systems. Proc Natl Acad Sci U S A 44:712–717

Navarro-Martín L, Blázquez M, Viñas J, Joly S, Piferrer F (2009) Balancing the effects of rearing at low temperature during early development on sex ratios, growth and maturation in the European sea bass (*Dicentrarchuslabrax*). Limitations and opportunities for the production of highly female-biased stocks. Aquaculture 296:347–358

Navarro-Martín L, Viñas J, Ribas L, Díaz N, Gutiérrez A, Di Croce L, Piferrer F (2011) DNA methylation of the gonadal aromatase (cyp19a) promoter is involved in temperature-dependent sex ratio shifts in the European sea bass. PLoS Genet 7:e1002447

Norouzitallab P, Baruah K, Vandegehuchte M et al (2014) Environmental heat stress induces epigenetic transgenerational inheritance of robustness in parthenogenetic *Artemia* model. FASEB J 28:3552–3563

Norouzitallab P, Baruah K, Biswas P et al (2016) Probing the phenomenon of trained immunity in invertebrates during a transgenerational study, using brine shrimp Artemia as a model system. Sci Rep 6:1–14

Norouzitallab P, Baruah K, Vanrompay D, Bossier P (2019) Can epigenetics translate environmental cues into phenotypes? Sci Total Environ 647:1281–1293

Ong-Abdullah M, Ordway JM, Jiang N, Ooi S-E, Kok S-Y, Sarpan N, Azimi N, Hashim AT, Ishak Z, Rosli SK, Malike FA, Bakar NAA, Marjuni M, Abdullah N, Yaakub Z, Amiruddin MD, Nookiah R, Singh R, Low E-TL, Chan K-L, Azizi N, Smith SW, Bacher B, Budiman MA, Van Brunt A, Wischmeyer C, Beil M, Hogan M, Lakey N, Lim C-C, Arulandoo X, Wong C-K, Choo C-N, Wong W-C, Kwan YY, Alwee SSRS, Sambanthamurthi R, Martienssen RA (2015) Loss of Karma transposon methylation underlies the mantled somaclonal variant of oil palm. Nature 525:533–537

Paneru B, Al-Tobasei R, Palti Y, Wiens GD, Salem M (2016) Differential expression of long non-coding RNAs in three genetic lines of rainbow trout in response to infection with Flavobacterium psychrophilum. Sci Rep 6:36032

Panserat S, Marandel L, Geurden I, Veron V, Dias K, Plagnes-Juan E, Pegourié G, Arbenoits E, Santigosa E, Weber G, Verlhac Trichet V (2017) Muscle catabolic capacities and global hepatic epigenome are modified in juvenile rainbow trout fed different vitamin levels at first feeding. Aquaculture 468(Part 1):515–523

Parker LM, Ross PM, O'Connor WA, Borysko L, Raftos DA, Pörtner H-O (2012) Adult exposure influences offspring response to ocean acidification in oysters. Glob Chang Biol 18:82–92

Peschansky VJ, Wahlestedt C (2014) Non-coding RNAs as direct and indirect modulators of epigenetic regulation. Epigenetics 9:3–12

Podgorniak T, Brockmann S, Konstantinidis I, Fernandes JM (2019) Differences in the fast muscle methylome provide insight into sex-specific epigenetic regulation of growth in Nile tilapia during early stages of domestication. Epigenetics 14(8):818–836

Potok ME, Nix DA, Parnell TJ, Cairns BR (2013) Reprogramming the maternal zebrafish genome after fertilization to match the paternal methylation pattern. Cell 153:759–772

Radford EJ, Ito M, Shi H, Corish JA, Yamazawa K, Isganaitis E et al (2014) In utero undernourishment perturbs the adult sperm methylome and intergenerational metabolism. Science 345:1255903

Romano G, Veneziano D, Nigita G, Nana-Sinkam SP (2018) RNA methylation in ncRNA: classes, detection, and molecular associations. Front Genet 9:1–9

Roth O, Beemelmanns A, Barribeau SM, Sadd BM (2018) Recent advances in vertebrate and invertebrate transgenerational immunity in the light of ecology and evolution. Heredity (Edinb) 121:225–238

Roy S (2020) Modulating innate immune memory in brine shrimp (Artemia Franciscana) and in giant freshwater prawn (MacrobrachiumRosenbergii). Faculty of Bioscience Engineering, University of Ghent, Ghent

Roy S, Kumar V, Bossier P et al (2019) Phloroglucinol treatment induces transgenerational epigenetic inherited resistance against Vibrioinfectionsand thermal stress in a brine shrimp (Artemia franciscana) model. Front Immunol 10:2745

Roy S, Bossier P, Norouzitallab P, Vanrompay D (2020) Trained immunity and perspectives for shrimp aquaculture. Rev Aquac 12(4):2351–2370

Salinas S, Brown SC, Mangel M, Munch SB (2013) Non-genetic inheritance and changing environments. Non Genet Inheritance 1:38–50

Sarda S, Zeng J, Hunt BG, Yi SV (2012) The evolution of invertebrate gene body methylation. Mol Biol Evol 29:1907–1916

Shao C, Li Q, Chen S, Zhang P, Lian J, Hu Q, Sun B, Jin L, Liu S, Wang Z, Zhao H, Jin Z, Liang Z, Li Y, Zheng Q, Zhang Y, Wang J, Zhang G (2014) Epigenetic modification and inheritance in sexual reversal of fish. Genome Res 24(4):604–615

Siggens L, Ekwall K (2014) Epigenetics, chromatin and genome organization: recent advances from the ENCODE project. J Internal Med 276:201–214

Skinner MK (2007) What is an epigenetic trangenerational phenotype? F3 or F2. Reprod Toxicol 25:2–6

Stadlbauer U (2017) Epigenetic and transgenerational mechanisms in infection-mediated neurodevelopmental disorders. Transl Psychiatry 7:e1113

Tahiliani M, Koh KP, Shen Y, Pastor WA, Bandukwala H, Brudno Y et al (2009) Conversion of 5-methylcytosine to 5-hydroxymethylcytosine in mammalian DNA by MLL partner TET1. Science 324:930–935

Talbert PB, Henikoff S (2010) Histone variants—ancient wrap artists of the epigenome. Nat Rev Mol Cell Biol 11:264–275

Uren Webster TM, Rodriguez-Barreto D, Martin SAM, Van Oosterhout C, Orozco-terWengel P, Cable J, Hamilton A, De Leaniz CG, Consuegra S (2018) Contrasting effects of acute and chronic stress on the transcriptome, epigenome, and immune response of Atlantic salmon. Epigenetics 13(12):1191–1207

Vastenhouw NL, Schier AF (2012) Bivalent histone modifications in early embryogenesis. Curr Opin Cell Biol 24:374–386

Verdone L (2006) Histone acetylation in gene regulation. Brief Funct Genomic Proteomic 5(3):209–221

Waddington CH (1942) The epigenotype. Endeavour 1:18–20

Wang Y, Liu H, Sun Z (2017) Lamarck rises from his grave: parental environment-induced epigenetic inheritance in model organisms and humans. Biol Rev 92(4):2084–2111

Weaver ICG, Cervoni N, Champagne FA, D'Alessio AC, Sharma S, Seckl JR et al (2004) Epigenetic programming by maternal behavior. Nat Neurosci 7:847–854

Wu CT, Morris JR (2001) Genes, genetics, and epigenetics: a correspondence. Science 293:1103–1105

Yu H, Zhao X, Li Q (2016) Genome-wide identification and characterization of long intergenic noncoding RNAs and their potential association with larval development in the Pacific oyster. Sci Rep 6:20796

Yue Y, Liu J, He C (2015) RNA N6-methyladenosine methylation in post-transcriptional gene expression regulation. Genes Dev 29:1343–1355

Zemach A, McDaniel IE, Silva P, Zilberman D (2010) Genome-wide evolutionary analysis of eukaryotic DNA methylation. Science 328:916–919

Zhao L, Schöne BR, Mertz-Kraus R, Yang F (2017) Sodium provides unique insights into transgenerational effects of ocean acidification on bivalve shell formation. Sci Total Environ 577:360–366

Zhi W, Fei Z (2017) MicroRNA-100 is involved in shrimp immune response to white spot syndrome virus (WSSV) and *Vibrio alginolyticus* infection. Sci Rep 7(1):42334

Zhu J-K (2009) Active DNA demethylation mediated by DNA glycosylases. Annu Rev Genet 43:143–166

Application of Stem Cell-Based Technologies in Management of Fisheries Resources

10

Sullip Kumar Majhi

Abstract

The vast worldwide fishery resources require urgent attention to safeguard the fish species from extinction. Recently, various anthropogenic activities like over harvesting, illegal fishing and habitat destruction have led to, unfortunately, extinction of several fish species. This is a serious issue and requires immediate attention; thus, raising the need to preserve the valuable genetic resources of endangered and commercially important fish species. It is noteworthy to mention that captive propagation technique is developed for only few species, especially those are into aquaculture. However, it is not developed for the species those are native to specific ecosystem and attract local consumers very much. To propagate such species and increase their population in nature, breeding technology is not available. Various assisted reproductive technologies have been developed over the years to propagate such fish species. Stem cell-based technologies are such approaches those can play vital role in safeguard such valuable native fish species from being extinct. In this article, all those approaches are discussed at length, those have evolved recently such as germ cell transplantation, in vitro gametogenesis and nuclear transfer technology to propagate such species beyond their reproductive lifespan and rejuvenate fishery of such fish species.

Keywords

Fisheries · Aquaculture · Reproduction · Germ cells · Nuclear transfer technology · Surrogacy

S. K. Majhi (✉)
ICAR-National Bureau of Fish Genetic Resources, Lucknow, Uttar Pradesh, India

© The Author(s), under exclusive license to Springer Nature Singapore Pte Ltd. 2021
P. K. Pandey, J. Parhi (eds.), *Advances in Fisheries Biotechnology*,
https://doi.org/10.1007/978-981-16-3215-0_10

10.1   Introduction

Various anthropogenic activities like overharvesting, illegal fishing and habitat destruction have led to extinction of several fish species. Although numerous efforts have been initiated for their conservation, like establishing closed areas and banned fishing seasons, but improvements in fisheries resources have been rather limited. This is a serious issue and requires immediate attention; raising the need to preserve the valuable genetic resources of endangered and commercially important fish species. An essential tool in this direction is the use of fish stem cells that hold immense potential in aquatic biotechnology and serves as a promising technique for protecting the genome and genetic diversity of aquatic fauna. They can be maintained under defined conditions to develop into stable cell lines.

Stem cells are biological undifferentiated cells that possess the ability to differentiate into specialized cells and divide through the process of mitosis to produce several stem cells (Mitalipov and Wolf 2009). Two major characteristics that make these cells unique from other cells of the body are self-renewal and pluripotency. The ability of self-renewal enables the stem cells to divide and form progeny cells, identical to the parent cells. Pluripotency is yet another unique property that has the potential to produce several specialized cells by undergoing differentiation. The stem cells are broadly categorized into (1) embryonic stem cells, those are isolated from the inner cell mass of blastocysts and, (2) adult stem cells those are found in the tissues and act as a repairing system for the body.

In fishes, research on stem cell culture started more than 20 years ago and most of the works were performed on model fish species such as medaka (*Oryzias latipes*) and zebrafish (*Danio rerio*). Recently, successful production of whole animals from haploid stem cells was reported in medaka. Among vertebrates, embryonic stem cell cultures were first attempted in fishes. Embryonic stem cell cultures have been reported in zebra fish and marine fish species like Asian sea bass, gilthead sea bream and sea perch by Parameswaran et al. (2007), Bejar et al. (2002) and Chen et al. (2007), respectively. Culture of fish germ cells and their transplantation into allogeneic and xenogeneic recipients have gained considerable interest for surrogacy and germline transmission.

In this chapter, some applications of fish stem cells and their importance in propagation of difficult to breed species and conservation of commercially important and endangered germ lines are discussed.

10.2   Germ Cell Transplantation

Germ cell transplantation (GCT) is a widely used reproductive tool that has been used recently for teleost fishes for propagation and conservation of vast fishery resources and elite germplasm. This technique has a tremendous advantage over the conventional induced breeding technique, as both the female and male individuals of the target fish species are not required to produce the offspring. Since the spermatogonial cell exhibits sexual plasticity, only the male germinal cell (spermatogonia) can be put to use to produce both the gametes (sperm and egg). This technology, very recently, has gained incredible scientific interest because of its immense potential for application in animal reproduction, production of transgenic animals, development of reproductive medicine and preservation of the genetic stock of endangered and valuable fish species (Khaira et al. 2005; McLean 2005; Brinster 2007).

In this technique, the donor-derived germ cells are transplanted into the sterile recipient's gonads through the genital papilla, using surgical and non-surgical interventions, where they proliferate, differentiate and produce functional surrogate gametes that are capable of producing viable offspring and transferring genetic information to subsequent generations through natural spawning or artificial insemination (Brinster 2002; Shinohara et al. 2006; Majhi et al. 2014). Several researchers have demonstrated germ cell transplantation in fishes (Takeuchi et al. 2004; Okutsu et al. 2007; Lacerda et al. 2008; Majhi et al. 2009). Primordial germ cells (PGCs) can also be converted into functional gametes by transplanting them into a recipient fish. On attaining maturity, the recipient fish is able to produce donor-origin gametes through artificial and natural spawning. Recent studies have shown that transplantation of PGC into the coelomic cavity of fish hatchlings and/or the blastodisc of embryos allows the donor cells to migrate towards and colonize the genital ridges of the recipient species (Takeuchi et al. 2003; Saito et al. 2008). The newly hatched larvae are usually preferred as recipients because immediately after hatching, they do not have a mature immune system (Manning and Nakanishi 1996) and hence, are unable to reject the donor germ cells, thus, allowing them to survive and then differentiate into mature gametes in their allogeneic gonads, producing normal and viable offspring. Takeuchi et al. (2004) successfully worked on the xenogeneic transplantation of PGCs of rainbow trout into gonads of newly hatched larvae of masu salmon which produced functional sperms derived from the donor rainbow trout. This technique has been applied on several fish species like nibe croaker (*Nibea mitsukurii*) (Yoshikawa et al. 2017), tilapia (*Oreochromis niloticus*) (Farlora et al. 2014) and yellowtail (*Seriola quiqueradiata*) (Morita et al. 2012). Considerable works related to transplantation of migration-stage PGCs into the blastodisc of embryos have also been performed by researchers (Saito et al. 2008; Yasui et al. 2011; Goto et al. 2012; Li et al. 2016).

This method of using fish hatchlings or embryos as recipients require sophisticated instruments and skills for cell transplantation into peritoneal cavity of the small sized larvae or the blastodisc of embryos. Furthermore, transplantation at such early stages takes considerably longer duration to reach adulthood and produce donor-derived functional gametes, adding heavily to the cost of offspring production from surrogate parents. To overcome this issue, development of surrogate broodstock, involving transplantation of gonial cells (oogonia or spermatogonia) into sexually mature hosts is found suitable where the endogenous germ cells are depleted naturally or experimentally (Majhi et al. 2008) to generate donor-derived gametes in a short duration.

Generally, the efficiency and success of GCT depends upon the recipient species where endogenous gametogenesis is naturally absent (such as triploids) or removed experimentally (Dobrinski 2005; Honaramooz et al. 2005; McLean 2005; Majhi et al. 2008), which provides niches for donor germ cells to colonize and develop. For this, several approaches like chemotherapy and X-ray or gamma-ray local radiation are used to deplete endogenous gametogenesis mainly in mammals (Ogawa et al. 1999; Brinster et al. 2003). Since the gonads in fishes are located in the coelomic cavity, application of local radiation is possible only by surgical intervention. Vilela et al. (2003) worked on tilapia and used different concentrations of busulfan drug along with different temperatures to deplete endogenous spermatogenesis. The study revealed that high temperatures help accelerate the spermatogenesis in the fish thus, speeding up the mitotic activity in germ cells and facilitating efficacy of busulfan drug. The invention of GCT technique in fish has created a new research avenue in reproductive biotechnology.

This novel technique can also be applied to various other fields such as conservation or recovery of endangered fish species and production of transgenic individuals. For conservation, apart from utilizing freshly isolated germline cell, it is also possible that the cells can be stored in cryogenic temperature for longer duration and then retrieved later for transplantation. Transplantation of germ cells from commercially important fishes to recipient species, those are prolific breeders and undergo sexual maturation more rapidly, can be performed, thereby, allowing the production of functional gametes and progeny of such species in a considerably short time.

10.3   In Vitro Gametogenesis

In vitro gametogenesis (IVG) is the technique of producing gametes of both male and female sexes in laboratory conditions. The fish gametes (sperm and egg), produced by in vitro gametogenesis through differentiation of female and male germ cell line, are fertilized to produce normal offspring. This reproductive technique offers potential benefits in research as well as in fisheries development. This approach of IVG aims to improve our current knowledge of gametogenesis process (development of gametes) and factors leading to infertility. IVG could also be used to expand and accelerate the genetic selection of off-springs with desirable traits. Successful in vitro sperm production is reported in several fish species. However, fish oocyte production remains a challenge for scientists. If oocyte development could be accomplished in vitro in the future, the same could simplify the unrestricted development of offspring from valuable fish species well beyond their capacity. The advantage of this technique is that gametes can be produced in vitro under human control and reliance on fish breeding season will not be necessary. Reproductive technologies like these will make the production of fish seeds very easy in the coming years and will produce the seeds as and when required for propagation of commercially important fishes and conservation of endangered germ lines.

Male germ cells, present in the testis, produce sperms which transmit genetic information between generations throughout the adult life of most animals. Male germ cells derived from PGCs that split during early developmental stage from somatic cells and then move towards the embryonic gonad, ultimately become gonocytes (Wylie 1999; Starz-Gaiano and Lehmann 2001). As the process of spermatogenesis starts, gonocytes of testis or pro spermatogonia resumes proliferation and becomes undifferentiated type-A spermatogonia, the male germ stem cells which are self-renewed to preserve the stem cell pool or differentiate into fertile sperm by meiosis (Hecht 1998; Brinster 2002). Since spermatogenesis presents a unique opportunity to conduct basic research and monitor animal reproduction and, more importantly, provides an excellent mechanism for genetic manipulation of the germ line in vertebrate organisms, efforts are increasingly being put into the in vitro recapitulation of this method.

Spermatogenesis is a complex process and is challenging to artificially rebuild it into in vitro. This includes creating a similar micro-environment to the testis environment, supporting endocrine and paracrine signals and ensuring the survival of somatic and germ cells from spermatogonial stem cells to mature sperm cells. In vitro spermatogenesis has the potential to produce and collect a large number of sperms, offering an opportunity to select the best sperm like cells both genetically and morphologically for fertilization. In addition, an in vitro sperm production system could help in determining the biological mechanisms in sperm production, such as hormones and growth factors, as well as epigenetic modifications, leading to problems in accessing and manipulating the testis environment in vivo. Specifically, in vitro production of sperm cells from spermatogonia has been observed in Japanese eel (*Anguilla japonica*) (Miura et al. 1996), medaka (*Oryzias latipes*) (Hong et al. 2004), and zebra fish (Sakai 2002; Kawasaki et al. 2016). However, such sperms were found to be fertile only for zebra fish, from freshly prepared spermatogonia as documented by Sakai (2002) or using SSCs grown for up to 1 month (Kawasaki et al. 2016). According to Miura et al. (1996), all stages involved in spermatogenesis of eels have been well established in organ culture of testes that are immature. In medaka species, fertile sperms were found at meiotic prophase during 10 days of primary culture of the spermatocytes (Sakai 1976). Success was achieved in deriving a normal spermatogonic cell line from adult medaka testis and in recapitulating spermatogenesis from that line to produce motile in vitro sperms. Medaka is one of the unique vertebrates where fertile sperm could be obtained in vitro condition even in the absence of any somatic cells as reported by Sakai (1976). This fish proved to be an excellent model for studying vertebrate development (Wittbrodt et al. 2002) as well as allowed for derivation of embryonic stem cell lines (Hong et al. 1996, 1998).

10.4 Induced Pluripotent Stem Cells (iPSCs)

Induced pluripotent stem cells are formed by direct molecular reprogramming from somatic cells and ES cells which are capable of limitless expansion and differentiation into multiple cell types. Due to rapidly increasing interest in genetic resource banking—cryopreservation, storage and usage of biomaterials (Wildt et al. 1997), the concept of iPSCs is appropriate. It is a unique way of reprogramming rare species, in order to preserve the genetic material of endangered animal species.

Since they are pluripotent and self-renewing, iPSCs obtained from endangered species have numerous applications (Trounson 2009) and can be means to rescue species from being extinct: preserving the genomes of individual animals, as pluripotent stem cells provide the possibility of producing iPSC-derived germ cells, which could then be used along with assisted reproduction methods to increase the diversity and size of the populations.

Although originally intended as repositories for germplasm (mostly sperm) or blood and entire tissues (i.e. genotyping, paternity analysis and forensic disease), genome resource banking also included developed somatic cells like fibroblasts. Since developed fibroblasts are senesce and may eventually become depleted, their availability is limited. iPSCs alternative (with suitable conditions for culture) may provide a self-renewing, inexhaustible resource of wildlife content. Thus, frozen stocks could be harvested sustainably, regenerating cells as required. iPSCs apparently can be produced relatively easily across species and is, therefore, potentially important for advances in assisted reproductive technologies. Ben-Nun et al. (2011) suggested that gametes may be derived from iPSCs in the future, though the amount of new knowledge needed to achieve this is formidable. Mouse ESCs were differentiated into oocytes and sperms, though with modest success (Daley 2007). The benefits of achieving this would be incalculable for iPSCs from endangered species, particularly in terms of the potential to create an inexhaustible supply of haploid gametes. Generating iPSC-derived sperm from samples of frozen somatic cells from long-dead animals will provide a way to infuse much-needed genetic diversity, using already validated methods of fertilization. An analogous method using iPSC-derived oocytes may provide an endless tool for basic research in in vitro fertilization, in vitro maturation of egg, intra-cytoplasmic sperm injection, and reproductive strategies, based on SCNT.

iPSCs were first developed by retrovirus-mediated induction from mouse somatic cells with four core reprogramming factors: Oct3/4, Sox2, Klf4, and c-Myc (Takahashi and Yamanaka 2006). The development of iPS cells paved way for fundamental and translational research. Efficient and robust reprogramming techniques have been developed and improved in mammals over the past 13 years, such as drug-inducing expression of defined factors (Bao et al. 2011), mRNA, miRNA or protein transfection as reprogramming factors (Kim et al. 2009; Zhou et al. 2009; Warren et al. 2010; Anokye-Danso et al. 2011), combination of chemical and genetic approaches (Huangfu et al. 2008; Shi et al. 2008; Lyssiotis et al. 2009) and direct reprogramming by small-molecule compounds (Li et al. 2017). However, the above mentioned iPSCs approaches are limited to mammals and there are few reports available for non-mammalian species (Rossello et al. 2013; Fuet and Pain 2017). Research conducted by Rossello et al. (2013) revealed that transcription factors from mammals may be used to produce iPSCs from embryonic fibroblasts of zebrafish. However, no iPS-like cell line has been produced from adult fish somatic cells in feeder-free environment for zebrafish. Several factors and pathways have recently been found to regulate cardiomyocyte regeneration in zebrafish, like Cdk9, Caveolin-1, nuclear factor kB, Neuregulin 1 (Nrg1), microRNA-101a, and bone morphogenetic protein signalling (Bednarek et al. 2015; Beauchemin et al.

2015; Gemberling et al. 2013; Matrone et al. 2015; Cao et al. 2016; Karra et al. 2015; Wu et al. 2016; Uygur and Lee 2016). Spontaneously, in vitro heart aggregates were generated from zebrafish larvae 3 days post fertilization with an efficiency of 0.4 heart aggregate per larva as reported by Grunow et al. (2015).

Several methods of pluripotent cell derivation exist. (a) ES cells can be obtained directly from the inner cell mass of pre-implanted blastocyst. (b) Somatic cell nuclear transfer is another technique in which adult somatic nuclei are microinjected into enucleated eggs. The eggs, that now contain the somatic cell nucleus, are stimulated by shock and the nucleus of somatic cell is then reprogrammed by the host egg cell that leads to formation of blastocyst. (c) Cellular hybridization involving formation of pluripotent hybrid cells via fusion of somatic cell with an embryonic stem cell. (d) Induced pluripotent stem cells which are developed from patient-specific SCs reprogrammed to an embryonic stem cell-like state.

iPSC technology is applicable as a revolutionary new fishing and aquaculture technology i.e. rejuvenating the IUCN red listed species, generating iPS cells from selected fish strains and creating clone individuals from selected lines, and cryopreserving significant fish strains as cell lines. There are several diseases modelled by using iPSCs to better understand their etiology, which may have been used further to improve putative fish disease therapies. Also, iPSCs might be used in the study of molecular mechanisms for many fish diseases. iPSC technology is a useful tool for some IUCN red listed species to preserve adequate biodiversity and encourage re-introduction of genetic material into populations to prevent natural extinction. So, it can be affirmed that iPS cells will be providing a powerful new tool for future fisheries and aquaculture production.

10.5   Somatic Cell Nuclear Transfer (SCNT)

Somatic Cell Nuclear Transfer (SCNT) is a cloning technique in which the nucleus of a SC is transferred to an enucleated egg's cytoplasm. The cytoplasmic factors affect the nucleus to become a zygote when this is done. The nucleus of the donor egg cell is removed, leaving it 'deprogrammed' now what are left are a somatic cell and an enucleated egg cell. They are then fused into the 'empty' ovum by inserting the somatic cell into it. The somatic cell nucleus is reprogrammed by its host egg cell, after being inserted into the egg. The ovum, which now contains the nucleus of the SC, is stimulated giving shock which then begins to divide. The egg is now viable and capable of creating an adult organism that contains all the genetic material, needed by only one parent. Development will normally ensure that this single cell forms a blastocyst (an early stage embryo of about 100 cells) with an identical genome to the original organism (i.e. clone) after several mitotic divisions (Wilmut et al. 1997).

Interspecies nuclear transfer (iSCNT) is a method of somatic cell nuclear transfer, used to promote the rescue of endangered species, or even after their extinction, to regenerate them. The technique is similar to SCNT cloning which usually is between farm animals and rodents, or where there is a ready supply of oocytes and surrogate animals. However, the cloning of highly endangered or extinct organisms involves the use of an alternate form of cloning. Nuclear fusion of interspecies involves a host and a donor of two separate organisms which are closely related species and are in the same genus. The theory of somatic cell nuclear transfer (SCNT) is that when transplanted into the stable environment of a recipient egg, the differentiated donor cell could be restored to full totipotency (Zhu and Sun 2000). However, the cloning success has been very limited in fish, even in model species like zebra fish, *Danio rerio*. Since the first adult cloned zebra fish, obtained by Lee et al. (2002) from long-term cultured fibroblast cells, the success of more than 2% has not been achieved using somatic cells. In general, fish SCNT is mostly unsuccessful in producing living adult clones (Luo et al. 2011). In addition, the iSCNT teleost only yields poor early embryonic growth. It is the case of gynogenetic bighead carp cloning, *Aristichthys nobilis* using gibel carp, *Carassius auratus gibelio* triggered eggs resulting in 27% progress before blastula stage (Liu et al. 2002). In order to perform the cloning technique, some basic steps must first be established. For instance, optimization of donor cell preparation is one crucial step because cell viability is the primary factor for a successful SCNT. A piece of fin has a high potential for regeneration (Akimenko et al. 2003) and is readily accessible, causing minimal damage to the fish (Labbé et al. 2013; Akimenko et al. 2003). In the case of critically endangered fish species such as sturgeon, this is particularly important. In addition, the fin-tissue can be harvested well before the individual's sexual maturation (Chenais et al. 2014), and this is particularly important for sturgeon species such as beluga, whose first maturation typically occurs at 20 years. Another important question for SCNT is whether an extender solution is needed for both egg washing and the working medium. In sturgeons, the eggs need to be washed before micromanipulation since they come with high viscosity coelomic fluid that includes loads of somatic ovarian and blood cells. To avoid the risk of transplanting an ovarian somatic cell into the recipient egg, the eggs must first be washed with a physiological saline solution. Placing both donor cells and recipient eggs in the same petridish with the extender solution is convenient during the SCNT process. The medium should therefore not be toxic to either of them, and should not cause or improve egg activation (Le Bail et al. 2010). Only after SCNT, it is important to incubate the nuclear transplants (NTs) for a certain amount of time before it is activated, so that the donor nucleus can be given the opportunity to stabilize and reprogram in the new environment (Chenais et al. 2014). It seems better to use non-enucleated eggs for sturgeon cloning, because enucleating sturgeon eggs can be dangerous and requires time and complicated procedures. Even putative spontaneous enucleation of the eggs (Wakamatsu 2008) may result in the restored embryo development normally (Meissner and Jaenisch 2006). According to Le Bail et al. (2010), micropyle transplantation with no activation of the eggs seems to increase the efficiency of cloned fish development. For this case, however, a micro needle has to be designed for penetration of the egg chorion. Finding the correct way of inserting the micro needle, and at which depth within the egg, is a real challenge due to the big size of sturgeon eggs; e.g. starlet egg, 1.8–2.8 mm, Russian sturgeon egg, 2.8–3.8 mm

(Hochleithner and Gessner 2012). Nevertheless, the position of the microinjection is very important, since the donor fin-nucleus must be at a favourable location in the host environment for reprogramming and the production of a cloned sturgeon. The Russian sturgeon listed as critically endangered (IUCN 2017) is the most favoured for caviar consumption within the Acipenseridae family (Hochleithner and Gessner 2012). Despite the advantage of a large amount of eggs per kg of body-weight (10,000–15,000), the drawback of the species is the late age at which the first gonad maturation is settled (Birstein et al. 1997; Vasilev et al. 2009; Gomez et al. 2006; Fontana et al. 1998; Hochleithner and Gessner 2012). As a recipient species, the starlet is promising. In fact, while it is listed as a vulnerable species, it is considered a model species for Acipenseridae as it is one of the smaller sturgeon species and can be easily evaluated in fish farms. The most significant advantage is that it achieves its sexual maturity faster than other sturgeon species (Gomez et al. 2006; Loi et al. 2001; Lanza et al. 2000). In addition, spawning occurs in both sexes in cycles of (Wildt 1992; Wells et al. 1998) years, which is even more common than other species within the same family (Hochleithner and Gessner 2012). Tung and colleagues in China initiated the study of nuclear transfer in fish in the 1960s (Di Berardino 1997; Zhu and Sun 2000) and since the 1970s, researchers in China have extensively documented nucleo-cytoplasmic hybrid development by transplanting the embryonic cell nuclei of one species into enucleated eggs of another species, primarily in cyprinid fish (Yan 1989). Gasaryan et al. (1979) obtained nuclear transplants which, by transferring embryonic cell nuclei into non-enucleated and enucleated loach eggs (*Misgurnus fossilis*), formed into feeding larvae. Diploid and fertile adult fish were formed from the nuclear fusion of blastula nuclei into enucleated (Wakamatsu et al. 2001) and non-enucleated (Bubenshchikovae Ju et al. 2005) unfertilized eggs in medaka. Adult zebra fish were cloned from cultured embryonic cell nuclei by Lee et al. (2002). In all of these studies, embryonic cell nuclei, but not adult somatic cell nuclei, were used as donors. Bubenshchikova et al. (2007) reported the first successful nuclear transfer using adult somatic cells in medaka.

Nuclear transfer has the potential to become one method for the management of fish genetic resources by enabling the regeneration of fish from cryopreserved somatic cells. However, survival rates are still low after nuclear transfer due to both biological and technical constraints. Among them, the difficulty to enucleate the recipient oocyte in fish, together with the need to alleviate some of the alterations associated with the enucleation methods, led several authors to inject the donor nucleus into non-enucleated oocytes (Bubenshchikovae Ju et al. 2005; Gasaryan et al. 1979; Siripattarapravat et al. 2009). Strikingly, diploid clones were produced which exhibited genetic characteristics of the donor fish. Wakamatsu and his team demonstrated, moreover, that diploidized activated eggs were a good template for the development of somatic cell nuclear transplants (Wakamatsu 2008), although the mechanism underlying a putative spontaneous erasure of the maternal genome remains to be explored. The application of interspecific somatic cell nuclear transfer (iSCNT) to endangered fish species has a great benefit, as reconstruction of critically endangered species can be achieved after transplantation of a single fin-cell into the egg-cytoplasmic environment of species whose eggs are easily accessible in farms.

References

Akimenko MA, Mari-Beffa M, Becerra J et al (2003) Old questions, new tools, and some answers to the mystery of fin regeneration. Dev Dyn 226(2):190–201

Anokye-Danso F, Trivedi CM, Juhr D et al (2011) Highly efficient miRNA-mediated reprogramming of mouse and human somatic cells to pluripotency. Cell Stem Cell 8 (4):376–388

Bao L, He L, Chen J et al (2011) Reprogramming of ovine adult fibroblasts to pluripotency via drug-inducible expression of defined factors. Cell Res 21(4):600–608

Beauchemin M, Smith A, Yin VP (2015) Dynamic microRNA-101a and Fosab expression controls zebrafish heart regeneration. Development 142(23):4026–4037

Bednarek D, González-Rosa JM, Guzmán-Martínez G et al (2015) Telomerase is essential for zebrafish heart regeneration. Cell Rep 12(10):1691–1703

Bejar J, Hong Y, Alvarez MC (2002) An ES-like cell line from the marine fish *Sparus aurata*: characterization and chimaera production. Trans Res 11(3):279–289

Ben-Nun IF, Montague SC, Houck ML et al (2011) Induced pluripotent stem cells from highly endangered species. Nat Methods 8(10):829

Birstein VJ, Bemis WE, Waldman JR (1997) The threatened status of acipenseriform species: a summary. In: Sturgeon biodiversity and conservation. Springer, Dordrecht, pp 427–435

Brinster RL (2002) Germline stem cell transplantation and transgenesis. Science 296 (5576):2174–2176

Brinster RL (2007) Male germline stem cells: from mice to men. Science 316(5823):404–405

Brinster CJ, Ryu BY, Avarbock MR et al (2003) Restoration of fertility by germ cell transplantation requires efficient recipient preparation. Biol Reprod 69:412–420

Bubenshchikova E, Kaftanovskaya E, Motosugi N (2007) Reprogramming of adult somatic cell nuclei to pluripotency in fish, medaka (*Oryzias latipes*), by a novel method of nuclear transfer using diploidized eggs as recipients. Develop Growth Differ 49:699–709

Bubenshchikovae Ju B, Pristyazhnyuk I, Niwa K et al (2005) Generation of fertile and diploid fish, medaka (*Oryzias latipes*), from nuclear transplantation of blastula and four-somite-stage embryonic cells into nonenucleated unfertilized eggs. Cloning Stem Cells 7:255–264

Cao J, Navis A, Cox BD et al (2016) Single epicardial cell transcriptome sequencing identifies Caveolin-1 as an essential factor in zebrafish heart regeneration. Development 143:232–243

Chen SL, Sha ZX, Ye HQ et al (2007) Pluripotency and chimera competence of an embryonic stem cell line from the sea perch (*Lateolabrax japonicus*). Mar Biotechnol 9(1):82–91

Chenais N, Depince A, Le Bail PY et al (2014) Fin cell cryopreservation and fish reconstruction by nuclear transfer stand as promising technologies for preservation of finfish genetic resources. Aquac Int 22(1):63–76

Daley GQ (2007) Gametes from embryonic stem cells: a cup half empty or half full? Science 316 (5823):409–410

Di Berardino MA (1997) Nuclear potential of mammalian cells. In: Genomic potential of differentiated cells. Columbia University Press, New York, pp 180–213

Dobrinski I (2005) Germ cell transplantation and testis tissue xenografting in domestic animals. Anim Reprod Sci 89(1–4):137–145

Farlora R, Hattori-Ihara S, Takeuchi Y et al (2014) Intraperitoneal germ cell transplantation in the Nile tilapia *Oreochromis niloticus*. Mar Biotechnol 16:309–320

Fontana F, Tagliavini J, Congiu L et al (1998) Karyotypic characterization of the great sturgeon, Huso huso, by multiple staining techniques and fluorescent in situ hybridization. Mar Biol 132 (3):495–501

Fuet A, Pain B (2017) Chicken induced pluripotent stem cells: establishment and characterization. In: Avian and reptilian developmental biology. Humana Press, New York, pp 211–228

Gasaryan KG, Hung NM, Neyfakh AA et al (1979) Nuclear transplantation in teleost *Misgurnus fossilis*. Nature 280:585–587

Gemberling M, Bailey TJ, Hyde DR, Poss KD (2013) The zebrafish as a model for complex tissue regeneration. Trends Genet 29(11):611–620

Gomez MC, Pope CE, Dresser BL (2006) Nuclear transfer in cats and its application. Theriogenology 66:72–81

Goto R, Saito T, Takeda T et al (2012) Germ cells are not the primary factor for sexual fate determination in goldfish. Dev Biol 370(1):98–109

Grunow B, Mohamet L, Shiels HA (2015) Generating an in vitro 3D cell culture model from zebrafish larvae for heart research. J Exp Biol 218:1116–1121

Hecht NB (1998) Molecular mechanisms of male germ cell differentiation. BioEssays 20 (7):555–561

Hochleithner M, Gessner J (2012) The sturgeons and paddle fishes (Acipenseriformes) of the world: biology and aquaculture. Createspace Independent Publishing Platform, North Charleston, pp 1–248

Honaramooz A, Behboodi E, Hausler CL et al (2005) Depletion of endogenous germ cells in male pigs and goats in preparation for germ cell transplantation. J Androl 26(6):698–705

Hong Y, Winkler C, Schartl M (1996) Pluripotency and differentiation of embryonic stem cell lines from the medakafish (Oryzias latipes). Mech Dev 60(1):33–44

Hong Y, Winkler C, Schartl M (1998) Production of medaka fish chimeras from a stable embryonic stem cell line. Proc Natl Acad Sci 95(7):3679–3684

Hong Y, Liu T, Zhao H et al (2004) Establishment of a normal medaka fish spermatogonial cell line capable of sperm production in vitro. PNAS 101(21):8011–8016

Huangfu D, Maehr R, Guo W (2008) Induction of pluripotent stem cells by defined factors is greatly improved by small-molecule compounds. Nat Biotechnol 26(7):795–797

International Union for Conservation of Nature (2017) The IUCN red list of threatened species. Version 2017–2. www.iucnredlist.org. Accessed 14 Sep 2017

Karra R, Knecht AK, Kikuchi K, Poss KD (2015) Myocardial NF-κB activation is essential for zebrafish heart regeneration. Proc Natl Acad Sci 112(43):13255–13260

Kawasaki T, Siegfried KR, Sakai N (2016) Differentiation of zebrafish spermatogonial stem cells to functional sperm in culture. Development 143(4):566–574

Khaira H, Mclean D, Ohl DA et al (2005) Spermatogonial stem cell isolation, storage, and transplantation. J Androl 26:442–450

Kim D, Kim CH, Moon JI et al (2009) Generation of human induced pluripotent stem cells by direct delivery of reprogramming proteins. Cell Stem Cell 4(6):472

Labbé C, Robles V, Herraez MP (2013) Cryopreservation of gametes for aquaculture and alternative cell sources for genome preservation. In: Advances in aquaculture hatchery technology. Woodhead Publishing, Cambridge, pp 76–116

Lacerda SM, Batlouni SR, Assis LH et al (2008) Germ cell transplantation in tilapia (Oreachromis niloticus). Cybium 32:115–118

Lanza RP, Cibelli JB, Diaz F et al (2000) Cloning of an endangered species (Bos gaurus) using interspecies nuclear transfer. Cloning 2(2):79–90

Le Bail PY, Depince A, Chenais N et al (2010) Optimization of somatic cell injection in the perspective of nuclear transfer in goldfish. BMC Dev Biol 10(1):64

Lee KY, Huang H, Ju B et al (2002) Cloned zebrafish by nuclear transfer from long-term-cultured cells. Nat Biotechnol 20:795–799

Li M, Hong N, Xu H et al (2016) Germline replacement by blastula cell transplantation in the fish medaka. Sci Rep 6(1):1–10

Li X, Liu D, Ma Y et al (2017) Direct reprogramming of fibroblasts via a chemically induced XEN-like state. Cell Stem Cell 21(2):264–273

Liu TM, Yu XM, Ye YZ et al (2002) Factors affecting the efficiency of somatic cell nuclear transplantation in the fish embryo. J Exp Zool 293:719–725

Loi P, Ptak G, Barboni B et al (2001) Genetic rescue of an endangered mammal by cross-species nuclear transfer using post-mortem somatic cells. Nat Biotechnol 19(10):962

Luo DJ, Hu W, Chen SP et al (2011) Critical developmental stages for the efficiency of somatic cell nuclear transfer in zebrafish. Int J Biol Sci 7(4):476

Lyssiotis CA, Foreman RK, Staerk J et al (2009) Reprogramming of murine fibroblasts to induced pluripotent stem cells with chemical complementation of Klf4. PNAS 106(22):8912–8917

Majhi SK, Hattori RS, Rahman SM et al (2008) Experimentally-induced depletion of germ cells in sub-adult Patagonian pejerrey (*Odontesthes hatcheri*). Theriogenology 71:1162–1172

Majhi SK, Hattori RS, Yokota A, Watanabe S, Strüssmann CA (2009) Germ cell transplantation using sexually competent fish: an approach for rapid propagation of endangered and valuable germline. PLoS One 4(7):e6132

Majhi SK, Hattori RS, Rahman SM, Strüssmann CA (2014) Surrogate production of eggs and sperm by intrapapillary transplantation of germ cells in cytoablated adult fish. PLoS One 9(4): e95294

Manning MJ, Nakanishi T (1996) The specific immune system: cellular defences. In: Iwama G, Nakanishi T (eds) The fish immune system. Academic Press, New York, pp 159–205

Matrone G, Wilson KS, Maqsood S et al (2015) CDK9 and its repressor LARP7 modulate cardiomyocyte proliferation and response to injury in the zebra fish heart. J Cell Sci 128 (24):4560–4571

Mclean D (2005) Spermatogonial stem cell transplantation and testicular function. Cell Tissue Res 322(1):21–31

Meissner A, Jaenisch R (2006) Mammalian nuclear transfer. Dev Dyn 235(9):2460–2469

Mitalipov S, Wolf D (2009) Totipotency, pluripotency and nuclear reprogramming. Adv Biochem Eng Biotechnol 114:185–199

Miura C, Miura T, Yamashita M et al (1996) Hormonal induction of all stages of spermatogenesis in germ-somatic cell coculture from immature Japanese eel testis. Develop Growth Differ 38 (3):257–262

Morita T, Kumakura N, Morishima K (2012) Production of donor-derived offspring by allogeneic transplantation of spermatogonia in the yellowtail (*Seriola quinqueradiata*). Biol Reprod 86 (6):176

Ogawa T, Dobrinski I, Brinster RL (1999) Recipient preparation is critical for spermatogonial transplantation in the rat. Tissue Cell 31:461–472

Okutsu T, Shikina S, Kanno M et al (2007) Production of trout offspring from triploid salmon parents. Science 317:1517

Parameswaran V, Shukla R, Bhonde R et al (2007) Development of a pluripotent ES-like cell line from Asian sea bass (*Lates calcarifer*)—an oviparous stem cell line mimicking viviparous ES cells. Mar Biotechnol 9:766–775

Rossello RA, Chen CC, Dai R et al (2013) Mammalian genes induce partially reprogrammed pluripotent stem cells in non-mammalian vertebrate and invertebrate species. eLife 2:e00036

Saito T, Goto-Kazeto R, Arai K et al (2008) Xenogenesis in teleost fish through generation of germ-line chimeras by single primordial germ cell transplantation. Biol Reprod 78:159–166

Sakai YT (1976) Spermiogenesis of the teleost, Oryzias latipes, with special reference to the formation of flagellar membrane. Dev Growth Differ 18:1–13

Sakai N (2002) Transmeiotic differentiation of zebrafish germ cells into functional sperm in culture. Development 129(14):3359–3365

Shi Y, Do JT, Desponts C et al (2008) A combined chemical and genetic approach for the generation of induced pluripotent stem cells. Cell Stem Cell 2(6):525–528

Shinohara T, Kato M, Takehashi M et al (2006) Rats produced by interspecies spermatogonial transplantation in mice and in vitro microinsemination. PNAS 103(37):13624–13628

Siripattarapravat K, Pinmee B, Venta PJ et al (2009) Somatic cell nuclear transfer in zebrafish. Nat Methods 6:733–735

Starz-Gaiano M, Lehmann R (2001) Moving towards the next generation. Mech Dev 105 (1–2):5–18

Takahashi K, Yamanaka S (2006) Induction of pluripotent stem cells from mouse embryonic and adult fibroblast cultures by defined factors. Cell 126:663–676

Takeuchi Y, Yoshizaki G, Takeuchi T (2003) Generation of live fry from intraperitoneally transplanted primordial germ cells in rainbow trout. Biol Reprod 69(4):1142–1149

Takeuchi Y, Yoshizaki G, Takeuchi T (2004) Surrogate broodstock produces salmonids. Nature 430:629

Trounson A (2009) Rats, cats, and elephants, but still no unicorn: induced pluripotent stem cells from new species. Cell Stem Cell 4:3–4

Uygur A, Lee RT (2016) Mechanisms of cardiac regeneration. Dev Cell 36(4):362–374

Vasilev VP, Vasileva ED, Shedko SV et al (2009) Ploidy levels in the Kaluga, Huso dauricus and Sakhalin sturgeon *Acipenser mikadoi* (Acipenseridae, Pisces). Biol Sci 426(1):228–231

Vilela DAR, Silva SGB, Peixoto MTD et al (2003) Spermatogenesis in teleost: insights from the nile tilapia (*Oreochromis niloticus*) model. Fish Physiol Biochem 28(1–4):187–190

Wakamatsu Y (2008) Novel method for the nuclear transfer of adult somatic cells in medaka fish (*Oryzias latipes*): use of diploidized eggs as recipients. Dev Growth Differ 50(6):427–436

Wakamatsu Y, Ju B, Pristyaznhyuk I (2001) Fertile and diploid nuclear transplants derived from embryonic cells of a small laboratory fish, medaka (*Oryzias latipes*). PNAS 98(3):1071–1076

Warren L, Manos PD, Ahfeldt T et al (2010) Highly efficient reprogramming to pluripotency and directed differentiation of human cells with synthetic modified mRNA. Cell Stem Cell 7 (5):618–630

Wells DN, Misica PM, Tervit HR et al (1998) Adult somatic cell nuclear transfer is used to preserve the last surviving cow of the Enderby Island cattle breed. Reprod Fertil Dev 10(4):369–378

Wildt DE (1992) Genetic resource banks for conserving wildlife species: justification, examples and becoming organized on a global basis. Anim Reprod Sci 28(1–4):247–257

Wildt DE, Rall WF, Critser JK (1997) Genome resource banks. Bioscience 47(10):689–698

Wilmut I, Schnieke AE, McWhir J et al (1997) Viable offspring derived from fetal and adult mammalian cells. Nature 385:810–813

Wittbrodt J, Shima A, Schartl M (2002) Medaka—a model organism from the far east. Nat Rev Genet 3(1):53–64

Wu C, Kruse F, Vasudevarao MD et al (2016) Spatially resolved genome-wide transcriptional profiling identifies BMP signaling as essential regulator of zebrafish cardiomyocyte regeneration. Dev Cell 36:36–49

Wylie C (1999) Germ cells. Cell 96(2):165–174

Yan SY (1989) The nucleo–cytoplasmic interactions as revealed by nuclear transplantation in fish. In: Malacinski GM (ed) Cytoplasmic organization systems: a primer in developmental biology. McGraw-Hill, New York, pp 61–81

Yasui GS, Fujimoto T, Sakao S et al (2011) Production of loach (*Misgurnus anguillicaudatus*) germ-line chimera using transplantation of primordial germ cells isolated from cryopreserved blastomeres. J Anim Sci 89:2380–2388

Yoshikawa H, Takeuchi Y, Ino Y (2017) Efficient production of donor-derived gametes from triploid recipients following intra-peritoneal germ cell transplantation into a marine teleost, Nibe croaker (*Nibea mitsukurii*). Aquaculture 478:35–47

Zhou H, Wu S, Joo JY (2009) Generation of induced pluripotent stem cells using recombinant proteins. Cell Stem Cell 4(5):381–384

Zhu ZY, Sun YH (2000) Embryonic and genetic manipulation in fish. Cell Res 10(1):17

Molecular Markers in Aquaculture

11

Ananya Khatei, Partha Sarathi Tripathy, and Janmejay Parhi

Abstract

The advent of molecular markers has revolutionized aquaculture through several genetic improvement programs. Since there is a growing demand for fish by the growing human population, the reduction in the fish genetic diversity can lead to its extermination. Molecular markers have proved beneficial not only in breeding programmes but also in planning conservation strategies based on genetic diversities of various fish species. These markers have been classified in various ways depending on the properties they possess such as dominant and codominant markers, type I and type II markers and linked and direct markers. The most commonly used markers in aquaculture include allozymes, Restriction Fragment Length Polymorphism (RFLP), Random Amplified Polymorphic DNA (RAPD), Amplified Fragment Length Polymorphism (AFLP), microsatellites, Expressed Sequence Tag (EST) and Single Nucleotide Polymorphism (SNP). SNPs are the most widely used markers in fish population genetics study since they are found in abundance. Microsatellite markers have also been developed in several commercially important fish species for desirable traits like growth enhancement, disease resistance, etc. Microsatellites have high Polymorphism Information Content (PIC) values for which they are considered very reliable. These molecular markers have application while comparing the wild and hatchery fish

The original version of this chapter has been updated: The authors Ananya Khatei and Janmejay Parhi affiliation has been updated. A Correction to this chapter is available at https://doi.org/10. 1007/978-981-16-3215-0_30

A. Khatei · J. Parhi (✉)
Fish Genetics and Reproduction, College of Fisheries, CAU (I), Lembucherra, Tripura, India
e-mail: Parhi.fgr.cof@cau.ac.in

P. S. Tripathy
Faculty of Biosciences and Aquaculture, Nord University, Bodø, Norway

© The Author(s), under exclusive license to Springer Nature Singapore Pte Ltd. 2021, corrected publication 2022
P. K. Pandey, J. Parhi (eds.), *Advances in Fisheries Biotechnology*,
https://doi.org/10.1007/978-981-16-3215-0_11

population. Hence for a sustainable management of fish resources it is imperative to utilize molecular markers.

Keywords

Allozymes · Breeding · Conservation · Diversity · Microsatellites

11.1 Introduction

Polymorphism is a consequence of certain inevitable mutations that take place during normal cellular development and environmental interactions. These polymorphisms bring about genetic diversity within a population. The types of variations include base substitutions, insertions, deletions or inversions within a locus of the interest. In the biological terms, markers are defined as any stable variation, which is heritable and can be measured or detected by an appropriate method. In genome, molecular marker occupies specific location and is linked to a specific gene or the inheritance of a particular character. Because of this role, it is widely used for the detection of the genetic variations within and between the populations.

A broad genetic diversity is a principal base for the selection of individuals or breeds. The detection of such diversity at an individual, family or population level has been possible due to the development of such molecular markers. These molecular markers are efficient aids to study the uniqueness of an individual as well as of population. These molecular approaches have made the selection of live stocks such as fishes, cattle, etc., easier. Molecular markers are broadly categorized into Type I and Type II markers. The former being a gene of known function and the latter is associated with random/unknown genomic regions. The details regarding these markers will be described subsequently in this chapter. The analysis of diversity and uniqueness, using various markers help in organizing various genetic improvement programmes via selection which is carried out using F2, backcrossing, near-isogenic lines and recombinant inbred lines. Considering the diverse range of traits some selection choices may have confusing outcomes like alleles of a particular gene may be beneficial for one trait but may have an adverse effect on others. Some genes of different traits are at times located so close that it may appear that they are controlled by the same locus because they are inherited together. Knowing such alleles can help in identifying individuals with a beneficial trait for selection.

Based on linked genes, the markers can be categorized as linked marker and unlinked or direct marker. The markers, that are so close to the gene of the interest and are inherited together, are called as linked markers. The association between the allele and the marker needs to be known in order to work at a population level. The second type of the marker is a functional polymorphic gene with variations in that particular trait. They are called direct markers.

Another major classification of the markers divides them into two groups' i.e. the codominant markers and the dominant markers. (a) **Codominant markers**: These are the markers that are able to distinguish between homozygous and heterozygous loci. When subjected to gel electrophoresis, it produces separate bands for dominant and recessive alleles and thereby determine the genotype at a loci. Codominant markers are more useful in population genetic analysis. Restriction Fragment Length Polymorphism (RFLP) and Allozymes are two major examples of the codominant markers.

(b) **Dominant markers**: The dominant markers exhibit a scenario that can be analyzed only as the presence or absence of fragments with a particular size. In this case, allele frequency can only be derived by presuming Hardy-Weinberg equilibrium. The examples of dominant markers include Random amplification of Polymorphic DNA (RAPD) and Amplified Fragment Length Polymorphism (AFLP).

A gene having a known function is considered as a **Type I marker**. Restriction Fragment Length Polymorphism (RFLP) and allozymes are the most common Type I markers since RFLP is obtained during the analysis of known genes and the protein of known function is identified as allozymes. Similarly, Expressed Sequence Tag (EST) is also considered as a Type I marker since it is a gene transcript (Raza et al. 2016). For a comparative study of gene transfer, identification of candidate species and genomic evolution, Type I markers are often used. In the course of evolution there are several genes that remain conserved within a group of organisms. Similarity and variations in such genes can be analyzed through comparative genomics. Many laboratories have developed EST markers for variety of fishes and shellfishes using their cDNA libraries. ESTs and known genes are widely used for developing microsatellite markers in different species and as these microsatellites are associated with the coding regions, they can produce high transferability between species and population. Such analysis has been carried out in case of common carp (Yue et al. 2004), rainbow trout (Rexroad et al. 2005) and Atlantic salmon (Ng et al. 2005).

Markers which are developed from anonymous genomic regions and their function are unknown are categorized under **Type II markers**. Type II markers are generally non-coding and are hence selectively neutral. The most common examples of Type II markers are Microsatellites, Randomly Amplified Length Polymorphism (RAPD) and (AFLP). SNPs come under Type II category unless it is developed using expressed sequences (Raza et al. 2016). These markers have found wide applications in population genetic studies that involve parameters such as population divergence and gene flow within and between populations. RAPD, a type II marker, has been widely used for polymorphism assays in Tilapia (Ahmed et al. 2004). Such analysis is also carried out in catfishes to develop a DNA fingerprint for detecting the pure strains and hybridizations (Huang et al. 2005).

11.1.1 Properties of Markers

1. Reproducibility: The quality of results obtained is depicted as reproducibility of the marker. To make it clear, let us consider RAPD marker as an example. The primers obtained for RAPD, are about 10 bp which is quite short. This can lead to nonspecific annealing. Hence, a proper standardized procedure needs to be followed in order to generate more authenticated data.

2. Polymorphic information content (PIC): Quantitative measurement of informativeness of a marker is called as PIC value. The PIC value can be obtained by deriving the informative off-springs from a pedigree after possible mating. While calculating this value it is often assumed that there is no recombination, since the marker locus is located very close to the gene locus.
3. Phenotypically neutral: If the marker is not affected by environment then it is said to be phenotypically neutral. Neutral mutations are not under any selection force. On the contrary, adaptive variations are analyzed through non-neutral phenotypic markers. Now a day, adaptive variations are gaining importance in case of diversity study in fisheries.

11.1.2 Allozymes

The proteins encoded by the allozymes exhibit a known enzymatic function and are, hence, categorized under Type I markers. Allozymes are variants of proteins or enzymes that are coded by different alleles at the same locus which is in contrast to isozymes that are coded by genes at different locus. When the protein is subjected to gel electrophoresis, they move according to charge and size. The relative frequencies of alleles are used to determine the genetic variation. For example, a typical phylogenetic analysis of Cyprinids was carried out using allozymes by Häunfling and Brandl (2000). After subjecting the muscle and liver tissues to gel electrophoresis, they obtained the allele frequency from the genotypes and they calculated mean distances within and between the taxa, from the distance matrices (Nei 1972). Their studies found allozymes to be an effective aid for phylogenetic analysis. The limitations of allozymes are that if some variations are masked in protein level or if there is presence of null alleles, then detection of variation becomes difficult. Allozymes are described as the workhorse of fish population genetics since 1970s and plays a key role even today in analyzing intraspecific genetic diversity among fish populations.

11.1.3 Restriction Fragment Length Polymorphism (RFLP)

This technique was invented by Allec Jeffreys in 1984 while studying about hereditary diseases. The unique patterns of DNA fragments obtained through these techniques are called as VNTR (Variable Number of Tandem Repeats). It is developed by cutting the DNA at highly specific sites using restriction enzymes. The basis of analysis is that the distance between two cleavage sites is variable for different individuals. Hence, the length of the fragments will differ across species. Since there is a large number of restriction sites in a genome, many fragments are produced which are separated by running it through the gel. Since heterozygosity can be detected by this marker o it can be categorized as codominant marker. RFLP has been used to differentiate between fish species as stated by Wolf et al. (2000).

11.1.4 Random Amplified Polymorphic DNA (RAPD)

This marker is developed using the polymerase chain reaction. Randomly constructed oligonucleotides are used as a primer for amplification using PCR. There is no need of prior knowledge of the genome to generate this marker. Various fragments are obtained and are then subjected to gel electrophoresis. Since there is no discrimination between coding and non-coding regions, the sampling is more random in this method which is a great advantage (Lynch and Milligan 1994). RAPD fingerprinting is a fast and effective method for developing DNA markers in fishes. The major disadvantage of RAPD is its dominant nature which doesn't produce any information on heterozygosity. Also, several studies provide evidence that they show very low reproducibility. However, the efficiency of RAPD in identifying the polymorphism both within and between fish population has been reported by Huang et al. (2005). It is used to differentiate geographically and genetically isolated populations (Fuchs et al. 1998). This helps in verifying the local adaptations within the species which could have been due to genetic drift or selection under various environmental conditions, RAPD was found to be highly beneficial. RAPD has also been used to determine heterosis of various strain combinations (diallele and reciprocal crosses) in case of Guppies (Shikano and Taniguchi 2002).

11.1.5 Amplified Fragment Length Polymorphism (AFLP)

It was first explained by Vos et al. (1995). AFLP is a PCR-based technique coupled with restriction endonuclease. Then ligating adapters to the sticky ends after which the complimentary of the primer is produced for amplifying a subset of the restriction fragment. Two most widely used restriction enzymes so far are EcoRI and MseI. The primers designed for AFLP comprises of a core sequence, an enzyme specific sequence and a selective extension. Since, the sequence is unknown; adapters of known sequence are ligated. This site is used for designing primers for amplification purpose. The amplified products are viewed after subjecting them to gel electrophoresis through autoradiography. Techniques of both RAPD and RFLP are combined in AFLP which thus overcomes the weaknesses of the both. This marker is efficient in differentiating individual strains. An example of such study by Botella et al. (2002) shows the characterisation of individual strains in *Photobacterium damselae*. Since AFLP follows Mendelian inheritance pattern, it is widely used in gene mapping and marker-assisted selection. Several AFLP markers have been developed for *Ictalurus punctatus* and *Ictalurus furcatus* (Liu et al. 1999). AFLP is an extremely powerful technique for the detection of genomic polymorphism. Like RAPD, AFLP are also considered as dominant markers. However, codominant scoring can be obtained by using softwares such as AFLP Quantapro, Key Genes, etc. Such results can be obtained by directly eluting the AFLP bands from gel, sequencing them and designing primers that are locus specific, for each band. The hybridization status of native and introduced trouts in California was analyzed using

AFLP marker. The major advantage of AFLP is its high reproducibility and it is relatively economical. However the PIC score for AFLP is less, being a disadvantage.

11.1.6 Sequence Characterized Amplified Region (SCAR)

SCAR is derived from RAPD or AFLP markers with an intention to convert the dominant score to a codominant score. The procedure involves simple PCR technique for which the primers are designed from the sequences obtained from the amplicons of RAPD or AFLP. The limitations of RAPD i.e. low reproducibility is overcome in this technique. Since it is very convenient and easy, it can be efficiently used in aquaculture and fisheries. It has been used for strain identification in the GIFT tilapia (Li et al. 2010). The genetic diversity of populations can be analyzed using SCAR marker utilizing the primers designed from the RAPD analysis. For example, a recent study by Chiu et al. (2012) has identified strains of Serranids and also analyzed the genetic diversity of their populations.

11.1.7 Microarray Techniques

Microarrays are designed probes against a surface to which cDNAs, derived from RNAs are hybridized. These microarrays can reveal the temporal and spatial patterns to which the organisms have been exposed. This is a technique to analyze several genes and their variations and expressions simultaneously. In case of fisheries and aquaculture the most widely used array technique is the high density arrays known as Expressed Sequence Tag (EST).

11.1.8 Expressed Sequence Tag (EST)

EST is generated from cDNA clones and is used in mapping and coding regions of genomic sequence. ESTs are the most useful tools in genome mapping and are also used to develop cDNA microarray which can be utilized for studying differential expression of genes. EST databases have been a very useful source of genomic information, linkage mapping and genomic mapping in aquaculture. ESTs were primarily used to identify zebra fish orthologous genes that proved highly beneficial in the study of different species of fishes. The primers can be designed from the cDNA sequences and length polymorphisms can be scored after amplification through PCR. An example of similar analysis is a study regarding the development of EST for gene mapping and Marker-Assisted Selection (MAS) in catfish which has carried out by Liu et al. (1999). ESTs are even more useful when radiation hybrid panels for species are available. A radiation panel consists of hybrid cell line. These cell lines consist of cells containing small fragments of irradiated chromosomes from certain species of interest. However in aquaculture, the development of such radiation panels for fish species is scarce, since there is already the availability of Bacteria Artificial Chromosome (BAC) library of important species which provide better results.

11.1.9 Microsatellites and Minisatellites

Microsatellites are a tandem repeat of sequences that range from 2 to 6 bp and are evenly distributed over the chromosomes occurring once in every 10 kbp in fish. Since, number of alleles is relatively higher in loci and it responses well to PCR amplification, it is considered to be a very good maker in genome study (Williams 2005). However, the use of microsatellites can be expensive and laborious. Minisatellites contain a longer repeat and are used in DNA fingerprinting. Sequences of the flanking regions are identified by PCR primers. In this method, allele differentiation can be detected even by 1 bp. Microsatellite has become the most commonly used tool for genetic mapping and studying population genetics in fish. Microsatellite markers utility fully depends on the polymorphic information content (PIC); higher the number of alleles higher will be the value of PIC. Thus, it follows a direct correlation with the number of alleles and their frequency is detected. Some evidences show that marine fishes exhibit weak population structuring at neutral loci and in these fishes nuclear microsatellite loci have proved beneficial in resolving decreased differentiation level in the fish population. It has been reported several times that there has been negative bias on estimation of differentiation due to mutational properties of microsatellite loci. Microsatellite has been widely used in aquaculture to develop markers linked to various important traits such as growth enhancement and disease resistance. A study on Tambaqui, *Colossoma macropomum* identified several microsatellites which comprised of both gene associated and anonymous markers. Being a codominant marker and exhibiting Mendelian inheritance pattern, this marker is quite useful.

11.1.10 Single Nucleotide Polymorphism (SNP) Marker

Single nucleotide polymorphism, as the name suggests, is formed by substitution of a single base. It represents a point mutation that gives rise to different alleles at a specific locus. The PIC value of this marker is lower than that of microsatellite. This marker gives the most accurate measure of genetic diversity. It is the most widely used marker for fish species since these are the most abundant form of genetic variations. Conventional methods used for SNP genotyping includes Ligation Chain Reaction (LCR), direct sequencing, Single strand conformational polymorphism assay (SSCP), etc. but with the advancements of technologies like pyrosequencing, Matrix-assisted laser desorption ionization time of flight (MALDI-TOF) mass spectrometry. Real time (RT) PCR, etc., the development of this marker is easier and provides precise accuracy of the results.

11.1.11 Mitochondrial DNA

The role of mtDNA is very crucial in the study of phylogenetics and bar-coding of fishes. With the exception of oysters, mt DNA is maternally inherited in all fishes and shellfishes. Hence, scientists are using the mutations over time to reconstruct the evolutionary relationships between and within the species. The size of this mtDNA in teleost fishes varies from 16,000 to 19,000 bps. Mostly Cytb, COI and Cyta genes are taken into consideration for phylogenetic analysis. Another very important region is the D-loop which is a hypervariable region, having tremendous utility in population genetics analysis. The analysis of mtDNA is mostly done through direct sequencing analysis. One of the major limitations in the approach is the non-Mendelian inheritance pattern of the mitochondrial genome.

11.1.12 Transcriptomics

Besides the conventional methods of interpreting genetic diversity, application of genomic tools to understand population structure, its response to environmental changes and diversity indices, are increasingly enhancing the advances in conservation biology. The interactions of population with its environment are governed by genomic variations between the populations. Hence transcriptomics can be used along with conventional population genetics in conservational strategies. For a species, which doesn't have prior molecular information can be subjected to RNA sequencing to identify the variation between the individuals or specific genotypes responsible for a phenotypic response. The mechanisms behind tolerance and adaptability to various environmental conditions can be understood by analyzing transcriptomics data. The basic transcriptomics approach includes collecting of tissues, RNA library preparation, cDNA library preparation and then high throughput sequencing. The differential expression is detected. This provides us a benchmark for managing the environmental requirements of the concerned fish species.

11.2   Selection of Markers and Their Utility in Aquaculture

Several markers have been developed for commercially important fish species such as catfishes, tilapias, trouts, carps and shellfishes that has benefited the aquaculture sector as well as setting strategies for fish population genetics study and conservation biology. The Genome Research on Atlantic Salmon Project (GRASP) is a Canadian government initiative that emphasizes on marker development and genome mapping. For fish species identification, mostly RAPD marker is used since it is inexpensive and the results obtained are reliable. However, genetic diversity analysis may require more advanced approaches. Another aspect of applicability of molecular markers is food chain traceability. In this field, AFLP has been used to detect seafood species in processed products (Maldini et al. 2006).

While planning any breeding programme or selecting the brooders, it is very important to evaluate the hybridization status of the species. Hybridization is brought about by crossing two strains or species, considering the improvement of desirable traits. Molecular markers can be utilized not only to detect the gene flow but also detect the hybridization. For this purpose mostly RAPD and isozymes have been used in aquaculture. The application of these markers could help in keeping track of the strains, lines thereby maintaining the stocks.

It is easier to select a marker for a monogenic trait since the variations within and between the groups can be analyzed on the basis of certain physiological performances and biochemical pathways. This approach requires thorough inquiry and knowledge about the species and trait. But this approach cannot be applied to the polygenic traits since they are governed by many genes. In such cases, the genes associated with the particular trait and their patterns of inheritance are needed to be studied. This method can locate the associated genes responsible for the expression of the trait by mapping and locating Quantitative Trait Loci (QTL). For QTL mapping, the most widely used markers are microsatellites and ESTs. The variations in the QTL can be more appropriately understood after analyzing population structure. Recently there has been emergence of sophisticated techniques for identifying QTL. Still the analysis of allelic frequencies at QTL in a much natural population remains a question. The development of markers had a tremendous impact in the field of conservation genetics. Determination of variations in a population to their inheritance pattern can be easily analyzed. Marker-Assisted Selection has wide application in organizing various breeding programmes to develop a superior variety of fish species. Assessment of variations between wild and captive populations and studying the genetic bottlenecks in a population has become easier due to markers. The markers have wide application in fisheries. It is not only for selection and breeding programmes but also for species identification, phylogenetic analysis and planning conservation strategies. Detection of diseases is another prospect of this advent which is an aid to deal with various disease outbreaks in fish.

References

Ahmed MM, Ali BA, El-Zaeem, S. Y. (2004) Application of RAPD markers in fish: part I—some genera (Tilapia, Sarotherodon and Oreochromis) and species (Oreochromis aureus and Oreochromis niloticus) of Tilapia. Int J Biotechnol 6(1):86–93

Botella S, Pujalte MJ, Macián MC, Ferrús MA, Hernández J, Garay E (2002) Amplified fragment length polymorphism (AFLP) and biochemical typing of Photobacterium damselae subsp. damselae. J Appl Microbiol 93(4):681–688

Chiu TH, Su YC, Pai JY, Chang HC (2012) Molecular markers for detection and diagnosis of the giant grouper (Epinephelus lanceolatus). Food Control 24(1–2):29–37

Fuchs H, Gross R, Stein H, Rottmann O (1998) Application of molecular genetic markers for the differentiation of bream (Abramis brama L.) populations from the rivers Main and Danube. J Appl Ichthyol 14(1–2):49–55

Häunfling B, Brandl R (2000) Phylogenetics of European cyprinids: insights from allozymes. J Fish Biol 57(2):265–276

Huang CF, Lin YH, Chen JD (2005) The use of RAPD markers to assess catfish hybridization. Biodivers Conserv 14(12):3003–3014

Li SF, Tang SJ, Cai WQ (2010) RAPD-SCAR markers for genetically improved NEW GIFT Nile Tilapia (Oreochromis niloticus niloticus L.) and their application in strain identification. Zool Res 31(2):147–153

Liu Z, Li P, Kucuktas H, Nichols A, Tan G, Zheng X et al (1999) Development of amplified fragment length polymorphism (AFLP) markers suitable for genetic linkage mapping of catfish. Trans Am Fish Soc 128(2):317–327

Lynch M, Milligan BG (1994) Analysis of population genetic structure with RAPD markers. Mol Ecol 3(2):91–99

Maldini M, Marzano FN, Fortes GG, Papa R, Gandolfi G (2006) Fish and seafood traceability based on AFLP markers: elaboration of a species database. Aquaculture 261(2):487–494

Nei M (1972) Genetic distance between populations. Am Nat 106(949):283–292

Ng SH, Chang A, Brown GD, Koop BF, Davidson WS (2005) Type I microsatellite markers from Atlantic salmon (Salmo salar) expressed sequence tags. Mol Ecol Notes 5(4):762–766

Raza S, Shoaib MW, Mubeen H (2016) Genetic markers: importance, uses and applications. Int J Sci Res Publ 6(3):221

Rexroad CE, Rodriguez MF, Coulibaly I, Gharbi K, Danzmann RG, DeKoning J et al (2005) Comparative mapping of expressed sequence tags containing microsatellites in rainbow trout (Oncorhynchus mykiss). BMC Genomics 6(1):54

Shikano T, Taniguchi N (2002) Using microsatellite and RAPD markers to estimate the amount of heterosis in various strain combinations in the guppy (Poecilia reticulata) as a fish model. Aquaculture 204(3–4):271–281

Vos P, Hogers R, Bleeker M, Reijans M, Lee TVD, Hornes M et al (1995) AFLP: a new technique for DNA fingerprinting. Nucleic Acids Res 23(21):4407–4414

Williams JL (2005) The use of marker-assisted selection in animal breeding and biotechnology. Revue Scientifique et Technique 24(1):379

Wolf C, Burgener M, Hübner P, Lüthy J (2000) PCR-RFLP analysis of mitochondrial DNA: differentiation of fish species. LWT Food Sci Technol 33(2):144–150

Yue GH, Ho MY, Orban L, Komen J (2004) Microsatellites within genes and ESTs of common carp and their applicability in silver crucian carp. Aquaculture 234(1–4):85–98

Microsatellite Markers for Fish Conservation

12

Jaya Kishor Seth, Anil Mohapatra, Swarup Ranjan Mohanty, and Sanmitra Roy

Abstract

The present book chapter deals with insight into the microsatellite DNA markers and their uses in biodiversity conservation in general and fish conservation in particular. As the chapter unfolds, the readers will know the fundamental aspects of microsatellite DNA markers, and their significance in managing and conserving biological wealth. Although there are many uses of microsatellite markers in biological sciences, for fish conservation, three essential utility viz., identification of fishes, identification of genetic stock, and population genetics analysis have been discussed.

Keywords

Integrated approaches · Management practices · Identification · Genetic stock · Population genetics analysis

12.1   Introduction

Human footprint across the globe has threatened biological diversity. Habitat degradation is the biggest threat to the loss of biological wealth. With the rise of the human population, we modify the landscape to meet our demand and increase the demand for energy consumption. Consequently, we are witnessing climate change worldwide and are on the verge of the sixth mass extinction process (Supple and Shapiro

J. K. Seth
Post-Graduate Department of Zoology, Berhampur University, Berhampur, Odisha, India

A. Mohapatra (✉) · S. R. Mohanty · S. Roy
Estuarine Biology Regional Centre, Zoological Survey of India, Gopalpur-on-Sea, Ganjam, Odisha, India

© The Author(s), under exclusive license to Springer Nature Singapore Pte Ltd. 2021

P. K. Pandey, J. Parhi (eds.), *Advances in Fisheries Biotechnology*,
https://doi.org/10.1007/978-981-16-3215-0_12

2018). According to the Millennium Ecosystem Assessment, there are around 1.75 million species reported worldwide, and the predicted number is varied from 5 to 100 million species. As per the Intergovernmental Science-Policy Platform on Biodiversity and Ecosystem Services (IPPBES) (2019) global biodiversity has witnessed the decline at an unprecedented rate in the human race history. Around one million species are threatened with extinction out of the planet's eight million species. This loss of biodiversity poses a serious threat to global food security. There are many international platforms/agreements like United Nation Convention of Biological Diversity (UNCBD), Convention on International Trade in Endangered Species of Wild Fauna and Flora (CITES), International Union for conservation of Nature and Natural Recourses (IUCN), Sustainable Development Goals (SDGs), and Aichi Biodiversity Targets which talk about the need and efforts related to conservation of biological diversity. For the effective management and conservation of biological diversity, integrating genomic technologies in identifying biological wealth have been given priority in recent decades (Supple and Shapiro 2018).

Across the globe, the aquatic environments are witnessing a severe threat due to numerous reasons, which is causing loss of natural genetic diversity, and fishes are not behind in the race (Patra et al. 2017). According to Food and Agriculture Organization (FAO), fishes are the most traded commodities globally, and they fulfil the protein demands of billions of people worldwide. The need for wild fishes for food security may be in higher demand in recent decades than earlier. Therefore, in this juncture of decline in fish genetic diversity and rise in demand, we need to have an integrated approach to understand the genetic diversity of fishes for proper management and conservation approaches. In this connection, this chapter describes the importance of one of the genomic technology i.e. microsatellite makers, for the conservation of fishes.

12.2 Molecular Markers

Molecular markers are the specific stretch/fragment of DNA, located at a specific location of the genome and used to flag the position of a particular gene or inheritance of particular characteristics.

12.3 Microsatellite Markers

The script of the genome is written in four nucleotides/chemical letters A, T, G, and C. Different combinations of these four letters, as a stretch of DNA molecule, provides uniqueness and specificity to each gene, cell and organism. The triplet codon/genetic code of these four letters encode specific protein. Any spelling variation in these four letter DNA stretch brings variation in the script of an individual genomes and in different organism, which ultimately generates biodiversity. Some of these variations may generate useful or diseased phenotypes. This variation in the stretch of the script of genome is very useful to differentiate various organisms at molecular level, their functional and adaptive role, and to understand the evolutionary path. One such spelling variation in a genome is the 5–50 fold repetition of 2–6 nucleotides base pairs stretch of DNA, called microsatellites DNA. Microsatellites are also called as short tandem repeats (STRs) or simple sequence repeats (SSRs) (Abdurakhmonov 2016). Another such variable tandem number repeats (VNTRs) are minisatellites DNA. Differences between minisatellites and microsatellite DNA are based on the numbers and nature of the tandem repeats. Minisatellites DNA are the heterogeneous fragments of 10–60/100 nucleotides sequences that have repeat size of 1–15 kb, whereas microsatellites DNA consists of a homogeneous fragment of mono, di, tri, tetra, penta, and hexa-nucleotides sequences of less then or around 1 kb (Abdurakhmonov 2016; Hodel et al. 2016; Saeed et al. 2016).

Microsatellites DNA occur in both prokaryotic and eukaryotic genome abundantly. These simple sequences repeat are distributed throughout the genome in non-random manners and can vary between region of the genome and different organisms or taxa. Microsatellites are found in noncoding regions, coding regions, other repetitive elements, transposons, and regulatory regions. The occurrence of these sequences is the highest in the untranslated regions (UTRs). They are known to involve in many biological functions like regulations of cell cycles, expression and silencing of various genes, organization of chromatin, and m-RNA splicing. Many phenotypic changes, morphological features, social behaviours in different organisms, various disease phenotypes, etc., are associated with microsatellite numbers, variation, and functions (Tridevi 2004; Hammock and Young 2005; Hodel et al. 2016; Saeed et al. 2016; Abdurakhmonov 2016).

12.4 Importance of Microsatellites as a Marker

Microsatellites are a marker of choice because they are highly polymorphic, multi-allelic, highly mutagenic, PCR-based marker, amplify from the low quantity of DNAs, cost-effective and easy to distinguish, co-dominant, suitable in detecting heterozygote from homozygote, useful in the construction of the genetic map, and are supposed to be a neutral marker (Abdul-Muneer 2014; Abdurakhmonov 2016)

12.5 Applications of Microsatellites in the Conservation of Fishes

There are number of application of Microsatellites DNA markers in biological sciences viz. taxonomic and phylogentic analysis, population genetics and conservation biology, diversity and cultivate analysis, hybridization and breeding, genetic tagging and QTL analysis, functional genomics, genomic mapping, forensic science, pedigree and gender identification, and diagnosis and identification of human disease (Abdurakhmonov 2016). However, following applications are useful in the conservation practices of fish and fisheries resources.

12.5.1 Identification of Fishes

The taxonomy of fishes is an ever changing subject with addition of different modern tools with classical taxonomic characters. The modern tools including modern photographic techniques, X-ray images, DNA barcoding etc. have made the science much easier in the modern taxonomic era. There are about 35,704 valid species of fishes belonging to 5208 genera documented worldwide (Fricke et al. 2020). The description of new species is an ongoing process and in the last 10 years (2011–2020) about 4117 fish species have been described (Fricke et al. 2020). Most of the fish species described in the recent past are either with the help of classical taxonomic tools or by the use of modern tools like DNA finger printing, recognized and authenticated as an established species. In recent years, several species are also described with a single specimen of collection, and ascertaining the validity of species is much dependant on DNA based studies for a more precise way of comparing that with its congeners. The collection gaps and the difficulty level for the collection of deep sea fishes had made DNA based studies more important for the rarely collected precious samples.

Identification of fishes is the gateway for any study related to fishes and their management and conservation practices. There are several tools and techniques available, along with the morpho-taxonomy for the proper identification of fishes. Microsatellite DNA markers are one such tool. Zhivotovsky et al. (2013) have demonstrated the use of microsatellite markers to identify Salmonid Fish. They have successfully provided insight into the inter-specific hybrids, genetic estimation, intra-specific variation, and individualization of biological specimens, using microsatellite DNA markers. Vanhaecke et al. (2012) have shown the use of microsatellite DNA markers and COI genes for species delimitation and identification of hybrids in galaxiid fishes. This study also justifies the use of microsatellite DNA markers and other conserved regions of the genome to solve the problems of overlapping traits shares between many species. The endemic fish species and data deficit need proper identification for their better conservation measures. Microsatellite DNA markers are useful to overcome these issues.

12.5.2 Identification of Stocks or Units

For fishes' management and conservation approaches, identification of discrete stocks or units is a critical element (Ihssen et al. 1981; Fetterolf 1981). As per Shaklee et al. (1990), a stock is defined as "a panmictic population of related individuals within a single species that is genetically distinct from other such population." The differentiation of species into the genetically distinct unit is the basis of evolution. This differentiation process depends on geographical and physical barriers, evolutionary forces like migration, selection, genetic drift, bottleneck effect, non-random matting, etc. If the size of a unit or stock declines, they will experience inbreeding and loss of genetic diversity, resulting in the loss of stock or units; consequently, there is a reduction in adaptability and survivability of the species. Microsatellite DNA markers help us to explore insight into the genetic diversity of the stock or the unit and can strengthen our management approaches.

12.5.3 Population Genetic Analysis

In population genetics, low genetic diversity due to any reason and deviation from the Hardy-Weinberg Genetic Equilibrium, coupling with inbreeding resulted in genetic bottleneck and reduces the individual stock's reproductive fitness of fish species. Therefore, frequent monitoring of the population genetic structure and its analysis is important for any conservation practices. Microsatellites DNA marker provides opportunities to keep track of the fish population's genetic structure (Abdul-Muneer 2014). Studies on the genetic diversity studies, using microsatellite DNA marker across the species, could be of great importance for understanding of population sub-structure and for biogeographical inference, which would further help us to understand the extent of diversity in the species and its management strategies (Hassanien and Gibley 2005). Microsatellites DNA markers are useful in evaluating heterozygosity, inbreeding coefficient and how the population is deviated from the Hardy-Weinberg Genetic Equilibrium (Nazish et al. 2018). Many studies provide insight into the usefulness of microsatellites DNA markers in genetic diversity and fish' population structure. Li et al. (2016) carried out population genetic analysis of *Percocypris pingi* and provided useful insight into the population structure by comparing the wild and hatchery populations of *P. pingi*. Shaw et al. (1999) have demonstrated microsatellite DNA analysis for population structure analysis in Atlantic herring (*Clupea harengus*). Liu et al. (2019) have revealed high gene flow between the population of *Liza affinis* in the coastal water along the eastern and southern China, using the six microsatellites loci. Landínez-García and Marquez (2018) have shown a high level of genetic diversity and gene flow among the population of one of the commercially important fish species *Curimata mivartii,* separated over a distance of 350 km in the lower section of the Colombian Magdalena-Cauca basin. Li et al. (2017) compared the population genetic structure of 4 aquaculture populations and 1 wild population of *Hemibarbus maculates* in China based on the 12 polymorphic loci of microsatellites DNA. The study also highlighted how anthropogenic activities reduced genetic diversity. Glover et al. (2011) revealed the genetic differentiation among the pelagic fish *Sprattus sprattus* of Northeast Atlantic, including Norwegian fjords, using microsatellites DNA markers.

12.6 Conclusion

In nature, the population of several species faced serious threat regularly due to both natural and anthropogenic pressure thus, pushing many species into the conservation important categories. The red list of IUCN is one of the most authoritative sources for assessment of the status of any plants and animals due to its strong scientific baseline of the assessment. Many fish species are now listed under the conservation categories of IUCN and many species of fishes are also on the verge of extinction as per IUCN assessment and are in the conservation important categories of IUCN. The Use of molecular markers in the identification of fish species are of much importance for the conservation biology (Arif et al. 2011) and management of any species either in wild or captivity. Thus, the genetic diversity of the fish population of particularly endangered species is very much essential to access the genetically weak population structure of either small populations or endemic population of any of the endangered species for the species specific management. The proper genetic study will also help in solving the wildlife crime related issues for the better identification of the species even from smaller body parts, thus, helping in conservation of particular species on regional and subregional basis.

Acknowledgement We thank Dr. Kailash Chandra, Director, Zoological Survey of India, for providing necessary working facilities. First author is thankful to authority of Berhampur University, Odisha, India for providing necessary facilities to work.

References

Abdul-Muneer PM (2014) Application of microsatellite markers in conservation genetics and fisheries management: recent advances in population structure analysis and conservation strategies. Genet Res Int 2014:691759. https://doi.org/10.1155/2014/691759

Abdurakhmonov IY (2016) Introduction to microsatellites: basics, trends and highlights, microsatellite markers. IntechOpen, London. https://doi.org/10.5772/66446. https://www.intechopen.com/books/microsatellite-markers/introduction-to-microsatellites-basics-trends-and-highlights

Arif IA, Khan HA, Bahkali AH, Homaidan AA, Farhan AHA, Sadoon MA, Shobrak M (2011) DNA marker technology for wildlife conservation. Saudi J Biol Sci 18:219–225

Fetterolf CM Jr (1981) Foreword to the stock concept symposium. Can J Fish Aquat Sci 38:4–5

Fricke R, Eschmeyer WN, Van der Laan R (eds) (2020) Eschmeyer's catalog of fishes: genera, species, references. http://researcharchive.calacademy.org/research/ichthyology/catalog/fishcatmain.asp. Electronic version accessed 14 Dec 2020

Glover KA, Skaala Ø, Limborg M, Kvamme C, Torstensen E (2011) Microsatellite DNA reveals population genetic differentiation among sprat (*Sprattus sprattus*) sampled throughout the Northeast Atlantic, including Norwegian fjords. ICES J Mar Sci 68:2145–2151

Hammock EAD, Young LJ (2005) Microsatellite instability generates diversity in brain and sociobehavioral traits. Science 308:1630–1634

Hassanien HA, Gibley J (2005) Genetic diversity and differentiation of Nile tilapia (*Oreochromis niloticus*) revealed by DNA microsatellites. Aquac Res 36:1450–1457. https://doi.org/10.1111/j.1365-2109.2005.01368.x

Hodel RG, Segovia-Salcedo MC, Landis JB, Crowl AA, Sun M, Liu X, Gitzndanner MA, Douglas NA, Germain-Aubrey CC, Chen S, Soltis DE, Soltis PS (2016) The report of my death was an exaggeration: a review for researchers using microsatellites in the 21st century. Appl Plant Sci 4:1600025. https://doi.org/10.3732/apps.1600025

Ihssen PE, Booke HE, Casselman JM, McGlade JM, Payne NR, Utter FM (1981) Stock identification: materials and methods. Can J Fish Aquat Sci 38:7838–7855

Landínez-García RM, Marquez EJ (2018) Microsatellite loci development and population genetics in Neotropical fish *Curimata mivartii* (Characiformes: Curimatidae). Peer J 6:e5959. https://doi.org/10.7717/peerj.5959

Li X, Deng Y, Yang K, Gan W, Zeng R, Deng L, Song Z (2016) Genetic diversity and structure analysis of *Percocypris pingi* (Cypriniformes: Cyprinidae): implications for conservation and hatchery release in the Yalong River. PLoS One 11(12):e0166769. https://doi.org/10.1371/journal.pone.0166769

Li L, Lin H, Tang W, Liu D, Bao B, Yang J (2017) Population genetic structure in wild and aquaculture populations of *Hemibarbus maculates* inferred from microsatellites markers. Aquacult Fish 2:78–73

Liu L, Zhang X, Sun D, Gao T, Song N (2019) Population genetic structure of *Liza affinis* (eastern Keelback mullet), reveals high gene flow inferred from microsatellite analysis. Ocean Sci J 54:245–256. https://doi.org/10.1007/s12601-019-0013-y

Nazish N, Abbas K, Abdullah S, Anjum Zia M (2018) Microsatellite diversity and population structure of *Hypophthalmicthys molitrix* in hatchery populations of Punjab. Turk J Fish Aquat Sci 18:1113–1122. https://doi.org/10.4194/1303-2712-v18_9_10

Patra BK, Kar A, Bhattacharya M, Parua S, Shit P (2017) Freshwater fish resource mapping and conservation strategies of West Bengal, India. Spatial Inform Res 25:635–645. https://doi.org/10.1007/s41324-017-0129-z

Saeed AF, Wang R, Wang S (2016) Microsatellites in pursuit of microbial genome evolution. Front Microbiol 6:1462. https://doi.org/10.3389/fmicb.2015.01462

Shaklee JB, Allendorf FW, Morizot DC, Whitt GS (1990) Gene nomenclature for protein-coding loci in fish. Trans Am Fish Soc 119:2–15

Shaw PW, Turan C, Wright JM, O'Connell M, Carvalho GR (1999) Microsatellite DNA analysis of population structure in Atlantic herring (*Clupea harengus*), with direct comparison to allozyme and mtDNA RFLP analyses. Heredity 83:490–499. https://doi.org/10.1038/sj.hdy.6885860

Supple AM, Shapiro B (2018) Conservation of biodiversity in the genomics era. Genome Biol 19:131. https://doi.org/10.1186/s13059-018-1520-3

Tridevi S (2004). Microsatellites (SSR): puzzles with in puzzle. Ind J Biotech 3:331–347

Vanhaecke D, Garcia de Leaniz C, Gajardo G, Young K, Sanzana J, Orellana G, Fowler D, Howes P, Monzon-Arguello C, Consuegra S (2012) DNA barcoding and microsatellites help species delimitation and hybrid identification in endangered galaxiid fishes. PLoS One 7(3):e32939. https://doi.org/10.1371/journal.pone.0032939. Epub 2012 Mar 6. PMID: 22412956; PMCID: PMC3295793

Zhivotovsky LA, Shaikhaev EG, Shitova MV (2013) Identification of salmonid fish using microsatellite markers with identical PCR-primers. Russ J Mar Biol 39(6):447–454

Cryopreservation in Aquaculture

13

Kavita Kumari and Praveen Maurye

Abstract

Cryopreservation technology has enhanced hatchery and aquaculture practice, valuable in selective breeding and provides greater control in breeding programmes. It is also helpful in production of interspecies hybrid and conservation of genome of domesticated and endangered species. The success of gamete cryopreservation depends on various factors such as the species selected, seasonality, and quality of gamete, collection techniques, diluents used, precise freezing and thawing regimes, storage conditions and post–thaw fertilization techniques. Till now cryopreservation of sperm is mostly practiced. Embryo and ova cryopreservation protocols are limited and constantly failed as reported by many authors due to the presence of thick chorion, impermeability of membrane, ample amount of yolk, toxicity of cryoprotectant and formation of intracellular ice. Cryopreservation protocols for semen of more than 200 fish species are available and many other fishes are still being optimized. The semen cryopreservation is till now at premature stage and with best preservation technique, only 50% survival is possible. Cell damage during cryopreservation occurs mainly during the freezing stage due to factors such as effect of solution, dehydration, extracellular and intracellular ice formation. Various studies have been carried out to study effect of cryoprotectants along with rate of cooling and thawing for different fish species to minimize the cell damage. Though cryopreservation protocol has been developed for sperm and oocyte of different species their commercialization is not successful. Further research is required for effect of freezing especially, cryoinjury at sublethal level for successful cryopreservation.

K. Kumari (✉) · P. Maurye
ICAR-Central Inland Fisheries Research Institute, Barrackpore, West Bengal, India

© The Author(s), under exclusive license to Springer Nature Singapore Pte Ltd. 2021
P. K. Pandey, J. Parhi (eds.), *Advances in Fisheries Biotechnology*,
https://doi.org/10.1007/978-981-16-3215-0_13

Keywords

Cryopreservation · Milt quality · Cryoprotectant · Extender

13.1   Introduction

In order to enhance aquaculture production and reduce the dependency on capture fisheries, cryopreservation of gametes becomes very important. Cryopreservation will support not only the increased fish production, but also help in the restoration of important genetic resources (Diwan et al. 2020). Cryopreservation is the process for preservation or storage of living materials such as cells, tissues, oocytes, embryos, spermatozoa, organs or any biological materials at low temperature (−196 °C), to arrest all biological activities to prevent DNA degradation and cell death and its storage for longer time for its retrieval and further use. In order to enhance fish production through aquaculture technologies and reduce the dependency on wild fish stock, cryobanking of gametes (sperm and ova) becomes very important. Blaxter (1953) reported the freezing of fish spermatozoa and subsequently more than 200 fish species spermatozoa have been cryopreserved (Benson et al. 2012; Diwan et al. 2020). In aquaculture cryopreservation of sperm has been done successfully. However embryo and ova cryopreservation protocols have meet limited success and there have been frequent failure as reported by many authors (Che-Zulkifli et al. 2019; Diwan et al. 2020).The failure in embryo preservation has been mainly due to the presence of thick chorion, impermeability of membrane, ample amount of yolk, toxicity of cryoprotectant and formation of intracellular ice (Morris et al. 2012). In the same manners the cryopreservation of oocyte has been difficult due to their cortical reaction with any preservative solution and quick activation after ovulation. Due to these problems, preservation of eggs and embryo did not gain much success and viability. Beside this commercialization of cryopreserved spermatozoa in aquaculture industry has failed due to lack of scaling up of the cryopreservation industry (Benson et al. 2012).

13.2   Present Status of Cryopreservation

Successful milt cryopreservation protocol is available for more than 200 species and cryobanks have been established (Diwan et al. 2020). Various factors such as cryoprotectant, cooling and thawing rate affect cryopreservation. Nahiduzzaman et al. (2012) studied effect of cryoprotectants along with rate of cooling and thawing on Indian Major Carp and *Labeo calbasu* spermatozoa. Recently, Jusdado et al. (2019) reported successful cryopreservation of the sperm of Japanese eel fish *Anguilla japonica* and European eel fish *Anguilla anguilla*. In methanol preserved sperm fertilization rate was higher which would be useful for further study (Diwan et al. 2020). The germ cell cryopreservation has been reported in fishes such as rainbow trout (Okutsu et al. 2006), Siberian sturgeon (Pšenička et al. 2016), *Tinca*

tinca (Linhartová et al. 2014), Catla and Rohu (Patra et al. 2016). However the successful cryopreservation is not available. In zebra fish embryo cryoinjury of mitochondria, plasma membrane and ribosomes occur even 1 min post preservation (Anchordoguy et al. 1987). Total mortality in some fishes such as *Labeo rohita*, Common carp have been reported (Harvey et al. 1983). Liu et al. (1998) identified methanol as a toxic cryoprotectant for zebra fish in early stage embryo only. Chen and Tian (2005) identified glycerol to be more toxic than DMSO for flounder embryo.

Fish embryo cryopreservation, is difficult compared to oocyte and therefore, recently more focus has been given to oocyte cryopreservation. Oocyte preservation is preferred due to their small size, more membrane permeability and lower sensitivity to chilling (Isayeva et al. 2004; Zhang et al. 2005). Various reports on fish oocytes cryopreservation particularly on zebra fish (Guan et al. 2010; Anil et al. 2011) and also on other fish species (Streit et al. 2014) are available. Isayeva et al. (2004) reported in zebra fish stage III oocyte was found to be more sensitive to chilling than stage IV. Tsai et al. (2009) reported stage III ovarian tissues in zebra fish was more sensitive to chilling than stage I and II. Guan et al. (2010) reported zebra fish oocytes viability in KCL buffer was higher comparable to L-medium. Marques et al. (2015) studied the viability of ovarian follicles in zebra fish after vitrification using different cryosolutions. Plachinta et al. (2004) evaluated toxicity of zebra fish embryo to different cryoprotectant and different life stage and found stage IV was more sensitive. Ponniah et al. (1999) reported unsuccessful cryopreservation of *L. rohita* egg and embryo whereas Kopeika et al. (2005) made effort to cryopreserve fertilized eggs of Loach.

13.3 Importance of Cryopreservation

Cryopreservation of gametes in aquaculture is carried out with an aim to synchronize the availability of both male and female gametes for the breeding programme, stable supply of sperm for hatchery production and sperm transfer among hatcheries when required (Suquet et al. 2000; Xin et al. 2020). Cryopreservation of fish gametes can also help in preservation of the genome of domesticated species (Labe et al. 2013). The cryopreservation of semen is especially important for those species in which the total volume of semen is very low (Suquet et al. 2000). The endangered genetic resources are also preserved through cryopreservation protocols (Brycon and Arias-castellanos 2004). Cryopreservation allows the sperm to be collected during the highest quality stages of its production and avoids ageing and improves quality of sperm (Che-Zulkifli et al. 2020). Cryopreservation technology has enhanced hatchery and aquaculture practice, valuable in selective breeding programmes by providing flexibility in spawning of females, greater control in breeding programmes and the ability to store favourable genes for extended periods (Tiersch and Mazik 2011). Nowadays, cryopreserved sperm are being used in production of interspecies hybrid (Che-Zulkifli et al. 2020).

13.4   Principle of Cryopreservation

A basic understanding of cryopreservation principle is essential for optimum result, cryopreservation, minimum cell damage and revival of large number of viable sperm after thawing (Morris 2000; Morris et al. 2012). The cryopreservation is till now at premature stage and with the best preservation technique, only 50% survival is possible. Damage to the cells during cryopreservation occur mainly during the freezing stage which might be due to the effect of solution, extracellular ice formation, dehydration and intracellular ice formation. Success of cryopreservation of fish gamete depends on various other factors such as the seasonality, quality of gamete, collection techniques, diluents used, precise freezing and thawing regimes, storage conditions and post-thaw fertilization techniques.

13.5   Protocol for Cryopreservation

Fish spermatozoa are unique in nature with distinctive characters. The quality and quantity of spermatozoa are species dependent and also individual specific (Kowalski and Cejko 2019). In general, they are immotile and only few species have motile spermatozoa. Cryopreservation of fish gamete has been well established. However cryopreservation of different fish species requires varying conditions which need to be standardized specifically (Jawahar and Betsy 2020). The main technical factor in achieving successful cryopreservation of fish spermatozoa or germplasm has been described as being dependent on extender, cryoprotectant and cooling rate (Gwo 2000). The different steps (Fig. 13.1) of cryopreservation of fish sperm include

1. Collection of milt
2. Evaluation of milt quality and quantity
3. Addition of extenders
4. Cryoprotectant addition
5. Dilution of Fish milt
6. Equilibration of Fish milt
7. Storage
8. Freezing of milt
9. Thawing of milt

13.5.1 Collection of Milt

The healthy brooders during breeding season should be selected for the good quality and quantity of milt. The brooders should be in matured condition. To obtain matured and good quality of the brooders, they should be fed with good quality diet having sufficient protein (Jawahar and Betsy 2020). The different stock of fish should be selected to avoid inbreeding (Uusi-Heikkilä 2020). The milt should be

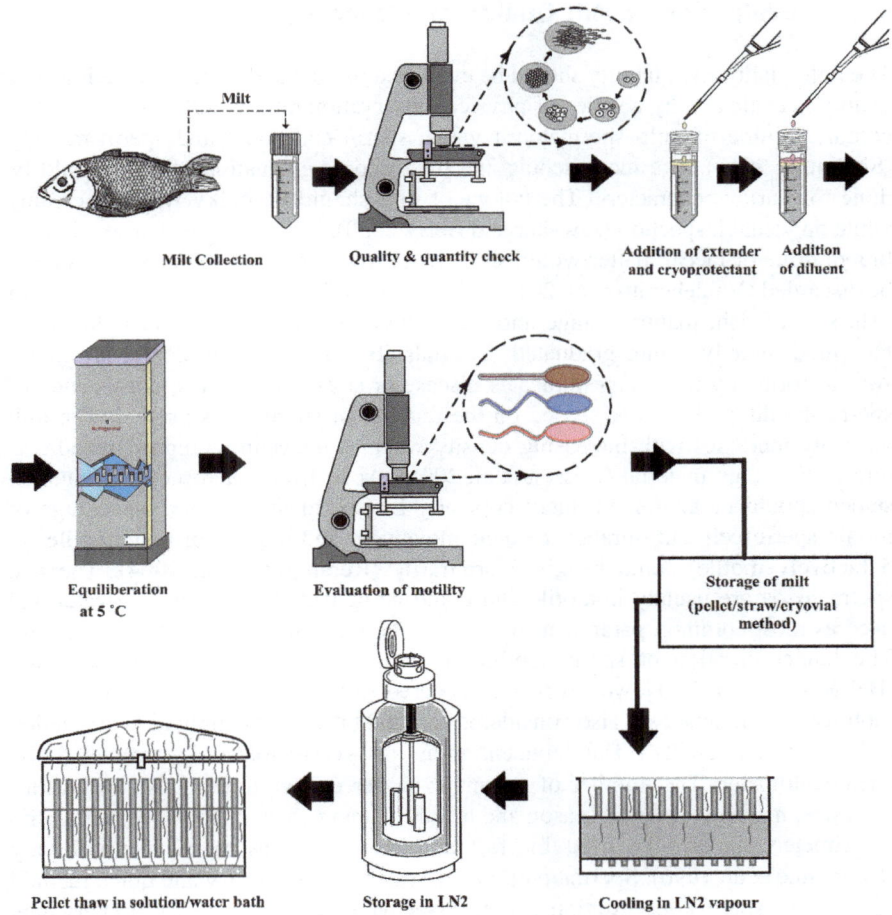

Fig. 13.1 Different steps involved in cryopreservation of fish gamete

collected in a clean and contamination free environment. The fish should be cleaned properly to avoid dirt, blood, faecal matter and mucus (Jawahar and Betsy 2020). The milt can be collected by using gentle pressure on abdominal region and stored immediately in ampoules or cryovials and kept at 0–4 °C temperature immediately (Nomura et al. 2018). The milt can be collected by surgical method from the fishes such as catfishes, murrels, etc., where the sperm cannot be collected through stripping. Frequency of milt collection can be decided based on the capacity of fish along with milt volume and density (Beirão et al. 2019).

13.5.2 Evaluation of Milt Quality and Quantity

The milt quality and quantity should be evaluated immediately after ejaculation. The quality is evaluated by a series of physical observation on the milt samples such as colour, volume of milt, spermatocrit value, sperm cell count and sperm motility (Shaliutina 2013). The macroscopic and microscopic evaluation of milt should be done for various characters. The colour of milt should be milky white to creamy white on visual inspection (Jawahar and Betsy 2020). The contaminated semen with blood, urine or faecal matter would show red, brown or green colour which should be discarded (Valdebenito et al. 2015). The volume of milt depends on the species of fish, size of fish, maturity stage and varies between individuals which should be measured directly, using graduated cryovials (Bustamante-González et al. 2016). Any abnormality in volume indicates disease or stress in fish. The consistency of sperm should be evaluated, based on the number of sperm cells present. The milt viscosity increases with increasing density whereas unevenness represents admixture with foreign material (Cosson et al. 2008). Apart from macroscopic evaluation semen should be examined microscopically for sperm motility i.e. percentage of motile sperm cell and duration of their movement and a score of 0 (immotile) to 5 (actively motile) could be given arbitrarily (Rurangwa et al. 2004). The fish spermatozoa are usually immotile and could be activated externally by water and used as most common parameter to evaluate the milt quality (Cabrita et al. 2009). The longer duration of sperm motility is considered good for cryopreservation (Beirão et al. 2011). However, it is considered to be species specific. Apart from motility, sperm density is also considered for qualitative and quantitative evaluation (Alavi et al. 2008). The concentration of spermatozoa is evaluated by haemocytometer. The number of spermatozoa per unit volume of fish milt varies with size, maturity, species, season and health (Rurangwa et al. 2004). Photoelectric colorimeter can also be used for fast and accurate measure of sperm density (Donoghue et al. 1996). Spermatocrit value, which is also an easy and quick method of assessing sperm concentration. It is accessed using microhematocrit capillary tube and expressed as packed cell volume (Sørensen et al. 2013). The percentage of live and dead spermatozoa could also be evaluated by using Eosin dye (Swanson and Bearden 1951). The dye will penetrate the dead cell only and will differentiate between dead and live cell. The dead cell should not exceed 25% for cryopreservation (Betsy and Kumar 2020). The milt should also be evaluated for biochemical properties such as pH. The pH of the milt varies from species to species and in general, the milt of the freshwater fish is slightly alkaline. For the good quality milt, the pH should be 6.5–7.0 (Betsy and Kumar 2020).

13.5.3 Addition of Extenders

The extender provides suitable dilution medium and inhibits the activation of spermatozoa for vitality and viability of sperm (Agarwal 2011). Extender is a solution of inorganic and organic chemicals, resembling that of blood or seminal plasma. Chemical formulations of extenders, used for cryopreservation, vary widely and depends on the physiological and chemical characteristics of spermatozoa (Cabrita et al. 2010). An extender is slightly hypertonic and prevents spermatozoa dehydration/exhaustion and act as medium that minimizes intracellular ice formation, prevents the change in intracellular solute concentration and prevents cryoinjuries (Xin et al. 2020). The sperm should be kept in an ideal extender with appropriate buffer, sugar, cryoprotectant, lipids and antibiotics (Rurangwa et al. 2004). The good buffer for extender should match with the properties of the seminal fluid of that particular species and composition varies from species to species (Holt 2000). The composition of a good extender should be equivalent to seminal fluid of the particular species (Suquet et al. 1994). Various media used for extenders are Fish Ringers solution, NaCl and Cortland's medium etc. (Jawahar and Betsy 2020). Different salts and compounds are used for formulating the extender medium. $NaHCO_3$, Tris and $KHCO_3$ are used as buffer, KCl prevents the sperm activation, NaCl provides tonicity, sugar act as a source of energy and osmotic balance, glycine for survival of sperm, mannitol protects toxicity effect, egg yolk, lecithin, etc. provides protection to membrane, apart from these Mg^{++} and Ca^{++} ions are used as a normal seminal fluid component (Diwan et al. 2010).

13.5.4 Cryoprotectant Addition

Cryoprotectants are chemicals that protect the cell and improve their survival. It minimizes ice crystal formation and changes the size and shape of ice crystal, prevents cell membrane rupture and provides more time for cell to dehydrate (Tiersch et al. 2007). Cryoprotectants may be permeating or non-permeating in nature. The permeating cryoprotectants such as DMSO, methanol, propylene glycol and glycerol are permeable to cell membrane and reduce the extracellular diffusion of water for ice crystal formation and prevent cell volume and solute change and also lower the homogenous nucleation temperature and ice crystal formation (Bagchi et al. 2008). Till now DMSO has given best result for cryopreservation of many species (Betsy and Kumar 2020). The permeating cryoprotectants may be toxic to cell and its concentration should not exceed 5–10% (Betsy and Kumar 2020). The non-permeating cryoprotectant such as sucrose, glucose, dextran, egg yolk serum, skim milk, etc. are not permeable to cell. It depresses the freezing point and increases the glass transformation temperature (Jawahar and Betsy 2020). Lipid and antibiotics are also added to the medium. Lipid minimizes the effect of cold shock and prevents plasma membrane rupture whereas antibiotics prevent any kind of microbial growth (Betsy and Kumar 2020).

13.5.5 Dilution of Fish Milt

For effective cryopreservation, the cryoprotectant and extender are mixed together to form suitable diluents (Irawan et al. 2010). This solution is a balanced salt buffer of specific pH and osmotic strength, sometimes with organic compounds. The diluent should be prepared fresh and mixed with milt in an appropriate ratio. The sperm diluent ration varies species to species and a ratio of 1:3–1:9 is used for different species to maintain the sperm viability (Betsy and Kumar 2020). The diluent should be slowly added to prevent any shock and should be checked for motility of sperm.

13.5.6 Equilibration of Fish Milt

Equilibration of fish milt is the pre-freeze storage period of fish spermatozoa in a cold handling cabinet or refrigerator for a particular period/time for cryoprotectant to bind with or permeate the fish spermatozoa and prevent cryoinjury (Irawan et al. 2010). The equilibration depends on species and even for the same species, it depends on the type and concentration of cryoprotectant and diluent. The diluted semen could be kept at 5 °C for brief period of time for equilibration and effective penetration of cryoprotectant (Betsy and Kumar 2020). It protects the sperm from freezing and thawing stress. The time of equilibration is also species specific and varies with the media too (Tiersch et al. 2007). The equilibration time should be brief and should not be longer than 60 min (Betsy and Kumar 2020). The samples showing more than 70% motility after equilibration should be taken for further use.

13.5.7 Storage

The semen mixed with diluent could be stored by pellet, straw or cryovials method. In pellet method semen is kept in liquid dry ice depression and stored in liquid nitrogen (-196 °C). The frozen pellet is thawed in warm tube with fresh diluent. The pellet method is economical, however, there is more chance of contamination due to open storage and also there is more chance of physical loss due to handling (Tomlinson 2005). Cryovials or ampoule is widely used method for cryopreservation (Jawahar and Betsy 2020). In this method the milt is kept inside cryovials and sealed for storage. The information can be marked on vials for retrieval. However, the method requires more storage space. The straw method is most popular for milt preservation where milt could be stored in 0.25, 0.5, and 1 mL straw (Viveiros and Godinho 2009). Straw method yields the best post-thaw survival of sperm due to better surface to volume ratio and thin film of milt (Betsy and Kumar 2020).

13.5.8 Freezing of Milt

Before freezing and storage, cooling is done at the rate of 10–45 °C per minute under liquid nitrogen vapour or programmable freezer until the temperature is reached at -70 °C to prevent any damage to sperm (Betsy and Kumar 2020). The slow cooling provides enough time for osmotic equilibrium then the straw or ampoules are transferred to liquid nitrogen at -196 °C in a cryocan for long time storage. The vials should be properly labelled with all the information such as date, batch, species, cryoprotectant, diluent, freezing method and post-thaw quality, etc. The cryocan should be periodically checked and filled. Apart from liquid nitrogen, solid carbon dioxide (-79 °C), liquid air (-183 °C) can also be used for storage (Hafez 2000; Arav et al. 2002). However, liquid nitrogen, being an inert, colourless and odourless, is most widely used for cryopreservation.

13.5.9 Thawing of Milt

Thawing is also considered an important variable in cryopreservation. Rapid thawing rates are used to avoid recrystallization whereas slow warming rate may result in recrystallization (Huebinger et al. 2016). The rapid and progressive thawing should be done to bring the spermatozoa at the fertilization temperature (Kopeika et al. 2007). The warm water thawing at 37–40 °C is widely used for the thawing (Betsy and Kumar 2020). The thawing temperature affects viability of sperm and it depends on storage method. The thawing could be done for 30 s to 5 min for French mini straw and cryovials, respectively (Betsy and Kumar 2020). Motility, fertilization and hatching rates, fry survival, etc. are the common criteria for judging the post-thaw viability/fertility of cryopreserved spermatozoa (Huebinger et al. 2016). After thawing, samples with >40% motility should be used further (Betsy and Kumar 2020). The fertilization rate and hatching percentage should be calculated and compared with the control treatments for evaluation of any cryopreservation protocol (Lee et al. 2020).

13.6 Changes in Sperm Quality During Cryopreservation

The freezing and thawing cause various stress to the spermatozoa and lead to reduced sperm motility, plasma membrane disruption and other damage (Yeste 2016). Sperm motility and viability are the most common parameters for quality evaluation. However, sometimes sub lethal injury also occurs which needs to be evaluated further (Herranz-Jusdado et al. 2019). Mitochondrial membrane potential has been a potential tool to access cryoinjuries (Pereira et al. 2019). Pereira et al. (2019) found changes in the membrane fluidity which could be useful in selecting optimal cryoprotectant for the cryopreservation. Excess reactive oxygen species production and lipid peroxidation are also useful to access cryoinjuries, especially for selection of suitable extenders. Cryoinjury also affect DNA of sperm such as degradation of transcript (Riesco et al. 2017). Cryopreservation also affects protein and lipid composition of spermatozoa of many fish species (Horokhovatskyi et al. 2016; Xin et al. 2018).

Martinez et al. (2016) in their review reported cryopreservation of sperm, oocytes, embryos, somatic cells, and primordial germ cells of early spermatogonia along with their advantages and disadvantages. Though cryopreservation protocol has been developed for sperm and oocyte of different species, the commercialization is not successful. Further research is required for effect of freezing especially at sub lethal level (Diwan et al. 2020).

References

Agarwal NK (2011) Cryopreservation of fish semen. In: Himalayan aquatic biodiversity conservation & new tools in biotechnology. Transmedia Publication, Srinagar

Alavi SMH, Psenicka M, Rodina M, Policar T, Linhart O (2008) Changes of sperm morphology, volume, density and motility and seminal plasma composition in Barbus barbus (Teleostei: Cyprinidae) during the reproductive season. Aquat Living Resour 21(1):75–80

Anchordoguy TJ, Rudolph AS, Carpenter JF, Crowe JH (1987) Modes of interaction of cryoprotectants with membrane phospholipids during freezing. Cryobiology 24(4):324–331

Anil S, Zampolla T, Zhang T (2011) Development of in vitro culture method for zebra fish ovarian tissue fragment. Cryobiology 63:311–312

Arav A, Yavin S, Zeron Y, Natan D, Dekel I, Gacitua H (2002) New trends in gamete's cryopreservation. Mol Cell Endocrinol 187(1–2):77–81

Bagchi A, Woods EJ, Critser JK (2008) Cryopreservation and vitrification: recent advances in fertility preservation technologies. Expert Rev Med Devices 5(3):359–370

Beirão J, Cabrita E, Pérez-Cerezales S, Martínez-Páramo S, Herráez MP (2011) Effect of cryopreservation on fish sperm subpopulations. Cryobiology 62(1):22–31

Beirão J, Boulais M, Gallego V, O'Brien JK, Peixoto S, Robeck TR, Cabrita E (2019) Sperm handling in aquatic animals for artificial reproduction. Theriogenology 133:161–178

Benson JD, Woods EJ, Walters EM, Critser JK (2012) The cryobiology of spermatozoa. Theriogenology 78(8):1682–1699

Betsy J, Kumar S (2020) Cryopreservation of fish gametes. Springer, Berlin

Blaxter JHS (1953) Sperm storage and cross-fertilization of spring and autumn spawning herring. Nature 172(4391):1189–1190

Brycon Y, Arias-castellanos JA (2004) Cryopreservation of Yamu. J World Aquacult Soc 35 (4):529–535

Bustamante-González JD, González-Rentería M, Rodríguez-Gutiérrez M, Cortés-García A, Ávalos-Rodríguez A (2016) Methodologies for spermatic evaluation in teleost. Int J Aquat Sci 7(2):95–106

Cabrita E, Robles V, Herráez P (2009) Sperm quality assessment. In: Cabrita E, Robles V, Herráez P (eds) Methods in reproductive aquaculture marine and freshwater species. CRC Press, Boca Raton, pp 93–147

Cabrita E, Sarasquete C, Martínez-Páramo S, Robles V, Beirao J, Pérez Cerezales S, Herráez MP (2010) Cryopreservation of fish sperm: applications and perspectives. J Appl Ichthyol 26 (5):623–635

Chen SL, Tian YS (2005) Cryopreservation of flounder (Paralichthys olivaceus) embryos by vitrification. Theriogenology 63(4):1207–1219

Che-Zulkifli CI, Chu KC, Mustafa S, Md Sheriff S, Ikhwanuddin MHD (2019) Use of dry ice for the shipping and packaging of cryopreserved giant grouper Epinephelus lanceolatus spermatozoa. Egyp J Aquat Res 46:85–90

Che-Zulkifli CI, Koh ICC, Shahreza MS, Ikhwanuddin M (2020) Cryopreservation of spermatozoa on grouper species: a review. Rev Aquacult 12(1):26–32

Cosson J, Groison AL, Suquet M. Fauvel C, Dreanno C, Billard R (2008) Studying sperm motility in marine fish: an overview on the state of the art. J Appl Ichthyol 24(4):460–486

Diwan AD, Ayyappan S, Lal KK, Lakra WS (2010) Cryopreservation of fish gametes and embryos. Indian J Anim Sci 80(4):109–124

Diwan AD, Harke SN, Panche AN (2020) Cryobanking of fish and shellfish egg, embryos and larvae: an overview. Front Mar Sci 7:251

Donoghue AM, Thistlethwaite D, Donoghue DJ, Kirby JD (1996) A new method for rapid determination of sperm concentration in turkey semen. Poult Sci 75(6):785–789

Guan M, Rawson DM, Zhang T (2010) Cryopreservation of zebra fish (Danio rerio) oocytes by vitrification. Cryoletters 31:230–238

Gwo JC (2000) Cryopreservation of aquatic invertebrate semen: a review. Aquacult Res 31:259–271

Hafez ESE (2000) Preservation and cryopreservation of gametes and embryos. In: Hafez ESE (ed) Reproduction in farm animals. Blackwell, Iowa, pp 431–442

Harvey B, Kelly RN, Ashwood-Smith MJ (1983) Cryobiology permeability of intact and dichorionated zebra fish embryos to glycerol and dimethyl sulfoxide. Aquaculture 20:432–439

Herranz-Jusdado JG, Gallego V, Morini M, Rozenfeld C, Pérez L, Kása E, Kollár T, Depincé A, Labbé C, Horváth Á, Asturiano JF (2019) Comparison of European eel sperm cryopreservation protocols with standardization as a target. Aquaculture 498:539–544

Holt WV (2000) Fundamental aspects of sperm cryobiology: the importance of species and individual differences. Theriogenology 53(1):47–58

Horokhovatskyi Y, Sampels S, Cosson J et al (2016) Lipid composition in common carp (Cyprinus carpio) sperm possessing different cryoresistance. Cryobiology 73:282–285

Huebinger J, Han HM, Hofnagel O, Vetter IR, Bastiaens PI, Grabenbauer M (2016) Direct measurement of water states in cryopreserved cells reveals tolerance toward ice crystallization. Biophys J 110(4):840–849

Irawan H, Vuthiphandchai V, Nimrat S (2010) The effect of extenders, cryoprotectants and cryopreservation methods on common carp (Cyprinus carpio) sperm. Anim Reprod Sci 122 (3-4):236–243

Isayeva A, Zhang T, Rawson DM (2004) Studies on chilling sensitivity of zebra fish (Danio rerio) oocytes. Cryobiology 49(2):114–122

Jawahar KTP, Betsy J (2020) Cryopreservation of fish gametes. In: Cryopreservation of fish gametes: an overview. Springer, Berlin, pp 151–175

Jusdado JG, Victor G, Marina M, Christoffer R, Luz P, Tamás M et al (2019) Eel sperm cryopreservation: an overview. Theriogenology 133:210–215

Kopeika J, Zhang T, Rawson DM, Elgar G (2005) Effect of cryopreservation on mitochondrial DNA of zebra fish (Danio rerio) blastomere cells. Mutat Res Fundament Mol Mech Mutagen 570:49–61

Kopeika E, Kopeika J, Zhang T (2007) Cryopreservation of fish sperm. In: Cryopreservation and freeze-drying protocols. Humana Press, Totowa, pp 203–217

Kowalski RK, Cejko BI (2019) Sperm quality in fish: determinants and affecting factors. Theriogenology 135:94–108

Labe C, Robles V, Herraez MP (2013) Cryopreservation of gametes for aquaculture and alternative cell sources for genome preservation. In: Allan G, Burnell G (eds) Advances in aquaculture hatchery technology. Woodhead Publishing, Cambridge, pp 76–116

Lee YH, Park JY, Lee IY, Zidni I, Lim HK (2020) Effects of cryoprotective agents and treatment methods on sperm cryopreservation of stone flounder, Kareiusbicoloratus. Aquaculture 531:735969

Linhartová Z, Rodina M, Guralp H, Gazo I, Saito T, Pšenička M (2014) Isolation and cryopreservation of early stages of germ cells of tench, Tincatinca. Czeh J AnimSci 59:381–390

Liu XH, Zhang T, Rawson DM (1998) Feasibility of vitrification of zebra fish (Danio rerio) embryos using methanol. Cryo Letters 19:309–318

Marques LS, Mikich AB, Leandro G, Laura AS, Daniel M, Zhang T et al (2015) Viability of zebra fish (Danio rerio) ovarian follicles after vitrification in a metal container. Cryobiology 71:1–7

Martinez PS, Horváth Á, Labbé C, Zhang T, Robles V, Herráez P et al (2016) Cryobanking of aquatic species. Aquaculture 472:156–177

Morris J (2000) Asymptote cool guide to cryopreservation, vol 44. Asymptote Ltd, Cambridge, pp 1–42

Morris GJ, Acton E, Murray BJ et al (2012) Freezing injury: the special case of the sperm cell. Cryobiology 64:71–80

Nahiduzzaman M, Hassan MM, Roy PK, Hossain MA, Hossain MAR, Tiersch TR (2012) Sperm cryopreservation of the Indian major carp, Labeo calbasu: effects of cryoprotectants, cooling rates and thawing rates on egg fertilization. Anim Reprod Sci 136:133–138

Nomura K, Koh ICC, Iio R, Okuda D, Kazeto Y, Tanaka H, Ohta H (2018) Sperm cryopreservation protocols for the large-scale fertilization of Japanese eel using a combination of large-volume straws and low sperm dilution ratio. Aquaculture 496:203–210

Okutsu T, Yano A, Nagasawa K, Shikina S, Kobayashi T, Takeuchi Y, Yoshizaki G (2006) Manipulation of fish germ cell: visualization, cryopreservation and transplantation. J Reprod Dev 52:685–693

Patra S, Mishra G, Dash SK, Verma D, Nandi S, Jayasankar P, Routray P (2016) Transplantation worthiness of cryopreserved germ cells of Indian major carp Rohu, Labeo rohita. Curr Sci 111:739–746

Pereira JR, Pereira FA, Perry CT, Pires DM, Muelbert JRE, Garcia JRE, Corcini CD, Varela Junior AS (2019) Dimethylsulfoxide, methanol and methylglycol in the seminal cryopreservation of Suruvi, Steindachneridion scriptum. Anim Reprod Sci 200:7–13

Plachinta M, Zhang T, Rawson DM (2004) Studies on cryoprotectant toxicity to zebra fish (Danio rerio) oocytes. Cryoletters 25:415–424

Ponniah AG, Sahoo PK, Dayal R, Barat A (1999) Cryopreservation of Tor putitora spermatozoa. Effect of extender composition, activation solution, cryoprotectant and equilibration time. Proc Natl Acad Sci U S A 1:53–59

Pšenička M, Saito T, Rodina M, Dzyuba B (2016) Cryopreservation of early stage Siberian sturgeon Acipenser baerii germ cells, comparison of whole tissue and dissociated cells. Cryobiology 72:199–122

Riesco MF, Raposo C, Engrola S, Martínez-Páramo S, Mira S, Cunha ME, Cabrita E (2017) Improvement of the cryopreservation protocols for the dusky grouper, Epinephelus marginatus. Aquaculture 470:207–213

Rurangwa E, Kime DE, Ollevier F, Nash JP (2004) The measurement of sperm motility and factors affecting sperm quality in cultured fish. Aquaculture 234(1–4):1–28

Shaliutina A (2013) Fish sperm motility parameters and total proteins profiles in seminal plasma during in vivo and in vitro storage. Doctoral dissertation, Thèse de Doctorat, Faculté des poisons et de la protection des eaux, Université du SudBohemie, République Czech, 96 p

Sørensen SR, Gallego V, Pérez L, Butts IAE, Tomkiewicz J, Asturiano JF (2013) Evaluation of methods to determine sperm density for the European eel, Anguilla anguilla. Reprod Domest Anim 48(6):936–944

Streit DP, Godoy LC, Ribeiro RP, Fornari DC, Digmayer M, Zhang T (2014) Cryopreservation of embryos and oocytes of South American fish species. In: Yamashiro H (ed) Recent advances in cryopreservation. IntechOpen, London

Suquet M, Billard R, Cosson J, Dorange G, Chauvaud L, Mugnier C, Fauvel C (1994) Sperm features in turbot (Scophthalmus maximus): a comparison with other freshwater and marine fish species. Aquat Living Resour 7(4):283–294

Suquet M, Dreanno C, Fauvel C et al (2000) Cryopreservation of sperm in marine fish. Aquacult Res 31:231–243

Swanson EW, Bearden HJ (1951) An eosin-nigrosin stain for differentiating live and dead bovine spermatozoa. J Anim Sci 10(4):981–987

Tiersch TR, Mazik PM (2011) Cryopreservation in aquatic species, 2nd edn. World Aquaculture Society, Baton Rouge

Tiersch TR, Yang H, Jenkins JA, Dong Q (2007) Sperm cryopreservation in fish and shellfish. Soc Reprod Fertil Suppl 65:493

Tomlinson M (2005) Managing risk associated with cryopreservation. Hum Reprod 20 (7):1751–1756

Tsai S, Rawson DM, Zhang T (2009) Development of cryopreservation protocols for early stage zebra fish (Danio rerio) ovarian follicles using controlled slow cooling. Theriogenology 71:1226–1233

Uusi-Heikkilä S (2020) Implications of size-selective fisheries on sexual selection. Evol Appl 13 (6):1487–1500

Valdebenito II, Gallegos PC, Effer BR (2015) Gamete quality in fish: evaluation parameters and determining factors. Zygote 23(2):177–197

Viveiros ATDM, Godinho HP (2009) Sperm quality and cryopreservation of Brazilian freshwater fish species: a review. Fish Physiol Biochem 35(1):137–150

Xin M, Niksirat H, Shaliutina-Kolešová A, Siddique MAM, Sterba J, Boryshpolets S, Linhart O (2020) Molecular and subcellular cryoinjury of fish spermatozoa and approaches to improve cryopreservation. Rev Aquac 12(2):909–924

Xin M, Shaliutina-Kolesova A, Sterba J et al (2018) Impact of cryopreservation on sterlet, Acipenser ruthenus sperm motility and proteome. Anim Reprod Sci 192:280–289

Yeste M (2016) Sperm cryopreservation update: cryodamage, markers, and factors affecting the sperm freezability in pigs. Theriogenology 85(1):47–64

Zhang T, Isayeva A, Adams SL, Rawson DM (2005) Studies on membrane permeability of zebra fish (Danio rerio) oocytes in the presence of different cryoprotectants. Cryobiology 50:285–293

Nanobiotechnology: Prospects and Applications in Aquaculture

14

Mohd Ashraf Rather, Deepak Agarwal, Sujit Kumar, and Jitendra Kumar Sundaray

Abstract

Application of nanotechnology in biological fields is known as Nanobiotechnology. Nanotechnology is fast emerging and most promising technology applied in all areas of science including aquaculture. Nanoparticles are those particles which have the dimension or size in the range of 1–100 nm. Different types of organic and inorganic nanoparticles have been developed for different purposes in fisheries and aquaculture research. Several methods have been used for the synthesis of nanoparticles, but the green synthesis of nanoparticles has emerged as an alternative to overcome the toxic effect of chemically synthesized nanoparticles. Nanoparticles have been used in many applications in the field of aquaculture medicine, including drug and gene delivery vaccination and diagnostics. Polymeric nanoparticles like chitosan, poly lactic-*co*-glycolic acid (PLGA) have been used for drug delivery because of their several key characteristics, like biocompatibility biodegradability, a stability of the drug, specificity of delivery etc. Use of nano-vaccines against many bacterial and viral pathogens has gained much interest in the last few years

M. A. Rather (✉)
Division of Fish Genetics and Biotechnology, Faculty of Fisheries, SKUAST-Kashmir, Srinagar, Jammu and Kashmir, India

D. Agarwal
Institute of Fisheries Post Graduation Studies, TNJFU, Chennai, Tamil Nadu, India

S. Kumar
Postgraduate Institute of Fisheries Education and Research, Kamdhenu University, Gandhinagar, Gujarat, India

J. K. Sundaray
ICAR-Central Institute of Freshwater Aquaculture, Bhubaneswar, Odisha, India

© The Author(s), under exclusive license to Springer Nature Singapore Pte Ltd. 2021
P. K. Pandey, J. Parhi (eds.), *Advances in Fisheries Biotechnology*,
https://doi.org/10.1007/978-981-16-3215-0_14

in aquaculture. Recently nanoparticles have been used specific and sensitive tool for diagnosis of various bacterial, fungal and viral diseases in aquaculture.

Keywords

Nanobiotechnology · Nanoparticles · Nanodelivery · Aquaculture

14.1 Introduction

Among all food producing sectors, aquaculture comes to be the fastest growing food production sector with a rate of 8.8% annually (FAO 2012). But the single unsolved question mark for aquaculture and its sustainability is—*"disease prevalence and the poor health of system."* In spite of the several strategies adopted on the national and international level, as improved laboratory facilities, diagnostic expertise and control and therapeutic strategies in order to handle disease outbreaks more effectively. The aquaculture industry is under uncertainty and the progress has not matched that of the rapidly developing aquaculture sector. The sector demand more technical innovation for the drug use, disease treatment, water quality management, production of tailored fish for suiting better health, productivity drive by epigenetic and nutrigenomic interaction, better breeding success by efficient delivery of maturation and spawning inducing agent, nutraceutical delivery for rapid growth promotion and culture time reduction, successful use of auto-transgenic and effective vaccines. For these multiple purposes the effort and importance of the nanotechnology and nanodelivery of drugs, vaccines, nutraceuticals, inducing hormones and growth-promoting anabolic bears the promise (Rather et al. 2011).

Nanotechnology has been defined by the US National Nanotechnology Initiative (NNI) as "understanding and control of matter at dimensions of roughly 1–100 nm where unique phenomena enable novel applications." More elaborately, it may be defined as "the study, design, creation, synthesis, manipulation and application of functional materials, devices, and systems through control of matter at the nanometer scale (1–100 nm, 1 nm being equal to 1×10^{-9} of a meter), that is at the atomic and molecular levels, and the exploitation of novel phenomena and properties of matter at that scale." Several applications of nanotechnology for aquaculture production are being developed. With a strong history of adopting new technologies, the highly integrated fish farming industry may be among the best to incorporate and commercialize nanotech products.

14.2 Types of Nanomaterials

As nanomaterials are small, they have a much greater surface area to volume ratio than the conventional forms. They can be produced in one dimension (surface film), two dimension (strand or fibers), or three dimension (particles) and in different irregular and regular shape such as sphere, rod tubes wires, etc. Because of their

Table 14.1 Classification of nanomaterials

Nature of material	Examples	Applications
Organic materials	Chitosan, alginates, gelatin	Drug and gene delivery
	Starch, liposomes, dextran	
Inorganic or metal based	Quantum dots, nanogold, nanosilver, metal oxides, etc.	Imaging, diagnostic
Carbon based	Fullerenes, dendrimers, nanotube	Drug and gene delivery
Polymeric carrier	Poly lactic-*co*-glycolic acid (PLGA), poly-alkylcyanoacrylates, polyethyleneamine	Drug, vaccine and gene delivery

Fig. 14.1 (**a**) Top-down method. (**b**) Bottom-down method

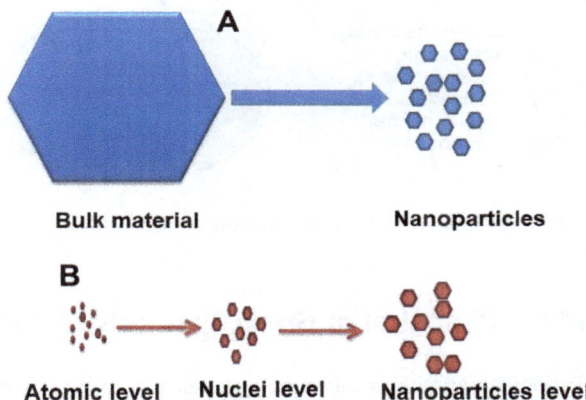

unique electrical, catalytic, magnetic and thermal features, these materials have received much attention among researcher in many fields of biological science including fisheries and aquaculture. Nanomaterials can be broadly classified based on nature or material used for their manufacture as given below Table 14.1.

14.3 Synthesis of Nanoparticles

Generally, there are two main methods for nanoparticle synthesis top-down and bottom-up. The top-down method involves mechanical breakdown, lithography techniques or grinding process of bulk metal to transform it from macroscale to nanoscale, which is followed by addition of stabilizing agents to confirm that the nanoparticles do not oxidize or congregate back to the microscale (Gaffet et al. 1996) (Fig. 14.1). On the other hand, bottom-up approaches include assembly of nanoparticles by various physical and chemical methods, including electrochemical reduction of metals, chemical vapor method or solution evaporation method or atoms self-assemble to new nuclei which grow into a particle of nanoscale (Fig. 14.2).

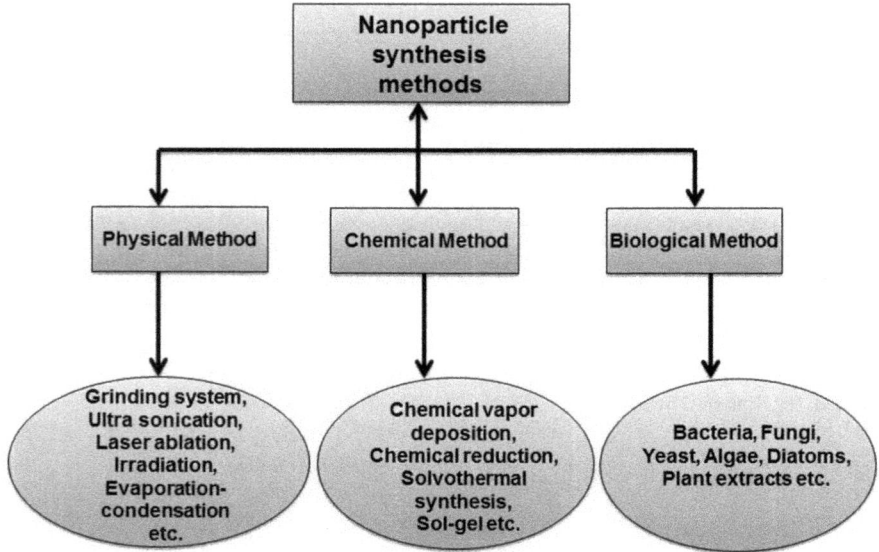

Fig. 14.2 Different methods for synthesis of nanoparticles

14.4   Biological or Green Synthesis of Nanoparticles

Biological synthesis of nanoparticles by using different biological agents has appeared as an alternative to overcome the toxic effect of chemically synthesized nanoparticles. In recent years, green synthesis and biological methods of metal nanoparticle formation has become a major area of interest for research across the globe. Biological synthesis of nanoparticles is a bottom-up approach that mostly involves reduction or oxidation reactions. Biologically synthesized nanoparticles are deduced from three main groups of organisms by:

1. Use of microorganisms like bacteria, algae (*Chlorophyceae, Phaeophyceae, Cyanophyceae, Rhodophyceae*), fungi, actinomycetes (prokaryotes), yeast, etc.
2. Use of plant extracts or microbial enzymes.
3. Use of templates like DNA, membranes, viruses, and diatoms.

The use of biological agents such as leaf extracts, bacteria, fungi and yeast for the synthesis of metallic nanoparticles is safe for biological applications with zero chemical toxicity (Antony et al. 2013; Mathur et al. 2014). The three main elements of a biosynthetic nanoparticle system are a solvent medium for synthesis, a reducing agent, and a nontoxic stabilizing agent. A flow diagram or procedure of synthesis of silver nanoparticles from neem plant is given in Fig. 14.3.

The leaf extracts acts both as reducing agent as well as a capping agent. Silver nitrate ions in aqueous form were reduced by neem leaf extract. The change in color

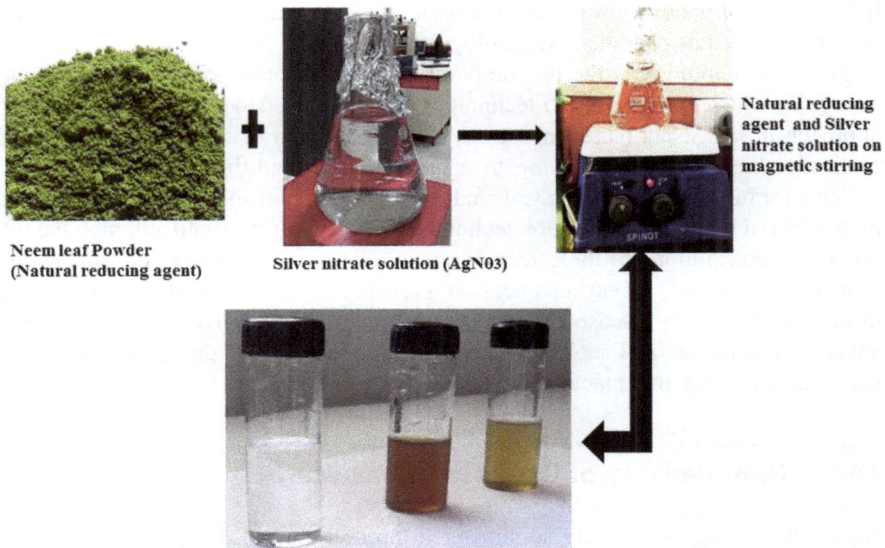

Neem leaf Powder
(Natural reducing agent)

Silver nitrate solution (AgN03)

Natural reducing
agent and Silver
nitrate solution on
magnetic stirring

Fig. 14.3 Green syntheses of silver nanoparticles at different time interval

of the reaction mixture from reddish yellow to brown indicated the formation of silver nanoparticles. Biosynthesis of metal nanoparticles is a kind of bottom-up approach for nanoparticle formation, where the central reactions like reduction/oxidation take place. Generally, two possible ways of the conversion of silver nitrate to silver nanoparticles by biological methods are found. The first reason may be the secondary metabolites (phenolic acid, terpenoids) existing in neem extract is believed to help in conversion of silver nitrate to the silver nanoparticles. The energy released during glycolysis process might be the second reason (Vignesh et al. 2013). Synthesis of metallic nanoparticles like silver, gold etc., using various plants and their extracts can be advantageous over other biological synthesis processes which involve the very complex events of maintaining microbial cultures.

Various algae such as *Spirulina platensis, Lyngbya majuscula,* and *Chlorella vulgaris* were used for the synthesis of metallic nanoparticles like silver, gold nanoparticles (Chakraborty et al. 2009; Niu and Volesky 2000). Among the biological resources, algae are called as bionano factories because both the live and dead dried biomasses were applied for the synthesis of metallic nanoparticles (Davis et al. 1998). Using *Ulva fasciata* extract as a reducing agent, synthesis of silver nanoparticles inhibited the growth of *Xanthomonas campestris pv. malvacearum* (Rajesh et al. 2012). Diatoms like *Navicula atomus* and *Diadesmis gallica* have been used to synthesize gold nanoparticles, and silica–gold bionanocomposites (Schröfel et al. 2011). Comparing with other biological material such as bacteria, fungi, and yeast, algae is correspondingly an essential organism in the synthesis of different nanoparticles. Hence, the studies of algae-mediated biosynthesis of nanometals have come under new branch of nanotechnology and it has been termed as phyconanotechnology (Sharma et al. 2016).

Nanobiotechnology is the unique combination of biotechnology and nanotechnology by which classical micro-technology can be merged to a molecular biological approach in real. Nanobiotechnology is the combination of small engineering and molecular biology that is leading to a new class of multifunctional devices and systems for biological and chemical analysis with better sensitivity, specificity and a higher rate of recognition. Nanobiotechnology is relatively new and although the full scope of contributions to these technological advances in the field of aquaculture remains unexplored, recent advances suggest that nanobiotechnology will have a profound impact on disease prevention, diagnosis, drug delivery, gene delivery, vaccine formation and delivery and so on. Some of the applications of nanobiotechnology in aquaculture are given in detail.

14.5 Nanodelivery of Drugs in Aquaculture

Aquaculture drugs are mostly administered through major three delivery routes as bath or immersion, second through in-feed or oral, and the third by injection. While the first mode of immersion or bath is more applicable, but it requires the drug in more amounts and the handling to fish cause unavoidable stress. So the most ease method come to be through in-feed formulation where the drugs are applied with normal feeding without stress and extra cost. And lastly, the injection seems to be impractical for fishes. The oral administration of drugs essentially needs the study of toxic kinetics, pharmacokinetics and pharmacodynamics of drugs in species and stage-specific manner before in-feed administration. In the in-feed delivery of drugs especially for the peptides, vaccine, and the DNA or RNA components the major problem is the gastric digestion or denaturation before reaching to the intestine for absorption (Florence et al. 1995). So the in-feed formulation and delivery of drugs through feed needs an immediate intervention of an innovative step. Here the nano-encapsulated or nano-coated drug delivery give a solution where the drugs, vaccines, adjuvant, enzymes, etc. as protected and passed to the intestine and made to stay for longer time in the intestine for better absorption and assimilation. One of the most promising and productive areas of nanotechnology application to animal research is the nano-pharmaceuticals (Tomlinson and Rolland 1996). One of the major classes of drug delivery systems is materials that encapsulate drugs to protect them during transit through the body. When encapsulation materials are produced from nanoparticles in the 1 to 100-nm size range instead of bigger micro particles, they have a larger surface area for the same volume, smaller pore size, improved solubility and different structural properties. This can improve both the diffusion and degradation characteristics of the encapsulation material. Another class of drug delivery system is nanomaterials that can carry drugs to their destination sites and also have functional properties. Certain nanostructures can be controlled to link with a drug, a molecule or an imaging agent, then attract specific cells and release their payload when required. The antibacterial properties of nanotubes are being studied.

Self-assembled stacking of cyclic peptides having an even number of alternating and L-amino acids forms the nanotubes. The nanotubes insert themselves readily into bacterial cell membranes and act as potent and selective antibacterial agents; both nanomaterials bucky balls and nanotube will undoubtedly become an important part of the total pharmaceutical tool kit over the next few years (Kannaki and Verma 2006). Targeting Ligands and receptor or ligand specific delivery of the drug is one key for targeted drug or gene or DNA delivery, which needs various ligands as small molecules galactose, glucose, and mannose, protein like transferins, antibodies, and low-density lipids (LDLS). Drug delivery is an area where nanotechnology has already had a significant impact (LaVan and McGuire 2003) and surely for this aquaculture drugs has the major share, because the drugs and nutraceuticals in aquaculture and their delivery methods need several concern as:

1. Cost of the drugs and nutraceutical used in culture.
2. Same time we need to target less waste as feed cost is the major cost and aquaculture is the margin sifting industry.
3. Efficacy and cost analysis of such drugs being used in aquaculture.
4. Environmental impact of these additives and drugs used.
5. Monitoring the residues level and its impact on food value.
6. Uncertain fate of the administered feeds in the water.
7. Feed palatability and acceptance by fish after use of drugs and nutraceuticals.
8. Toxicity of drugs to fishes at higher dose.

The factors mentioned above strongly suggest the need of the effective delivery of the drugs where the efficacy can be achieved at low dose in a sustained manner, without any toxicity to fish and residue of drug in the flesh. Nanodeliveries—has satisfaction for several of these issues in aquaculture drugs like drug safety, residue level in flesh and sustain release to reduce the frequency of dosing and overall cost of drug or treatment. There are several natural polymeric and other nano carriers which can easily be applied for aquaculture drug without adding much cost. Goal of more sophisticated drug delivery techniques like nanodelivery is to

1. Deploy to a target site to limit side effects.
2. Shepard drugs through specific areas of the body without degradation.
3. Maintain a therapeutic drug level for prolonged periods of time.
4. Predictable controllable release rates.
5. Reduce dosing frequent and increase fish compliance without frequent handling and stress.

Therefore, the delivery of drugs and nutraceuticals in aquaculture will have to find the way through nano-delivery (Rather et al. 2011). Nanomedicine is a rapidly growing aspect of nanotechnology and there is an opportunity to use these techno-logical advances to monitor and improve fish health. The poor stability of pharmaceuticals in natural water has inevitably led to many fish medicines being delivered via the food or accepting that much of any aqueous treatment may be simply washed away. Nano carriers have been exploited to make new drug delivery systems for humans, and these may also be used for veterinary medicines including those for fishes. The approaches include solid core drug delivery systems (SCDDS), which involve coating a solid NP with a fatty acid shell to contain the drug of interest. This methodology works at relatively low temperature and pressure, making it especially useful for heat sensitive or labile pharmaceuticals. Porous NMs can also be used as a drug delivery matrix. For example, mesoporous silica particles can be used for the controlled release of substances. Oral delivery is the easiest and practical through in-feed formulations and only option for the nutraceuticals delivery. The various in-feed formulation forms like pelleted form, micro encapsulated, microparticulate form and nano particulate or encapsulated forms are the means of delivery. As the immersion pose handling stress and more amounts of drugs required which raise the treatment cost, impact on the environment and may cause possible threat on other organism and resistance building. While the Injection method becomes unrealistic for field application, the oral/ in-feed delivery with add of nano carriers becomes excellent mean of delivery for drugs and nutraceuticals (Rather et al. 2011).

14.6 Gene or DNA Delivery

There is a relation between nanoparticles and gene or DNA delivery, Polynucleotide vaccines work by delivering genes encoding relevant antigens to host cells where they are expressed, producing the antigenic protein within the vicinity of professional antigen-presenting cells to initiate immune response. Such vaccines produce both humoral and cell-mediated immunity because intracellular production of protein, as opposed to extracellular deposition, stimulates both arms of the immune system (Gurunathan and Freidag 2000). The key ingredient of polynucleotide vaccines, DNA, can be produced cheaply and has much better storage and handling properties than the ingredients of the majority of protein-based vaccines. Hence, polynucleotide vaccines are set to supersede many conventional vaccines, particularly for viral diseases. However, there are several issues related to the delivery of polynucleotide which limits their application. These issues include efficient delivery of the polynucleotide to the target cell population and its localization to the nucleus of these cells, and ensuring that the integrity of the polynucleotide is maintained during delivery to the target site (Mohanraj and Chen 2006). Nanoparticles loaded with plasmid DNA could also serve as an efficient sustained release gene delivery system due to their rapid escape from the degradative endolysosomal compartment to the cytoplasmic compartment (Panyam et al. 2002). Hedley et al. (1998) reported that following their intracellular uptake and endolysosomal escape, nanoparticles could release DNA at a sustained rate resulting in sustained gene expression. This gene delivery strategy could be applied to facilitate the transgenic production, for modulating the expression of a gene, say GH gene in fishes to boost growth, produce the tailored fish product or health food. The DNA vaccines administered as liposomal complexes also improve the antibody response over that seen with free

DNA (Gregoriadis et al. 1997). These include a reduction in the rapid clearance of cationic liposomes and the production of efficiently targeted liposomes. At the cellular level, the problems may be overcome by improving receptor-mediated uptake using appropriate ligands, the endowment of liposomes with endosomal escape mechanisms, and a more efficient translocation of DNA to the nucleus and the efficient dissociation of the liposome complex just before the entry of free DNA into the nucleus. If it is delivered though in-feed formulations, then the protection of the chitosan or alginate nano-capsulation in a single layer or a layer over another carrier will ensure the gastric protection. Promising results were reported in the formation of complexes between chitosan and DNA. Although chitosan increases transformation efficiency, the addition of appropriate ligands to the DNA–chitosan complex seems to achieve a more efficient gene delivery via receptor-mediated endocytosis (Eldridge et al. 1990). These results suggest that chitosan has comparable efficacy without the associated toxicity of other synthetic vectors and therefore, can be an effective gene delivery vehicle in vivo (Murata et al. 1998).

In aquaculture, several gene expression studies have been done through nanoparticle administration in different fish species. The first report on use of chitosan-conjugated nanodelivery of gonadotropic hormone in fish was done in *Cyprinus carpio* trough administration of chitosan-gold nanoconjugates LHRH to investigate the surge of gonadotropins and reproductive output in female fish (Rather et al. 2013). Later, Rather et al. (2016), analyzed the expression of kisspeptin gene through nano-delivery of chitosan encapsulated nanoparticle of kisspeptin-10 in *Catla catla*. In fishes, several genes have been reported which affect the gonadal development and maturity. One of the gene sox9 plays a crucial role in determining the fate of several cell types and is a primary factor in regulation of gonadal development. In *Clarias batrachus*, the expression of sox9 was analyzed through the nanodelivery of Chitosan-conjugated LHRH (Rather et al. 2016). Expression vector have also been used in the nanodelivery of specific gene constructs. A chitosan nanoparticle conjugated StAR gene construct was prepared in a eukaryotic expression vector and administered in the *Clarias batrachus* through intramuscular injection. The StAR gene was detected in several tissues to get the regulating rate and timing of steroidogenesis (Rather et al. 2017). Identification of key genes responsible for any physiology in fishes is a crucial step which can be done through transcriptomics analysis. RNA sequencing of the cell or tissue before or after some stimulation leads to the identification of several differential expressed genes (Agarwal et al. 2020). These genes can be conjugated with nanoparticles and administered in the fish to analyze the phonotypical changes, though the use of different nanoparticles might affect the physiology of the animal through accumulation in the body.

14.7 Vaccine-Adjuvant

Vaccination is one of the important methods of prevention of disease in advance by developing antibody agents the particular pathogen. Most of the vaccines are applied as a fluid form and generally injected into blood stream. These vaccines require the cool temperature to be stored and they also have limited life span within which they are to be utilized. These two limitations have prevented the utility of vaccines, particularly in the fishes and shrimp. Therefore, more robust and durable vaccines are the only solution for the successful eradication of the particular disease.

Many organisms, particularly microorganisms, have novel and interesting structures that could be exploited, for example, the lattice-type crystalline arrays of bacterial S-layers and bacterial spore coats both of which have protective prosperities. In principle, the spore coat could be used not only as a delivery vehicle for a variety of different molecules but also as a source of new and novel self-assembling proteins. Spore coats are comprised of protein, have ordered arrays of photometric subunits, exhibit self-assembly and have protective prosperities. A spore-based display system provides several advantages with respect to systems based on the use of conventional vaccines; these include the robustness of the bacterial spore allowing storage in the desiccated form, ease of production and safety. And it can be suitably used for aquaculture bacterial vaccines. Similarly, in the liposomes as vaccine adjuvants, liposomes have been firmly established as immune-adjuvants (enhancers of the immunological response), potentiating both cell-mediated and humoral immunity (Gregoriadis et al. 1996). Liposomal vaccines can be made by associating microbes, soluble antigens, cytokines (Gregoriadis et al. 1996) or deoxyribonucleic acid (DNA) (Gregoriadis et al. 1997) with liposomes, the latter stimulating an immune response on the expression of the antigenic protein (Gregoriadis 1997). Liposomes encapsulating antigens which are subsequently encapsulated within alginate lysine microcapsules (Cohen et al. 1991) or to chitin or chitosan or other polymeric microcapsules to protect the gastric digestion and to control antigen release for improving the antibody response. Liposomal vaccines may also be stored dried at refrigeration temperatures for up to 12 months and still retain their adjuvanticity (Kim and Jeong 1995). In DNA nano-vaccines using nanocapsules and ultrasound methods, the United States Department of Agriculture (USDA) is completing trials on a system for mass vaccination of fish in fishponds using ultrasound. Nanocapsules containing short strands of DNA are added to a fishpond where they are absorbed into the cells of the fish. Ultrasound is then used to rupture the capsules, releasing the DNA and eliciting an immune response from the fish. This technology has so far been tested on rainbow trout by Clear Springs Foods (Idaho, US) a major aquaculture company that produces about one-third of all U.S. farmed trout (Mongillo 2007; ETC 2003).

Advances in nanotechnology have also proved to be beneficial in therapeutic fields such as drug discovery, drug delivery, and gene/protein delivery. Now a day synthetic siRNA is considered as a highly promising therapeutic agent for a viral disease like WSSV. However, clinical use of siRNA has been hampered by instability in the body and inability to deliver sufficient RNA interference compounds to the tissues or cells. To address this challenge, we present here a single siRNA nanocapsule delivery technology, which is achieved by encapsulating a single siRNA molecule within a degradable polymer nanocapsule with a diameter around 20 nm and positive surface charge. Several works provide a potential novel platform for siRNA delivery that can be developed for therapeutic purposes.

14.8 Management of Animal Breeding

Aquaculture today is growing vertically and horizontally, growing rate of mariculture indicates the kind of horizontal expansion in the sea where the immediate need felt is for potentiating the maturation of the fishes for early and effective breeding same time easy artificial spawning to ensure uninterrupted supply of quality seed which is the backbone for aquaculture growth. The nano-delivery of the hormones, anabolic agents, hormone analogue and the spawning inducing agent is very important for it where the effective delivery without much wastage can be targeted through in-feed formulation in marine cages. Same time the oral delivery of the nano carrier-based or nano-encapsulated hormones will have an option of mass breeding of the fishes in confined water in natural or artificial habitat (Rather et al. 2011).

14.9 Nano Smart Delivery System in Cell vs. Transgenic

Today the worldwide discussion is about the uses and abuse of transgenics. Human need and greed over the luring enhanced production and desirable trait attainment for diseases resistance and environmental tolerance seems necessary but same time the impact on biodiversity and food safety bring a big question mark even on survival and sustainability. Here the big dilemma can be solved by the nano-delivery by using smart nano cell delivery, which will ensure only the cell specific or tissue specific modification of expression leading production of particular hormone say growth hormone in a specific tissue. The delivery will be on the basis of cell surface-based receptor interaction and multiple nanocoating materials, this will be able to release the gene or DNA construct in the nucleus of decided cell. It will also be used for triggering the protein- protein interaction channel to trigger the various synthetic cascade of the cell leading to expression of the particular gene by various factors or cofactors activation and triggering binding to genes response element. This particular mean will be effective as it will lead to temporary and even long lasting change in the somatic cell only. The epigenetic or protein-protein based triggering will have a definite life of triggering, while the genetic integration or transgenics will be a lifelong modification of expression in that animal. Besides that, it will have the chance of genetic contamination and environmental impact. As the somatic cell, alteration will not be carried to the germplasm. Oral delivery if achieved it becomes super simple, overall it will be the least tedious than transgenic gene transfer to embryo where the success rate is less and it requires the sophisticated techniques or equipment.

14.10 Disease Diagnosis

Unlike higher animals, this application of nanobiotechnology seems to be meaningless in aquaculture. To prevent the mass mortality of commercially important fish and shellfish due to diseases at an early stage needs an important attention. The technologies that are used in higher animals can be mimicked and applied in aquaculture for example in animals the tumor detection is possible now at an early stage by RNA nanoparticles (Guo et al. 2012). Similarly, there are some other techniques that can be applied in aquaculture like nanobodies, a new generation of antibody-based therapeutics for disease diagnosis (Jain 2005).

14.11 Nano-biosensors

Nanotechnology based biosensors can be used in the aquaculture industry for microbe control. Researchers at NASA, USA have developed a sensitized carbon nanotubes based biosensor that is capable of detecting minute amount of microbes including bacteria, virus, parasite, and heavy metals from water and foods sources. Nano colloidal sliver is one the most beneficial product of nanotechnology and act as a catalyst and work on a wide spectrum of bacteria, fungi, parasites and virus by rendering an enzyme inoperative, which is used for their metabolism. Unlike antibiotic resistant strains of bacteria, no such strain is known to develop by using colloidal sliver. Even these sliver nanoparticles are able to kill methicillin-resistant *S. aureus*. Tracking nano-sensors are being developed. "Smart fish" may be fitted with sensors and locators that relay data about their health and geographical location to a central computer. Such technology may be used to control cognitive cage systems or individual fish (ETC Group Report 2003).

14.12 Tagging and Nano-barcoding

Radio frequency ID (Rfid) is a chip with a radio circuit incorporating nanoscale component with an identification code embedded in it. These tags can hold more information, scanned from a distance and embedded in the product to identify any object anywhere automatically. These tags may be used as a tracking device as well as device to monitor the metabolism, swimming pattern and feeding behavior of fish. A nano-barcode is a monitoring device consisting of metallic stripes containing nanoparticle where variations in the striping provide the method of encoding information. By incorporating the nano-barcoding, processing industry and exporters can monitor the source or track the delivery status of their aqua product till it reaches the market. Further, coupled with nanosensors and synthetic DNA tagged with color-coded probes, nanobarcode device could detect pathogen and monitor for temperature change, leakage etc.

14.13 Conclusion and Future Direction

Nanotechnology is a global business enterprise impacting industries, universities, industry regulation agents Worldwide. Nanobiotechnology in all sectors including aquaculture is still in its early stages of expansion; however, the development is multidirectional and fast-paced. Although there are many exhilarating potential biological applications of nanomaterials, one needs to distinguish definite scientific promises from the hype and to constantly improve the understanding of their interactions with intracellular structures, the process, and the environment.

References

Agarwal D, Gireesh-Babu P, Pavan-Kumar A, Koringa P, Joshi CG, Chaudhari A (2020) Transcriptome analysis of *Clarias magur* brain and gonads suggests neuro-endocrine inhibition of milt release from captive GnRH-induced males. Genomics 112(6):4041–4052

Antony JJ, Nivedheetha M, Siva D, Pradeepha G, Kokilavani P, Kalaiselvi S, Sankarganesh A, Balasundaram A, Masilamani V, Achiraman S (2013) Antimicrobial activity of Leucas aspera engineered silver nanoparticles against Aeromonas hydrophila in infected Catla catla. Colloids Surf B Biointerfaces 109:20–24

Chakraborty N, Banerjee A, Lahiri S, Panda A, Ghosh AN, Pal R (2009) Biorecovery of gold using cyanobacteria and an eukaryotic alga with special reference to nanogold formation—a novel phenomenon. J Appl Phycol 21:145–152

Cohen S, Bernstein H, Hewes C, Chow M, Langer R (1991) The pharmacokinetics of, and humoral responses to, antigen delivered by microencapsulated liposomes. Proc Natl Acad Sci U S A 88:10440–10444

Davis SA, Patel HM, Mayes EL, Mendelson NH, Franco G, Mann S (1998) Brittle bacteria: a biomimetic approach to the formation of fibrous composite materials. Chem Mater 10:2516–2524

Eldridge JH, Hammond CJ, Meulbroek JA, Staas JK, Gilley RM, Tice TR (1990) Controlled vaccine release in gut-associated lymphoid tissues I. Orally administered biodegradable microspheres target the Peyer's patches. J Control Release 11:205–214

ETC (Action Group on Erosion, Technology and Concentration) (2003) Down on the farm: the impact of nanoscale technologies on food and agriculture. http://www.etcgroup.org/en/materials/publications.html?pub_id=80

FAO (2012) Status and trends, world review of fisheries and aquaculture

Florence AT, Hillery AM, Hussain N, Jani PU (1995) Nanoparticles as carriers for oral peptide absorption: studies on particle uptake and fate. J Control Release 36:39–44

Gaffet E, Tachikart M, El Kedim O, Rahouadj R (1996) Nanostructural materials formation by mechanical alloying: morphologic analysis based on transmission and scanning electron microscopic observations. Mater Charact 36(4):185–190

Gregoriadis G (1997) Genetic vaccines: strategies for optimization. Pharm Res 15:661–669

Gregoriadis G, Gursel I, Gursel M, McCormack B (1996) Liposomes as immunological adjuvants and vaccine carriers. J Control Release 41:49–56

Gregoriadis G, Saffie R, de Souza JB (1997) Liposome mediated DNA vaccination. FEBS Lett 402:107–110

Guo P, Shu Y, Binzel D, Cinier M (2012) Synthesis, conjugation, and labeling of multifunctional pRNA nanoparticles for specific delivery of siRNA, drugs, and other therapeutics to target cells. Methods Mol Biol 928:197

Gurunathan SC, Freidag WB (2000) DNA vaccines: a key for inducing long-term cellular immunity. Curr Opin Immunol 12:442–447

Hedley ML, Curley J, Urban R (1998) Microspheres containing plasmid-encoded antigens elicit cytotoxic T-cell responses. Nat Med 4:365–368

Jain KK (2005) The role of nanobiotechnology in drug discovery. Drug Discov Today 10:1435

Kannaki TR, Verma PC (2006) The challenges of 2020 and the role of nanotechnology in poultry research. In: Proceedings of National Seminar Central Avian Research Institute, Izzatnagar (India), pp 273–277

Kim CKE, Jeong J (1995) Development of dried liposome as effective immuno-adjuvant for hepatitis B surface antigen. Int J Pharm 115:193–199

LaVan DAT, McGuire R (2003) Langer small-scale systems for in vivo drug delivery. Nat Biotechnol 21:1184–1191

Mathur A, Kushwaha A, Dalakoti V, Dalakoti G, Singh DS (2014) Green synthesis of silver nanoparticles using medicinal plant and its characterization. Pharm Lett 5(5):118–122

Mohanraj VJ, Chen Y (2006) Nanoparticles—a review. Trop J Pharm Res 5(1):561–573

Mongillo JF (2007) Nanotechnology 101. Greenwood Press, West Port

Murata J, Ohya Y, Ouchi T (1998) Design of quaternary chitosan conjugate having antennary galactose residues as a gene delivery tool. Carbohydr Polym 32(2):105–109

Niu H, Volesky B (2000) Gold-cyanide biosorption with L-cysteine. J Chem Technol Biotechnol 75:436–442

Panyam J, Zhou WZ, Prabha S, Sahoo SK, Labhasetwar V (2002) Rapid endo-lysosomal escape of poly (DL-lactide-co-glycolide) nanoparticles: implications for drug and gene delivery. Fed Am Soc Exp Biol J 16:1217–1226

Rajesh S, Raja DP, Rathi JM, Sahayaraj K (2012) Biosynthesis of silver nanoparticles using Ulva fasciata (Delile) ethyl acetate extract and its activity against Xanthomonas campestris pv. Malvacearum. J Biopest 5:119–128

Rather MA, Sharma R, Aklakur M, Ahmad S, Kumar N, Khan M, Ramya VL (2011) Nanotechnology: a novel tool for aquaculture and fisheries development. A prospective mini-review. Fish Aquacult J 16:1–15

Rather MA, Sharma R, Gupta S, Ferosekhan S, Ramya VL, Jadhao SB (2013) Chitosan-nanoconjugated hormone nanoparticles for sustained surge of gonadotropins and enhanced reproductive output in female fish. PLoS one 8(2):e57094

Rather MA, Bhat IA, Gireesh-Babu P, Chaudhari A, Sundaray JK, Sharma R (2016) Molecular characterization of kisspeptin gene and effect of nano–encapsulted kisspeptin-10 on reproductive maturation in Catla catla. Domest Anim Endocrinol 56:36–47

Rather PK, Bhat IA, Rather MA, Gireesh-Babu P, Kumar K, Purayil SBP, Sharma R (2017) Steroidogenic acute regulatory protein (StAR) gene expression construct: development, nanodelivery and effect on reproduction in air-breathing catfish, Clarias batrachus. Int J Biol Macromol 104:1082–1090

Sharma N, Rather MA, Ajima MN, Gireesh-Babu P, Kumar K, Sharma R (2016) Assessment of DNA damage and molecular responses in Labeo rohita (Hamilton, 1822) following short-term exposure to silver nanoparticles. Food Chem Toxicol 96:122–132

Schröfel A, Kratošová G, Bohunická M, Dobročka E, Vávra I (2011) Biosynthesis of gold nanoparticles using diatoms—silica-gold and EPS-gold bionanocomposite formation. J Nanopart Res 13(8):3207–3216

Tomlinson E, Rolland AP (1996) Controllable gene therapy pharmaceutics of non-viral gene delivery systems. J Control Release 39(2-3):357–372

Vignesh V, Anbarasi KF, Karthikeyeni S, Sathiyanarayanan S, Subramanian P, Thirumurugan T (2013) A superficial phyto-assisted synthesis of silver nanoparticles and their assessment on hematological and biochemical parameters in Labeo rohita (Hamilton, 1822). Colloids Surf A Physicochem Eng Asp 39:184–192

Application of Nanotechnology for Abiotic Stress Management in Aquaculture

15

Neeraj Kumar, Shashi Bhushan, Dilip Kumar Singh, Prem Kumar, and Nitish Kumar Chandan

Abstract

The aquatic ecosystem is one of the largest ecosystems which has several interfering abiotic components impacting aquatic organisms. Climate change and pollution are the main drivers of the abiotic factors; due to such impact, production of fish is tremendously affected. However, a new technology in the form of nano is considered to be a boon for aquatic systems and aquaculture production. The aquaculture sector is the fastest blooming food industry globally which plays a major role in meeting the increased demands for animal protein. Climate change, pollution and environmental factors are the major hindrances for this sector in the attempts to achieve global food security. But nanotechnology has a tremendous potential to enhance the production of the aquaculture through mitigation of such abiotic factors. Nanotechnology can be applied to aquaculture in feed formulation, water management, drug delivery, vaccination, antimicrobial application, reproduction and functional feeding. This technology is also being implemented in the analysis of biomolecules, clinical diagnosis, protein or cells, development of nonviral vectors for gene therapy, as transport vehicles for DNA, disease therapeutics, etc. Application of different forms of nanotechnology can reduce the impact of abiotic stress in aquatic ecosystems. This chapter attempts to

N. Kumar (✉)
ICAR-National Institute of Abiotic Stress Management (NIASM), Pune, India

S. Bhushan · D. K. Singh
ICAR-Central Institute of Fisheries Education, Mumbai, India

P. Kumar
ICAR-Central Institute of Brackishwater Aquaculture, Kakdwip Research Centre, Kakdwip, India

N. K. Chandan
ICAR-Central Institute of Freshwater Aquaculture, Bhubaneshwar, India

© The Author(s), under exclusive license to Springer Nature Singapore Pte Ltd. 2021
P. K. Pandey, J. Parhi (eds.), *Advances in Fisheries Biotechnology*,
https://doi.org/10.1007/978-981-16-3215-0_15

211

give an overview of the application of nanotechnology for abiotic stress mitigation.

Keywords

Abiotic stress · Nanotechnology · Fish · Nutraceuticals · Pollution

15.1 Introduction

Sustainable and productive agriculture is the need of the time to ensure the existence of humanity, and recent agriculture practices can produce enough food as per our needs, but it will not be sustainable without technological interventions (Kundu et al. 2013; Hadioui et al. 2013). In view of ensuring food security, nanotechnology has to play a very crucial and promising role for sustainable food production (Gagné et al. 2016). It has to be done through agriculture including crops, animal husbandry, poultry and fisheries. The protein source from the aquatic ecosystem is very important and therefore, global aquaculture has to play a key role in ensuring food and nutritional security. The global production of fish was estimated to be 171 million ton during 2016 which has estimated cost of US\$362 billion (Shah and Mraz 2020). The global fish consumption has increased from 9.0 kg in 1961 to 20.2 kg in 2015 (FAO 2018). The fisheries sector is still experiencing uncertainty with regard to sustainability as aquaculture waste has negative impact on productivity inside aquaculture systems. In addition to this, drastic climate change has affected the environment throughout the globe. It is a fact that today, most of the aquatic ecosystem are contaminated with other contaminants; it may be pesticides (Kumar et al. 2011) or metal (Kumar et al. 2018a, b, 2019) or some other non-biodegradable materials. All these instances belong to abiotic factors, which is very deadly and disastrous to humans, plants and animals including fish. The aquatic animals are under threat because of climate change and hence the production of safe aquatic organism (fish and shrimp) is a dubious proposition due to contamination of aquatic environment reaching to the top level (Bashir et al. 2020). The toxicity of organic and inorganic contaminates has increased with higher temperature and the same has occurred in the aquatic systems (Kumar et al. 2017a, 2018b, 2019). Therefore, it is essential to introduce nanotechnology to minimize such impact of abiotic stress related to temperature, low dissolved oxygen, low and high pH, metal and pesticide pollutions, heat wave, cold wave and other abiotic factors. Nanotechnology has a very broad application such as in ponds management, water management, identification and treatment of fish diseases, optimum delivery of nutrients and drug (Huang et al. 2015). Therefore, it is an attempt made in this chapter to inform the readers of new opportunities that has arisen in the form of nanotechnology for management of abiotic stress in aquaculture system.

15.2 Significance of Nanoparticles in Aquaculture

The nanoparticles play both a direct and indirect role in aquaculture systems. It can be used as nutraceutical, pharmaceuticals, drug delivery, vaccine development, nano-enzymes, nano delivery systems, pond water purifications and waste management, pond and cage sterilization, biofilm and fouling control, packing and barcode and tagging. Nanoparticles are also important for nano drug delivery, reproduction, immune-modulatory, growth promoter, antioxidant, feed digestion, cancer prevention, RNAi application and diseases control (Figs. 15.1 and 15.2).

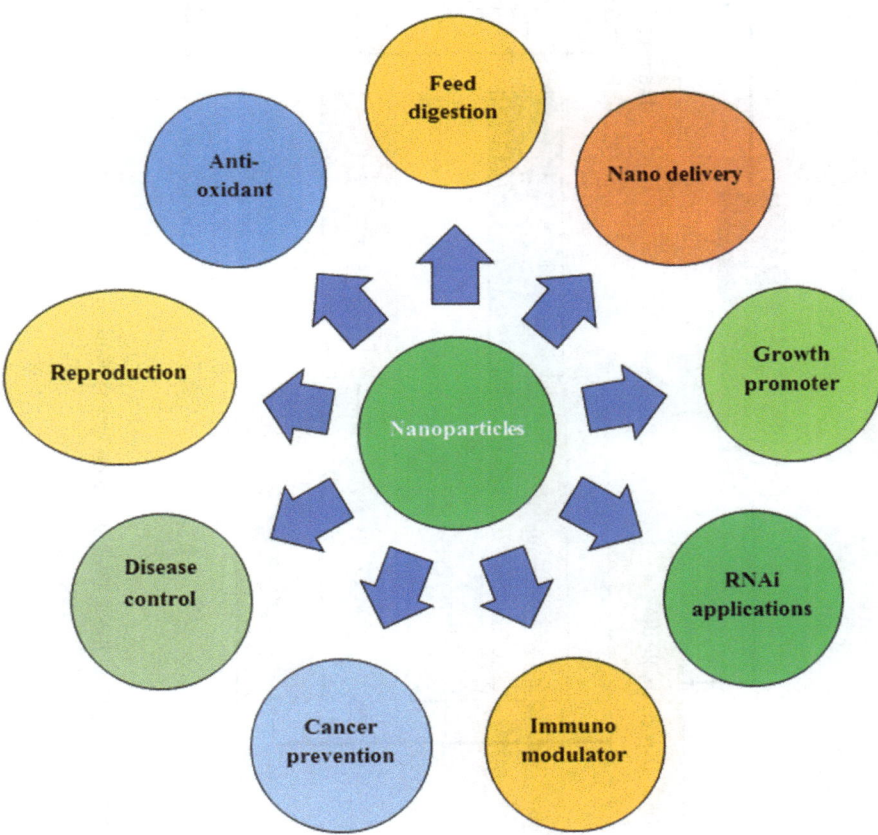

Fig. 15.1 Role of nanoparticles in aquaculture systems

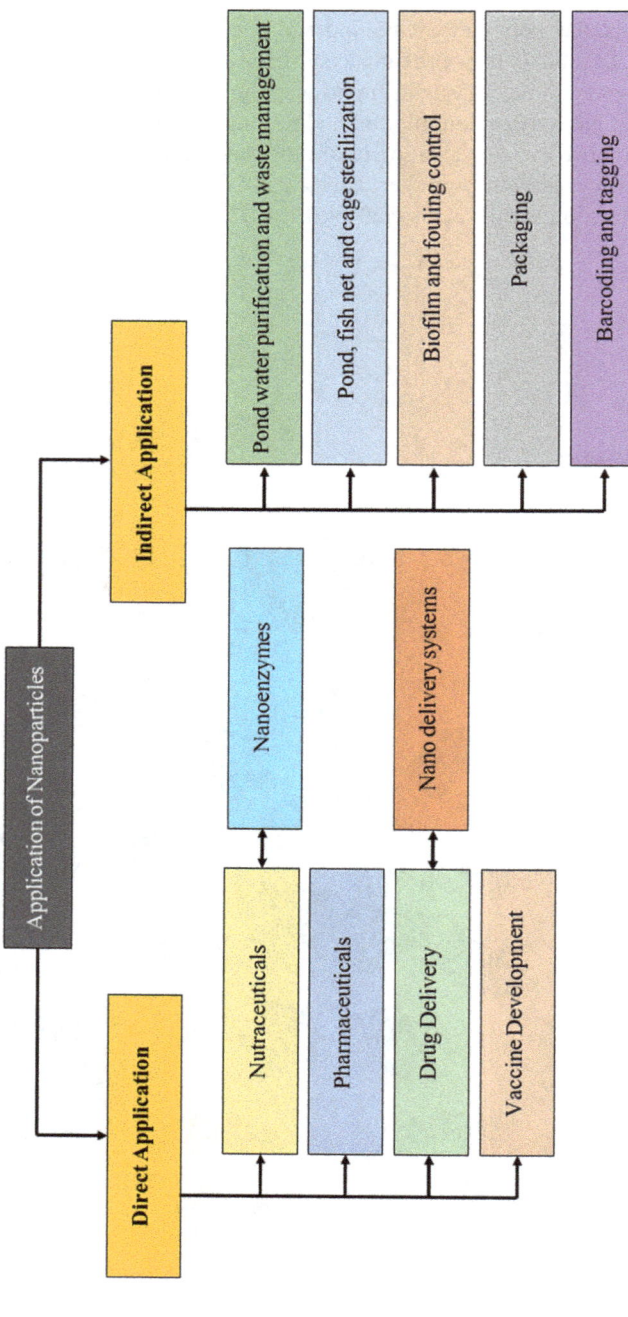

Fig. 15.2 Application of nanoparticles in aquaculture systems

15.3   Type of the Nanomaterial

Nanomaterial can be divided in two groups, namely the ultrafine nano-sized particles and engineered nanoparticles. The ultrafine nano-sized particles generally occur in nature and engineered nanoparticles are produced in controlled environments in an intended way (Oberdörster et al. 2005). The nanomaterials can be used in the aquaculture and other sectors such as carbon nanotubes (single-walled or multi-walled), nanosphere, nanocapsules, liposomes, dendrimers and polymeric nanoparticles (Table 15.1).

15.4   Role of Zinc Nanoparticles (Zn-NPs)

Zinc is an essential element for fish and shellfish to regulate metabolic activities. It has a major role in fish physiology and biological process such as enzymatic activities, hormone biosynthesis and metabolism of carbohydrate and protein (Kumar et al. 2017b, 2018a, 2020a; Wang and Wang 2015). It also has an important role in transcription factors and more enzymatic process in animal and fish (Chen et al. 2015; Wang and Wang 2015; Swain et al. 2016). It is essential for metabolic activities viz. lactate dehydrogenase, alcohol dehydrogenase, alkaline phosphatase, superoxide dismutase, aldolase, DNA and RNA polymerases, reverse transcriptase and carboxypeptidase (Prasad 1991; Swain et al. 2016) as well as for functioning in protein kinases and phosphatases (Chen et al. 2015). Deficiency of zinc is the major cause of growth retardation in fish and shellfish. However, Zn can be supplemented in the diet of animal and fish in the form of inorganic salts and organic chelates viz. zinc sulphate ($ZnSO_4$), zinc oxide (ZnO) and zinc propionate, zinc gluconate, zinc acetate, Zinc methionine (ZnMet), zinc lysine (ZnLys) and zinc glycine (ZnGly), respectively. Zinc in the form of organic chelates showed higher bioavailability compared to inorganic Zn but it has higher cost (Zhao et al. 2014). Zinc deficiency could be the reason for slower growth performance, increased mortality, cataracts,

Table 15.1 Type and use of different forms of nanoparticles

S. no	Type of nanoparticles	Structure	Application
1	Nanospheres	Spherical	Aquaculture, drug delivery
2	Nanocapsules	Combination of core and shell	Aquaculture, drug delivery
3	Carbon nanotubes	Cylindrical tube	Aquaculture, drug delivery and anticancer
4	Liposomes	Lipid-bi-layer globules	Aquaculture, reproduction and drug delivery
5	Dendrimers	Highly branched end and central core	Drug deliver
6	Polymeric nanoparticles	Polymers as chitosan, PLGA	Aquaculture and drug delivery

skin erosion, find and dwarfism (Wang and Wang 2015). The bioavailability of zinc may reduce due to the presence of tri-calcium phosphate in fish meal phytate, or phytic acid in soybean meal, and other oil seeds and grain (Hossain et al. 2003). Generally, absorption of Zn in animal is least and differs with the sites of the gastrointestinal tract and age of the animal (Swain et al. 2016). Nano Zn has an advantage over inorganic form of Zn. Generally, nano particles size varies from 1 to 100 nm but sometime it may go up to 100–1500 nm. Smaller size of nanoparticles can easily reach to the gastrointestinal tract and it is more effective than the larger sized nanoparticles (Feng et al. 2009). Nano form of Zn is more effective as compared to organic and inorganic Zn due to their larger surface area (Zaboli et al. 2013). However, it can easily pass through the small intestine and move into blood circulation, brain, lung, kidney, heart, liver, intestine, spleen and stomach (Hillyer and Albrecht 2001).

15.4.1 Characteristic of Zn-NPs

Zinc oxide/acetate is the basic compound for synthesis of zinc nanoparticles, which is generally found in the powdered form which is white and yellow-white crystalline in colour. The wurtzite (hexagonal) and zinc blende are common crystalline forms, which are generally soluble in water (Sirelkhatim et al. 2015). Zinc nanoparticle (Zn-NPs) has an advantage over ZnO due to its most important properties viz. increases surface area and hence speeds up the chemical reactions, smaller size of nanoparticles compared to ZnO, that can penetrate the bacterial cell wall which leads to cell death. Further, ROS (reactive oxygen species), produced due to nanoparticles, can attack the bacterial cell wall and hence, antibacterial activities may increase due to the large surface area. Due to Zn-NPs the many defects observed in the bacterial cell wall viz. edges and corners, which showed the abrasive nature of ZnO (Stoimenov et al. 2002). Zinc nanoparticles (Zn-NPs) are very potent for growth enhancement, feed utilization (Kumar et al. 2018c) and also used as antibacterial agent. It also increases the immunity and the ability to reproduce in fishes and animals. Zn-NPs can work in lower and higher doses which showed several effects on animal performance and other physiological functions. Zn-NPs can be supplemented in lower concentrations for better production of animal and fish which is better than conventional Zn and may also reduce the contamination load in environments. It can be also supplemented in the salmon diet in the form of $ZnSO_4$ (zinc sulphate) and zinc gluconate and both are equally efficient (Maage et al. 2001). Moreover, the Zn-NPs come in the third position globally, in terms of the highest rate of production after SiO_2-NPs and TiO_2-NPs (Piccinno et al. 2012). However, Zn-NPs can supplement in the feed and it can replace the conventional Zn in the fish and animal diet at lower concentration and safeguard the environment. Some of the Zn-NPs size and zeta potential has been determined in the laboratory which is shown in Fig. 15.3.

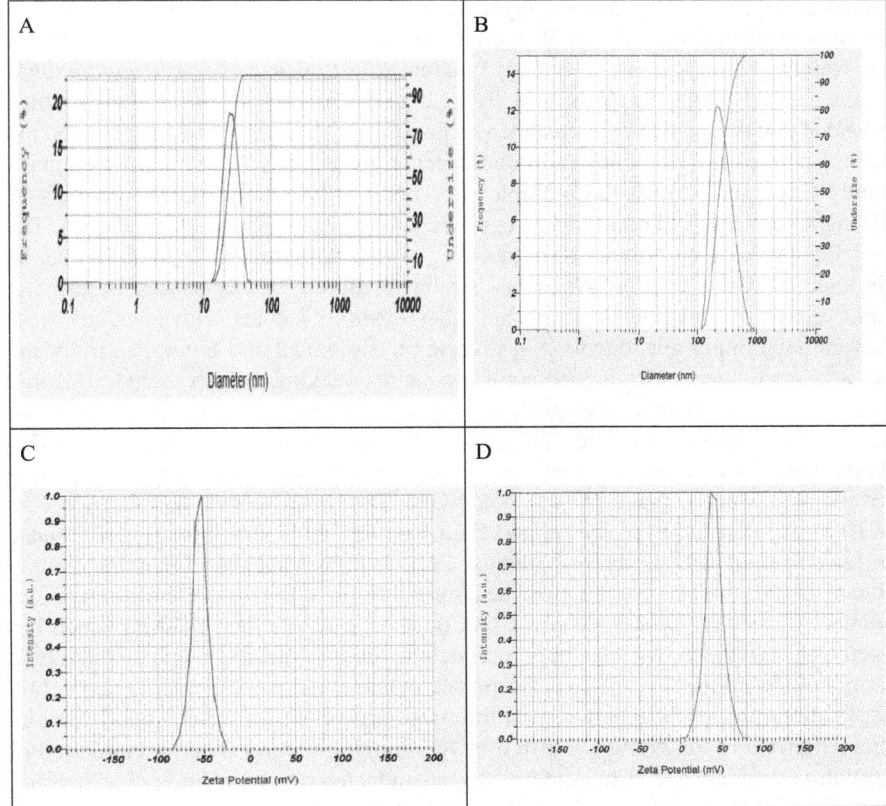

Fig. 15.3 Size of Zn-NPs (**a**) (23 nm) and (**b**) (206 nm); zeta potential (**c**) (−53.7 mV) and (**d**) (−41 mV)

15.5 Role of Nanoparticles in Growth Performance Against Abiotic Stress

It is a proven technology that the dissolution process of Zn-NPs generates abundant Zn-NPs which possess high bioavailability to the aquatic animal including fish (Hogstrand 2011). However, the application of Zn-NPs has an advantage over inorganic zinc and organic zinc and, therefore, are highly beneficial to aquatic animals and other farmed animals. A study conducted on *Macrobrachium rosenbergii* fed with Zn-NPs (10–60 mg/kg) concluded that Zn-NPs noticeably enhanced the growth and survival (Muralisankar et al. 2014) of the fish. It is universally proven that zinc is an essential element for the growth and development for humans, animals, and fish when present in a lower concentration. However, at higher concentrations, it is believed to be toxic and harmful for the food web and subsequently, the food chain (Hayat et al. 2007; Hao et al. 2013), as studies demonstrated that Zn-NPs @ 30 mg/kg diet showed significant improvements in the weight gain of the fish. The study conducted in *Pangasianodon hypophthalmus* supplemented diet with Zn-NPs at the rate of 10 mg/kg diet for 75 days which revealed that the growth performance viz. final weight gain (%), low feed conversion ratio, high protein efficiency ratio and specific growth rate of the fish was significantly improved against the multiple stressors. Further application of Zn-NPs @ 20 mg/kg diet showed retardation in growth performance (Kumar et al. 2018d). The beneficial role of Zn-NPs @ 10 mg/kg diet in growth performance might be due to the characteristics of zinc which has strong relation with carbohydrates, protein, lipid, enzymes and nucleic acid and other macromolecules. However, all these factors have important functions in diverse physiological and biological roles and act as electron acceptor, as a cofactor for metalloenzymes which includes alcohol dehydrogenase, alkaline phosphatase, carbonic anhydrase and DNA and RNA polymerase (Muralisankar et al. 2014). Thangapandiyan and Monika (2020) demonstrated that application of Zn-NPs in fish diet improved the growth performance viz. weight gain, FCR, PER, SGR. The study concluded that Zn-NPs @10 mg/kg diet followed by 7.5 and 5 mg/kg diet had higher growth performance in *Labeo rohita* after 45 days of feeding. Several studies suggested that the application of Zn-NPs in fish diet enhanced healthy condition of the fish. A similar study has been conducted by Mondal et al. (2020) on inorganic zinc and Zn-NPs which has been concluded that among the experimental groups, dietary supplementation of inorganic Zn improved growth performance in fish in a dose-dependent manner but, application of Zn-NPs in diet noticed improved growth performance up to 20 mg/kg diet. Higher dose of Zn-NPs in fish diet @ 30 mg/kg diet resulted in retardation of growth performance in fish. Another study conducted by Onuegbu et al. (2018) on Zn-NPs as feed supplements in African Catfish for enhancement of growth performance concluded that supplementation of Zn-NPs in the diet significantly enhanced the growth performance in term of higher weight gain, higher SGR and lowest FCR. The application of Zn-NPs in diet improves the feed intake and due to the smaller size of nanoparticles, the nutrient easily passes through the member and improves bioavailability and gastrointestinal tract absorption of the Zn in nano form.

Selenium is an essential nutrient for aquatic animals and has a fundamental role to play in the biological process for growth and development which is required for homeostatic functions. Kumar et al. (2018e) studied on *P. hypothalamus* fed with Se-NPs for mitigation of lead and high-temperature stress. The experiment was conducted for 60 days to evaluate the growth performance in the form of weight gain percentage, specific growth rate, feed conversion rate and protein efficiency ratio in *P. hypothalamus*. Application of Se-NPs through diet is shown to have significantly enhanced the growth performance of the fish which might be due to its biological function such as the role it plays in enzymatic oxidation-reduction, nucleic acid metabolism and in promoting the activity of easily oxidized substances like carotenoids and vitamin A which may increase protein and water in the cells (Tayeb and Qader 2012). Our laboratory finding also confirmed that application of nutrient in nanoparticles form reduced the impact of abiotic stress in fish (Kumar et al. 2019, 2020a, 2018e).

15.6 Role of Nanoparticles in Reduction of Oxidative Stress Against Abiotic Stress

The oxidative stress in the form of catalase (CAT), superoxide dismutase (SOD), glutathione-*S*-transferase (GST) and glutathione peroxidase (GPx) has been elevated with abiotic stress in aquatic animal (fish). Application of Zn-NPs at the rate of 10 mg/kg diet significantly reduced ($p < 0.01$) the oxidative stress in fish reader under arsenic, lead and temperature stress (Kumar et al. 2018a). The ability of Zn-NPs in reduction of oxidative stress might be due to the role of zinc in transcription factor, antioxidant defence system, and DNA repair (Dhawan and Chadha 2010; Verstraeten et al. 2004). It also protects the cell against lipid peroxidation as it acts as an antioxidant to minimize cell damage due to the formation of free radicals in the cell (Prasad 1991). The production of free radicals developed oxidative stress which modified the protein structure, cellular damages and disease pathogenesis and eventually results in the alteration in equilibrium between antioxidant level and cellular pro-oxidants (Halliwell 2006). However, selenium acts as an essential nutrient for humans and animals and it showed strong antioxidative properties through selenoproteins via selenocysteine (Fig. 15.4). The selenocysteine is the active component of glutathione peroxidase and supports in neutralization of the adverse effects of reactive oxygen species with the help of CAT, SOD, GST and GPx (Reeves and Hoffmann 2009). Generally, the biological systems are more prone to oxidative damage which interferes with their native structure and leads to induce oxidative stress (OS). To reduce and or neutralized the oxidative stress the biological systems have antioxidative defence systems through catalase, superoxide dismutase, glutathione-*S*-transferase and peroxidase. Selenium is also a strong antioxidative component which helps in neutralization of reactive oxygen species during stress condition (Kumar et al. 2020a). The study conducted in sheep and fed with Se-NPs through diet reduced the thiobarbituric acid reactive substances (TBARS) in plasma which indicates reduction of lipid peroxidation (Sadeghian et al. 2012). Another

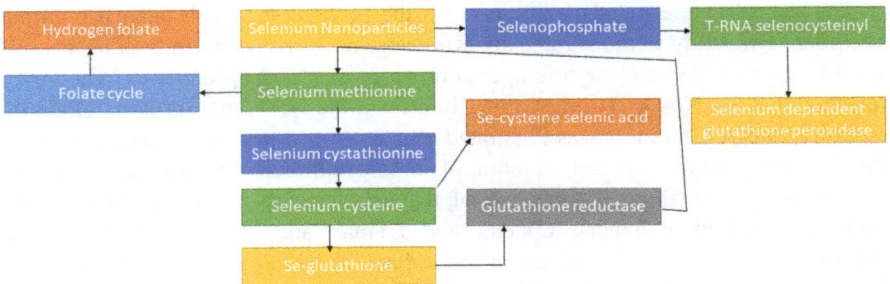

Fig. 15.4 Role of selenium nanoparticles in anti-oxidative potential

study conducted in the weaned Taihang black male for application of Se-NPs through diet has shown the reduction of oxidative stress (Shi et al. 2011). Zhou et al. (2009), demonstrated that supplementation of Se-NPs improved the antioxidative status through glutathione peroxidase in crucian carp (*Carassius auratus*). Further, Se-NPs also has a protective effect against chromium-induced oxidative stress and cellular damage in rat thyroid (Hassanin et al. 2013). It also has a very important role in maintenance of glutathione, catalase, superoxide dismutase, and malondialdehyde during thyrotoxicity from exposure to hexavalent chromium (Hassanin et al. 2013). Selenium has important role in anti-oxidative as it formed selenomethionine which leads to formed Se cysteine from the precursor through Se-cystathionine.

15.7   Role of Nanoparticles in Improvement of Immune System Against Abiotic Stress

In an aquaculture system, many nanoparticles have been used for the enhancement of fisheries production. These nanoparticles can enhance the immune system of the fish but the functioning of the mechanistic parts is not yet clear. The nanoparticles can pass through the aquatic animal and interact with many cells inside the body, including immune cell. The appropriate dose response of nanoparticles may stimulate and trigger the immune cells in the fish. Nanoparticles help in secretion of chemokines and cytokines like signalling molecules which provide the platform for coordination between molecular and immune cells. Nanoparticles have positive and negative charge which influences the immune cell. The positive charge showed higher inflammatory response compared to negative charged (Dobrovolskaia and McNeil 2007). It might be due to the nature of macrophage as negatively charged sialic acid on their outer surface and react with cationic substances (Dwivedi et al. 2009). Macrophages can recognize foreign antigens through different cells and functions viz. toll like receptors which can bind with and activate the antigen and antibody through signal transduction pathways and inflammatory response (Dwivedi et al. 2009). The study has been conducted to evaluate the different aspects of innate immune systems in fish which are protected by different layers and barriers, which is mentioned in Table 15.2.

The nanostructured diet of zinc ensures the protection of fish against multiple stressors. The study demonstrated by Kumar et al. (2018d) on mitigating role of Zn-NPs against lead and high-temperature stress in *P. hypophthalmus*. The study revealed that Zn-NPs @ 10 mg/kg diet improved the NBT, total protein, albumin, globulin, and A/G ratio. It was illustrated that Zn-NPs potentiates innate immunity through cell phagocytosis and secretion of soluble antimicrobial molecules. Zn-NPs also enhanced the thermal tolerance of the fish through improving their innate immunity through enhanced CTmax and LTmax and CTMin and LTmin in Pangasius species. The mechanism of enhancing thermal tolerance by using Zn-NPs might be due to the role of Zn-NPs in immune-modulations through enhancement of non-specific immunity (Rink and Gabriel 2000; Fraker et al.

Table 15.2 Role of nanoparticles in innate immune defenses with different mechanism

Innate immune defence	Study in fish	Nanoparticles	References
Cellular response			
Neutrophils	*Carassius auratus*	Zinc	Ortega et al. (2015)
Internalization of nanoparticles	*Oreochromis niloticus*	Zinc	Kaya et al. (2016)
	Oreochromis niloticus	Zinc	Kaya et al. (2016)
	Catla catla	Zinc	Asghar et al. (2018)
	Carassius auratus L.	Zinc	Ortega et al. (2015)
Antioxidant defenses			
SOD	*Catla catla*	Zinc	Asghar et al. (2018)
	Carassius auratus	Zinc	Benavides et al. (2016)
	Embryo-larval zebrafish	Zinc	Zhao et al. (2016)
	Embryo-larval zebrafish	Zinc	Zhao et al. (2013)
	Channa striatus	Ag-NPs	Kumar et al. (2018c)
	Pangasianodon hypophthalmus	Se-NPs	Kumar et al. (2018e)
Catalase (CAT)	Embryo-larval zebrafish	Zinc	Zhao et al. (2013)
Glutathione-sulfotranferase (GST)	*Carassius auratus*	Zinc	Benavides et al. (2016)
	Catla catla	Zinc	Asghar et al. (2018)
Glutathione (GSH)	*Pangasianodon hypophthalmus*	Zinc	Kumar et al. (2020b)
	Pangasianodon hypophthalmus	Zinc	Kumar et al. (2018d)
	Pangasianodon hypophthalmus	Zinc	Kumar et al. (2017b)
Biomarkers of oxidative stress			
Lipid peroxidation (LPO)	*Carassius auratus*	Zinc	Benavides et al. (2016)
	Catla catla	Zinc	Asghar et al. (2018)
	Embryo-larval zebrafish	Zinc	Zhao et al. (2016)
	Pangasianodon hypophthalmus	Zinc	Kumar et al. (2020b)
Cytokines	*Carassius auratus* L	Zinc	Ortega et al. (2015)
Lysozyme	*Oreochromis niloticus*	Zinc	Kaya et al. (2015)
External barriers			
Gill	*Oreochromis niloticus*	Zinc	Kaya et al. (2016)

(continued)

Table 15.2 (continued)

Innate immune defence	Study in fish	Nanoparticles	References
	Catla catla	Zinc	Asghar et al. (2018)
	Carassius auratus	Zinc	Benavides et al. (2016)

2000). The study also concluded that a Zn deficient diet was responsible for hypothermic in fish and animals; however, it is clear that Zn has a very important role in haemostasis of the fish (Topping et al. 1980). Zn is also important for the augmentation of the heat shock protein, which is noted for cofactor in more hundreds of protein which is essential for several enzymes in transcription of nucleic protein as DNA (Arslan et al. 2006). Zinc nanoparticles are also effective for marine fish. The work highlighted by Sallama et al. (2020) on *Siganus rivulatus* fed diet with zinc and zinc nanoparticles concluded that Zn-NPs could improve the immune system of the fish through enhancement of the IgM titer. The study also revealed that Zn-NPs increased phenoloxidase and lysozyme activities. Mondal et al. (2020) demonstrated that albumin, globulin, total protein and A:G ratio was improved with supplementation of Zn-NPs. In this study, the level of peroxidase activity was observed to be significantly higher with Zn-NPs. Apart from peroxidase, the phagocytosis is also very important for developing better immunity against pathogenic infections. Swain et al. (2018) emphasized that supplementation of Zn-NPs in diet significantly improved the respiratory burst, myeloperoxidase, haemagglutination activities and enhanced the agglutination for the foreign bodies. In fish blood, many proteins showed bactericidal activity, which is important for innate immune in fish. Zn-NPs enhanced production of B-lymphocyte which is responsible for improvement in lysozyme activity and also enhanced the bacterial agglutination capacity to control pathogen in fish and animals. Se-NPs also have a very important immune-modulator for living organism including fish, in which NBT total protein and A:G ratio are important components of innate immunity (Qin et al. 2016). The study conducted by Kumar et al. (2018e) showed that on supplementation of Se-NPs at the rate of 1 mg/kg diet and it has ability to enhance the immune systems of the fish against multiple stressors. Similarly, Se-NPs in combination with riboflavin also has the potential to enhance the immune system of the fish against arsenic pollution and high temperature in fish (Kumar et al. 2020a). The efficiency of the Se-NPs has been enhanced in combination of riboflavin as the immune-stimulatory properties have been shown in the Se-NPs at the rate of 0.5 mg/kg diet. Se-NPs have immunomodulatory properties might be due to similar nature and function of Se-NPs with selenite, selenomethionine, and methyl selenocysteine. It has an important role in upregulating selenoenzymes and demonstrated higher bioavailability and lower toxicity in animals (Wang et al. 2007).

15.8 Role of Nanoparticles in Reproduction Against Abiotic Stress

Selenium and zinc are the important nutrients for reproduction in aquatic animals. Selenium and zinc have both beneficial and toxic effects on fish reproduction as it depends upon the dose of nutrients. A few studies have been conducted on zinc nanoparticles on reproductive efficiency. Zn-NPs has been demonstrated in Rainbow trout to evaluate milt quality and breeding (Kazemia et al. 2020). Zinc may help in the maturity, activation, capacitation of sperm and finally productive performance in animals and fish (Thompson et al. 2002; Yamaguchi et al. 2009). Zinc has the potential to improve reproductive efficiency which might be due to the role it has in the maintenance of sperm motility and other parameters such as rate, viability, sperm movement (%) (Jiang et al. 2016). Gammanpila et al. (2007) demonstrated that zinc sulphate has the ability to enhance reproductive efficiency in term of sperm motility, sperm viability, hatching rate, total number of spawns, seed production and fecundity. The study conducted by Kazeto et al. (2006) demonstrated that the application of Zn-NPs improve the regulations of reproductive hormones such as gonadotropins, follicular stimulating hormones (FSH) and luteinizing (LH), Estradiol (E2) and 11-ketotestosterone (11-KT) in fish. Even though the application of higher concentration of Zn-NPs significantly reduced the FSH and LH in fish and had an effect on gonadal function that represented in down alteration of both Estradiol (E2) and testosterone levels. Deepa et al. (2019) demonstrated that exposure to Zn nanopowered to common carp effect on testicular steroidogenesis which is essential for androgen production and spermatogenesis. The results were expected to have a positive effect on the progression of spermatogenesis (Yamaguchi et al. 2009; Foresta et al. 2014). In this present study, however, exposure of Zn showed a toxic response to the reproductive status of fish even at the lowest desired concentration. The study also showed that exposure of Zn-NPs to zebrafish embryos deteriorates hatching, tail malformations and shortened body length of the larvae at lower concentrations and embryonic mortality at higher concentrations (Zhu et al. 2008; Bai et al. 2010). The alteration in embryonic stage in term of mortality might be due to higher concentration of Zn-NPs induced hypoxia and alteration in hatching enzyme. Generally, the hatching enzymes constitute choriolysin H (HCE), proteases and choriolysin L (LCE) which belong to the family of astacin protease. In hatching process, several steps are involved in embryos as hatching enzymes are digested to chorion which is secreted by hatching gland cells and it is proteolytic in nature (Inohaya et al. 1997). It is difficult to measure the effect of Zn-NPs due to transitory existence of the hatching enzyme. In addition to this mechanism, larger Zn-NPs might be aggregated and block the channel pore of the chorion cells which depleted in oxygen supply to embryos which effect on development in primitive stage (Cheng et al. 2007). In the study conducted by Zhu et al. (2008), it was observed that the zinc nanoparticles altered the embryos and larva structure deformities such as pericardial edema and body arcuation, tissue ulceration in Zebra fish. The mechanism in alteration of the structure of embryos and larva might be due to the exposure to Zn-NPs bulk suspensions. Even though Zn-NPs accumulated in the surface of zebrafish larvae which might have direct interaction with the surface cell of zebrafish larvae, causing a disruption of the cell membrane and lead to ulceration and eventually result in death (Dawson et al. 1988). Fish breeding is the key process for enhancement of fisheries production. Selenium has a very important role to play in reproduction as it determines the development of spermatozoa and is also incorporated into the mitochondrial capsule protein (Boitani and Puglisi 2008; Hansen and Deguchi 1996). It is also important for spermatogenesis, normal testicular development, and spermatozoa motility and function in fish (Moslemi and Tavanbakhsh 2011). Selenoprotein is the important regulatory protein which is responsible for transporting of selenium to the testis during spermatogenesis (Olson et al. 2005).

15.9   Summary

The application of nanotechnology in aquaculture is still in an infancy stage. The documentation of nanotechnology in aquaculture is mainly used in feed material, packaging, drug delivery, coating of the ship, etc. However, great attention must be paid to the use the nanotechnology for enhancement of fisheries production, since today, the world is facing great challenges in the form of climate change, pollution and other abiotic stress factor in aquatic ecosystems. Application of nanotechnology may boost and alleviate such abiotic stress and enhance production. Researchers are engaged in devising methods to achieve higher production through the use of nano-based tools in aquaculture in areas such as waterborne food, growth, reproduction, and culturing of species, their health, and water treatment, in order to increase aquaculture production. It is observed that the use of nanoparticles, however, often create toxicity in the systems, hence its safety must be ensured before their full-scale implementation. Its toxicity depends upon several factors such as: diameter, form, surface charge, and chemistry concentration, time of exposure, nature of the NPs, medium composition, route of particle administration, and target species immune systems. It is also observed that the concentration of nanoparticles may be lower (via feed) or higher (via water or surface treatments). Thus, it is not possible to draw inferences on the potential adverse effects that it may have on the end consumer. However, before using the nanoparticles, one should ensure its safety through moderation of the dose. Studies on various nanoparticles suggested that some NPs may be less available to the body and exist easily through excreta and cause environmental problems. In addition to this, it plays a very important role in the enhancement of growth performance, reproduction, reduction of cellular metabolic stress, immune-modulation, reduction of genotoxicity, DNA damage, etc. Moreover, nanotechnology in the form of mineral nutrition is still in its infancy and more studies need to be conducted to understand the role of nanoparticles in animal and fisheries production systems.

References

Arslan A, Csermely P, Soti C (2006) Protein homeostasis and molecular chaperones in aging. Biogerontology 7:383–389

Asghar MS, Qureshi NA, Jabeen F, Khan MS, Shakeel M, Chaudhry AS (2018) Ameliorative effects of selenium in ZnO NP-induced oxidative stress and hematological alterations in *Catla catla*. Biol Trace Elem Res 186(1):279–287. https://doi.org/10.1007/s12011-018-1299-9

Bai W, Zhang Z, Tian W, He X, Ma Y, Zhao Y (2010) Toxicity of zinc oxide nanoparticles to zebrafish embryo: a physicochemical study of toxicity mechanism. J Nanopart Res 12:1645–1654

Bashir I, Lone FA, Bhat RA, Mir SA, Dar ZA, Dar SA (2020) Concerns and threats of contamination on aquatic ecosystems. Bioremediat Biotechnol 27:1–26

Benavides M, Fernandez-Lodeiro J, Coelho P, Lodeiro C, Diniz MS (2016) Single and combined effects of aluminum (Al_2O_3) and zinc (ZnO) oxide nanoparticles in a freshwater fish, *Carassius auratus*. Environ Sci Pollut Res Int 23:24578–24591

Boitani C, Puglisi R (2008) Selenium, a key element in spermatogenesis and male fertility. Adv Exp Med Biol 636:65–73

Chen L, Yang J, Zheng M, Kong X, Huang T, Cai YD (2015) The use of chemical-chemical interaction and chemical structure to identify new candidate chemicals related to lung cancer. PLoS One 10(6):e0128696

Cheng JP, Flahaut E, Cheng SH (2007) Effect of carbon nanotubes on developing zebrafish (*Danio rerio*) embryos. Environ Toxicol Chem 26(4):708–716

Dawson DA, Stebler EF, Burks SL, Bantle JA (1988) Evaluation of the developmental toxicity of metal-contaminated sediments using short-term fathead minnow and frog embryo-larval assays. Environ Toxicol Chem 7(1):27–34

Deepa S, Kumar RM, Gupta YR, Gowda MKS, Senthilkumaran B (2019) Effects of zinc oxide nanoparticles and zinc sulfate on the testis of common carp, *Cyprinus carpio*. Nanotoxicology 13(2):240–257

Dhawan DK, Chadha VD (2010) Zinc: a promising agent in dietary chemoprevention of cancer. Indian J Med Res 132:676–682

Dobrovolskaia MA, McNeil SE (2007) Immunological properties of engineered nanomaterials. Nat Nanotechnol 2:469–478

Dwivedi PD, Misra A, Shanker R, Das M (2009) Are nanomaterials a threat to the immune system? Nanotoxicology 3:19–26

FAO (2018) The state of world fisheries and aquaculture 2018—meeting the sustainable development goals. Rome. Licence. In: CC BY-NC-SA 3.0 IGO

Feng M, Wang ZS, Zhou AG, Ai DW (2009) The effects of different sizes of nanometer zinc oxide on the proliferation and cell integrity of mice duodenum-epithelial cells in primary culture. Pak J Nutr 8:1164–1166

Foresta C, Garolla A, Cosci I, Menegazzo M, Ferigo M, Gandin V, Toni DL (2014) Role of zinc trafficking in male fertility: from germ to sperm. Hum Reprod 29(6):1134–1145

Fraker PJ, King LE, Laakko T, Vollmer T (2000) The dynamic link between the integrity of the immune system and zinc status. J Nutr 130:1399S–1406S

Gagné F, Auclair J, Pilote M, Turcotte P, Gagnon C (2016) Neurotoxicity of zinc oxide nanoparticles and municipal effluents to fathead minnows. Front Nanosci Nanotechnol 2 (1):48–57

Gammanpila M, Yakupitiyage A, Bart A (2007) Valuation of the effects of dietary vitamin C, E and zinc supplementation on reproductive performance of Nile tilapia (*Oreochromis niloticus*). Sri Lanka J Aquat Sci 12:39–60

Hadioui M, Leclerc S, Wilkinson KJ (2013) Multimethod quantification of Ag+ release from nanosilver. Talanta 105:15–19

Halliwell B (2006) Oxidative stress and neurodegeneration: where are we now? J Neurochem 97 (6):1634–1658

Hansen JC, Deguchi Y (1996) Selenium and fertility in animals and man—a review. Acta Vet Scand 37:19–30

Hao L, Chen L, Hao J, Zhong N (2013) Bioaccumulation and sub-acute toxicity of zinc oxide nanoparticles in juvenile carp (*Cyprinus carpio*): a comparative study with its bulk counterparts. Ecotoxicol Environ Saf 91:52–60

Hassanin KM, Abd El-Kawi SH, Hashem KS (2013) The prospective protective effect of selenium nanoparticles against chromium-induced oxidative and cellular damage in rat thyroid. Int J Nanomedicine 8:1713–1720

Hayat S, Javed M, Razzaq S (2007) Growth performance of metal stressed major carps viz. *Catla catla*, *Labeo rohita* and *Cirrhinus mrigala* reared under semi-intensive culture system. Pak Vet J 27:8–12

Hillyer JF, Albrecht RM (2001) Gastrointestinal persorption and tissue distribution of differently sized colloidal gold nanoparticles. J Pharm Sci 90:1927–1936

Hogstrand C (2011) Zinc. Academic, New York

Hossain MA, Matsui S, Furuichi M (2003) Effect of zinc and manganese supplementation to tricalcium phosphate rich diet for tiger puffer (*Takifugu rubripes*). Bangladesh J Fish Res 7 (2):189–192

Huang S, Wang L, Liu L, Hou Y, Li L (2015) Nanotechnology in agriculture, livestock, and aquaculture in China. A review. Agron Sustain Dev 35:369–400

Inohaya K, Yasumasu S, Araki K, Naruse K, Yamazaki K, Yasumasu I, Iuchi I, Yamagami K (1997) Species dependent migration of fish hatching gland cells that commonly express astacin-like proteases in common. Develop Growth Differ 39(2):191–197

Jiang M, Wu F, Huang F, Wen H, Liu W, Tian J, Yang C, Wang W (2016) Effects of dietary Zn on growth performance, antioxidant responses, and sperm motility of adult blunt snout bream, Megalobrama amblycephala. Aquaculture 464:121–128

Kaya H, Aydin F, Gurkan M, Yilmaz S, Ates M, Demir V, Arsalan Z (2016) A comparative toxicity study between small and large size zinc oxide nanoparticles in tilapia (*Oreochromis niloticus*): organ pathologies, osmoregulatory responses and immunological parameters. Chemosphere 144:571–582

Kazeto Y, Ijiri S, Adachi S, Yamauchi K (2006) Cloning and characterization of a cDNA encoding cholesterol side-chain cleavage cytochrome P450 (CYP11A1): tissue-distribution and changes in the transcript abundance in ovarian tissue of Japanese eel, Anguilla japonica, during artificially induced sexual development. J Steroid Biochem Mol Biol 99:121–128

Kietzmann M, Braun M (2006) Effects of the zinc oxide and cod liver oil containing ointment Zincojecol in an animal model of wound healing. Dtsch Tierarztl Wochenschr 113(9):331–334

Kumar N, Singh NP (2019) Effect of dietary selenium on immuno-biochemical plasticity and resistance against Aeromonas veronii biovar sobria in fish reared under multiple stressors. Fish Shellfish Immunol 84:38–47

Kumar N, Jesu AP, Pal AK, Remya S, Aklakur A, Rana RS, Gupta S, Raman RP, Jadhao SB (2011) Anti-oxidative and immuno-hematological status of Tilapia (*Oreochromis mossambicus*) during acute toxicity test of endosulfan. Pestic Biochem Physiol 99(1):45–52

Kumar N, Krishnani KK, Meena KK, Gupta SK, Singh NP (2017a) Oxidative and cellular metabolic stress of Oreochromis mossambicus as biomarkers indicators of trace element contaminants. Chemosphere 171:265–274

Kumar N, Krishnani KK, Kumar P, Singh NP (2017b) Zinc nanoparticles potentiates thermal tolerance and cellular stress protection of *Pangasius hypophthalmus* reared under multiple stressors. J Therm Biol 70:61–68

Kumar N, Krishnani KK, Brahmane MP, Gupta SK, Kumar P, Singh NP (2018a) Temperature induces lead toxicity in Pangasius hypophthalmus: an acute test, antioxidative status and cellular metabolic stress. Int J Environ Sci Technol 15(1):57–68

Kumar N, Krishnani KK, Gupta SK, Singh NP (2018b) Effects of silver nanoparticles on stress biomarkers of Channa striatus: immuno-protective or toxic? Environ Sci Pollut Res 25 (15):14813–14826

Kumar N, Krishnani KK, Kumar P, Sharma R, Baitha R, Singh DK, Singh NP (2018c) Dietary nano-silver: does suort or discourage thermal tolerance and biochemical status in air-breathing fish reared under multiple stressors? J Therm Biol 77:111–121

Kumar N, Krishnani KK, Singh NP (2018d) Effect of dietary zinc-nanoparticles on growth performance, anti-oxidative and immunological status of fish reared under multiple stressors. Biol Trace Elem Res 186(1):267–278

Kumar N, Krishnani KK, Gupta SK, Sharma R, Baitha R, Singh DK, Singh NP (2018e) Immuno-protective role of biologically synthesized dietary selenium nanoparticles against multiple stressors in Pangasinodon hypophthalmus. Fish Shellfish Immunol 78:289–298

Kumar N, Gupta SK, Bhushan S, Singh NP (2019) Impacts of acute toxicity of arsenic (III) alone and with high temperature on stress biomarkers, immunological status and cellular metabolism in fish. Aquat Toxicol 214:105233

Kumar N, Chandan NK, Wakchaure GC, Singh NP (2020a) Synergistic effect of zinc nanoparticles and temperature on acute toxicity with response to biochemical markers and histopathological attributes in fish. Comp Biochem Physiol Part C 229:108678. https://doi.org/10.1016/j.cbpc.2019.108678

Kumar N, Gupta SK, Chandan NK, Bhushan S, Singh NP, Kumar P, Wakchaure GC, Singh NP (2020b) Mitigation potential of selenium nanoparticles and riboflavin against arsenic and elevated temperature stress in *Pangasianodon hypophthalmus*. Sci Rep 10(1):17883. https://doi.org/10.1038/s41598-020-74911-2

Kundu P, Anumol EA, Ravishankar N (2013) Pristine nanomaterials: synthesis, stability and applications. Nanoscale 5:5215–5224

Maage A, Julshamn K, Berge GE (2001) Zinc gluconate and zinc sulphate as dietary zinc sources for Atlantic salmon. Aquac Nutr 7:183–187

Mondal AH, Behera T, Swain P, Das R, Sahoo SN, Mishra SS, Das J, Ghosh K (2020) Nano zinc vis-à-vis inorganic zinc as feed additives: effects on growth, activity of hepatic enzymes and non-specific immunity in rohu, *Labeo rohita* (Hamilton) fingerlings. Aquac Nutr 26 (4):1211–1222

Moslemi MK, Tavanbakhsh S (2011) Selenium–vitamin E supplementation in infertile men: effects on semen parameters and pregnancy rate. Int J Gen Med 4:99–104

Muralisankar T, Bhavan PS, Radhakrishnan S, Seenivasan C, Manickam N, Srinivasan V (2014) Dietary supplementation of zinc nanoparticles and its influence on biology, physiology and immune responses of the freshwater prawn, *Macrobrachium rosenbergii*. Biol Trace Elem Res 160:56–66

Oberdörster G, Oberdörster E, Oberdörster J (2005) Nanotoxicology: an emerging discipline evolving from studies of ultrafine particles. Environ Health Perspect 113:823–839

Olson GE, Winfrey VP, Nagdas SK, Hill KE, Burk RF (2005) Seleno-protein P is required for mouse sperm development. Biol Reprod 73:201–211

Onuegbu CU, Aggarwal A, Singh NB (2018) ZnO nanoparticles as feed supplement on growth performance of cultured African catfish fingerlings. J Sci Ind Res 77(4):213–218

Ortega VA, Katzenback BA, Stafford JL, Belosevic M, Goss GG (2015) Effects of polymer-coated metal oxide nanoparticles on goldfish (*Carassius auratus L.*) neutrophil viability and function. Nanotoxicology 9:23–33

Piccinno F, Gottschalk F, Seeger S, Nowack B (2012) Industrial production quantities and uses of ten engineered nanomaterials in Europe and the world. J Nanopart Res 14(9):1–11

Prasad AS (1991) Discovery of human zinc deficiency and studies in an experimental human model. Am J Clin Nutr 53:403–412

Qin F, Shi M, Yuan H, Yuan L, Lu W, Zhang J, Tong J, Song X (2016) Dietary nano-selenium relieves hypoxia stress and, improves immunity and disease resistance in the Chinese mitten crab (*Eriocheir sinensis*). Fish Shellfish Immunol 54:481–488

Reeves MA, Hoffmann PR (2009) The human selenoproteome: recent insights into functions and regulation. Cell Mol Life Sci 66:2457–2478

Rink L, Gabriel P (2000) Zinc and the immune system. Proc Nutr Soc 59(4):541–552

Sadeghian S, Kojouri GA, Mohebbi A (2012) Nanoparticles of selenium as species with stronger physiological effects in sheep in comparison with sodium selenite. Biol Trace Elem Res 146:302–308

Sallama AE, Mansour AT, Alsaqufi AS, Salem M, El-Feky MM (2020) Growth performance, antioxidative status, innate immunity, and ammonia stress resistance of Siganus rivulatus fed diet supplemented with zinc and zinc nanoparticles. Aquacult Rep 18:100410

Shah BR, Mraz J (2020) Advances in nanotechnology for sustainable aquaculture and fisheries. Rev Aquac 12:925–942

Shi L, Xun W, Yue W, Zhang C, Ren Y, Shi L, Wang Q, Yang R, Lei F (2011) Effect of sodium selenite, Se-yeast and nano-elemental selenium on growth performance, Se concentration and antioxidant status in growing male goats. Small Ruminant Res 96:49–52

Sirelkhatim A, Mahmud S, Seeni A, Kaus NHM, Ann LC, Bakhori SKM, Hasan H, Mohamad D (2015) Review on zinc oxide nanoparticles: antibacterial activity and toxicity mechanism. Nano Micro Lett 7:219–242

Stoimenov PK, Klinger RL, Marchin GL, Klabunde KJ (2002) Metal oxide nanoparticles as bactericidal agents. Langmuir 18:6679–6686

Swain PS, Rao SBN, Rajendran D, Dominic G, Selvaraju S (2016) Nano zinc, an alternative to conventional zinc as animal feed supplement: a review. Anim Nutr 2:134–141

Swain P, Das R, Das A, Padhi SK, Das KC, Mishra SS (2018) Effects of dietary zinc oxide and selenium nanoparticles on growth performance, immune responses and enzyme activity in rohu, *Labeo rohita* (Hamilton). Aquac Nutr 25:1–9

Tayeb IT, Qader GK (2012) Effect of feed supplementation of selenium and vitamin E on production performance and some hematological parameters of broiler. KSU J Nat Sci 15:46–56

Thangapandiyan S, Monika S (2020) Green synthesized zinc oxide nanoparticles as feed additives to improve growth, biochemical, and hematological parameters in freshwater fish Labeo rohita. Biol Trace Elem Res 195:636–647

Thompson ED, Mayer GD, Walsh PJ, Hogstrand C (2002) Sexual maturation and reproductive zinc physiology in the female squirrelfish. J Exp Biol 205:3367–3376

Topping DL, Clark DG, Dreosti IE (1980) Impaired thermoregulation in cold exposed zinc deficient rats: effect of nicotine. Nutr Rep Int 24:643–648

Verstraeten SV, Zago MP, MacKenzie GG, Keen CL, Oteiza PI (2004) Influence of zinc deficiency on cell-membrane fluidity in Jurkat, 3T3 and IMR-32 cells. Biochem J 378:579–587

Wang J, Wang WX (2015) Optimal dietary requirements of zinc in marine medaka Oryzias melastigma: importance of daily net flux. Aquaculture 448:54–62

Wang H, Zhang J, Yu H (2007) Elemental selenium at nano size possesses lower toxicity without compromising the fundamental effect on selenoenzymes: comparison with selenomethionine in mice. Free Radic Biol Med 42:1524–1533

Yamaguchi S, Miura C, Kikuchi K, Celino FT, Agusa T, Tanabe S, Miura T (2009) Zinc is an essential trace element for spermatogenesis. Proc Natl Acad Sci U S A 106(26):10859–10864

Zaboli K, Aliarabi H, Bahari AA, Abbasalipourkabir R (2013) Role of dietary nano-zinc oxide on growth performance and blood levels of mineral: a study on in Iranian angora (Markhoz) goat kids. J Pharm Health Sci 2:19–26

Zhao X, Wang S, Wu Y, You H, Lv L (2013) Acute ZnO nanoparticles exposure induces developmental toxicity, oxidative stress and DNA damage in embryo-larval zebrafish. Aquat Toxicol 137:49–59

Zhao CY, Tan SX, Xiao XY, Qiu XS, Pan JQ, Tang ZX (2014) Effects of dietary zinc oxide nanoparticles on growth performance and antioxidative status in broilers. Biol Trace Elem Res 160:361–367

Zhao X, Ren X, Zhu R, Luo Z, Ren B (2016) Zinc oxide nanoparticles induce oxidative DNA damage and ROS-triggered mitochondria-mediated apoptosis in zebrafish embryos. Aquat Toxicol 180:56–70

Zhou X, Wang Y, Gu Q, Li W (2009) Effect of different dietary selenium source (selenium nanoparticle and selenomethionine) on growth performance, muscle composition and glutathione peroxidase enzyme activity of crucian carp (*Carassius auratus* gibelio). Aquaculture 29:78–81

Zhu X, Zhu L, Duan Z, Qi R, Li Y, Lang Y (2008) Comparative toxicity of several metal oxide nanoparticle aqueous suspensions to zebrafish (*Danio rerio*) early developmental stage. J Environ Sci Health 43:278–284

Nutritional Biotechnology to Augment Aquaculture Production

16

Bijay Kumar Behera

Abstract

Aquaculture is the fastest growing sector globally and the demand of food fishes is also increasing day by day. Nutritional biotechnology has great role in aquaculture production enhancement. Selection of proper feed ingredients, advanced technologies for feed formulation and preparation, feed processing and feeding techniques, anti-nutritional factors, their detoxification are very essential aspects of fish feed. Carbohydrate utilization as protein sparing effect in fish nutrition, nutraceuticals in fish nutrition, suitable feed attractants, fish biochemical and physiological responses in extreme environmental stresses are some of the priority areas of research in fish nutritional biotechnology. Further, probiotics, prebiotics and immunostimulants are very much useful in preventing disease outbreak and growth. This chapter deals with various developments in nutritional biotechnology for enhancing aquaculture production.

Keywords

Aquaculture · Immunostimulant · Nutritional biotechnology · Prebiotics · Probiotics

16.1 Introduction

Globally, consumption of food fishes is projected to reach 165 million tonnes by 2030. As capture fisheries are now approaching full exploitation worldwide, a large part of fish requirement will have to come from aquaculture. The identification of

B. K. Behera (✉)
Aquatic Environmental Biotechnology and Nanotechnology Division, ICAR-Central Inland Fisheries Research Institute, Kolkata, India

© The Author(s), under exclusive license to Springer Nature Singapore Pte Ltd. 2021

P. K. Pandey, J. Parhi (eds.), *Advances in Fisheries Biotechnology*,
https://doi.org/10.1007/978-981-16-3215-0_16

231

alternative fish species and suitable feed ingredients for their diet formulations has therefore become very important. The main goal of fish nutrition as a scientific discipline is to produce feeds that support good growth rates while maintaining proper fish health, resulting in a safe and healthy product for the consumer at least cost. In this regard, many scientists are studying about the safe utilization of nutrients and their interactions when alternative feed ingredients from plants are used as substitute to the traditional and expensive fish meal and oil. The present day research focuses on issues such as characterization of nutrient effects on brooders fish, eggs, larvae, juveniles, and at different stages of growth. Measurements include nutrients' effect on growth, feed utilization, digestibility, alterations in metabolic pathways, fish health and welfare parameters, nutrient bioavailability and retention. Modern tools within genomics and proteomics are gradually being used to discover novel pathways for effective utilization of nutrients.

The aquaculture feed sector of Asia has made tremendous developments during the last two decades. At present about 20 million ton of manufactured aqua feed are being used in aquaculture sector of India, of which the majority is consumed in shrimp culture. If the rapid growth of aquaculture persists, the feed requirement will increase many fold. Hence, more scientific understanding and interventions are required for sustainable development of the sector. Sustainable commercial carp feed production has become a challenge to the aquaculture nutritionists. It has been estimated that a feed with an FCR of 1.3 could make a commercial carp feed sustainable. Exploration of novel genes related to growth enhancement and use of different nutraceuticals have raised the hope of achieving that target. Isolation of a potential growth hormone gene and subsequent transfer of that gene to enhance fish production is approaching to reality. Addition of new immunostimulants in aqua feed has increased the possibility of safer production through high-tech aquaculture. Strategies for increasing utilization of cheaper nutrients like carbohydrate by various technological interventions and addition of feed attractants paved the way for developing low cost feeds for sustainable carp culture (Gopakumar 2003). Currently, quality enhancement of fish flesh by dietary and gene manipulation is a focused area of research in fish nutrition.

Besides feed development, feed quantity and its management are critical factors for profitability of fish farming, especially in intensive aquaculture. A biologically ideal feed may not be economically viable if feed management is poor. A proper knowledge about fish nutrition and feed management is of paramount importance for sustainable fish production. The following are the priority areas of research in fish nutrition:

- Nutritional requirement of fishes with respect to growth.
- Selection of proper feed ingredients.
- Energy requirements of fish.
- Feed formulation and preparation.
- Feed processing and feeding techniques.
- Carbohydrate utilization as protein sparing effect in fish nutrition.
- Omega-3 fatty acids in nutrition and health.

- Nutraceuticals in fish nutrition.
- Fish nutrition, biochemical and physiological responses in extreme environmental stresses.
- Quality control and storage of feed ingredients.
- Anti-nutritional factors, their detoxification and pathology.
- Left-over feed and waste *vis-a-vis* water quality management.
- Molecular strategies to enhance fish flesh quality and quantity.

Biotechnology is being used to answer some of the technical and environmental concerns of fish farming including aquaculture nutrition. Over the last decade, world has witnessed spectacular growth in fish feed industry of many developing countries. It is further anticipated that world aquaculture production will continue to increase and consequently the need of fish feed. Nutrition and feeding play a pivotal role in sustainable aquaculture and feed constitutes about 40–50% of the total cost of aquaculture production. Use of nutritionally balanced and complete formulated feed will continue to play a dominant role in finfish and shellfish production. Hence, alternative and biotechnologically improved feed ingredients should be sought along with improvements in pond management and productivity.

Presently, the most common protein source for many fish diets is fish meal, a by-product of fish processing, because of its high quality protein content. However, it has some disadvantages. In addition to its high cost, it contains phosphorus far beyond the requirement for optimal growth in fish, leading to potential environmental concerns. The excess phosphorus, released into the water, can cause problems such as eutrophication and consequently excess algal growth.

16.2   Plant Proteins as Alternatives to Fish Meals

Plant proteins have the potential to reduce the problem of phosphorus pollution, since plants do not contain such high phosphorus levels. Moreover, use of plant protein as a feed ingredient can help to reduce the burden on fish meal supplies. As a result of these concerns, researchers are using biotechnological tools to produce alternative plant-based protein sources, suitable for use in aquaculture. However, use of plant-derived materials such as legume seeds, different types of oilseed cake, leaf meals, leaf protein concentrates and root tuber meals as feed ingredients is often limited due to the presence of a wide variety of anti-nutritional substances (Fournier et al. 2004). Plant proteins often require processing to remove these substances, which are produced by plants as natural defence mechanisms. Important among these are the protease inhibitors, phytates, glucosinolates, saponins, tannins, lectins, oligosaccharides and non-starch polysaccharides, phytoestrogens, alkaloids, antigenic compounds, gossypols, cyanogens, mimosine, cyclopropenoid fatty acids, canavanine, antivitamins and phorbol esters. These compounds must be destroyed during processing to prevent harmful effect to the fish. Problems associated with anti-nutritional factors can also be solved by producing feed enzymes to counteract them. Phytase is one such example, which can help the fish to make the best use of phosphorous available in a plant-protein-based feed. One of the most exciting technological developments has come from the ability to manipulate the plant genome to produce products suitable for use in aquaculture. The use of genetically modified crops to eliminate anti-nutritional factors and increase specific nutrients (limiting amino acids, n-3 fatty acids, etc.) is now possible.

16.3 Feed Additives

Adding specific nutrients to feed can improve animal digestion and thereby reduce feed costs. A lot of feed additives are being currently used and new concepts are continuously developed with the help of biotechnology.

16.4 Utilization of Plant Fibers in Fish Feed Through Enzymes

Although enzymes have been in use for a long time in other areas such as detergents, textiles, baking and brewing, their use in the animal and fish feed has been a recent phenomenon. This is because enzymes need to be designed to suit the appropriate application. Industrial enzymes are mainly produced from microorganisms by a process of fermentation and extraction. These enzymes can be produced in large quantities from genetically modified microbes with desired properties to make them economically viable. Feed enzymes need to be robust to stand variations in pH and temperatures. They need to have high temperature stability to withstand pelletization and also have a long shelf life. Over the years, feed enzymes have gone through an evolution from liquid to powder to granules in their product forms in order to make them more heat stable. Lately, the granulation technology has been developed in a way that the enzyme molecules are coated with inert materials like cellulose and wax to give the enzyme full protection. This technology helps the enzyme to achieve longer shelf life and is most suitable for pelletized feed. In Europe and Australia, feed enzymes have been used for nearly a decade but their usage in Indian fish feed have been only in recent years.

Several alternate feed sources like sunflower, rapeseed and safflower can provide protein, and sorghum, millets or rice bran for energy could be used as feed ingredients if supplemented by substrate-specific enzymes. Cell wall composition of this high fiber feed ingredients show that they contain large amount of arabino-xylans, pectic polysaccharides and some cellulose. Use of specific enzymes like xylanase, pectinase, and cellulase could allow breakdown of fiber releasing energy as well as increase the protein digestibility due to better accessibility of the protein (Alford et al. 1996). In this way the feed cost can be reduced and the protein levels in the feed can be increased. There is vast scope for increasing the use of enzymes in near future with the availability of high quality enzymes with more predictable performances.

16.5 Phytase in Aquafeed

One of the enzymes that have really caught the attention of the researchers in India is phytase, an enzyme that breaks down the indigestible phytic acid (phytate) in cereals and oilseeds and releases digestible phosphorus. It has been known for several decades. However, the feed industry could not use it economically due to its high production costs. Microbial phytase became commercially available in 1990s as the result of biotechnological improvements.

Phytase reduces the use of expensive supplemental inorganic phosphorus such as dicalcium phosphate (Jongbloed et al. 2000). Phytase releases minerals (Ca, Mg, Zn and K), amino acids and proteins, which are complexed with the phytate molecule. Phytate itself is an anti-nutritive factor, when hydrolyzed gives better performance in animals. Presently, a substantial number of farmers in India are using phytase to reduce the cost of their feed. Use of phytase in aquaculture industry has helped farmers in getting better control over algal blooms due to the reduction in phosphorus levels.

Aqua feed manufacturers are also looking into the possibility of using phytase to release the nonavailable phosphorus from deoiled rice bran, soya and wheat (Boling et al. 2000). There are various enzyme formulations, available in the market with varied activities. Because enzymes are produced by different microorganisms, the characteristics as well as the composition of enzymes would be different. The manufacturer by experience and good quality control can guarantee consistent results. However, comparison of enzyme products is difficult unless in vivo testing is carried out. Each microorganism produces enzymes at different optimum pH values, different optimum temperature of operation and different affinity to the substrate in feed. It has not yet been possible to develop an in vitro method that can predict in vivo performance.

Improving expression of the phytase gene in plants is underway as a means to commercially produce phytase. Plants such as alfalfa are used as "bioreactors" and also by developing plant cultivars that would produce enough transgenic phytase so that their additional supplementation in grain or meals is not necessary.

16.6 Micro-nutrients and Vitamins in Fish Feed

The absorption and availability of inorganic trace minerals such as zinc, cobalt, calcium etc., vary depending upon the nature of the minerals (sulfate, oxide or carbonate), their solubility and ionization. Trace minerals are now being attached to oligopeptides to make them more bioavailable. Commercial preparations of proteinated selenium and chromium are used in poultry feed production. In the case of vitamins, due to varying availability and stability of vitamins in ingredients, supplemental vitamins are widely incorporated in diets. Stability of the vitamins is achieved through the application of advanced technologies which involve preparation of biologically active derivatives, coating technologies, carriers and diluents

(Bamji and Lakshami 1998). These vitamins are much more stable than naturally occurring forms.

16.7   Nutraceuticals

Nutraceutical implies that the extract of food is demonstrated to have a physiological benefit or provide protection against a disease and/or improve growth. Functional foods are defined as being consumed as part of a usual diet but are demonstrated to have physiological benefits and/or reduce the risk of disease beyond basic nutritional functions. Nutraceuticals are often used in nutrient premixes or nutrient systems in the food. In aquaculture, application of nutraceuticals includes inclusion of feed additives such as antioxidants, vitamins, minerals and carotenoids etc. in the feed (Brower 1998).

Nutraceuticals are also extracted from many fishes which include omega-3 oil, chitosan and glucosamine, originally derived from waste products. Hundreds of tons of marine by-products are available annually which are driving force for both research and commercialization in the area of marine nutraceuticals.

16.8   Fish Feed Attractants

Color, smell, odor, taste, and flavor of several food and feed stuffs play an important role in attracting animals toward a particular feed (Kozasa 1986). Attractants are used in the aquafeed industry for identification and consumption of feed efficiently. Marine by-products such as mollusk meal, squid meal, squid liver meal, shrimp head meal, shrimp meal, krill meal, krill hydrolysates and artemia are used as attractants in aquafeeds.

16.9   Probiotics in Fish Nutrition

Microorganisms are naturally present in the digestive tract of the animals. Some microbes' aid to digestion, others can potentially cause pathogenesis. The microbial ecology of the gut merits greater attention due to its implications for nutrition, feed conversion and disease control. Use of antibiotics disturbs the microbiological balance of gut flora, eliminating most of the beneficial flora microbes. When we stop antibiotic treatment, pathogens begin to re-establish themselves in the intestine. Overgrowth of these organisms and subsequent invasion of the system by pathogenic organisms cause inflammatory, immunological, neurological and endocrinological problems. Using probiotics can help build up the beneficial bacteria in the intestine and competitively exclude the pathogenic bacteria (Gatesoupe 1999). Probiotic bacteria also release enzymes, which help in the digestion of feed. Concept of using probiotics in animal feed, particularly poultry and aquaculture, is slowly becoming popular. Common organisms in probiotic products are *Aspergillus oryzae*,

Lactobacillus acidophilus, *L. bulgaricus*, *L. plantarium*, *Streptococcus lactis*, and *Saccharomyces cerevisiae*. These products can be administered through water or incorporated in the feed. The most important quality parameter of probiotics is that the vegetative or the spore forms have to be viable to be able to multiply in the gut. Secondly, they should be resistant to antibiotics. Genetic engineering would help develop probiotics with special properties like secreting enzymes and vitamins in large quantities. Such products would be the future generation feed additives.

16.10 Prebiotics in Fish Nutrition

The concept of prebiotics in aquafeed is fairly recent. Prebiotics are basically feed for probiotics that are resistant to attack by endogenous enzymes and hence, reach the site for proliferation of gut microflora. Some of the prebiotics, which are currently used in animal feed, are mannan-oligosaccharides (MOS), fructo-oligosaccharide, and mixed oligo-dextran. Mannan-oligosaccharides are mainly obtained from the cell walls of yeasts. Other sources of MOS are copra or palm kernel meal. MOS interferes with the colonization of the pathogens. Cell surface carbohydrates are primarily responsible for cell recognition. Bacteria have lectins (glycoprotein) on the cell surface that recognize specific sugars and allow the cell to attach to that sugar. Binding of *Salmonella*, *E. coli* and *Vibrio* sp. is shown to be mediated by a mannose specific lectin like substance, present on the bacterial cell surface. Similarly, fructo-oligosaccharides from chicory have been used as prebiotics to competitively exclude pathogenic bacteria (Xu et al. 2003). The pH of the lumen gets reduced, thus, preventing the entry of pathogenic bacteria. The concept of using prebiotics has not yet been accepted but the advantages of prebiotics are that it can stand high pelletizing temperatures in the feed and also have a long shelf life.

16.11 Quality Management and Detection of Contaminants in Fish Feeds

Modern and biotechnological tools like PCR, biomonitoring and DNA/gene-based diagnostics could be developed and used for detection of *Aspergillus* fungi, *Salmonella* species, mycotoxins and other contaminants in fish feeds and feeding environment.

16.12 Growth Improvement and Enhancement of Quantitative Traits in Fish

Enhancement of natural growth rates of fish in aquaculture has been extensively explored, with gains arising from improvements in husbandry, nutrition and genetic selection. It can be advantages to aquaculture by shortening production times, enhancing feed conversion efficiency, and controlling product availability.

Endocrine approaches to manipulate growth have also been extensively explored, principally by applications of somatotropins such as growth hormone (GH), prolactin and placental lactogen, insulin-like growth factor-1, thyroid hormones and sex steroids (Mclean and Devlin 2000). In the recent past, there have been instances both in India and abroad where growths of fish have been increased by incorporation of growth hormone gene.

16.13 Highly Unsaturated Fatty Acids (HUFA) and Qualitative Enhancement of Fish

Highly unsaturated fatty acids (HUFA) are receiving considerable attention due to their involvement in human health. Fish is an important dietary source of long-chain C20 and C22 highly unsaturated fatty acids (HUFA) (Ackman 1980), which are crucial to the health of higher vertebrates also. They play pivotal roles in number of biological functions including cardiovascular functions, neural development, eicosanoid signaling, regulation of gene expression and ion channel modulation (Kang et al. 2004).

Traditional emphasis has been on dietary lipids and oils because of their effects on lipid composition, in particular that of the lipoproteins. The classical way by which lipids were evaluated, however, gradually changed in the last decade as it became evident that fatty acids themselves regulate lipid homeostasis not only at the level of the lipids interaction with proteins, but also on the genetic level by affecting gene expression. Changes in the lipid content at high concentration are the major cause for several diseases, i.e., obesity, coronary heart disease and cancer (Kushi et al. 1997). Earlier studies clearly showed that risk of coronary heart disease was greatly reduced in populations where fish was major portion of the food. In the recent years, researches have been initiated to conduct molecular analysis and characterization of delta-6 fatty acyl desaturase genes from several fin fishes which are involved in biosynthesis of omega-3 and omega-6 PUFA (Clarke 1998).

16.14 Application of Feed Probiotics

The concept behind the composition of feed probiotics is to apply the beneficial bacterial strains in feed, using binders such as eggs and cod liver oil to obtain the beneficial microbial effects with more efficacies and at less environmental cost. The majority of commercial preparations contain either *Lactobacillus* or *Saccharomyces cerevisiae*, nitrifying bacteria, *Streptococci*, *Roseobacter* and *Bacillus* sp. Beneficial effects of regular use of probiotics in fish feed in the UK and other European countries have been reported. In aquaculture, probiotics can also be encapsulated in feed or in live food such as rotifers and Artemia. Another efficient application of probiotics to aquatic animals is through bioencapsulation or infusions in diets. According to the FAO and WHO guidelines, probiotic organisms, used in food, must be capable of surviving passage through the gut. They must have the ability to resist gastric juices and exposure to bile. In addition, probiotics must be able to proliferate and colonize the digestive tract to be safe, effective and maintain their effectiveness and potency for the duration of the shelf life of the product. The benefits of inclusion of bacterial strains into feed ingredients include improvements in feed values, contributions to enzymatic digestion, inhibition of pathogenic microorganisms, anti-mutagenic and anti-carcinogenic activity, growth-promoting factors and enhanced immune response. The effects of probiotic, *B. subtilis* on intestinal microbial diversity and immunity of the orange-spotted grouper *Epinephelus coioides* has been studied. The innate cellular response and respiratory burst activity of the supplemental groups were significantly higher. Probiotic *B. subtilis* increased the intestinal microbial diversity by stimulating the bacterial populations of *Paenibacillus* sp., *Lactobacillus oeni* strain 59 b and *Methylacidiphilum infernorum* strain V4, which are beneficial for *E. coioides*. Use of feed probiotics in aquaculture has opened the window onto the possibility of sustainable commercial aquaculture. Feed probiotics were found to provide better growth results, an improved feed conversion ratio and enhanced. It will soon be necessary to conduct studies relating to probiotic resistance to antibiotics and the possibility of transmission of genetic elements to other microorganisms in the fish GIT, and thus, to humans while consuming the aquaculture product.

16.15 Future Perspectives of Probiotics

In aquaculture, probiotics are administered through feed and water additives. However, supplementation of probiotics through feed is a better method to ensure the efficiency of the probiotic bacteria in the gut of fish without interacting with the surrounding medium. However, their use in fish feed production is still rare. Generally, probiotics have proven their promising growth results in fish by enhancing the feed conversion efficiency, as well as conferring protection against harmful bacteria by competitive exclusion, production of organic acids, hydrogen peroxide and several other compounds.

Although the role of probiotics in aquaculture is well known, the route of administration that is feasible in organisms other than fish and shellfish needs to be examined. For probiotic doses, the process of transmission is yet to be standardized for better economic and sustainable uses and to augment the aquaculture production. Bacterial strains used as probiotics are also equally significant. As described by several authors, *Bacillus* spp. is commonly used as probiotic bacteria in animal nutrition, such as *B. cereus, B. toyoi, B. licheniformis,* and *B. subtilis*. Even though, probiotic applications in mariculture have been found to be limited, there has been increasing interest in the possible use of probiotics in marine species such as summer flounder, *Paralichthys dentatus*, common dentex, *Dentex dentex*, and Japanese flounder, *P. olivaceus* Exploration is essential to elucidate the beneficial effects of probiotic bacteria as stress probiotics, which may be of a great importance for eco-friendly production and natural resource management. Probiotics can also effectively trigger the piscine immune system. For all these reasons, their use will definitively increase in the future. All probiotics must be evaluated for their safety before being used in fish nutrition. Also, dose–response relationships need to be established. In the same way, the application of probiotics for fish has garnered much interest, and today a large number of studies have demonstrated their potential benefits to aquatic hosts. However, these studies were generally laboratory based or conducted in small-scale aquarium facilities, and thus, efficacy at the farm level needs to be determined. Additionally, many of the underlying molecular mechanisms and signaling pathways, as well the impact on indigenous microbes are poorly understood, the reproducibility of these applications is often problematic. Alternatively, there is a need to increase studies on microbial ecology in aquaculture systems, correlating it to microbial communities.

Nowadays, a variety of probiotic strains, present in the gastrointestinal tract (GIT) of aquatic animals have been isolated and characterized using biochemical, morphological and molecular techniques. The development of molecular techniques such as PCR (polymerase chain reaction), FISH (fluorescent in situ hybridization), DGGE (denaturing gradient gel electrophoresis) and generation of genomic libraries have started to unveil the diversity present in aquaculture systems. Currently, next-generation sequencing methodologies offer great potential for phylogenetic identification of probiotic microorganisms without using conventional cultivation techniques. The application of up-to-date molecular procedures to study the gut microbiota and the development and validation of research methods, in vitro, ex vivo and in vivo models, have provided important information to help in understanding the mechanisms of action behind the effects. Synbiotics, as a combination of probiotics and prebiotics, have been studied to determine the synergistic effects. However, further investigations related to the interaction between probiotics and other functional additives at the molecular level are warranted in aquaculture. Although probiotic preservation is essential, the techniques for the same are yet to be established and standardized in aquaculture sector.

16.16 Immunostimulants for Aquaculture

Aquaculture is one of the fastest growing sectors in the world and Asia. However, disease outbreaks are a significant constraint to aquaculture production and trade and are affecting adversely economic development in many countries in the Asia-Pacific region.

Problems relating to the use of antibiotics, drugs, and chemical treatments to prevent diseases in fish have led to new concept i.e., disease prevention through immunostimulants. The adverse effects of antibiotics viz., development of resistant strains, residual problems in fish and human carry over, poor knowledge of farmers and their cost have been realized in many countries by putting strong stricture on their use. The use of drugs/chemicals has also achieved partial success only. The vaccination in fish, particularly in Indian context, will take a long time to come up. Above all, vaccination only protects from a single or few diseases for which it contains the antigen. It only provides specific from immunity to fish. On the other hand, fish rely more on the nonspecific defense mechanisms than mammals. Although, Indian farmers have started intensive or super-intensive system of culture practices, many farmers are still following semi-intensive culture methods with few ponds. We cannot recommend them for practicing vaccination. Above all, there is not a single commercial vaccine available for fish in India. Thus, the only alternative is to use immunostimulants for sustainable farming.

The intensive fish culture system is a highly stressful environment for fish. This stress in fish suppresses the immune responses, and the fish, kept under these conditions, become highly susceptible to diseases. Thus, the increase of nonspecific immunity in stressed fish is important for resistance against many diseases, for which the immunostimulants will play a major role. By definition, an immunostimulant is a chemical, drug, stressor, or action that elevates the nonspecific defense mechanisms or the specific immune response. Immunostimulants can activate nonspecific defense mechanism or they may be administered with a vaccine to activate nonspecific defense mechanisms as well as heightening a specific immune response.

Immunostimulants will be most effective for short-lived fish living in cool and cold waters because the development of a specific immune response is temperature-dependent. Nonspecific defense mechanisms, important in the fish, are the barriers in places such as the skin and scales, and lytic enzymes of mucus and sera; cellular aspects include monocytes, macrophages, neutrophils and cytotoxic cells. The specific immune response is induced by immunogenic stimulation with production of antibodies, resulting from the increase in numbers of antibody-producing cells. The nonspecific factors become activated by injury or stress. The mechanism of action of various immunostimulants is diverse in nature or may be poorly understood. Broadly, they may act through following mechanisms:

(a) Stimulators of T-lymphocytes: Levamisole, Freund's complete adjuvant, glucans, muramyl dipeptide.
(b) Stimulators of B-Cells: Lipopolysaccharides.
(c) Inflammatory agents inducing chemotaxis-Silica and Carbon particles.
(d) Cell membrane modifiers: Detergents, sodium dodecyl sulfate, Quarternary ammonium compounds, saponins.
(e) Nutritional factors: vitamins C and E.
(f) Cytokines: Leukotriene, Interferon.
(g) Heavy metals: Cadmium.
(h) Animal and fish extracts: mitogens.

In general, immunostimulants enhance the activity of macrophages, complements, phagocytes, lymphocytes, and nonspecific cytotoxic cells, resulting in resistance and protectoin to various diseases (Tables 16.1 and 16.2).

A good number of natural or synthetic substances have been tried in channel catfish salmon and rainbow trout, mostly. Out of which β-1, 3 glucan, levamisole, chitosan, ascorbic acid and α-tocopherol have already been evaluated in the laboratory for carps. Ascorbic acid, α-tocopherol and quarternary ammonium compounds have also been tried in carps. It has been observed that glucan, a long-chain

Table 16.1 Immunostimulants evaluated in fish and shrimp farming

A	Synthetic chemicals	Levamisole, FK 565 (lactoyl tetrapeptide from *Streptomyces, Olivaceogriseus*)
B	Biological substances	
	(a) Bacterial derivatives	MDP (muramyl dipeptide from *Mycobacterium* species), LPS (lipopolysaccharide), FCA, Vibrio vaccine, *Achromobacter stenohalis* and *Clostridium butyricum* cells, EF203 (fermented egg product), peptidoglucan (from *Brevibacterium lactofermentum* and *Vibrio* sp.)
	(b) Yeast derivatives	β-1,3 glucan, β-1,6 glucan
	(c) Nutritional factors	Vitamins C, E and A
	(d) Hormone	Growth hormone, prolactin
	(e) Cytokines	Interferon, interleukin
	(f) Polysaccharides	Chitosan, chitin, lentinan, schizophyllan, oligosaccharide
	(g) Animal and plant extract	Ete (tunicate), Hde (Abalone), Saponin, glycyrrhizin
	(h) Others	Lactiferin, Soyabean protein, Quil A, Saponin, Spirulina

Table 16.2 List of pathogens successfully controlled by immunostimulants exposure in fish/shrimp

A	Bacteria	*Aeromonas hydrophilla, A. salmonicida, Edwardsiella trada, E. ictaluri, Vibrio anguillarum, V. vulnificus, V. salmonicida, Yersinia ruckeri, Streptococcus* sp.
B	Virus	Infectious hematopoietic necrosis, viral hemorrhagic septicemia, yellow head bacculo virus/white spot virus.
C	Parasite	*Ichthyopthirius multifiliis*

polysaccharide extracted from yeast cell wall, is the best immunostimulant followed by levamisole, ascorbic acid, and α-tocopherol in carps. These could work efficiently, even in immunosuppressive conditions, to raise the nonspecific immunity level as well as specific resistance against two common bacterial pathogens viz., *Aeromonas hydrophila* and *Edwardsiella tarda*. These two pathogens cause heavy economic loss in fish farming in India by producing disease such as *Aeromonas septicaemia* and *Edwardsiellosis* or emphysematous putrefactive disease.

16.17 Conclusion

As the feed cost in the aquaculture is around 50–60% of the cost of production, it influences the profit of the fish farmers/entrepreneurs greatly. Therefore, low cost feed with improvement FCR is the need of the hour to maximize the profit from the aquaculture industry. Hence, advanced research in nutritional biotechnology is required to develop low cost feed, probiotics, prebiotics and immunostimulants etc. for aquaculture production enhancement.

References

Ackman RG (1980) The Iatroscan TLC-FID system. Methods Enzymol 72:205–252

Alford BB, Liepa GU, Vanbeber AD (1996) Cottonseed protein: what does the future hold? Plant Foods Hum Nutr 49:1–11

Bamji MS, Lakshami AV (1998) Less recognized micronutrient deficiencies in India. Bull Nutr Found India 19(2):5–8

Boling SD, Webel DM, Mavromichalis I, Parson CM, Baker DH (2000) The effects of citric acid on phytate-phosphorus utilization in young chicks and pigs. J Anim Sci 78:682–689

Brower V (1998) Nutraceuticals: poised for a healthy slice of the healthcare market? Nat Biotechnol 16:728–731

Clarke BV (1998) Zinc fingers in *Caenorhabditis elegans*: finding families and probing pathways. Science 282:2018–2022

Fournier V, Huelvan C, Desbruyeres E (2004) Incorporation of a mixture of plant feedstuffs as substitute for fish meal in diets of juvenile turbot (*Psetta maxima*). Aquaculture 236:451–465

Gatesoupe FJ (1999) The use of probiotics in aquaculture: a review. Aquaculture 180:147–165

Gopakumar K (2003) Indian aquaculture. J Appl Aquac 13(1/2):1–10

Jongbloed AW, Mroz Z, Weij-Jongbloed RV, Kemme PA (2000) The effects of microbial phytase, organic acids and their interaction in diets for growing pigs. Livest Prod Sci 67:113–122

Kang M, Morsy N, Jin X, Lupu F, Akbarali HI (2004) Protein and gene expression of Ca2+ channel isoforms in murine colon: effect of inflammation. Pflugers Arch 449:288–297

Kozasa M (1986) Toyocerin (*Bacillus toyoi*) as growth promotor for animal feeding. Microbiol Aliment Nutr 4:121–135

Kushi LH, Fee RM, Folsom AR (1997) Physical activity and mortality in post-menopausal women. J Am Med Assoc 277:128792

McLean E, Devlin RH (2000) Application of biotechnology to enhance growth of salmonids and other fish. In: Fingerman M, Nagabhushnam R (eds) Recent advances in marine biotechnology. CRC Press, Boca Raton, pp 17–55

Xu ZR, Hu CH, Xia MS, Zhan XA, Wang MQ (2003) Effects of dietary fructo oligosaccharide on digestive enzyme activities. Intestinal microflora and morphology of male broilers. J Anim Sci 82:1030–1036

Pigmentation in Fishes

17

Archana Sinha

Abstract

Fish as well as different species of prawns and crabs contains variety of carotenoids, the pigment molecules in their body, exhibiting brilliant coloration in fins, scales, muscle, gonads, and carapaces. Carotenoids viz. astaxanthin, cryptoxanthin, β-carotene, derivatives of β-carotene, canthaxanthin, zeaxanthin, lutein, etc. are the predominant molecules available in their bodies. Form maintenance and enhancement of the color in captivity, fish need an adequate level of carotenoids in their feed. The metabolic transformation of the molecules into vitamin A through either central or terminal cleavage has been shown after taking β-carotene, lutein, and cryptoxanthin. The different metabolites and their ultimate conversion into either retinol or dehydroretinol have been described. In the present communication, different types of carotenoids present, the problems associated with the designing of utilization of carotenoids as feed additives, understanding the mechanism of absorption, and sources of different pigments for the farmers and entrepreneurs have been discussed.

Keywords

Carotene · Ornamental fish · Astaxanthin · Color pigments

17.1 Introduction

Carotenoids are the red yellow coloration of many plant and animal tissues. More often it is the characteristic color of these products that the consumer uses for identification of quality. This association of quality and color is probably the most

A. Sinha (✉)
ICAR-CIFRI, Kolkata, India

© The Author(s), under exclusive license to Springer Nature Singapore Pte Ltd. 2021
P. K. Pandey, J. Parhi (eds.), *Advances in Fisheries Biotechnology*,
https://doi.org/10.1007/978-981-16-3215-0_17

important aspect of marketing aquaculture products. The pigments or coloration of fish and crustaceans in both wild and cultured conditions in aquarium or farm have attracted all humans and aqua culturists in particular, from time immemorial. The magnificent coloration such as red, yellow, green, blue, pink, orange, and other colors in fish and several species of prawns and crabs are due to the presence of a class of lipid or fat-soluble compounds known as carotenoids. These animals accumulate the pigments through various foods and synthesize the same through a series of mechanisms and ultimately accumulate in their muscles, scales, fins, and gonads. An example may be cited that of the red coloration in fish (such as salmonids, *Puntius,* and other fish) and prawns which is synthesized from a molecule or compound known as astaxanthin, available in crustaceans, molluscs, echinodermata, insects, copepod, etc. which are some of the favorite food of fish and other crustaceans. We have seen in our kitchen that when we fry prawn or shrimp, it turns red this is owing to the presence of the astaxanthin or astacene molecule. Herbivorous fish accumulate their pigmentation from various such molecules, present in blue-green algae, aquatic vegetation etc.

17.2　Types of Carotenoid Pigments

There are varieties of carotenoid pigments which have been identified from different tissues as well as scales and fins of different fish. The most common types of carotenoids are astaxanthin, β-carotene, metabolites of β-carotene such as 8′-apo, 10′-apo, 12′-apo and 14′-apo β-carotenals, α-carotene, echinenone, cryptoxanthin, lutein, taraxanthin, zeaxanthin, etc. The structure of the above-mentioned carotenoids is shown in Fig. 17.2. These carotenoids are available in algae, aquatic plants, higher angiosperm plants, coelenterates, annelids, crustaceans, insects, mollusks, echinoderms, fish etc. Fish accumulate their pigmentation from the above-mentioned carotenoids. It may be mentioned that the carotenoid molecules after entering the intestinal lumen, the dioxygenase and monooxygenase enzyme system break at least 35–50% of its concentration into different forms of vitamin A (Goswami and Bhattacharjee 1982; Goswami and Sarma 2005), while rest of the molecule is transported into the different parts of the body. The accumulation in the concerned regions are regulated through carotenoids binding protein, the lymph and the overall circulation play a significant role in the distribution and accumulation of the carotenoid's molecule. Further, role of different hormones from the pituitary, adrenal, thyroid gland, and gonads play important role.

Structure of Pro Vitamin A - Carotenoids [1-VII] and Retinol (Vitamin A₁) and Dehydroretinol (Vitamin A₂) [VIII-IX]

β-carotene: Present in higher angiosperm plants, aquatic plants, blue-green algae and fish.

Lutein: Present in higher angiosperm plants, aquatic plants, blue-green algae and fish.

Cryptoxanthin: Present in higher angiosperm plants, aquatic plants, blue-green algae and fish.

Astaxanthin: Present in coelenterata, annelids, crustaceans, molluscs, echinoderms and fish.

Zeaxanthin: Higher plants, aquatic plants, algae, fungi and scales of some marine fish

Echnenone: Find in Annelids, molluscs, echinodermata, proto chordates, and fish

Canthaxanthin: Available in mushroom and fish.

Carotenoid containing organisms, found in the aquatic food chain, are the main source of these carotenoids. However, the commercial feed ingredients like corn,

corn gluten meal, and alfalfa are being used as important source of pigments such as zeaxanthin and lutein. Alternatively, common carotenoid-rich ingredients like marigold meal (lutein), red pepper extract (capsanthin) and krill or crustacean meals (astaxanthin) are used as fish feed. Canthaxanthin is a proven effective pigmentary additive for the tropical fish *Trichogaster leeri*, especially to enhance reproductive processes (Frank and Cogdell 1993).

17.3 Absorption of Carotenoids

The feed of fish and crustaceans which contains a wide spectrum of carotenoid pigments is ingested by them. More than 600 carotenoids have been identified in nature (Pfander 1993). Approximately, 50 out of the 600 carotenoids available in the nature can be converted into the essential nutrient and vitamin A. The carotenoids, released from the food matrices are solubilized in the gut in presence of fat and conjugated bile acids, after that they get absorbed into intestinal mucosal cells which is a critically important process. After metabolization of the carotenoids within the cell by oxidative cleavage through one or more routes, they form retinal, retinoic acid, and small amount of intermediate breakdown products named β-apocarotenoids. Major portion of unutilized carotenoids are transported to muscles, scales and fins. They possess many conjugated double bonds—usually 9–13, each one can form several geometric isomers. β-carotene contains 9 double bonds in its polyene chain that are free to assume cis/trans configurations, can theoretically form 275 isomers whereas, its asymmetric isomer, α-carotene can form 512. Thus, the total possible number of compounds in the class, including all possible isomers, easily exceeds 200,000. Approximately 50 carotenoids that are largely in the all-trans configuration are in focus for scientific attention.

The function and action of carotenoids are determined by their physical and chemical properties. Carotenoids have ability to absorb into visible light. The delocalized Π-electrons of carotenoids can undergo to a photon-induced transformation, yielding an excited singlet state (S2) that can efficiently transfer its energy to chlorophyll, forming an excited singlet state of chlorophyll of slightly low energy. The central polyene chain of the carotenoids must be in direct overlapping contact with the porphyrin ring of chlorophyll. In nature, the physical structure of the chloroplasts facilitates the transfer of energy, absorbed by carotenoids to chlorophyll. Thus, carotenoids gather energy that is used in photosynthesis. The manifold function of carotenoids is regulated through its structural peculiarizes and the associated β-ionone ring.

Carotenoids also serve as an essential photoprotective function in photosynthetic organisms. As they are highly unsaturated, they either extract or donate electrons from suitable core actants, leading to the free-radical *anions* and *cation*, which in turn can react with oxygen as well as other molecules. Carotenoids, therefore, show both antioxidant and pro-oxidant activities under various conditions. They are nonpolar and largely associated with the lipophilic parts of the cell; including action of carotenoids which can be attributed directly to their physicochemical properties.

Their antioxidant and pro-oxidant properties are useful for aquatic animals including fish and crustaceans (Goswami 2007).

The metabolism of carotenoids is highly dependent on their chemical nature. Carotenoids are transported by lipoproteins in the plasma, with the hydrocarbon carotenoids mainly associated with low density lipoproteins and the xanthophyllous distributed more evenly between low density and high-density lipoproteins. Carotenoids are taken up differentially by tissues, but little is known of the factors regulating this process. The absorbed quantity of carotenoids is distributed and deposited in the skin, scale, fins, muscle, and gonads in different stages of maturation. The color changing phenomenon in discus fish is not only fascinating but also an indicator of its health and condition. Various factors govern the color change from stress to lighting color to the types of plants in the aquarium.

17.4   Application of Carotenoids in Crustacean Culture

The crustaceans contain astaxanthin as the predominant carotenoid and it has been studied extensively in the shrimp *Penaeus japonicus* (Bird and Savage 1990), It was observed that rate of survival of *P. japonicus* on astaxanthin-supplemented diets was higher compared with those containing β-carotene or algal meal. The suggested oxidative pathway for metabolism of dietary carotenoids, include that of β-carotene and zeaxanthin to astaxanthin. Prawns have ability to deposit astaxanthin in their tissue directly. Astaxanthin is the most effective carotenoid for prawn pigmentation. The three carotenoids namely β-carotene, zeaxanthin and astaxanthin lead to the deposition of mainly astaxanthin esters. Dietary astaxanthin gets stored in the integument, carapace, epidermis, and hepatopancreas of the prawns. It is assumed that astaxanthin in the prawn serve as an intracellular oxygen reserve and allows the crustaceans to survive even under the hypoxic conditions of culture systems. Howel and Matthews (1991) explained that the discoloration in blue diseases of cultured tiger prawn *Penaeus monodon* was due to nutritional lack of carotenoids. Astaxanthin, astaxanthin esters and β-carotene are the main carotenoids present in *P. monodon* exoskeletons. Pond-raised prawns, showing blue coloration, had a total carotenoids concentration of 4.3–7 ppm compared with 26.3 ppm in the exoskeleton of wild caught prawns. Mayers (1994) demonstrated effective carapace pigmentation with Spirulina supplemented diets.

Carotenoid pigments perform or regulate manifold activities in fish. They help in growth, reproduction, developing immunity, maintaining epithelial tissues and hematological parameters including the maintenance of hemopoietic tissues and in particular vitamin A synthesis (Goswami and Dutta 1991).

17.5 Application of Carotenoids in Feed

Although there have been continuous discussions on the addition of pigment producing materials in the feed as an additive, for enhancing the coloration yet the important point is the understanding the mechanism of action as well as status of its absorption and assimilation. Several feed additives are prepared without understanding its physical and chemical properties. An important example is pigmenting a fish species *Polyacanthus fasciatus*. Based on the knowledge about carotenoids available and their ratio in ingredients, one can select appropriate additive which will serve the purpose of providing coloration. Further, preparation of an additive requires the knowledge of photolabile and thermolabile properties and lipid materials to be used or preparation of water-soluble matrix and use of specific lipid as vehicle. The concept of binding with actinomyosin and carrier protein i.e., carotenoids binding protein requires a thorough idea. Several feeds are prepared with pigmented materials, which do not serve the specific purpose. Preparation and storage as well as feeding are a delicate technology where all these above points have to be taken into account.

17.6 Hormonal Methods to Enhance Coloration

Hormonal methods have been used by the aquarists to induce and enhance colors in fish since long. For decades, guppy breeders have been using this method to enhance the female fish colors to make them look more attractive. As it is believed that a colorful female most probably engenders colorful mates. This statistic is well-documented for fancy Guppies. The method is as follows:

1. Dissolve one-tenth of a gram of methyl testosterone in a half cup of 70% alcohol.
2. Next, take a quart bottle and transfer this alcoholic solution to it.
3. Add water to the solution in the quart bottle and shake it well. The stock solution is ready.
4. Six drops of this stock solution should be added to three gallons of water on alternate day.

The treatment is recommended to be continued for 4 weeks in order to bring out the hidden colors. This method is used by the breeders to make discus more sellable.

17.7 Problems in Maintenance of Brightness of Pigments

It is important to know about carotenoids and their occurrence as well as their metabolism by converting them into vitamin A. Considering all these facts, figures, and background of the pigments, we must answer some of the queries raised by the entrepreneurs, with reference to the retaining restoration of pigments. It has been found when ornamental fish are reared in an aquarium they start to lose coloration.

Table 17.1 Content of the natural carotenoids in different sources

Animal source	Content (mg/kg)	Plant source	Content (mg/kg)
Crab meal	75–1300	Marigold (petal meal)	7000
Crayfish meal	30–800	Chlorella	4000
Shrimp meal	100–130	Yeast	1000
Shrimp oil	25–125	Sea weed	390–900
		Corn gluten	290
		alpha alpha	280

Sometimes, the common fish feed would not be competent enough for the pigmentation. There is no doubt that fish with faded color, has no demand or less market value.

The plants synthesize carotenoids naturally which get modified into animal tissues. The researchers observed that the fish cannot synthesize pigments and therefore, they rely on a dietary supply of carotenoids to exhibit their natural skin pigmentation. Fish must obtain them from their diet to maintain and enhance the color (Boonyaratpalin and Lovell 1977). Pigmentation in aquatic organisms attracted many researchers and they have shown that pigmentation in fish can be achieved by inclusion of crustacea and or crustacean processing wastes in their diets (Torrissen 1991; Spinelli and Mahnken 1978). Application of plant sources as carotenoid pigments such as paprika, *Capsicum annuum* L. (Bitzer 1963; Ellis 1979), yeasts, *Rhodotorula sanneii* and *Ciferri et Redaelli* (Salvolainen and Gyllenberg 1970), yeasts, *Phaffia rhodozyma* (Johnson et al. 1977), chestnut flowers (Neamtu et al. 1976), dried flowers (Pailan et al. 2012), hippophage oil, *Hippophae rhamnoides* Linne (Kamata et al. 1977), spirulina algae (Alagappan et al. 2004) and *Hibiscus rosa-sinensis* (Sinha and Asimi 2007) for ornamental fish diets has been attempted (Table 17.1).

17.8 Feed Additives for Coloration

17.8.1 China Rose Petals (*Hibiscus rosasinensis*)

Chinese rose (*Hibiscus rosa-sinensis*), an ancestor of many cultivated hybrid roses, is having mostly red, pink, or white flowers. Hibiscus flowers yield a dark purplish dye, which is used for making shoe polish. Its leaves are also being used as an emollient, anodyne (reliving pain) and laxative in medicines. It is used as a potent natural carotenoid source to enhance color of goldfish *(Carassius auratus L.)*.

17.8.2 Dried Marigold (*Tagatef patula*) Petals

The marigold flower contains lutein, cryptoxanthin, β-carotene, zeaxathin, and other pigments derivative. The fine dried powder of the colored pellets is a good source and device which, after the addition in the feed, would restore the coloration. The faded pigments start rejuvenation after depositing the different carotenoids molecules in their muscles, scales, fin, etc.

17.8.3 Gulmohar (*Puya)*

The petals, androecium and gymnasium of gulmohar, *Caesalpinia pulcherrima* are good source of several carotenoids. A large amount of pigments could be isolated from the anthers. A simple study showed that 200 g of anthers will produce about 5–8 mg of pigments. The experiment conducted in the laboratory showed that by simple dipping of anther acetone in a test tube one can isolate the pigments.

17.8.4 Waste from Crustaceans

The waste produced from the shrimp, prawn and crab processing industries give rise to a large amount of pigments and astaxanthin in particular. Salmonids eat shrimps, prawns, lobsters, carbs which contain a large amount of astaxathin, zeaxanthin, taraxanthin, echinenone etc. The fish, after ingesting the foods containing the above-mentioned carotenoids, converts about 30–40% into vitamin A as the rest 60–70% are deposited in the muscle (Storebakken and Goswami 1996). The crustaceans waste, the shell, carapace, cephalothorax, etc. contain a large amount of astaxanthin which is used in pigmenting both marine and freshwater fish and crustaceans. There are several species of prawns and carbs which provide some affordable impetus for small aqua culturists of ornamental fish.

17.8.5 Chironomids

Chironomids contain a large amount of carotenoid pigment. The culture of chironomids, tubifex should be encouraged so that entrepreneurs can adopt a mass culture program. The culture of chironomids is a simple process where a continuous flow of water is maintained. There is a good market and demand of chironomids/tubifex which are available in the big markets, public places of the metros of India.

17.8.6 Aquatic Angiosperms

Several species of aquatic angiosperms such as *Azolla* sp. provide a good source of several pigments. The dry powder or the fresh culture of species like *Azolla* could be added to the feed formulations of various species of fish. Besides this, *Azolla* also fix a considerable amount of nitrogen in aquatic environment.

17.8.7 Blue-Green Algae and Algal Powder

Various species of blue-green algae and algal powder are efficient precursors of several molecules of pigments. Mass culture of algae and preparation of dried powder from it could be used for the preparation of a diet containing algal powder. *Spirulina* is being used in several places of India as a precursor of pigments as well as a high calorie fish feed.

17.8.8 Zooplanktons

Several species of zooplankton such as *Daphnia, Cyclops, and Bosmia* accumulate astaxanthin, idoxanthin, cryptoxanthin in their body. These are potent and main food of several species and particularly in their larval forms. There should be a mass culture of the above-mentioned species which could be utilized for restoration of the pigments.

17.9 Requirement of Carotenoids

It is understood that the fish cannot synthesize the carotenoid molecules in their body and obtain or accumulate the same from food (Goodwin 1984). For example, it is well established that the red coloration of Salmonids comes from food rich in crustaceans (Storebakken and Goswami 1996) which contains astaxanthin. These lipid soluble pigments are found in it and color differs in different parts of the body of the fish, such as scales, fins, muscles, and gonads. The multifaceted role of carotenoids has been illustrated by a series of workers (Goodwin 1984; Krinsky 1993; Torissen and Naevdal 1984). In gold fish (*Carrasius auratus*) the optimum level of astaxanthin is found to be 36–37 mg/kg of diet for intense coloration and better survival. Carotenoids play a significant role in sexual dimorphism, growth of fries and fingerlings. They also function as a potent antioxidant, immune modulator, scavenger of singlet oxygen and in the metamorphosis of gametes (Lygren et al. 1999).

The most significant role of pigment molecules is its ability to be converted into vitamin A. As mentioned earlier, out of 600 carotenoids, hardly 50 such carotenoid molecules have the ability to be converted into different forms of vitamins A, such as retinol (present in marine fish, to some extent in freshwater fish, adult amphibian, reptiles, aves, mammals including humans) and dehydroretinol (in freshwater fish and larval amphibian).

17.10 Extraction and Estimation of Carotenoids from Fish and Crustaceans

The carotenoid pigments are extracted from the dried scales, fins and the tissues of liver, muscles, gonads etc. of fish. It is easy to extract the lipids from dried samples. Depending upon the amount of moisture, the application of anhydrous sodium sulfate needs to be measured. Acetone, light petroleum, benzene, diethyl ether, etc. are the best solvents for the extraction of various pigments from lipid. Tissues such as liver or muscle from the concerned species are minced with anhydrous Na_2SO_4 and carotenoid pigments are extracted with light petroleum (40–60 °C) and diethyl ether etc. The extract, containing the esters, is pooled together, evaporated and is estimated from their UV-visible absorption maxima of the respective compounds. The esters are saponified with methanolic KOH (10%) under refluxed conditions and later extracted with light petroleum (40–60 °C). Alkali can be removed after 5–7 washings with distilled water. It is recommended to extract again with light petroleum 40–60 °C, dry over anhydrous Na_2SO_4 and measured for alcoholic form. Different carotenoids and vitamin A analogues can be extracted with this method.

17.11 Estimation

The most commonly used procedures for estimation of different carotenoids are methods like (a) assay based on UV-visible spectroscopy, later followed by $SbCl_3$ reaction (b) colorimetry, (c) fluorometric methods, (d) bioassay procedure etc. Chromatography techniques such as thin layer column and HPLC are widely used methods which give correct estimation. The compounds are separated through either thin layer chromatography or column chromatography. UV light is used in detecting the site of occurrence from the chromatography as well as calculation of rf values. However, use of the HPLC technique has revolutionized the estimation as well as characterization of the compounds. The emergence of various columns and eluents as well as different software packages have helped in estimating the amount of both carotenoids and different vitamin A compounds (Guillou et al. 1993). Further total carotenoids can be measured by using the UV-Visible absorption spectroscopy. Capacity factor, Retention time, and maximum absorption wavelength of some carotenoids and retinoids found in freshwater fish is mentioned in Table 17.2.

Liver oil extracts from *Cirrhinus mrigala* estimated using HPLC has been shown in Fig. 17.1a, b. The separation of carotenoids and retinoids are shown there. The amounts of various compounds are measured from the peak and areas.

The light petroleum (40–60 °C) extracted liver oil and with the addition of internal standards was dried in the rotatory vacuum evaporator and later dissolved in the HPLC solvent and injected (50 µL) to the pump. The detected peaks of each compound were measured from the area covered. Besides the HPLC the ordinary TLC and column chromatography are also used to identify and quantify the carotenoids available in scaly fins and tissues and carapace.

Table 17.2 Capacity factor, retention time, and maximum absorption wavelength of some carotenoids and retinoids found in freshwater fish

Compound	Capacity factor (K)	Retention time (min)	Absorption maxima (nm)		
Crustaxanthin (3,4,3′,4′-tetrahydroxy-β-carotene)	0.66	3.65	460	–	488
Astaxanthin (3′,3′-dixydroxy-4,4′-diketo-β-carotene)	0.99	4.37	–	480	–
Isozeaxanthin (4,4′-dehydroxy-β-carotene)	1.30	5.05	452	–	478
Lutein (3,3′-dehydroxy-α-carotene)	1.53	5.57	448		476
Phoenicoxanthin (3-hydroxy-4,4′-diketo-β-carotene)	1.70	5.95		475	
Zeaxanthin (3,3′-dehydroxy-β-carotene)	1.94	6.47	454		482
Canthaxanthin (4′,4′-diketo β-carotene)	2.81	8.39		472	
CAEE (β-apo-8′-carotenoic acid ethyl ester)	4.30	11.65		445	
Citranaxanthin (β-apo-8′-carotenal)	4.46	12.02		468	
Isocryptoxanthin (4-hydroxy-β-carotene)	5.60	14.53	427		450
Cryptoxanthin (3-hydroxy-β-carotene)	6.26	15.98	454		480
Echinenone (4-keto β-carotene)	7.97	19.73		460	
α-carotene	19.60	45.33	448		476
β-carotene	21.64	49.80	454		482
Retinol$_2$ (3,4-dehydroretinol)	0.51	3.33		352	
Retinal$_2$ (3,4-dehydroretinal)	0.66	3.65		390	
Retinol$_1$	0.75	3.85		326	
Retinal$_1$	0.95	4.30		377	
Retinyl acetate	1.15	4.72		326	
Retinyl propionate	1.42	5.32		326	
Retinyl palmitate	26.06	59.54		326	

Operating system = 300 mm × 3.9 mm Nova-Pal c18 column (4μm): acetonitrile/dichloromethane/methanol/water/propionic acid (71/22/4/2/1, v/v/v/v/v) as mobile phase and spectrophotographic solution; flow rate = 1 mL min^{-1}; temperature = 0 °C. (Source: Guillou et al. 1993)

17.12 Conversion of Carotenoids into Vitamin A in Freshwater Fish

Provitamin A carotenoids are responsible for the formation of vitamin A in all the animals, where carotenoids such as β-carotene, cryptoxanthin, astaxanthin, lutein are converted into retinol and dehydroretinol. However, owing to the presence of a substantial amount of dehydroretinol in freshwater fish (Abdullah et al. 1954;

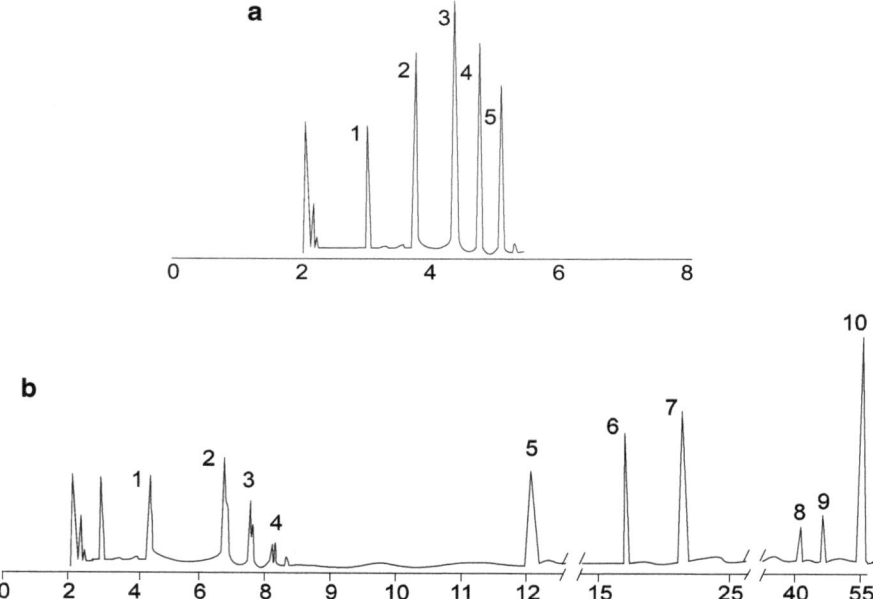

Fig. 17.1 (a) Chromatogram of standard retinoids at a concentration of 3 mg. *1*. Dehydroretinol, *2*. dehydroretinol, *3*. retinol, *4*. retinal, *5*. retinyl propionate. (**b**) HPCL chromatogram of *Cirrhinus mrigala*: showing the carotenoids profiles, five carotenoids were identified and the remaining peaks were not identified. 1. Astaxanthin, 2–4. unknown carotenoids, 5. CAEE, 6. cryptoxanthin, 7. lutein, 8, 9. unknown carotenoids, 10. β-carotene

Balasundaram et al. 1956), the biogenesis of vitamin A_2 in fish is still obscure. Carotenoids such as β-carotene, cryptoxanthin act as precursors of retinol in fresh-water fishes rich in retinol. In case of the dehydroretinol rich fish, lutein, astaxanthin and cryptoxanthin act as precursors of dehydroretinol (Goodwin 1984). Although more than 600 carotenoids and their structures are known, only 50 of them are having provitamin A status.

The mechanism for absorption, transportation, storage and uptake of carotenoids and vitamin A has been reviewed by Blomhoff (1994). The uptake of β-carotene from the intestine is stimulated by bile, the bile acids and glycocholate influenced by the dioxygenase systems. The conversion of β-carotene takes place in the intestinal mucous by the enzyme β-carotene 15, 15′ dioxygenase. β-carotene converting enzymes have been found in a number of mammals as well as in lower vertebrate. The cleavage of provitamin A carotenoids to vitamin A follows two paths. The molecule is oxidatively divided by central or eccentric cleavage (Fig. 17.2), the later forming one short and one long β-apocarotenal. Central cleavage yields two molecules of vitamin A, while eccentric cleavage yields one (Bachmann et al. 2002; During et al. 1996). Besides β-carotene, other carotenoids such as cryptoxanthin, astaxanthin etc. get converted into retinol and dehydroretinol.

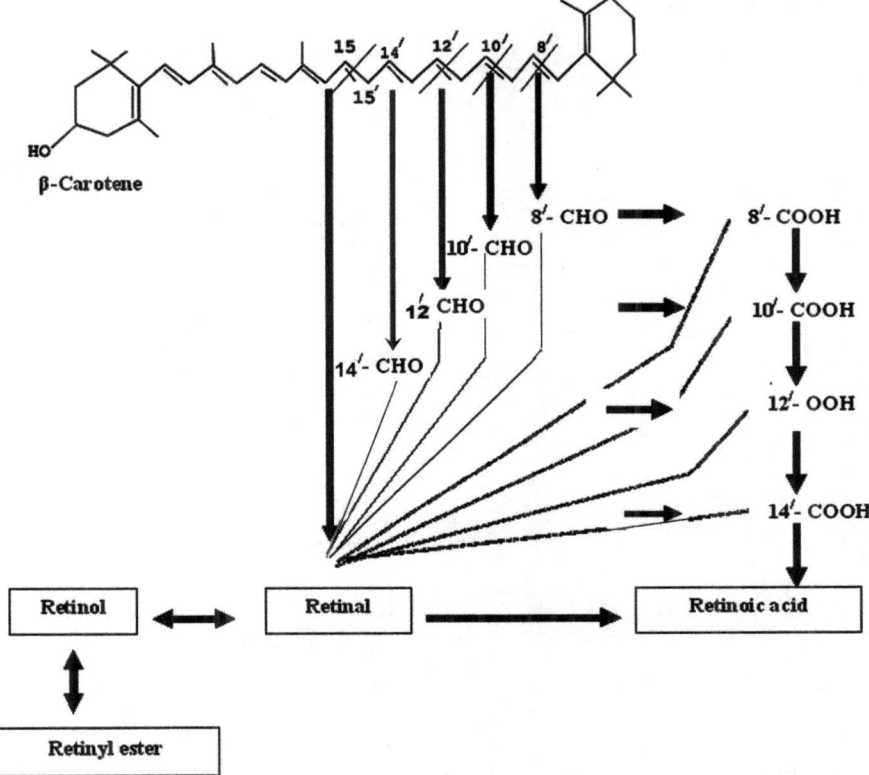

Fig. 17.2 Eccentric cleavage of ß-carotene

The amount of β-carotene in zooplankton has been estimated to be insufficient to account for the amount of vitamin A which is found in zooplankton-eating fish (Goodwin 1984)ˑ It is, therefore, unlikely that β-carotene is an important vitamin A source for these fish. In general, fish also have a poor ability to absorb β-carotene in salmonids. The predominant carotenoids in the marine environment are astaxanthin, lutein and zeaxanthin. It was suggested early that astaxanthin is a vitamin A precursor for fish (Torissen and Naevdal 1984; Storebakken and Goswami 1996). Goodwin (1984) considered it unlikely that carotenoids with a hydroxy and keto substitute β-ionone might have a provitamin A activity. In the early sixties, however, Granguad et al. (1962) published papers on the conversion of astaxanthin to vitamin A in the intestines of several species of freshwater fish.

The metabolism of Lutein and cryptoxanthins is shown in Figs. 17.3 and 17.4. These carotenoid pigments are responsible to exhibit color in the muscle, scale, fins, gonads and carapace as well as conserving into retinol and dehydroretinol. The optimum requirements of a balanced diet for aquarium fish are different, and actually more demanding, than those for commercial food fish. A feed for aquarium fish

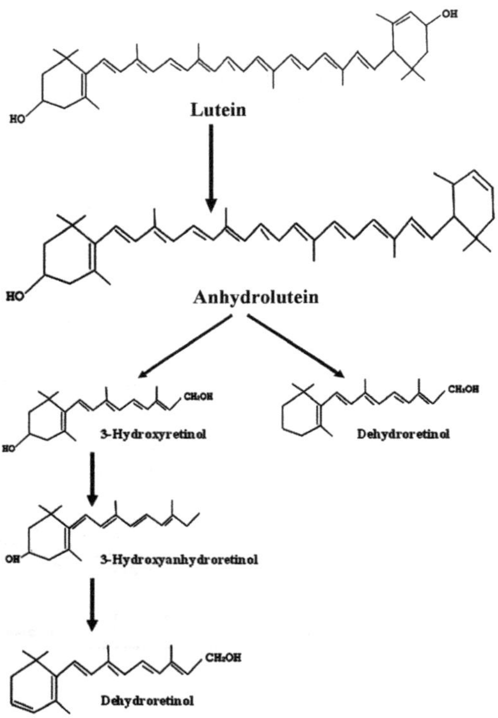

Fig. 17.3 Metabolism of
lutein

should be nutritionally balanced, palatable, and resistant to crumbling, water stable, buoyant, and capable to enhance pigmentation in ornamental fish and crustaceans. The pigmentation of cultured fish is lower than wild one even after feeding carotenoid-rich diet (Tripathy et al. 2019). Mandal et al. (2020) explained the genetics of color variation in cultured and wild fish to prove the same.

17.13 Genetic Basis of Carotenoid Coloration in Fish

Advances in DNA sequencing discovered the novel genes associated with carotenoid-based phenotypes at a rapid rate. New gene families are characterized, including those coding for scavenger receptors, β-carotene oxygenases, and ketolases. Carotenoid processing genes are conserved across deep phylogenetic time scale and they are associated with gene and genome duplication. Sometimes in parallel in deeply diverged taxa, they have been co-opted for various functions. Hybridization, as a tool contributes to the movement of genes across deep and recent timescales. Natural and artificial hybridization, combined with whole-genome sequence comparisons, has been used to identify several candidate genes for carotenoid-based avian feather coloration.

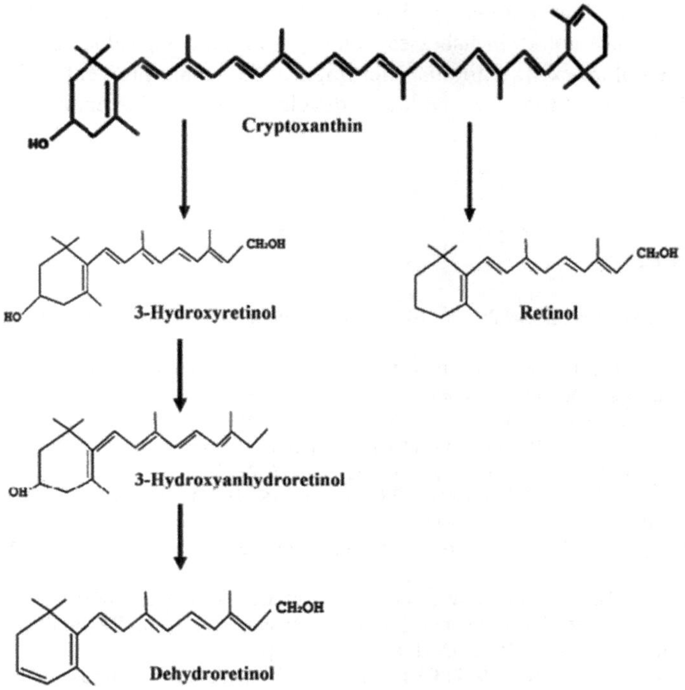

Fig. 17.4 Metabolism of cryptoxanthin

Pigmentation is one of the most conspicuous traits due to which varied colors are observed in the natural organisms. Identification of genes that are important for carotenoid transport, deposition, and processing is difficult, in contrast to the well-characterized genes involved in the melanogenesis pathways. Genetic basis is poorly understood in heritability of carotenoid-based coloration which has been demonstrated in different vertebrate species (Olsson et al. 2013). A list of 11 candidate genes with potential roles in the uptake, deposition and degradation of carotenoids in vertebrates was developed by Walsh et al. (2012). Transporter proteins and enzymes, involved in the uptake and metabolism of dietary carotenoids, are under genetic, rather than environmental, control (Yonekura and Nagao 2007; Hill and Johnson 2012). Magalhaes and Seehausen (2010) found that the genetic factors play the main role in generation of discrete variation, such as color polymorphisms or differentiation between closely related cichlid taxa. Koi are majestic, elegant, bright, and colorful ornamental fish that descended from the well-known carp. Decades ago, there were only blue, black, white, and red koi fish. Today, these aquatic creatures have been bred in different combinations to create a diverse variety of patterns and colors. Based on the facts, that chromatophores of fish skin has a type of cell that contains color pigments and these pigments utilize carotenoids to bring forth shades of yellow (Xanthophylls), red and orange (Carotenoids), and brown and black (Melanin), there are ample scopes to enhance colors in fishes especially of ornamental value. Genetics dictate where these colors exist, while the diet impacts the actual pigments. These are the important facts for further research and development in color enhancement of the fishes.

References

Abdullah MM, Morcos SR, Salah MK (1954) Content of vitamin A_2 in some Nile fishes. Biochem J 56:567–571

Alagappan M, Vijula K, Sinha A (2004) Utilization of spirulina algae as a source of carotenoid pigment for blue gouramis (*Trichogaster trichopterus* Pallas). J Aquaricult Aquat Sci 10:1–11

Bachmann H, Desbarats A, Pattison P, Sedgewick M, Riss GWA, Cardinault N, Duszka C, Goralczyk R, Grolier P (2002) Feedback regulation of carotene 15, 15-monooxygenase by retinoic acid in rats and chickens. J Nutr 132:3616–3622

Balasundaram SC, Cama HR, Sundaresan PR, Varma TNR (1956) Vitamin A $_2$ in Indian freshwater fish liver oils. Biochem J 64:150–154

Bird JN, Savage GP (1990) Carotenoid pigmentation in aquaculture. Proc Nutr Soc New Zealand 15:45

Bitzer RA (1963) The coloring of future hatchery trout. U.S. Trout News 8(2):13

Blomhoff R (1994) Overview of vitamin A metabolism and function. In: Blomhoff R (ed) Vitamin A in health and disease. Mercel Dekker, New York, pp 1–35

Boonyaratpalin M, Lovell RT (1977) Diet preparation for aquarium fish. Aquaculture 12:53–62

During A, Nagao A, Hoshino C, Terao J (1996) Assay of β-carotene 15, 15′- dioxygenase activity by reverse-phase high-pressure liquid chromatography. Anal Biochem 241:199–205

Ellis JN (1979) The use of natural and synthetic carotenoids in the diet to color the flesh of salmonids. In: Halver JE, Tiews K (eds) Proceedings of the world symposium on finfish and fish feed technology, Hamburg 20-23 June, 1978, vol 2, pp 353–364

Frank HA, Cogdell RJ (1993) The photochemistry and function of carotenoids in photosynthesis. In: Young A, Britton G (eds) Carotenoids in photosynthesis. Chapman and Hall, London, pp 252–326

Goodwin TW (1984) The biochemistry of carotenoids, animals, vol 2. Chapman and Hall, London, 122p

Goswami UC (2007) Vitamin-A in freshwater fish. In: Presidential address in the section—animal, veterinary & fishery sciences, 94th session of Indian Science Congress, Annamalai University, Annamalai. Poc. 94th session Indian Science Congress III, pp 1–26

Goswami UC, Bhattacharjee S (1982) Biosynthesis of 3-dehydroretinol: metabolism of lutein (β-ε -carotene-3,3′ diol) in *Clarias batrachus* and *Ompok pabda*. Biochem Int 5(4):545–552

Goswami UC, Dutta NK (1991) Vitamin A-deficient diet and its effects on certain hematological parameters of *Heteropneustes fossilis*, 3-4 dehydroretinol- rich freshwater fish. Int J Vit Nutr Res 61:205–209

Goswami UC, Sarma N (2005) Efficiency of a few retinoids and carotenoids *in vivo* in controlling benzo[a]pyrene-induced forestomach tumor in female Swiss mice. Br J Nutr 94:540–543

Granguad R, Massonet R, Conquy T, Ridolfo J (1962) In vitro conservation of astaxanthin into vitmin A by the intestinal mucosa of Gambusia holbruki. Comp Rendus Hebd Acad Sci Paris 254:479–581

Guillou A, Choubert G, de la Noue J (1993) Separation and determination of carotenoids, retinol, retinal and their dehydro forms by isocratic reverse-phase HPLC. Food Chem 476:93

Hill GE, Johnson JD (2012) The vitamin A-redox hypothesis: a biochemical basis for honest signaling via carotenoid pigmentation. Am Nat 180:E127–E150

Howel BK, Matthews P (1991) Discoloration and blue disease in formed figure prawn. Comp Biochem Physiol 98B:375–379

Johnson EA, Conklin DE, Lewis MJ (1977) The yeast *Phaffia rhodozyma* as a dietary pigment source for salmonids and crustaceans. J Fish Res Board Can 34(12):2417–2421

Kamata T, Neamtu GC, Simpson KL (1977) The pigmentation of rainbow trout (*Salmo gairdneri*) with *Hyppophae rhamnoides* oil. Rev Roum Biochim 14(4):253–258

Krinsky NI (1993) Actions of carotenoids in biological systems. Annu Rev Nutr 13:561–587

Lygren B, Hamre K, Waagbø R (1999) Effects of dietary pro-and antioxidants on some protective mechanisms and health parameters in Atlantic salmon. J Aquat Anim Health 11:211–221. https://doi.org/10.1577/1548-8667(1999)011<0211:EODPAA>2.0.CO;2

Magalhaes IS, Seehausen O (2010) Genetics of male nuptial colour divergence between sympatric sister species of a Lake Victoria cichlid fish. J Evol Biol 23:914–924

Mandal SC, Tripathy PS, Khateia A, Behera DU, Ghosh A, Pandey PK, Parhi J (2020) Genetics of colour variation in wild versus cultured queen loach, *Botia dario* (Hamilton, 1822). Genomics 112(5):3256–3267

Mayers SP (1994) Developments in world aquaculture feed formulations and role of carotenoids. Pure Appl Chem 66(5):1069–1076

Neamtu G, Weaver CM, Wolke RE, Simpson KL (1976) The pigmentation of rainbow trout with extracts of floral parts from Aesculus. Rev Roum Biochim 13:25–30

Olsson M, Stuart-Fox D, Ballen C (2013) Genetics and evolution of colour patterns in reptiles. Semin Cell Dev Biol 24(6–7):529–541

Pailan GH, Sinha A, Kumar M (2012) Rose petal meal as natural carotenoid source in pigmentation and growth of rosy barb (*Puntius conchonius*). Indian J Anim Nutr 29(3):291–296

Pfander H (1993) Carotenoids: an overview. In: Packer L (ed) Methods in enzymology, vol 213. Academic Press, San Diego, pp 3–31

Salvolainen JET, Gyllenberg HG (1970) Feeding of rain bow trouts with *Phodotorula sanneii* preparations. III. Amounts and qualities of carotenoids. Lebensm Wiss Technol 3:18–20

Sinha A, Asimi OA (2007) China rose (*Hibiscus rosasinensis*) petals: a potent natural carotenoid source for goldfish (*Carassius auratus* L.). Aquacult Res 38(11):1123–1128. https://doi.org/10.1111/j.1365-2109.2007.01767.x

Spinelli J, Mahnken C (1978) Carotenoid deposition in pen-reared salmonids fed diets containing oil extracts of red crab (*Pleuronocondes planipes*). Aquaculture 13:213–223

Storebakken T, Goswami UC (1996) Plasma carotenenoid concentration indicates the availability of dietary astaxanthin for Atlantic Salman, *Salmo salar*. Aquaculture 146:147–153

Torrissen KR (1991) Genetic variation in growth rate of Atlantic salmon with different Trypsin-like isozyme pattern. Aquaculture 93:299–312

Torissen OJ, Naevdal G (1984) Pigmentation of salmonid: genetical variation in carotenoid deposition in rainbow trout. Aquaculture 38:59–66

Tripathy PC, Devi NC, Parhi J, Priyadarshi H, Patel AB, Pandey PK, Mandal SC (2019) Molecular mechanisms of natural carotenoid-based pigmentation of queen loach, *Botia dario* (Hamilton, 1822) under captive condition. Sci Rep 9(1):1–12. https://doi.org/10.1038/s41598-019-48982-9

Walsh N, Dale J, McGraw KJ, Pointer MA, Mundy NI (2012) Candidate genes for carotenoid coluration in vertebrates and their expression profile in the carotenoid-containing plumage and bill of a wild bird. Proc R Soc B Biol Sci 279:58–66

Yonekura L, Nagao A (2007) Intestinal absorption of dietary carotenoids. Mol Nutr Food Res 51:107–115

Immunoprophylactic Measures in Aquaculture

Akshaya Panigrahi, R. Naveenkumar, and R. R. Das

Abstract

High intensification of aquaculture practices leads to diseases of microbial etiology of economic significance. These diseases have surfaced in rearing systems and constitute a significant threat to sustainable aquaculture which calls for proper investigation and appropriate solutions. Immunoprophylaxis focuses on tackling the effects of high intensification in aquaculture. Immunoprophylaxis is pragmatic immunology which is concerned with preventing infectious diseases through the use of vaccines, immune sera, immunostimulants, plant as well as animal products, and gamma globulins. This confers immunity in fish/shellfish during intensification of aquaculture thereby benefiting farmers and can possibly even double the output. Intensification often encourages many existing and emerging diseases and needs drugs, synthetic chemicals, and vaccination for treatments both as prophylaxis as well as curative. Immunoprophylaxis could be both active as well as passive. Vaccines (live, attenuated, inactivated,, subunit, recombinant vector, DNA-based and synthetic/peptide) are classified as active, and the microbial derivatives viz. probiotics, prebiotics as well as plant and animal extracts that boost the nonspecific immunity in both fishes and crustaceans are considered passive immunoprophylactics. Immunoprophylaxis, also known as prophylactic immunization, is one of the alternate, highly productive strategies in the primary prevention of diseases.

A. Panigrahi (✉) · R. Naveenkumar · R. R. Das
ICAR-Central Institute of Brackishwater Aquaculture, Chennai, India
e-mail: akshaya.panigrahi@icar.gov.in

© The Author(s), under exclusive license to Springer Nature Singapore Pte Ltd. 2021
P. K. Pandey, J. Parhi (eds.), *Advances in Fisheries Biotechnology*,
https://doi.org/10.1007/978-981-16-3215-0_18

18.1 Introduction

World aquaculture produces constitute around 60% of food fish supply globally which fulfills over 20% of the increasing population's animal protein requirements. Intensification of the production system through sustainable approaches is required to optimize the returns. With high stocking density, supplemental feeding and scientific pond management, following BMPs and biosecurity norms are integral to fish/shrimp farming systems' husbandry practices. Aquaculture uses various methods and techniques to cultivate aquatic organisms which involves farming of aquatic species ranging from freshwater to saltwater. As per Food and Agriculture Organization, aquaculture is a booming industry representing 46% of the total global fish production (about 172.6 million tonnes) which is worth USD 238 billion in 2017. The annual escalation in global fish consumption (3.2%) surpasses meat produced from all terrestrial farm animals (2.8%) between 1961 and 2016 (FAO 2018). Furthermore, fish accounted for about 20% of animal protein consumed by humans globally (FAO 2016). This massive global demand for fish was met by aquaculture as capture fishery which has been plateaued since the late 1980s (FAO 2018). Like freshwater aquaculture, brackish water aquaculture has a long history of the employment of traditional practices from a way of life in bheries (enclosed paddy fields) and pokkali (unique salt tolerant rice variety) fields in India, to the export-oriented aquaculture industry. In this journey of enhanced production, it is seen in India that there is a phenomenal increase of more than 25-fold from a measly 28,000 tonnes in 1988–1989 to 7,47,694 tonnes in 2019–2020; the most critical turning point was the regulated introduction and successful establishment of SPF (Specific Pathogen Free) *P. vannamei* in 2009.

Intensification in aquaculture often brings with it many existing and emerging diseases that need drugs, synthetic chemicals and vaccination for their prevention and control. However, due to the indiscriminate use of antimicrobial compounds, chemical disinfectants, anti-parasitic and other synthetic drugs, there has been a negative impact on the host and environmental health. Since there is only partial success seen in these practices, an alternative approach to stimulating the immune system of farmed aquatic animals is practised. In practicing the alternative prophylactic method in intensive sustainable aquaculture, zero-tolerance to antibiotics can be imposed while mitigating the risk of infectious diseases. Such immunoprophylactic approaches explores the beneficial role of vaccination and immunostimulant compounds, i.e., microbial components (MDP, LPS, beta-1-3-glucan, bacterin, peptidoglycan), plant products (polysaccharides, levamisole, etc.), and yeast derivatives. They are known to elicit the innate and adaptive immune systems of fish and shrimp.

18.2   Concept of Immunoprophylaxis

Immunoprophylaxis, otherwise known as prophylactic immunization, is one of the highly efficacious, alternate strategies in the primary prevention of disease (Eisenhauer et al. 2013).

General prophylactic measures followed in aquaculture operations include:

- Better management practices (BMPs) in culture operations.
- Probiotics and bioremediation process to improve host-gut microbiome and environmental health and their interaction.
- Enhancing host resistance to infectious diseases by immunostimulants.
- Nutraceutical diets for immunomodulating the host.
- Vaccination of the host against specific pathogens.
- Genetic selection/selective breeding programs for resistance to diseases.

Immunoprophylaxis is the mainstay of preventing infections via vaccines or antibody-containing preparations or immunostimulants to provide immune protection against disease. It is a branch of practical immunology which is concerned with preventing infectious diseases by injecting immunological preparations such as vaccines, immune sera, immunostimulants, and gamma globulins, to elicit immunity (Fig. 18.1). It is quite evident from the human point of view that it has greatly helped in preventing several infectious diseases, viz. poliomyelitis, yellow fever, diphtheria, tetanus, and smallpox.

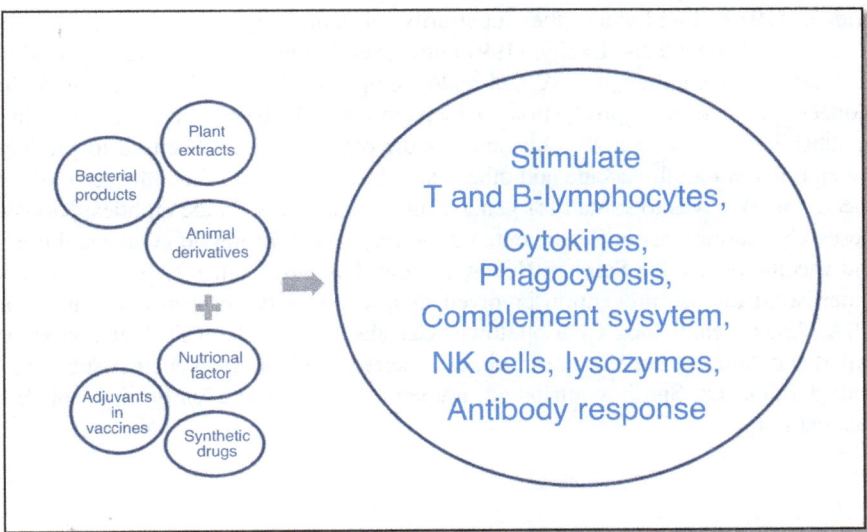

Fig. 18.1 The concept of the immune response of prophylactic treatment

Vaccination may reduce the risk of introducing and establishing any infectious agent and containing its horizontal/vertical spread to other populations/farms, thereby aiding internal biosecurity. Biosecurity is a relatively new term used to describe measures for preventing and controlling diseases (FAO 2013; OIE 2011). A commonly used definition of biosecurity mentions "a set of measures that reduce the risk of introducing and spreading disease agents." Biosecurity may be external (or bio-exclusion) or internal (or biocontainment) depending on the spread/control of pathogens to a system/population or within a population. Since aquaculture involves many individuals, boosting innate and adaptive immunity through vaccination is better as herd immunity ensures protection against diseases even in few nonvaccinated animals and poor responders.

18.3   Historical Perspective

Fish farming is entering a new era for sustainable intensification in many countries. In earlier times, antibiotics or chemotherapeutics were primarily used for disease prevention and treatment. Snieszko et al. (1938) first reported on the protective immunity in carps immunized with *Aeromonas punctata*, a manuscript in Polish. However, since this manuscript was not available elsewhere globally, the first report in English was written by Duff (1942), who demonstrated the protection in trout against *A. salmonicida* by parenteral inoculation of oral administration. Until the 20–30 year duration of World War II, very few reports on disease prevention by vaccination were published. In those days, immunoprophylaxis was disregarded because of the easy availability and use of antimicrobial compounds. Snieszko and Friddle (1949) highlighted the superiority of sulfamerazine to a vaccine for controlling furunculosis. Evelyn (1997) described the post-World War II period as "the era of chemotherapy." A Colorado company, Wildlife Vaccines, was the pioneer in vaccine production which produced bacterins. Tavolek Inc. (a subsidiary of Johnson & Johnson) was the other company licensed to produce the enteric red mouth vaccine and other bacterins. During that time, the scale of the vaccine market was too small for generating decent profits. In the eighties, Biomed Research Laboratories in Seattle entered the fray. Many professional aquaculturists and vaccinologists like Stephen Newman, Tony Novotny, James Nelson, with their expertise in disease prevention involved themselves in bacterin production in the USA. Aqua Health Inc., Charlottetown, Canada, with William D. Paterson as an expert and Aquaculture Vaccines Ltd., UK, started as a Wildlife Vaccines subsidiary with Patrick D. Smith contributed immensely during the early days of fish vaccination.

18.4   Concept of Immunoprophylaxis in Aquaculture

Intensive aquaculture is associated with severe disease outbreaks caused by various pathogenic microorganisms, resulting in loss of production due to the reduction of growth and increased mortality (Sivasankar and Kumar 2017). Different types of chemotherapeutic agents, drugs and biological products are used for preventing diseases in aquaculture. Indiscriminate use of chemical agents and natural products has potentially negative impact on the environment and human health (Sivasankar and Kumar 2017). Using antibiotics to prevent diseases could lead to Antimicrobial resistance (AMR) in microorganisms (George et al. 2006). Immunoprophylaxis is a protective management tool aiding in the prevention and treatment of diseases. The prophylactic measures for disease prevention are implemented with prebiotics, probiotics, vaccines, and immunostimulants (Panigrahi and Azad 2007; Panigrahi et al. 2011; Sivasankar and Kumar 2017).

18.5   Classification of Immunoprophylaxis

Immunity can be achieved by either active or passive immunization.

1. *Active Immunoprophylaxis (Vaccines):*
   —through vaccines (inactivated, live, attenuated, subunit, recombinant vector, DNA-based and synthetic/peptide-based)
2. *Passive Immunoprophylaxis (Immunostimulants):*
   —through immunoglobulin preparation or immunostimulants (pro-prebiotic compounds, nutritional factors, animal and plant derivatives, etc.)

18.5.1 Active Immunoprophylaxis Measure (Vaccines)

Presently, there are several commercial vaccines available that offer prevention of infectious bacterial and viral diseases (Table 18.1 and 18.2). Generally, for disease prevention, vaccines are classified as inactivated whole-cell, inactivated bacterin, live, attenuated, DNA-based, Nano-based, subunit based,, genetically modified and polyvalent.

18.5.2 Passive Immunoprophylaxis

Passive immunoprophylaxis can be achieved through immunomodulatory preparations like immunostimulants (Pro and prebiotic compounds, nutritional factors, animal and plant derivatives, etc.) or immunoglobulins (Vijayan et al. 2017).

Table 18.1 Available commercial bacterial vaccines for infectious diseases in aquaculture

S. no	Types of vaccine	Name of vaccines	Diseases prevention	Vaccinated host	References
1.	Inactivated or "killed" vaccine	*Vibrio anguillarum*-Ordalii	Vibriosis	Salmonids, rainbow trout	Sommerset et al. (2005)
2.	Inactivated or "killed" vaccine	*Vibrio salmonicida* Bacterin	Vibriosis	Salmonids Coldwater	Sommerset et al. (2005)
3.	DNA vaccine	*Vibrio anguillarum*	Vibriosis	Seabass	Kumar et al. (2007)
4.	Formalin-killed bacterin vaccine	*Aeromonas salmonicida* Bacterin	Furunculosis	Rainbow trout	Gudmundsdóttir and Björnsdóttir (2007)
5.	Live attenuated vaccine	*Aeromonas hydrophila* vaccine	Motile Aeromonas Septicemia	Common carp	Jiang et al. (2016)
6.	Live attenuated vaccine	*Edwardsiella ictalurii* vaccine	Edwardsiellosis	Catfish	Kordon et al. (2019)
7.	Recombinant GAPDH vaccine	*Edwardsiella tarda* vaccine	Edwardsiella tarda	Japanese flounder	Liu et al. (2005)
8.	Modified live vaccine	*Flavobacterium columnare* vaccine	Columnaris disease	Channel catfish	Shoemaker et al. (2011)
9.	Live attenuated vaccine	*Flavobacterium psychrophilum* vaccine	Flavobacteriosis	Salmonids	Sudheesh and Cain (2016)
10.	Live attenuated vaccine	*Renibacterium salmoninarum* vaccine	Bacterial kidney disease	Salmonids	Toranzo et al. (2009)
11.	Formalin-killed bacterin vaccine	*Yersinia ruckeri* Bacterin	Enteric redmouth disease	Salmonids	Jaafar et al. (2019)
12.	Formalin-killed bacterin vaccine	*Streptococcus agalactiae* vaccine	Streptococcosis	Tilapia	Evans et al. (2004)

Table 18.2 Available commercial viral vaccines for infectious diseases in aquaculture

S. no.	Types of vaccine	Name of vaccines	Diseases prevention	Host vaccination	References
1.	DNA vaccine	Infectious hematopoietic necrosis virus vaccine	Infectious hematopoietic necrosis	Salmonids	Garver et al. (2005)
2.	DNA vaccine	Infectious pancreatic necrosis virus vaccine	Infectious pancreatic necrosis	Salmonids	Cuesta et al. (2010)
3.	Formalin-killed vaccine	Iridoviral disease vaccine	Iridoviral disease	Red Sea bream	Nakajima and Kunita (2005)
4.	DNA vaccine	Spring viremia of carp vaccine	Spring viremia of carp	Common carp	Embregts et al. (2019)
5.	Recombinant vaccine	Nodavirus vaccine	Viral nervous necrosis	European seabass	Gonzalez-Silvera et al. (2019)

18.6 Immunostimulants

An immunostimulant is a substance that enhances the innate or nonspecific immune response by interacting directly with the cells of the immune system and activating them. Immunostimulants can be grouped under different agents based on their source, such as bacterial preparations, polysaccharides, animal or plant extracts, nutraceuticals, and cytokines (Sakai 1999). In nature, the defense mechanism in all living organisms (vertebrates and invertebrates) are common, but there still exists some dissimilarities. The vertebrate has a more complex mechanism whereas the invertebrates generally lack adaptive immunity and exclusively rely on their innate immunity. Therefore, it is necessary to inculcate products that can enhance host immunity and resistance to infectious diseases in shrimps disease prevention and control (Farzanfar 2006). Cellular and humoral immunities occur both in vertebrates and invertebrates. Immunostimulants as adjuvants (helper compounds) are effective when used with fish vaccines and as additives in aquafeeds. As immunostimulants generate passive immunity, they are used as prophylactic agents prior to the recognition of the elevated risk of disease (Raa 1996).

18.6.1 General Aspects of Immunostimulants

The success story of immunostimulants in fish/shrimp involves bacteria causing infectious diseases such as *A.salmonicida, A. hydrophila, Vibrio anguillarum, V. vulnificus, V. salmonicida, Yersinia ruckeri, Streptococcus* spp.; viruses causing infectious diseases such as hematopoietic necrosis, yellow head virus, viral hemorrhagic septicemia and the parasite like *Ichthyophthirius multifiliis*. Immunostimulants are known to elicit innate (nonspecific) defense mechanisms and improve resistance to specific pathogens (Sakai et al. 1999; Patil et al. 2014). In invertebrates, the memory factor is not fully developed and being very feeble, the memory is short-lived. There are chemical substances that activate leukocytes and lymphocytes (Lunden and Bylund 2000). Adjuvant (FCA) is one of the essential immunomodulators used in animals to enhance specific immune response and is used in conjunction with the injection of fish bacterins (Anderson 1992). The β-glucans, present in cell walls of plants, fungi, and bacteria, appear to be most promising among the immunostimulants in fish and shrimp culture ponds, although an oral application was found to be the preferred route of choice.

18.6.2 Role of Immunostimulant

Immunostimulants increase immunocompetence and disease resistance in aquatic animals (Sakai 1999). Many studies have been conducted on immunostimulants in fish and shellfish, and these include chemical agents, polysaccharides, extracts of plant and animal origins, components of bacteria, cytokines, and nutritional agents (Sakai 1999). In crustaceans, three circulating hemocytes are generally identified: hyaline cells, semigranular cells, and large granular cells. These cells form the basis for cellular immune responses, including phagocytosis and eliminating foreign bodies or microbial agents (Hose et al. 1990). Hemocytes form an integral part of the crustaceans' immune system. They are associated with enzymes prophenoloxidase (proPO), and ProPO activates systems involved in encapsulation, melanin formation and function as non-self-recognition systems (Johansson and Soderhall 1989). Phenoloxidase (PO) is activated from prophenoloxidase by a serine proteinase in the presence of a small number of microbial components such as peptidoglycan, lipopolysaccharides, or b-1.3-glucan through the recognized receptors. The PO catalyzes the stepwise oxygenation and reactions of monophenols through O-diphenols to O-quinones, which subsequently leads to melanin formation (Johansson and Soderhall 1989).

Phagocytosis is an essential mechanism in crustaceans to get rid of microorganisms or foreign bodies. Reactive oxygen intermediates such as superoxides, hydroxide radicals, and peroxides are produced during the process of phagocytosis. This activity is known as respiratory burst and these products have microbicidal activities (Song and Hsieh 1994). Many of these parameters such as total hemocyte count (THC), phenoloxidase activities, superoxide dismutase (SOD), an enzyme that catalyses the two steps of rapid dismutation of the toxic superoxide anion to non-toxic molecular oxygen and hydrogen peroxide, etc. have been used to evaluate the immune status of shrimps (Maftuch et al. 2013).

Several studies have revealed the enhancement in immunological parameters in many fish species after administration of medicinal plants or extracts, including phagocytic activity, respiratory burst activity, nitrogen oxide, myeloperoxidase content, complement activity, lysozyme activity, total protein (globulin and albumin), and antiprotease activity (Dugenci et al. 2003; Talpur, 2014).

18.7 Classification of Immunostimulants

18.7.1 Synthetic Chemicals

These are the synthetic immunomodulatory compounds that are known to stimulate the immune response. Levamisole is an antihelminthic compound that combats nematodes infection in humans and animals as well. It can increase the phagocytic activity, the respiratory burst activity and increase antibody-producing cells and other immune parameters (Siwicki 1989).

18.7.2 Microbial Product as Immunoprophylaxis Measure

18.7.2.1 Bacterial Derivatives
MDP (Muramyl dipeptide): MDP (*N*-acetylmuramyl-L-alanyl-D-isoglutamine), obtained from Mycobacterium increases the phagocytic activities, respiratory burst, and migration activities of kidney leucocytes and resistance to *A. salmonicida*. Various bacterial products used for immunoprophylaxis in aquaculture is shown in Table 18.3.

Table 18.3 Bacterial products used for immunoprophylaxis in aquaculture

Lactobacillus plantarum	Increased THC, PO and SOD activities, phagocity activity, and resistance to *V. alginolyticus* infection.	Chiu et al. (2007)
Beta-1-3-glucan (*Schizophyllum commune*)	Increased PO and SOD activities, PA, superoxide anion, survival against WSSV	Chang et al. (2003)
Beta-glucan	Increased mRNA expression of lysozyme and SOD	Wang et al. (2008)
Bacterin (Vibrio)	Increased immune response	Powell et al. (2011)
Bacterial LPS	Increased immune gene expression and disease resistance, proPO activity	Rungrassamee et al. (2013)
Bacterial genomic DNA extract	Increased THC, PTP, HLAT	Amar and Faisan (2012)
Peptidoglycan	Increased survival against Yellowhead baculovirus infection	Boonyaratpalin (1995)
Mannan oligosaccharide (MOS)	Increased growth and survival, PO, SOD activities, resistance to NH_3 stress	Zhang et al. (2012)
Peptidoglycan+ MOS	Increased growth, THC, respiratory burst activity, protection against WSSV infection	Apines-Amar et al. (2014)

18.7.2.2 LPS (Lipopolysaccharide)

LPS being a cell wall component of gram-negative bacteria, is effective against *A. hydrophilla* infection in rainbow trout and stimulates the innate immune response (Nya and Austin 2010). Atlantic salmon (Salati et al. 1987), in red sea bream *Pagrus major*, in goldfish lymphocytes (Neumann et al. 1995). It is also known to elicit phagocytic, microbicidal activity, and hemocyte proliferation in shrimp (Karunasagar et al. 1996).As these substances are very potent in low doses, they can be used for fish immunizing programs.

18.7.2.3 FCA (Freund's Complete Adjuvant)

As attenuated *Mycobacterium butyricum* is found in the FCA mineral oil, it improves immune response in fish. The FCA can increase cellular and humoral immune response in rainbow trout and protect it against *V. anguillarum* infection (Kajita et al. 1992). The effect of FCA on a *P. piscicida* vaccine was observed in fish (Kawakami et al. 1998).

18.7.2.4 Vibrio Bacterin

Vibrio anguillarum bacterin (inactivated whole-cell vaccine) is the most successful vaccine for salmonid fish, administered through injection, oral, and immersion methods (Sakai et al. 1999). Immunostimulation of *V. anguillarum* bacterin was seen in fish and shellfish. Hemocyte proliferation and migration were observed in *P. monodon* treated with vibrio bacterin (Horne et al. 1995; Chou et al. 1995).

18.7.2.5 *Clostridium butyricum* Cells

C. butyricum bacterin activates leucocytes and can enhance resistance to vibriosis disease in rainbow trout and also activates phagocytosis and increased superoxide anion production when introduced orally (Sakai et al. 1995; Young et al. 1987). It also revealed that *C. butyricum* bacterin shows immunostimulatory effects like the stimulation of macrophages and NK cells and improves further protection against Candida infection.

18.7.2.6 EF203

It is a fermented product of chicken eggs and oral administration of it to rainbow trout stimulates the immune response and ensures increases protection against *Streptococcus* infection (Yoshida et al. 1993).

18.7.3 Yeast Derivatives

Glucan: Glucan is a yeast-derived bioactive compound (long-chain polysaccharide extract) that stimulates the nonspecific defense mechanism in fishes and shellfishes to combat bacterial pathogens. Yeast glucan (β-1,3- and β-1,6-linked glucan) and β-1,3 glucan (VST) is derived from cell walls of baker's yeast like *Saccharomyces cerevisiae* and *Schizophyllum commune*, respectively.

18.7.4 Polysaccharides

18.7.4.1 Chitin and Chitosan
These polysaccharides are naturally abundant in the exoskeleton of crustaceans, insects, and cell walls of few fungi. They are known to elicit cellular and humoral immune responses like macrophage activity and protect the host from individual pathogenic organisms (Kawakami et al. 1998; Anderson and Siwicki 1994).

18.7.4.2 Lentinan, Schizophyllan, and Oligosaccharide
These compounds are reported to elicit cellular and noncellular defense mechanisms in fish and shrimp.

18.7.5 Medicinal Plant as Immunostimulant

Nowadays, antibiotics and chemotherapeutic agents have negative impact such as immunosuppression, development of antimicrobial resistance and bioaccumulation in the tissues and environment health. An alternative way of prophylactic measure is using a medicinal plant derivative (Table 18.4). It has rich bioactive compounds like polyphenolics compounds, flavonoids, saponins, tannins, terpenoids, alkaloids, curcumin, quinone, polypeptides, colchicine, capsaicin, and flavopiridol. These bioactive compounds trigger or modulate the immune system (Nair et al. 2019). That enhances the phagocytosis and macrophage activity, produce complementary compounds, cytokines, T lymphocytes, lysozyme activity which act on innate (nonspecific), and adaptive defense mechanisms (Farooqi and Qureshi 2018).

18.7.6 Animal Derivatives as Immunostimulant

Marine tunicate, *Ecteinascidia turbinata* (Ete) and abalone, Haliotis, *Discus hannai* extracts, and fractions (Hde) are known to have immunomodulatory roles and inhibits tumor growth in vivo. *Ete* (Tunicate) can enhance the phagocytosis and increases the rate of survival of Eel when injected against *A. hydrophila*. Rainbow trout injected with Hde against *V. anguillarum* infection showed increased survival along with enhanced phagocytic activities (Sakai et al. 1991).

18.7.6.1 Firefly Squid
Firefly squid, *Watasenia scintillans*, can stimulate the immunity of fish as observed in superoxide anion, macrophages proliferation, and the lymphoblastic transformation of lymphocytes in vitro.

Table 18.4 Report on the importance of medicinal plant extracts in enhancing growth and immune response in fish

Plants	Used parts	Route of administration	Culture species	Immune response	References
Viscum album	Leaf	Oral	*Oncorhynchus mykiss* (rainbow trout)	Enhances the phagocytic activity and respiratory burst, total protein	Dugenci et al. (2003)
Capparis spinosa	Leaf	Oral	*Oncorhynchus mykiss* (rainbow trout)	Cytokine genes (IL-1b, IL-8, TNF-a, TGF-b, -IL-10) against *A. hydrophila*	Bilen et al. (2016)
Zingiber officinale	Root	Oral	*Oncorhynchus mykiss* (rainbow trout)	Enhances phagocytic, respiratory burst, total protein	Dugenci et al. (2003)
Astragalus radix	Root	Oral	*Oreochromis niloticus* (Nile tilapia)	Increases the lysozyme activity, phagocytic, respiratory burst	Yin et al. (2006)
Solanum trilobatum	Leaf	Injection	*Oreochromis mossambicus*	Rising respiratory burst, lysozyme, serum bactericidal, serum protein against *A. hydrophila*	Divyagnaneswari et al. (2007)
Zingiber officinale	Root	Oral	*Oreochromis mossambicus*	Increases lysozyme production, Phagocytic, total protein, albumin, globulin against *V. vulnificus*	Immanuel et al. (2009)
Ocimum sanctum	Leaf	Oral	*Epinephelus tauvina*	Phagocytic, bactericidal against *V. harveyi*	Sivaram et al. (2004)
Euphorbia hirta	Leaf	Oral	*Cyprinus carpio* (common carp)	Phagocytic, respiratory burst, lysozyme, peroxidase against *A. hydrophila*	Pratheepa and Sukumaran (2014)
Lawsonia inermis	Leaf	Injection	*Cyprinus carpio* (common carp)	Lysozyme and bactericidal activity, phagocytic and respiratory burst activity against *A. hydrophila*	Soltanian and Fereidouni (2016)
Azadirachta indica	Leaf	Injection	*Carassius auratus*	Phagocytic, respiratory burst against *A. hydrophila*	Harikrishnan et al. (2009)
Cynodon dactylon	Leaf	Oral	*Catla catla* (Indian carp)	Lysozyme, antiprotease, complement, respiratory burst, nitrogen species, myeloperoxidase against *A. hydrophila*	Kaleeswaran et al. (2011)

18.7.7 Nutritional Factors

18.7.7.1 Nutrient Based Immunoprophylactic Measures

Nutritional immunostimulating compound plays a vital role in the survival, growth, physiology functions and immune response of fish/shellfish. It plays an essential role in maintaining epithelial barriers, nonspecific cellular factors such as phagocytosis by macrophages and neutrophils, nonspecific humoral factors such as lysozyme, complement, and transferrin and specific humoral and cellular immunity.

18.7.7.2 Vitamins and Minerals

Vitamins and minerals in negligible amount play a vital role in physiological functions and immunological responses (Table 18.5). Physiological processes such as growth, disease resistance to infections, wound healing, stress reduction and other biological processes are enhanced by vitamins. It also acts as an immunomodulator, enhances antioxidant activity, enhances calcium, phosphorus and iron absorption and regulate moult phase in shrimp and fish growth and also increases survival.

Vitamin C has a vital antioxidant and anti-stress characteristic that also induces different physiological processes, including growth, metabolism, reproduction, and protection against tissue damage, and a cofactor in cellular functions related to neuroendocrine modulation, hormone and immune systems. Ascorbic acid is found to have a significant role. Tewary and Patra (2008) revealed a positive correlation between vitamin C in diet and resistance against *A. hydrophila* in *Labeo rohita*.

Vitamin E can elicit specific and cell-mediated immunity against infection in Japanese Flounder *Paralichthys olivaceus* (Villegas et al. 2006) and macrophage phagocytosis in fish such as channel catfish Ictalurus punctatus (Wise et al. 1993) and turbot *Scophthalmus maximus* (Fernández et al. 2002). Vitamin E deficiencies in trout result in reduced protection against *Y. ruckeri* (Blazer and Wolke 1984).

18.7.8 Hormones

Growth hormones have a powerful influence on immunocompetent cells like macrophages, lymphocytes, and Natural Killer cells. Exogenous growth hormone (GH) has several activities in fish, including lymphocytes, NK cellproliferation and SOD production. Prolactin has multiple immunomodulatory roles, including the activation of immunocompetent cells and production of superoxide anions (Sakai et al. 1996).

18.7.9 Cytokines

Cytokines are polypeptides, or glycoproteins which plays a significant role as modulators in the immune system. Recombinant cytokines as immunostimulants

Table 18.5 Vitamins and minerals involved in enhancing the disease resistance and immune response in aquaculture species

Immunostimulants	Effect on host	Species	References
Vitamin C supplement with diet	Increases the growth, survival, THC, superoxide anion, PO activity, respiratory burst activity (NBT cells) resistance to *Aeromonas hydrophila* infections	Rohu *(L. rohita)*	Tewary and Patra (2008)
	Stimulates innate immune response and enhance resistance against high pH stress and protection against *Aeromonas hydrophila* infection	Megalobrama amblycephala	Liu et al. (2016)
Vitamin E	Increased growth, THC, SOD activity, and decrease TBA value	Flatfish	Pulsford et al. (1995)
	PO activity, respiratory burst activity, protect oxidative damage of tissues	Atlantic salmon	Lygren et al. (2001)
Folic acid (vitamin B9)	Improves the growth, antioxidant activity, enhance the ProPO activity and increase digestive enzymes, and better survival	*Macrobrachium rosenbergii*	Asaikkutti et al. (2016)
Copper (Cu)	Increased growth, THC, superoxide production	Tiger shrimp	Lee and Shiau (2002)
Zinc (Zn)	Increased growth, survival, THC, PO activity, phagocytic activity, alkaline phosphates, and SOD activities	White shrimp	Lin et al. (2013)
Selenium (se)	Increases superoxide anion, PO activity, respiratory burst activity, and protection against *Edwardsiella ictaluri* infection	Catfish	Paripatananont and Lovell (1997), Wang et al. (1997)
Chitin	Increased proPO activity		Sivakumar and Felix (2011)
Sodium alginate	Increased respiratory burst, PO activity, resistance to *V. alginolyticus* infection		Cheng et al. (2004)

can have a lot of utility as Immunoprophylactic agents provided they are characterized.

18.7.10 Algal Derivatives

Laminarin is a component of brown algae, e.g., Phaeophyceae, and its structure is with β(1, 6)-branched β(1, 3)-D-glucan. Laminarin extracted from *Laminaria*

Table 18.6 Other derivatives used as immunostimulants

Immunostimulants	Types	Functions	References
Levamisole	Drugs	Induction of B and T lymphocytes, monocytes, and macrophages	Patil et al. (2012), Biswajit et al. (2014)
Recombinant cytokines	Cytokines/ adjuvants	Generation of interferons and interleukins to stimulate effective immune responses	Galeotti (1998), Sirko et al. (2011)
Glucans	Carbohydrates	Stimulation of anti-tumor mechanisms and enhancement of host resistance to a variety of microbial pathogens in mammalian	Sahoo and Mukherjee (2001), Vetvicka (2011), Sajeevan et al. (2009)
Trehalose	Carbohydrates	Production of antibody, stimulation of specific immunity against different bacterial infections	Parant et al. (1978), Oswald et al. (1997)
Chitosan	Animals originated	Activating the production of cytokines such as IL-1β, TNF-α, and reactive oxygen intermediates to promote the defense system against microbial infections	Mastan (2015), Chen et al. (2009)
CpG oligonucleotides and imiquimod	Drugs/ adjuvants	Maturation and migration of DCs, and enhancement of humoral and cellular immune responses	Mizumoto et al. (2005)
Sodium alginate	Seaweed	Increased respiratory burst, PO activity, resistance to *V. alginolyticus* infection	Cheng et al. (2004)

hyperborea has immunomodulatory in aquatic animals as well. Water solubility is an issue for β-(1,3) D-glucan, making them less easy to handle than aqueous soluble laminaria.

Details of different other derivatives used as immunostimulants in aquaculture are shown in Table 18.6.

18.8   Combating Diseases Through Immunoprophylaxis Measures

Shellfishes lack adaptive immune systems and instead, have nonspecific immune system to combat pathogenic infections. Immunostimulants are dietary additives that enhance the innate (non-specific) defense mechanisms and increase resistance to pathogens in shrimp. Beta-glucans are increasingly used in aquaculture and are seemingly very promising in immune enhancement in aquaculture (Ajadi et al. 2016; Barman et al. 2013). Mode of action of various immunoprophylactic agents are summarized in Table 18.7.

Table 18.7 Mechanism of action of immunoprophylactic agents

Immunostimulants	Mechanism of action
Freund's complete adjuvant (FCA), glucans, FK-565 (Lactoyl tetrapeptide from *Streptomyces olivaceogriseus*)	Stimulators of T lymphocytes
Bacterial endotoxins, lipopolysaccharides, and chitin and chitosan	Stimulates of B cells and macrophage activity
Cytokines	Leukotriene, interferon
Detergents and sodium dodecyl sulfate, quarternary ammonium compounds (QAC), saponins	Cell membrane modifiers
Muramyl dipeptide (a purified form of mycobacteria)	Enhances the antibody activity, stimulation of polyclonal activation of lymphocytes, and activation of macrophages
Lentinan, Schizophyllan, and oligosaccharide	Increases the lysozyme activity, phagocyte activity, and complement activity
Levamisole	Enhancement of cell-mediated cytotoxicity, lymphokine production, and suppressor cell function and stimulation of the phagocytic activity of macrophages and neutrophils.
Yeast derivatives glucans (β1-3- and β1-6-linked glucan)	Increased lysozyme activity, phagocyte activity, complement activity, and bactericidal activity of macrophages
Plant based immnunostimulant (e.g., *Ocimum sanctum*, *Emblica officinalis*, *Cynodon dactylon*, and *Adathoda vasica*	Improved the phagocytic activity, serum bactericidal activity, albumin–Globulin (A/G) ratio
Nutritional factor—vitamin C	Collagen synthesis and cellular functions related to neuromodulation, hormone, and immune systems

18.8.1 Combating Vibriosis

Shellfish/finfish have always been affected by vibriosis, and it has a globally high mortality rate. Vibrio species' distribution is worldwide, and infections due to Vibrios are frequently encountered in hatcheries, but outbreaks are widespread in pond-reared shrimps (Lavilla-pitogo et al. 1998). Vibriosis is caused by a group of gram-negative bacteria, especially members of the family Vibrioceae.

Pores or wounds on the exoskeleton may also be a good source of entry of vibrios into the host system (Alday-Sanz et al. 2002). Bacterial penetration through the gills is typical because they are covered by a thin exoskeleton (Taylor and Janney 1992). The intestines which are not lined by exoskeleton also predispose the shrimp to Vibrio attack. The bacteria present in food, water and the environment can easily penetrate the system (Jayabalan et al. 1982). Over 280 shrimp hatcheries along the coastline of Thailand suffered a setback from seed production due to luminescent bacterial diseases caused by *V. harveyi*. Many researchers have reported vibriosis infections in penaeid shrimps and at least 14 species of *Vibrio* have been implicated as the cause of the bacteria diseases. They include *Vibrio harveyi, Vibrio parahaemolyticus, Vibrio alginolyticus, Vibrio splendidus, Vibrio mimicus, Vibrio vulnificus, Vibrio anguillarum, Vibrio damsela, Vibrio cambelli, Vibrio fischeri, Vibrio ordalli, Vibrio orientalis, Vibrio logei,* and *Vibrio mediterranei* (Eaves and Ketterer 1994; Lavilla-pitogo et al. 1998).

18.8.2 Combating *Aeromonas hydrophila*

The effect of *Astragalus membranaceus, Portulaca oleracea, Flavescent sophora,* and *A. paniculata* on immunomodulation in *Cyprinus carpio* was studied by Wu et al. (2000) and the effect of two Chinese medicinal herbs (*A. membranaceus* and *L. japonica*) on the immune response of Nile tilapia (*O. niloticus*) has been reported by Ardo et al. (2008).

18.9 Immunostimulation Act on Nonspecific Defense Mechanisms

Many immunomodulatory materials elicit the nonspecific immune response in crustaceans, which is regarded as the first defense line against challenging pathogens. The hemocytes present in hemolymph of crustaceans (granular and semigranular) and the prophenoloxidase (Propo) system are the primary immune mechanisms (Johansson and Soderhall 1989). Phenoloxidase being the terminal enzyme in this activation system, is activated by the cell wall components like lipopolysaccharides or peptidoglycans from bacteria and β-1, 3 glucan from fungi pattern recognition system (Smith et al. 1984).

18.10 Diet and Husbandry Practices Toward Immunoprophylaxis

The health of the animal is a reflection of its diet and farming practices. Domesticated animals resistant to specific pathogens can be reared in a proper environment; feed and health management form an integral part of aquaculture. The interaction between the pathogens, stress factor, microbiota and the immune response can be superimposed to design healthy feed and additives. Three main strategies, (a) combating pathogens by following hygiene practices, (b) providing a stress-free environment, and (c) efficient feed to optimize fish/shrimp immune system, need to be addressed by fish farmers to enhance the defense of fish and shellfish.

18.10.1 Husbandry Practices Followed for Better Prophylaxis

(a) Fighting the pathogens.
(b) Preserving the intestinal vital microbiota.

(c) Improving strict hygiene to avoid infection.
(d) Antimicrobial preparations and compounds from plant and animal origin.
(e) Improving the welfare of fish through healthy rearing module.
(f) Addressing interaction between neuroendocrine and immune functions in fish and shellfish.
(g) Improving water quality and bioremediation.
(h) Improving rearing conditions through carrying capacity approach.
(i) Fatty acid and antioxidants and their importance for health.

18.10.2 Probiotics and Prebiotics

The microorganisms in the cultural environment play a fundamental role in maintaining balance in the living creatures and in the environment by recycling natural resources. The utilization of microbial community in improving the health of the host and the environment has been investigated quite intensively. Microbes in probiotics have been extensively used globally in medicine, animal husbandry and agriculture.

18.10.3 Nutraceuticals

Nutraceuticals have been used in the last decade due to the increasing cost and side effects of therapeutic pharmaceutical agents. This concept of food for health and well-being was the primary basis of the Indian system of medicine for ages, gaining importance following the drawbacks of modern medicine systems. Stress is the common factor affecting humans (due to contemporary lifestyles) and plants and animals (due to farm intensification). These products containing dietary fiber, prebiotics, probiotics, poly-unsaturated fatty acids, antioxidants and other different types of herbal/ natural foods generally improve the antioxidant defense mechanism and innate immunity. Though the beneficial effects of nutraceuticals have been well established in humans and terrestrial farm animals, similar reports in aquaculture are scarce.

18.10.4 Genetic Selection

Genetic selection for disease resistance has been a milestone in agriculture/animal husbandry to meet the global hunger for food. Breeding for disease resistance is a well-established science in crop protection and has shown mixed success in the livestock sector. However, such programs for cultured aquatic species are limited. The ever-increasing cost of disease prevention and control has shifted the focus of health management from prophylactic and therapeutic intervention to genetic selection. Since the genetic selection programs are time-consuming, involving huge costs, the essential factor to be considered is the economic cost of the disease in question.

Selection for production traits can be easily achieved by measuring growth and survival. Selection for disease resistance is complicated due to the possibility of increased susceptibility to non-target pathogen and the loss of production traits as observed in dairy/beef cattle and poultry. Hence, such selection programs may ideally employ a straight forward multi-disciplinary approach with biologists, microbiologists, immunologists, epidemiologists, virologists, pathologists, environmental experts, and specialists in production systems management. At times, it may not be possible to select for resistance to certain diseases; a classic example being WSSV and TSV in shrimp. It is indeed possible to select for families having resistance to TSV whereas it is not possible to select for resistance to WSSV due to its low heritability.

18.11 CIBAstim, a Shrimp Immunostimulant: A Case Study

CIBASTIM, an immunostimulant product prepared from whole-cell *Vibrio,* has been proven to improve growth, survival and immunity in shrimps. This microbial product (MP) was obtained from heat-killed beneficial bacterial isolate developed and refined by ICAR–CIBA through ICAR- & NFDB-funded research projects. The product was extensively field-tested in different shrimp farming regions of India and the results indicated that it improved immunity in shrimps there by providing better growth, survival and production benefits. Field trial studies revealed substantial growth, survival (6–15%) and net profit returns upto Rs. 40,000/ha.

Benefits/uniqueness
- The product improves the efficacy of feed consumption, which results in higher growth and survival in shrimps and also improves immunity factors in shrimp.
- Field trial studies revealed that CIBASTIM fed shrimps exhibited uniform size, reduced tail rot incidence and other bacterial problems.
- The product maintains good water quality during culture operation.
- The product usage by shrimp farmers revealed higher production with the minimum cost of application.

The product has been commercialized and Technology transfer of CIBASTIM has generated scientific publications, revenue and employment to several entrepreneurs.

18.12 Biofloc-based Farming System

Beneficial bacteria are recognized as a useful tool to fight diseases, minimizing the rampant use of drugs and avoiding usage of antibiotics. Biofloc is a consortium of particulate matter formed predominantly by a biota of aerobic and heterotrophic bacteria, protozoa, microalgae, metazoan, exoskeletons, feces, and detritus. The diverse microbial community present in the biofloc system acts like natural probiotics and stimulates non-specific immune activity. The biofloc system acts as a natural source of probiotic bacteria (Ferreira et al. 2015; Panigrahi et al. 2019d, 2020b) which act as an immunostimulant source (Xu and Pan 2014, Anand et al. 2017; Kumar et al. 2018). This improves the immunity of fish/shellfish and reduces the prevalence of disease. The complex microbial interaction in the biofloc enhances the immune response of the cultured animals (Kim et al. 2014; Panigrahi et al. 2019a, b, c, 2020a, b). Biofloc technology can be seen more as a mechanism, which provides shrimp/fish a chance to keep the immune system active at all times as they get exposed to various microbes.

These microbes facilitate competitive strategies in complex communities, like competitive exclusion or harm the other microbes either by challenging or producing toxins (Panigrahi et al. 2018; Hibbing et al. 2010). Further, they can inhibit the proliferation of pathogens along with immunomodulation in the host.

18.13 Conclusion

Immmunoprophylatic approaches in aquaculture are to be followed in a most judicious way, exploring the beneficial role of vaccination and immunostimulant compounds. This is more important in present-day aquaculture as it is evolving toward sustainable intensification. We have standardized various eco-based, economically feasible models like Organic farming technology, Biofloc-based farming technology, Low input sustainable farming technology, Probiotic based seed, and grow-out production system, Bio-secured Zero water exchange shrimp Farming technology, Green water technology and Polyculture and Zero stocking model where a healthy system of farming is practiced often with immunoprophylactic measures. The sustainable intensification in aquaculture is undoubtedly feasible, thus increasing productivity through eco-friendly approaches to incorporate immunoprophylactic modules.

References

Ajadi A, Sabri MY, Dauda AB, Ina-Salwany MY, Hasliza AH, Malaysia P (2016) Immunoprophylaxis: a better alternative protective measure against shrimp vibriosis—a review. PJSRR 2:58–69
Alday-Sanz V, Roque A, Turnbull JF (2002) Clearing mechanisms of Vibrio vulnificus biotype I in the black tiger shrimp Penaeus monodon. Dis Aquat Org 48(2):91–99
Amar EC, Faisan JP Jr (2012) Induction of immunity and resistance to white spot syndrome virus (WSSV) in shrimp Penaeus monodon (Fabricius) by synthetic oligodeoxynucleotide and bacterial DNA. Philippine Agric Sci 95(3):267–277
Anand PS, Kumar S, Kohli MP, Sundaray JK, Sinha A, Pailan GH, Dam RS (2017) Dietary biofloc supplementation in black tiger shrimp, Penaeus monodon: effects on immunity, antioxidant and metabolic enzyme activities. Aquac Res 48(8):4512–4523
Anderson DP (1992) Immunostimulants, adjuvant, and vaccine carriers in fish: applications to aquaculture. Annu Rev Fish Dis 2:281–307

Anderson DP, Siwicki AK (1994) Duration of protection against Aeromonas salmonicida in brook trout immunostimulated with glucan or chitosan by injection or immersion. Progress Fish Cult 56(4):258–261

Apines-Amar MJ, Andrino KG, Amar EC, Cadiz RE, Corre VL Jr (2014) Improved resistance against White Spot Virus (WSV) infection in tiger shrimp, Penaeus monodon by combined supplementation of peptidoglycan and mannan oligosaccharide (MOS). Extreme Life Biospeol Astrobiol 6(1)

Ardo L, Yin G, Xu P, Váradi L, Szigeti G et al (2008) Chinese herbs (Astragalus membranaceus and Lonicera japonica) and boron enhance the non-specific immune response of Nile tilapia (Oreochromis niloticus) and resistance against Aeromonas hydrophila. Aquaculture 275:26–33

Asaikkutti A, Bhavan PS, Vimala K (2016) Effects of different levels of dietary folic acid on the growth performance, muscle composition, immune response and antioxidant capacity of freshwater prawn, Macrobrachium rosenbergii. Aquaculture 464:136–144

Barman D, Nen P, Mandal SC, Kumar V (2013) Immunostimulants for aquaculture health management. J Marine Sci Res Dev 3:134

Bilen S, Altunoglu YC, Ulu F, Biswas G (2016) Innate immune and growth promoting responses to caper (Capparis spinosa) extract in rainbow trout (Oncorhynchus mykiss). Fish Shellfish Immunol 57:206–212

Biswajit D, Suvakanta D, Chandra CR, Jashabir C (2014) An overview of levamisole hydrochloride with immunostimulant activity. Am J Pharm Health Res 2(4):1–9

Blazer VS, Wolke RE (1984) The effects of α-tocopherol on immune response and non-specific resistance factors of rainbow trout (Salmo gairdneri Richardson). Aquaculture 37:19

Boonyaratpalin S (1995) Effects of peptidoglycan (PG) on growth, survival, immune responses, and tolerance to stress in black tiger shrimp, Penaeus monodon. Fish Health Section

Chen WR, Sarker A, Liu H, Naylor MF, Nordquist RE (2009, February) Effects of immunostimulants in phototherapy for cancer treatment. In: Biophotonics and immune responses IV, vol 7178. Int Soc Opt Photon, p 71780A

Cheng W, Liu CH, Yeh ST, Chen JC (2004) The immune stimulatory effect of sodium alginate on the white shrimp Litopenaeus vannamei and its resistance against Vibrio alginolyticus. Fish Shellfish Immunol 17(1):41–51

Chiu CH, Guu YK, Liu CH, Pan TM, Cheng W (2007) Immune responses and gene expression in white shrimp, Litopenaeus vannamei, induced by Lactobacillus plantarum. Fish Shellfish Immunol 23(2):364–377

Chou H, Huang C, Wang C, Chiang H, Lo C (1995) Pathogenicity of a baculovirus infection causing white spot syndrome in cultured penaeid shrimp in Taiwan. Dis Aquat Org 23(3):165–173

Cuesta A, Chaves-Pozo E, de Las Heras AI, Saint-Jean SR, Perez-Prieto S, Tafalla C (2010) An active DNA vaccine against infectious pancreatic necrosis virus (IPNV) with a different mode of action than fish rhabdovirus DNA vaccines. Vaccine 28(19):3291–3300

Divyagnaneswari M, Christybapita D, Michael RD (2007) Enhancement of nonspecific immunity and disease resistance in Oreochromis mossambicus by Solanum trilobatum leaf fractions. Fish Shellfish Immunol 23(2):249–259

Duff DCB (1942) The oral immunization of trout against bacterium salmonicida. J Immunol 44:87–94

Dugenci SK, Arda N, Candan A (2003) Some medicinal plants as immunostimulant for fish. J Ethnopharmacol 88(1):99–106

Eaves LE, Ketterer PJ (1994) Mortalities in red claw crayfish Cheraxquadricarinatus associated with systemic Vibrio mimicus infection. Dis Aquat Org 19(3):233–237

Eisenhauer M, Smith DJ, Backflow Solutions Inc (2013) Methods and systems for tracking backflow assemblies. US Patent 8,463,823

Embregts CW, Rigaudeau D, Tacchi L, Pijlman GP, Kampers L, Veselý T, Pokorová D, Boudinot P, Wiegertjes GF, Forlenza M (2019) Vaccination of carp against SVCV with an oral DNA vaccine or an insect cells-based subunit vaccine. Fish Shellfish Immunol 85:66–77

Evans JJ, Klesius PH, Shoemaker CA (2004) Efficacy of Streptococcus agalactiae (group B) vaccine in tilapia (Oreochromis niloticus) by intraperitoneal and bath immersion administration. Vaccine 22(27–28):3769–3773

Evelyn TP (1997) A historical review of fish vaccinology. Dev Biol Stand 90:3–12

FAO (2013) FAO yearbook. Fishery and aquaculture statistics-2011, Rome, 76 pp

FAO (2016) The state of world fisheries and aquaculture 2016. Contributing to food security and nutrition for all, Rome, 200 pp

FAO (2018) The state of world fisheries and aquaculture 2018 - Meeting the sustainable development goals. Rome, p 2

Farooqi FS, Qureshi WUH (2018) Immunostimulants for aquaculture health management. J Pharmacognosy Phytochem 7(6):1441–1447

Farzanfar A (2006) The use of probiotics in shrimp aquaculture. FEMS Immunol Med Microbiol 48 (2):149–158

Fernández AB, de Blas I, Ruiz I (2002) Immunological system in teleost. Cells and organs. Aquatic magazine, 16, Spanish

Ferreira GS, Bolívar NC, Pereira SA, Guertler C, do Nascimento Vieira F, Mouriño JLP, Seiffert WQ (2015) Microbial biofloc as source of probiotic bacteria for the culture of Litopenaeusvannamei. Aquaculture 448:273–279

Galeotti M (1998) Some aspects of the application of immunostimulants and a critical review of methods for their evaluation. J Appl Ichthyol 4(3–4):189–199

Garver KA, LaPatra SE, Kurath G (2005) Efficacy of an infectious hematopoietic necrosis (IHN) virus DNA vaccine in Chinook Oncorhynchus tshawytscha and sockeye O. nerka salmon. Dis Aquat Org 64(1):13–22

George MR, Maharajan A, John KR, Jeyaseelan MJ (2006) Shrimps survive white spot syndrome virus challenge following treatment with Vibrio bacterin

Gonzalez-Silvera D, Guardiola FA, Espinosa C, Chaves-Pozo E, Esteban MÁ, Cuesta A (2019) Recombinant nodavirus vaccine produced in bacteria and administered without purification elicits humoral immunity and protects European sea bass against infection. Fish Shellfish Immunol 88:458–463

Gudmundsdóttir BK, Björnsdóttir B (2007) Vaccination against atypical furunculosis and winter ulcer disease of fish. Vaccine 25(30):5512–5523

Hibbing ME, Fuqua C, Parsek MR, Peterson SB (2010) Bacterial competition: surviving and thriving in the microbial jungle. Nat Rev Microbiol 8(1):15–25

Horne MT, Poy M, Pranthanpipat P (1995) Control of vibriosis in black tigershrimp, Penaeus monodon, by vaccination. In: Chou LM et al (eds) The third Asian fisheries forum. Asian Fisheries Society Manila, Philippines, pp 459–467

Hose JE, Martin GG, Gerard AS (1990) A decapod hemocyte classification scheme integrating morphology, cytochemistry, and function. Biol Bull 178(1):33–45

Immanuel G, Uma RP, Iyapparaj P, Citarasu T, Punitha Peter SM, Michael Babu M, Palavesam A (2009) Dietary medicinal plant extracts improve growth, immune activity and survival of tilapia Oreochromis mossambicus. J Fish Biol 74(7):1462–1475

Jaafar RM, Al-Jubury A, Dalsgaard I, Mohammad Karami A, Kania PW, Buchmann K (2019) Effect of oral booster vaccination of rainbow trout against Yersinia ruckeri depends on type of primary immunization. Fish Shellfish Immunol 85:61–65

Jayabalan N, Chandran R, Sivakumar V, Ramamoorthi K (1982) Occurrence of luminescent bacteria in sediments. Curr Sci 51(14):710–711

Jiang X, Zhang C, Zhao Y, Kong X, Pei C, Li L, Nie G, Li X (2016) Immune effects of the vaccine of live attenuated Aeromonas hydrophila screened by rifampicin on common carp (Cyprinus carpio L). Vaccine 34(27):3087–3092

Johansson MW, Soderhall K (1989) Cellular immunity in crustaceans and the proPO system. Parasitol Today 5:171–176

Kajita Y, Sakai M, Atsuta S, Kobayashi M (1992) Immunopotentiantion activity of Freund's complete adjuvant in rainbow trout Oncorhynchus mykiss. Nippon Suisan Gakkaishi 58:433–437

Karunasagar I, Otta SK, Karunasagar I, Joshna K (1996) Application of Vibrio vaccine in shrimp culture. Fishing Chimes 16:49–50

Kawakami H, Shinohara N, Sakai M (1998) The non-specific immunostimulation and adjuvant effects of Vibrio anguillarum bacterin, M-glucan, chitin or Freund's complete adjuvant in yellowtail Seriola quinqueradiata to Pasteurella piscicida infection. Fish Pathol 33:287–292

Kim SK, Pang Z, Seo HC, Cho YR, Samocha T, Jang IK (2014) Effect of bioflocs on growth and immune activity of Pacific white shrimp, L itopenaeus vannamei postlarvae. Aquac Res 45 (2):362–371

Kordon AO, Abdelhamed H, Ahmed H, Baumgartner W, Karsi A, Pinchuk LM (2019) Assessment of the live attenuated and wild-type Edwardsiella ictaluri-induced immune gene expression and langerhans-like cell profiles in the immune-related organs of catfish. Front Immunol 10:392

Kumar SR, Parameswaran V, Ahmed VI, Musthaq SS, Hameed AS (2007) Protective efficiency of DNA vaccination in Asian seabass (Lates calcarifer) against Vibrio anguillarum. Fish Shellfish Immunol 23(2):316–326

Kumar VS, Pandey PK, Anand T, Bhuvaneswari GR, Dhinakaran A, Kumar S (2018) Biofloc improves water, effluent quality, and growth parameters of Penaeus vannamei in an intensive culture system. J Environ Manag 215:206–215

Lavilla-Pitogo CR, Leaño EM, Paner MG (1998) Mortalities of pond-cultured juvenile shrimp, Penaeus monodon, associated with dominance of luminescent vibrios in the rearing environment. Aquaculture 164(1–4):337–349

Lee MH, Shiau SY (2002) Dietary copper requirement of juvenile grass shrimp, Penaeus monodon, and effects on non-specific immune responses. Fish Shellfish Immunol 13(4):259–270

Lin YC, Chen JC, Morni WZW, Putra DF, Huang CL, Li CC, Hsieh JF (2013) Vaccination enhances early immune responses in white shrimp Litopenaeus vannamei after secondary exposure to Vibrio alginolyticus. PLoS One 8(7):e69722

Liu Y, Oshima SI, Kurohara K, Ohnishi K, Kawai K (2005) Vaccine efficacy of recombinant GAPDH of Edwardsiella tarda against edwardsiellosis. Microbiol Immunol 49(7):605–612

Liu B, Wan J, Ge X, Xie J, Zhou Q, Miao L, Ren M, Pan L (2016) Effects of dietary Vitamin C on the physiological responses and disease resistance to pH stress and Aeromonas hydrophila infection of Megalobrama amblycephala. Turk J Fish Aquat Sci 16(2):421–433

Lunden T, Bylund G (2000) The influence of in vitro and in vivo exposure to antibiotics on mitogeninduced proliferation of lymphoid cells in rainbow trout (Oncorhynchus mykiss). Fish Shellfish Immunol 10:395–404

Lygren B, Hjeltnes B, Waagbø R (2001) Immune response and disease resistance in Atlantic salmon (Salmo salar L.) fed three levels of dietary vitamin E and the effect of vaccination on the liver status of antioxidant vitamins. Aquac Int 9(5):401–411

Maftuch, Prasetio E, Sudianto A, Rozik M, Nurdiyani R, Sanusi E, Nursyam H, Fariedah F, Marsoedi, Murachman (2013) Improvement of innate immune responses and defense activity in tiger shrimp (Penaeus monodon Fab.) by intramuscular administration of the outer membrane protein Vibrio alginolyticus. Springerplus 2(1):432

Mastan SA (2015) Use of immunostimulants in aquaculture disease management. Int J Fish Aquat Stud 2(4):277–280

Mizumoto N, Gao J, Matsushima H, Ogawa Y, Tanaka H, Takashima A (2005) Discovery of novel immunostimulants by dendritic-cell–based functional screening. Blood 106(9):3082–3089

Nakajima K, Kunita J (2005) Red sea bream iridoviral disease. Uirusu 55(1):115–125

Nair A, Chattopadhyay D, Saha B (2019) Plant-derived immunomodulators. In: New look to phytomedicine. Elsevier, Amsterdam, pp 435–499

Neumann NF, Fagan D, Belosevic M (1995) Macrophage activating factors secreted by mitogen stimulated goldfish kidney leucocytes synergies with bacterial lipopolysaccharide to induce nitric oxideproduction in teleost macrophages. Dev Comp Immunol 19:475–482

Nya EJ, Austin B (2010) Use of bacterial lipopolysaccharide (LPS) as an immunostimulant for the control of *Aeromonas hydrophilla* infections in rainbow trout, Oncorhynchus mykiss (Walbaum). J Appl Microbiol 108:686–694

OIE (World Organisation for Animal Health), Paris (2011) Access at www.oie.int/biosecurity. Accessed 16 Oct 2012

Oswald IP, Dozois CM, Petit JF, Lemaire G (1997) Interleukin-12 synthesis is a required step in trehalose dimycolate-induced activation of mouse peritoneal macrophages. Infect Immun 65 (4):1364–1369

Pal S (2015) Phage therapy an alternate disease control in aquaculture: a review on recent advancements. IOSR J Agr Vet Sci 8:68–81

Panigrahi A, Azad IS (2007) Microbial intervention for better fish health in aquaculture: the Indian scenario. Fish Physiol Biochem 33(4):429–440

Panigrahi A, Viswanath K, Satoh S (2011) Real-time quantification of the immune gene expression in rainbow trout fed different forms of probiotic bacteria Lactobacillus rhamnosus. Aquac Res 42(7):906–917

Panigrahi A, Saranya C, Sundaram M, Kannan SV, Das RR, Kumar RS et al (2018) Carbon: nitrogen (C: N) ratio level variation influences microbial community of the system and growth as well as immunity of shrimp (Litopenaeusvannamei) in biofloc based culture system. Fish Shellfish Immunol 81:329–337

Panigrahi A, Sundaram M, Chakrapani S, Rajasekar S, SyamaDayal J, Chavali G (2019a) Effect of carbon and nitrogen ratio (C: N) manipulation on the production performance and immunity of Pacific white shrimp Litopenaeusvannamei (Boone, 1931) in a biofloc-based rearing system. Aquac Res 50(1):29–41

Panigrahi A, Sundaram M, Saranya C, Kumar RS, Dayal JS, Saraswathy R (2019b) Influence of differential protein levels of feed on production performance and immune response of pacific white leg shrimp in a biofloc–based system. Aquaculture 503:118–127

Panigrahi A, Sundaram M, Saranya C, Swain S, Dash RR, Dayal JS (2019c) Carbohydrate sources deferentially influence growth performances, microbial dynamics and immunomodulation in Pacific white shrimp (Litopenaeus vannamei) under biofloc system. Fish Shellfish Immunol 86:1207–1216

Panigrahi A, Esakkiraj P, Jayashree S, Saranya C, Das RR, Sundaram M (2019d) Colonization of enzymatic bacterial flora in biofloc grown shrimp Penaeus vannamei and evaluation of their beneficial effect. Aquac Int 27(6):1835–1846

Panigrahi A, Saranya C, Ambiganandam K, Sundaram M, Sivakumar MR (2020a) Evaluation of biofloc generation protocols to adopt high density nursery rearing of Penaeus vannamei for better growth performances, protective responses and immuno modulation in biofloc based technology. Aquaculture 522:735095

Panigrahi A, Das RR, Sivakumar MR, Saravanan A, Saranya C, Sudheer NS (2020b) Bio-augmentation of heterotrophic bacteria in bioflocSystem improves growth, survival, and immunity of Indian white shrimp Penaeus indicus. Fish Shellfish Immunol 98:477–487

Parant M, Audibert F, Parant F, Chedid L, Soler E, Polonsky J, Lederer E (1978) Nonspecific immunostimulant activities of synthetic trehalose-6, 6'-diesters (lower homologs of cord factor). Infect Immun 20(1):12–19

Paripatananont T, Lovell RT (1997) Comparative net absorption of chelated and inorganic trace minerals in channel catfish Ictalurus punctatus diets. J World Aquacult Soc 28(1):62–67

Patil US, Jaydeokar AV, Bandawane DD (2012) Immunomodulators: a pharmacological review. Int J Pharm Pharm Sci 4(Suppl 1):30–36

Patil PK, Gopal C, Panigrahi A, Rajababu D, Pillai SM (2014) Oral administration of formalin killed *Vibrio anguillarum* cells improves growth and protection against challenge with *Vibrio harveyi* in banana shrimp. Lett Appl Microbiol 58(3):213–218

Powell A, Pope EC, Eddy FE, Roberts EC, Shields RJ, Francis MJ, Smith P, Topps S, Reid J, Rowley AF (2011) Enhanced immune defences in Pacific white shrimp (Litopenaeus vannamei) post-exposure to a vibrio vaccine. J Invertebr Pathol 107(2):95–99

Pratheepa V, Sukumaran N (2014) Effect of Euphorbia hirta plant leaf extract on immunostimulant response of Aeromonas hydrophila infected Cyprinus carpio. PeerJ 2:671

Pulsford AL, Crampe M, Langston A, Glynn PJ (1995) Modulatory effects of disease, stress, copper, TBT and vitamin E on the immune system of flatfish. Fish Shellfish Immunol 5(8):631–643

Raa J (1996) The use of immunostimulatory substances in fish and shellfish farming. Rev Fish Sci 4 (3):229–288

Rungrassamee W, Maibunkaew S, Karoonuthaisiri N, Jiravanichpaisal P (2013) Application of bacterial lipopolysaccharide to improve survival of the black tiger shrimp after Vibrio harveyi exposure. Dev Comp Immunol 41(2):257–262

Sahoo PK, Mukherjee SC (2001) Effect of dietary β-1, 3 glucan on immune responses and disease resistance of healthy and aflatoxin B1-induced immunocompromised rohu (Labeo rohita Hamilton). Fish Shellfish Immunol 11(8):683–695

Sajeevan TP, Philip R, Singh IB (2009) Dose/frequency: a critical factor in the administration of glucan as immunostimulant to Indian white shrimp Fenneropenaeus indicus. Aquaculture 287 (3–4):248–252

Sakai M (1999) Current research status of fish immunostimulants. Aquaculture 172:63–92

Sakai M, Kamiya H, Atusta S, Kobayashi M (1991) Immunodulatory effects on rainbow trout, Oncorhynchus mykiss, injected with the extract of abalone, Haliotis discus hannai. J Appl Ichthyol 7:54–59

Sakai M, Yoshida T, Atsuta S, Kobayashi M (1995) Enhancement of resistance to vibriosis in rainbow trout, Oncorhynchus mykiss (Walbaum.), by oral administration Clostridium butyricumbacterin. J Fish Dis 18:187–190

Sakai M, Kobayashi M, Kawauchi H (1996) In vitro activation of fish phagocytic cells by GH, prolactin and somatolactin. J Endocrinol 151:113–118

Salati F, Hamaguchi M, Kusuda R (1987) Immune response of red sea bream to Edwardsiella tarda antigens. Fish Pathol 22:93–98

Shoemaker CA, Klesius PH, Drennan JD, Evans JJ (2011) Efficacy of a modified live Flavobacterium columnare vaccine in fish. Fish Shellfish Immunol 30(1):304–308

Sirko A, Vanek T, Gora-Sochacka A, Redkiewicz P (2011) Recombinant cytokines from plants. Int J Mol Sci 12(6):3536–3552

Silva YJ, Moreirinha C, Pereira C, Costa L, Rocha RJ, Cunha Â, Gomes NC, Calado R, Almeida A (2016) Biological control of Aeromonas salmonicida infection in juvenile Senegalese sole (Solea senegalensis) with phage AS-A. Aquaculture 450:225–233

Sivaram V, Babu MM, Immanuel G, Murugadass S, Citarasu T, Marian MP (2004) Growth and immune response of juvenile greasy groupers (Epinephelus tauvina) fed with herbal antibacterial active principle supplemented diets against Vibrio harveyi infections. Aquaculture 237(1–4):9–20

Sivasankar P, Kumar GS (2017) Influence of pH on dynamics of microbial enhanced oil recovery processes using biosurfactant producing Pseudomonas putida: mathematical modelling and numerical simulation. Bioresour Technol 224:498–508

Siwicki AK (1989) Immunostimulating influence of levamisole on non-specific immunity in carp (Cyprinus carpio). Dev Comp Immunol 13:87–91

Smith VJ, Söderhäll K, Hamilton M (1984) β 1, 3-glucan induced cellular defence reactions in the shore crab, Carcinusmaenas. Comp Biochem Physiol A Physiol 77(4):635–639

Snieszko SF, Friddle SB (1949) Prophylaxis of furunculosis in brook trout (Salvelinus fontinalis) by oral immunization and sulfamerazine. Progressive Fish Culturist 11(3):161–168

Snieszko S, Piotrowska W, Kocylowski B, Marek K (1938) Badaniabakteriologiczneiserogicznenadbakteriamiposocznicykarpi. Memoires de l'Institutd'Ichtyobiologie et Pisciculture de la Station de Pisciculture Experimentale a Mydlniki de l'UniversiteJagiellonienne a Cracovie 38

Soltanian S, Fereidouni MS (2016) Effect of Henna (Lawsonia inermis) extract on the immunity and survival of common carp, Cyprinus carpio infected with Aeromonas hydrophila. Int Aquatic Res 8(3):247–261

Sommerset I, Krossøy B, Biering E, Frost P (2005) Vaccines for fish in aquaculture. Expert Rev Vaccines 4(1):89–101

Song YL, Hsieh YT (1994) Immunostimulation of tiger shrimp (Penaeus monodon) hemocytes for generation of microbicidal substances: analysis of reactive oxygen species. Dev Comp Immunol 18(3):201–209

Sudheesh PS, Cain KD (2016) Optimization of efficacy of a live attenuated Flavobacterium psychrophilum immersion vaccine. Fish Shellfish Immunol 56:169–180

Talpur AD (2014) Mentha piperita (peppermint) as feed additive enhanced growth performance, survival, immune response and disease resistance of Asian seabass, Lates calcarifer (Bloch) against Vibrio harveyi infection. Aquaculture 420:71–78

Taylor GR, Janney RP (1992) In vivo testing confirms a blunting of the human cell mediated immune mechanism during space flight. J Leukoc Biol 51(2):129–132

Tewary A, Patra BC (2008) Use of vitamin C as an immunostimulant-effect on growth, nutritional quality, and immune response of Labeorohita (Ham.). Fish Physiol Biochem 34:251–259

Toranzo AE, Romalde JL, Magariños B, Barja JL (2009) Present and future of aquaculture vaccines against fish bacterial diseases. Options Mediterr 86:155–176

Vetvicka V (2011) Glucan-immunostimulant, adjuvant, potential drug. World J Clin Oncol 2(2):115

Vijayan, K.K., Makesh, M., Otta, S.K., Patil, P.K., Poornima, M. and Alavandi, S.V., 2017. Prophylaxis in aquaculture

Villegas JG, Fukada H, Masumoto T, Hosokawa H (2006) Effect of dietary immunostimulants on some innate immune responses and disease resistance against Edwardsiella tarda infection in Japanese flounder (Paralichthyso livaceus). Aquacult Sci 54:153–162

Wang C, Lovell RT, Klesius PH (1997) Response to Edwardsiella ictaluri challenge by channel catfish fed organic and inorganic sources of selenium. J Aquat Anim Health 9(3):172–179

Wang YC, Chang PS, Chen HY (2008) Differential time-series expression of immune-related genes of Pacific white shrimp Litopenaeus vannamei in response to dietary inclusion of ß-1,3-glucan. Fish Shellfish Immunol 24:113–121

Wise DJ, Tomasso JR, Gatlin DM III, Bai SC, Blazer VS (1993) Effects of dietary selenium and vitamin E on red blood cell peroxidation, glutathione peroxidase activity, and macrophage superoxideanion production in channel catfish. J AquatAnim Health 5:177–182

Wu T et al (2000) The SRP authentication and key exchange system. RFC 2945

Xu WJ, Pan LQ (2014) Evaluation of dietary protein level on selected parameters of immune and antioxidant systems, and growth performance of juvenile Litopenaeus vannamei reared in zero-water exchange biofloc-based culture tanks. Aquaculture 426:181–188

Yin G, Jeney G, Racz T, Xu P, Jun X, Jeney Z (2006) Effect of two Chinese herbs (Astragalus radix and Scutellaria radix) on non-specific immune response of tilapia, Oreochromis niloticus. Aquaculture 253(1–4):39–47

Yoshida T, Sakai M, Kitao T, Khlil SM, Araki, et al. (1993) Immunodulatory effect of the feremented products of chicken egg, EF203, on rainbow trout, Oncorhynchus mykiss. Aquaculture 109:207–214

Young CH, Kaneda S, Mikami Y, Arai T, Igarashi K et al (1987) Protection activity induced by the bacterial vaccine, heat-killed Closteridiumbutyricum against Candida albicans infections in mice. Jpn J Med Mycol 28:262–269

Zhang J, Liu Y, Tian L, Yang H, Liang G, Xu D (2012) Effects of dietary mannan oligosaccharide on growth performance, gut morphology and stress tolerance of juvenile Pacific white shrimp, Litopenaeus vannamei. Fish Shellfish Immunol 33(4):1027–1032

DNA Vaccines for Fish

Megha Kadam Bedekar and Sajal Kole

Abstract

Aquaculture has been playing an important role in global food security for the last three decades. To meet the increasing consumer demands, fish farming has shifted from extensive to intensive industrial-scale production. However, this intensification has led to frequent occurrences of infectious diseases, hampering the development and profitability of fish farms. To control the persistent and emerging diseases and sustain the intensive fish production, aquaculture industry requires a wide number of effective vaccines. Although there have been significant progress in the field of fish vaccinology, only few traditional inactivated vaccines, mostly against pathogenic bacteria, got licenses till date. Thus, there is urgent need for effective prophylaxis against other serious fish pathogens, viz., viruses, parasites, and intracellular bacteria. In this regard, DNA vaccination has come as a promising alternative strategy in fish vaccinology. It involves delivery of plasmid DNA, encoding a vaccine antigen to the host, which in turn expresses the antigenic protein and induces strong and long-lasting immune responses inside the host. In this chapter, the key aspects relating to DNA vaccination in aquaculture, viz., action mechanisms, current status, underlying challenges, existing opportunities, and future directions regarding the use of DNA vaccines in commercial fish farming are reviewed and discussed.

M. K. Bedekar (✉)
Department of Aquatic Animal Health, ICAR-Central Institute of Fisheries Education, Mumbai, Maharashtra, India
e-mail: megha.bedekar@cife.edu.in

S. Kole
Department of Aqualife Medicine, Chonnam National University, Yeosu, Republic of Korea

© The Author(s), under exclusive license to Springer Nature Singapore Pte Ltd. 2021
289
P. K. Pandey, J. Parhi (eds.), *Advances in Fisheries Biotechnology*,
https://doi.org/10.1007/978-981-16-3215-0_19

Keywords

Vaccines · DNA vaccination · Fish · Infectious diseases

19.1 Introduction

Disease prevention is an essential criterion for the continued progress of aquaculture around the world. Among the prevalent measures available in combating the disease problem in aquaculture, vaccination is regarded as the most efficient prophylaxis to control infectious pathogens. However, in the aquaculture industry, developments of effective vaccines are still in a relatively early phase and most of the commercial fish vaccines available are against bacterial diseases which are simply formulated by inactivated bacteria, applied by either immersion or injection with an oil adjuvant (Heppell and Davis 2000). These traditional vaccine technologies have long proven their inefficiency in providing prophylactic treatment against several other important diseases, mainly of viral and parasitic origin, thereby an urgent need was felt to develop alternative vaccines for farmed fishes. In this context, during the last decade of the previous century, DNA vaccination came as a viable and promising alternative vaccine technology for aquaculture.

The very idea of this promising technology first gained worldwide attention in 1990, when Wolff and his co-researchers showed that injecting DNA plasmid in mouse muscle resulted in a significant expression of the protein encoded by the plasmid (Wolff et al. 1990). Following year, the technology was tested in fish, wherein expression of reporter gene was observed in carp muscle after injection of plasmid DNA (Hansen et al. 1991). However, only after 1996, the technology was first implemented in fish vaccinology studies where the developed plasmid constructs, encoding the viral glycoprotein, proved to be extremely efficient for protecting fish against rhabdoviruses (Anderson et al. 1996; Lorenzen et al. 1998). Since then, various studies have been or are being carried out to develop DNA vaccines against wide range of fish pathogens including viruses, bacteria, and parasites. But the only success story, regarding DNA vaccination in aquaculture, lies in the commercialization and licensing of APEX-IHN (Novartis/Elanco) in 2005, for protecting Atlantic salmon (*Salmo salar*) against Infectious Hematopoietic Necrosis Virus (IHNV) in British Colombia and CLYNAV (Elanco) in 2017, a polyprotein-encoding DNA vaccine against Salmon Pancreas Disease Virus (SPDV) infection in Atlantic salmon for use within the European Union (EU).

Apart from these, the progress of DNA vaccines for aquaculture remains at experimental level due to various reasons, viz., DNA vaccines against many non-rhabdoviral diseases have not yet yielded sufficient results in terms of protection whereas the DNA vaccines which gave encouraging results in eliciting improved immune responses (both humoral and cellular adaptive immunity) and protection against the targeted pathogens are yet to succeed in field trial. Moreover, the safety concerns and environmental impacts remained the most critical limitation for their worldwide acceptance. With this background, the present chapter aims to discuss the mechanisms of action of DNA vaccines. In addition, the chapter also gives an update on the recent development of DNA vaccines for aquaculture and the associated challenges in the DNA vaccine developmental processes and their futuristic solutions.

19.2   Working Mechanism of DNA Vaccines

DNA vaccines are basically composed of circular DNA plasmids that contain the gene, encoding the vaccine antigen from pathogen of interest. Structurally, the plasmid contains a eukaryotic expression vector with different transcriptional elements, viz., promoter and enhancer sequences, the gene of interest, a poly-adenylation sequence, transcriptional termination sequence, antibiotic resistance gene, and origin of replication. These transcriptional elements allow the amplification of the plasmid to large quantities inside the bacterial cells. Following administration of the plasmid DNA into host fish, the promoter helps in flanking of antigenic gene, whereas other termination elements facilitate expression of antigenic protein using the host's cell machinery (Fig. 19.1).

Sequentially, after intramuscular administration of pDNA, the professional antigen presenting cells (APCs) such as macrophages and dendritic cells (DC), at the administration site, take up the pDNA in their nucleus, followed by transcription of the antigenic gene into mRNA transcripts with the help of promoter. The mRNA transcripts are then transported to the cytoplasm wherein, translation process takes place using host cell ribosome resulting in synthesis of polypeptide into the endoplasmic reticulum (ER). Subsequently, the transmembrane protein, on its way from ER to Golgi cisternae, subjected to folding, glycosylation, and non-covalent association into trimers and finally reaches the cell membrane in its mature form. Post expression of antigenic protein, the host immune responses elicited in 3 major ways, viz., innate immune response by interferon induction, cell-mediated immune response by MHC class I antigen presentation pathway, and humoral immune response by MHC class II antigen presentation pathway.

19.2.1 Innate Immune Response by DNA Vaccination

The induction of innate immune responses by DNA vaccination is mainly associated with the recognition of DNA motifs, mRNA transcripts and/or expressed foreign protein antigen by TLRs or other pattern recognition receptors on neighboring cells and thereby activating the interferon type I response (Tonheim et al. 2008; Collins et al. 2019). The early IFN response is believed to transient which in turn activates the antigen presenting cells (APCs) and makes way for structured adaptive immune responses involving T cells (Utke et al. 2008; Chang et al. 2015; Embregts et al. 2017) and B cells (Lorenzen and LaPatra 1999, 2005; LaPatra et al. 2000; Tian et al. 2008a; Embregts et al. 2017).

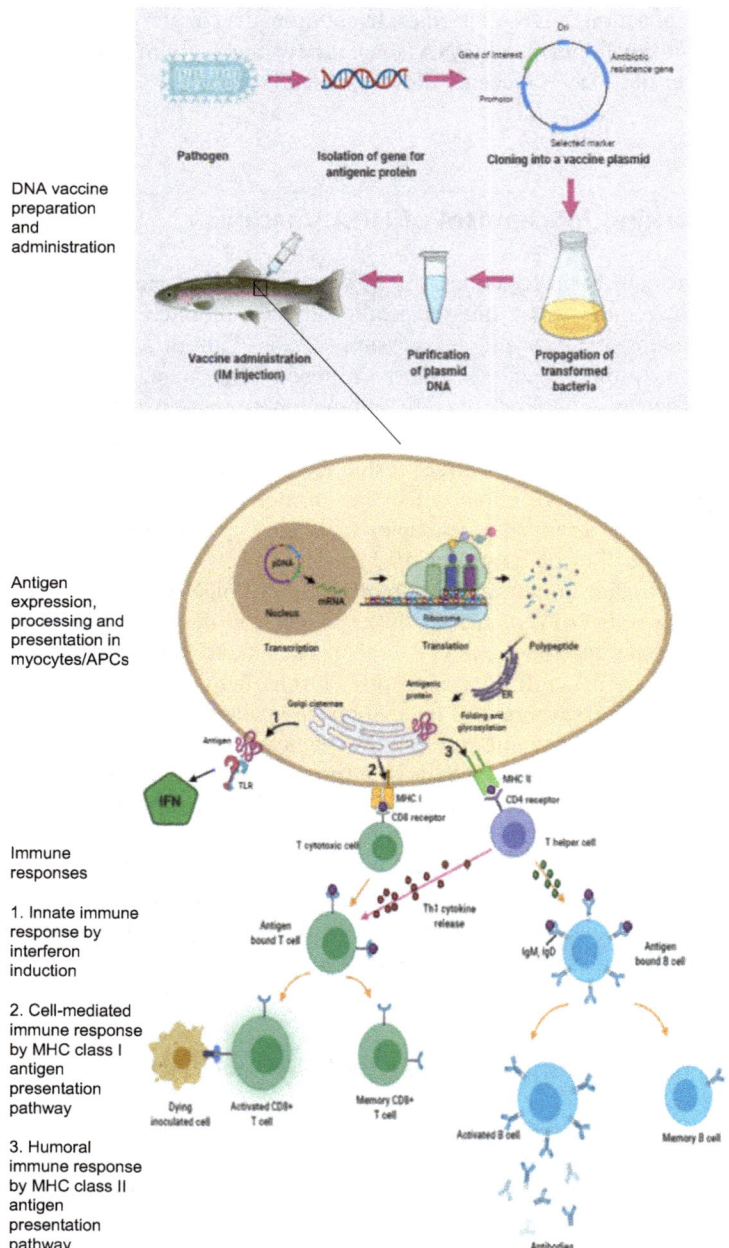

Fig. 19.1 Working mechanism of DNA vaccine

19.2.2   Cell-Mediated Immune Response by DNA Vaccination

The cell-mediated immune (CMI) responses are the unique feature of DNA vaccines as compared to conventional inactivated or subunit vaccines, where only humoral immune response takes place (Liu 2003). The CMI responses are initiated by presentation of endogenously expressed antigenic proteins into the extracellular space by MHC class I molecules, mimicking the transfected cells appear like the pathogen infected cell. Subsequently, the T-cell receptors (TCR) recognize the peptides presented by MHC class I on the activated APCs which in turn stimulate CD 8^+ T-cell (cytotoxic T cell) response. Following this, the CMI responses are represented in two ways, viz., secretion of pro-inflammatory cytokines by the activated Th1 cells and killing of the transgene presenting cells by the CD 8^+ T cell (Collins et al. 2019).

19.2.3   Humoral Immune Response by DNA Vaccination

Beside the MHC class I antigen presentation pathway, the APC also takes up soluble antigens released from another transgene producing cell (e.g., a myocyte), processes it and presents the peptide on MHC class II molecules at the cell surface (Shirota et al. 2007). Following this, naive CD4+ T helper cells interact with specific antigen-MHC class II complexes on APCs, leading to the differentiation of B cells into antibody producing plasma cells (Boudinot et al. 1998; Utke et al. 2008; Embregts et al. 2017). In addition, the DNA vaccine encoded antigenic protein can directly stimulate the adaptive humoral immune response, wherein the native antigenic proteins are recognized by B cell receptors and circulating antibodies, followed by activation of B lymphocytes and antibody production.

Furthermore, in terms of long-term protective mechanisms of DNA vaccination, the published literatures in majority of cases demonstrated generation of long-lived activated T or B cell populations (immune memory), but the exact mechanism is yet to be established. Among these studies, antibody responses were reported to be important for long-term protection. However, further analysis of antibodies showed that apart from antigen-specific antibody or neutralizing antibodies, which are considered to be pre-requisites for protection (Boudinot et al. 1998; Kurath 2008), non-neutralizing antibodies may also confer protection (Traxler et al. 1999; Jones 2001; Bootz et al. 2017). Likewise, the immunoglobulin (IgM, IgT, and IgD) gene expression analysis also showed to contribute for antigen-specific B cell differentiation and proliferation, post DNA vaccination (Kaattari et al. 2002; Ye et al. 2011), which in turn provided long-term immunity.

19.3 Development of DNA Vaccines for Aquaculture

Although it is ascertained that DNA vaccine confers an improved immune response and protection as they stimulate both humoral and cellular adaptive immunity (Munang'andu and Evensen 2015), in addition to its role in some innate immune response mechanisms (Liu 2011; Dalmo 2018), most DNA vaccines developed against fish pathogens are in experimental phase barring APEX-IHN (Novartis/ Elanco) and CLYNAV (Elanco) as mentioned earlier. Thus, it is important to review scientific publications on experimental DNA vaccines for fish in order to get an overall idea of the development process of DNA vaccines till date along with the efficacies and constraints associated with them. In this context, this section summarizes a detailed review on the experimental DNA vaccines with five subsections categorized according to the type of pathogen.

19.3.1 DNA Vaccine Against RNA Viruses (Table 19.1)

19.3.1.1 Rhabdovirus

Rhabdovirus is negative sense single stranded RNA virus. Fish rhabdovirus belongs to 2 genera, viz., Novirhabdoviruses which include viral hemorrhagic septicemia virus (VHSV), infectious hematopoietic necrosis virus (IHNV), Hirame rhabdovirus (HIRRV) and Vesiculoviruses which include spring viremia of carp virus (SVCV). DNA vaccines against fish rhabdoviruses are maximally studied as they showed most promising results in terms of protective efficacies. Among the various genes of rhabdoviruses, gene encoding the glycoprotein (G) proved to be very effective antigenic gene for construction of DNA vaccine in providing long-lasting protection, irrespective of the individual virus. From the reported studies involving DNA vaccines against VHSV/IHNV, it was observed that a dual-phase immune response was orchestrated in vaccinated fish with a rapidly activated transient nonspecific interferon (IFN) related antiviral protection, followed by a long-lasting specific immunity (LaPatra et al. 2001; Lorenzen et al. 2002) in salmonids, resulting in 80–100% protection. However, recent reports of limited protection of Pacific herring and muskellunge against VHSV following DNA vaccination suggested that the conferred protection depends on the host-pathogen combination (Hart et al. 2017; Millard et al. 2017). In case of flounder, DNA vaccines encoding the G protein of HIRRV (Takano et al. 2004; Seo et al. 2006; Yasuike et al. 2007) and VHSV (Byon et al. 2005; Lazarte et al. 2017; Lim et al. 2019) conferred effective protections. For SVCV, barring the study of Embregts et al. (2017) and Emmenegger and Kurath (2008), wherein, 95–100% and 88% protection was observed in common carp and koi carp, respectively, other studies elicit low to moderate (30–60%) protection (Kanellos et al. 2006; Zhang et al. 2017, 2018, 2019b). Further, the studies also indicated that protective efficacies of SVCV DNA vaccines are dependent on the promoter, plasmid dose and/or mode of challenge (immersion/injection).

Table 19.1 Status report of experimental DNA vaccine against RNA viruses

Virus	Fish (size in g)	Vaccination Vaccine	Antigenic gene	Modality	Dose (μg/fish)	Challenge Mode (dpv)	Dose	RPS	Reference
Rhabdovirus: (−)ssRNA virus									
Viral hemorrhagic septicemia virus (VHSV)	Rainbow trout (3–4 g)	pcDNA3-vhsG	Glycoprotein (G) gene of VHSV	IM (Prime only)	0.1, 1	Immersion (51 and 103 dpv)	5×10^5 TCID$_{50}$/mL for 2 h	51 dpv: 97–100% 103 dpv: 100%	Lorenzen et al. (2000)
	Rainbow trout (10 and 100 g)	pcDNA3-vhsG	Glycoprotein (G) gene of VHSV	IM and IP (Prime only)	0.1, 1	Immersion (7 dpv)	10 g: 1×10^5 100 g: 5×10^5 TCID$_{50}$/mL for 2 h	10 g: IM: 80–90%; IP: 0–35% 100 g: IM: 92–96%; IP: 0–48%	McLauchlan et al. (2003)
	Rainbow trout (2 g)	pcDNA3-vhsG pcDNA3-ihnG	Glycoprotein (G) gene of VHSV & IHNV	IM (Prime only)	1	Immersion (40 dpv)	1×10^6 TCID$_{50}$/mL for 6 h	pcDNA3-vhsG: 98–100% pcDNA3-ihnG: 95–97%	Lorenzen et al. (2009)
	Rainbow trout (3.5 g)	pcDNA3-CMV-vhsG pcDNA3-Mx-vhsG	Glycoprotein (G) gene of VHSV (CMV/Mx promoter)	IM (CMV: Prime only; Mx: Prime-Boost (1μg Poly(I:C) 7dpv)	1	Immersion (14 and 60 dpv)	2–3×10^5 TCID$_{50}$/mL	14 dpv: CMV—100% Mx—67.8% 60 dpv: CMV—95.7% Mx—75.8%	Sepúlveda et al. (2019)
	Pacific Herring (age 1.5 years)	pcDNA3-vhsG	Glycoprotein (G) gene of IVa & Ia VHSV genotype	IM (Prime only)	4 g	Immersion (70 dpv)	6–8×10^3 PFU/mL for 1 h	IVa: 40.4–51.1% Ia: 11.1–27.3%	Hart et al. (2017)
	Muskellunge (56 g)	pVHSivb-G	Glycoprotein (G) gene of IVb VHSV genotype	IM (Prime only)	10	Immersion (70 dpv)	10^2 PFU/mL for 90 min	45%	Millard et al. (2017)
	Flounder (3 g)	pCMV-VHSg	Glycoprotein (G) gene of VHSV	IM (Prime only)	10	IP (30 dpv)	Low: 1×10^2 TCID$_{50}$/fish High: 1×10^3 TCID$_{50}$/fish	Low: 100% High: 93%	Byon et al. (2005)

(continued)

Table 19.1 (continued)

Virus	Fish (size in g)	Vaccination				Challenge		RPS	Reference
		Vaccine	Antigenic gene	Modality	Dose (μg/fish)	Mode (dpv)	Dose		
	Flounder (10 g)	pCMV-VHSg	Glycoprotein (G) gene of VHSV	IM (Prime only)	10	IP (30 dpv)	Low: 1×10^2 TCID$_{50}$/fish High: 1×10^3 TCID$_{50}$/fish	Low: 96% High: 90%	Byon et al. (2006)
	Flounder (3.62 g)	pFC-vG-miR155	Glycoprotein (G) gene of VHSV + flounder miR-155	IM (Prime only)	1	IM (30 dpv)	Low: 5×10^3 PFU/fish High: 5×10^4 PFU/fish	Low: 100% High: 91.7%	Lim et al. (2019)
	Flounder (11.4 g)	pEF-GD	Glycoprotein (G) gene of VHSV + flounder DDX41	IM (Prime only)	1	IP (15 and 30 dpv)	10^5 TCID$_{50}$/fish	15 dpv: 70% 30 dpv: 64%	Lazarte et al. (2017)
Infectious hematopoietic necrosis virus (IHNV)	Rainbow trout (1.8 g)	pIHNVw-G	Glycoprotein (G) gene of IHNV WRAC strain	IM (Prime only)	0.1	Immersion (29 dpv)	2.8×10^4 PFU/mL for 1 h	100%	Corbeil et al. (2000a)
	Rainbow trout (1.8 g)	pIHNVw-G	Glycoprotein (G) gene of IHNV WRAC strain	IM (Prime only)	0.1	Immersion (42 dpv)	IHNV strain, PFU/mL WRAC: 10^3 220–90: 10^6 AK-14: 10^5 RB-1: 10^7 Col-85: 10^7 Shizuoka: 10^5 32–87: 10^6	WRAC—90% 220–90—56.2% AK-14—75% RB-1—72.2% Col-85—66.7% Shizuoka 71.8% 32–87—82.3%	Corbeil et al. (2000b)
	Rainbow trout (120 g)	pIHNVw-G	Glycoprotein (G) gene of IHNV WRAC strain	IM (Prime only)	1, 10, 25	IP (42 dpv)	10^5 PFU/fish	100%	La Patra et al. (2000)

Rainbow trout (0.56 g)	pcDNA3-IHNV-G	Glycoprotein (G) gene of IHNV	IM (Prime only)	10	Immersion (30 and 70 dpv)	10^5 PFU/mL for 5 h in 2 L water	30 dpv: 93% 70 dpv: 87%	Kim et al. (2000)
Rainbow trout (2 g)	pIHNVw-G	Glycoprotein (G) gene of IHNV WRAC strain	IM (Prime only)	0.1, 1.0, 2.5	Immersion (4 dpv)	10^5 PFU/mL for 60 min in ($10 \times$ total g of fish) L water	82–96%	La Patra et al. (2001)
Rainbow trout (2.5 g)	pIHNVw-G	Glycoprotein (G) gene of IHNV WRAC strain	IM (Prime only)	0.1	IP (90 dpv)	10^4–10^8 PFU/fish	100%	Kurath et al. (2006)
Rainbow trout (1.2–4 g)	pM and pU	Glycoprotein (G) gene of IHNV WRAC strain (M & U genogroup)	IM (Prime only)	1	Immersion/ IP (28 dpv)	Immersion: 2×10^5 PFU of IHNV-M/ mL in 1 L water for 1 h IP injection: 5×10^6 PFU IHNV-U/fish	IHNV-M challenge: pM: 88%; pU: 58% IHNV-U challenge: pM: 93%; pU: 96%	Peñaranda et al. (2011)
Rainbow trout (0.4 g)	pIRF1A-G-pMT-M	Glycoprotein (G) gene of IHNV (inducible fish promoter IRF1A)—Matrix (M) gene of IHNV (ZnCl$_2$ inducible metallothionein promoter pMT)	IM (Prime only)	3	Immersion (30 dpv)	10^5 PFU/mL for 5 h	42%	Alonso et al. (2011)
Rainbow trout (5 g)	pcDNA-G (PLGA NPs encapsulation)	Glycoprotein (G) gene of IHNV	Oral (via feed for 4 and 8 days)	22, 43	IP (42 dpv)	2×10^3 PFU/fish	11–22%	Adomako et al. (2012)
Rainbow trout (3–4 g)	pIRF1A-G (Alginate encapsulation for oral)	Glycoprotein (G) gene of IHNV (inducible fish promoter IRF1A)	Oral gavage IM (Prime only)	100 (oral) 5 (IM)	Immersion (30 dpv)	10^5 TCID$_{50}$/ mL for 3 h	Oral: 56% IM: 70%	Ballesteros et al. (2015a)

(continued)

Table 19.1 (continued)

Virus	Fish (size in g)	Vaccination		Dose (μg/ fish)	Challenge			RPS	Reference
		Vaccine	Antigenic gene	Modality		Mode (dpv)	Dose		
	Rainbow trout (3 g)	pIHNch-G	Glycoprotein (G) gene of IHNV SD12 strain	IM (Prime only)	0.1, 1, 5, 10	Immersion/ IP (28 dpv)	Immersion: 10^5 PFU/mL for 60 min in ($10 \times$ total g of fish) L water IP: 10^2 PFU/fish	Immersion: 87–93% Injection: 86.7–95.6%	Xu et al. (2017a)
	Atlantic salmon (Pre-smolt: 57 g; Smolt: 71 g)	pCMV4-G	Glycoprotein (G) gene of IHNV	IM (Prime only)	25	Immersion/ Co-habitation (60 dpv)	Immersion: 5×10^3 PFU/mL for 1 h in 20 L water Co-habitation: 10 fish injected with 4.9×10^3 with 50 Pre-smolt/ 40 smolt	Immersion: Pre-smolt: 96% Smolt: 90% Co-habitation: Pre-smolt: 100% Smolt: 93%	Traxler et al. (1999)
	Chinook salmon (3 g) Sockeye Salmon (3 g)	pIHNVw-G	Glycoprotein (G) gene of IHNV WRAC strain	IM (Prime only)	0.1, 1	IP Chinook: 45 dpv Sockeye: 35 dpv	Chinook: 3.6×10^7 PFU/fish Sockeye: 1.7×10^7 PFU/fish	Chinook: 71–78% Sockeye: 61–86%	Garver et al. (2005)

		Construct	Antigen	Dose (µg)	Route (Prime/Boost)	Challenge route	Challenge dose	Protection	Reference
VHSV/IHNV	Rainbow trout (2 g)	pcDNA3-vhsG + pcDNA3-ihnG	Glycoprotein (G) gene of VHSV/IHNV	1 (each)	IM (Prime only)	Immersion (80 dpv)	1×10^4 TCID$_{50}$/mL for 2 h VHSV/ IHNV/ VHSV +IHNV	VHSV: 90% IHNV: 88% VHSV+IHNV: 85%	Einer-Jensen et al. (2009)
Hirame rhabdovirus (HIRRV)	Flounder (2 g)	pCMV-HRVg	Glycoprotein (G) gene of HIRRV	1, 10	IM (Prime only)	IM (28 dpv)	4.0×10^3 TCID$_{50}$/fish	70.5–90.1%	Takano et al. (2004)
	Flounder (3 g)	pcDNA-G, pcDNA-N, pcDNA-G + N	Glycoprotein (G) Nucleoprotein (N) gene of HIRRV	5	IM (Prime only)	Immersion (21 dpv)	1.2×10^4 TCID$_{50}$/mL for 90 min	pcDNA-G: 94.7% pcDNA-N: 26.3% pcDNA-G + N: 97.4%	Seo et al. (2006)
	Flounder (10 g)	pHRV-G, pHRV-N	Glycoprotein (G) Nucleoprotein (N) gene of HIRRV	10	IM (Prime only)	IM (28 dpv)	3.2×10^3 TCID$_{50}$/fish	pHRV-G-86.5% pHRV-N-1.5%	Yasuike et al. (2007)
Spring viremia of carp virus (SVCV)	Common carp (11.4 g)	pCMV*i*/ GF + pVR1223/GF	Glycoprotein (G) gene of SVCV (CMV-Intron A promoter)	25	IM (Prime only)	IP (42 dpv)	$2.5 \times 10^{3.6}$ PFU/fish	48%	Kanellos et al. (2006)
	Common carp (4 g)	pEGFP-G (Single-walled carbon nanotubes, SWCNTs)	Glycoprotein (G) gene of SVCV	Imm: 20, 40 mg/ L IM:12, 20µg	Immersion (6 h) IM (Prime only)	IM (28 dpv)	4.0×10^5 TCID$_{50}$/fish	Immersion: 45–57.5% Injection: 55–65%	Zhang et al. (2017)
	Common carp (4 g)	pcDNA-M (Single-walled carbon nanotubes, SWCNTs)	Matrix-protein (M) gene of SVCV	10	IM (Prime only)	IP (28 dpv)	3.0×10^6 TCID$_{50}$/fish	51.3%	Zhang et al. (2018)
	Common carp (4 g)	pEGFP-M (Single-walled carbon nanotubes, SWCNTs)	Matrix-protein (M) gene of SVCV	35 mg/L	Immersion (10 h) (Prime-Boost, 3 dpv)	IP (21 dpv)	3.0×10^3 TCID$_{50}$/fish	46.3%	Zhang et al. (2019b)

(continued)

Table 19.1 (continued)

Virus	Fish (size in g)	Vaccination				Challenge			Reference
		Vaccine	Antigenic gene	Modality	Dose (µg/fish)	Mode (dpv)	Dose	RPS	
	Common carp (1.5–2 g)	pcDNA3-SVCV-G	Glycoprotein (G) gene of SVCV	IM (Prime only)	0.1, 1	Immersion (75 dpv)	8×10^6 PFU/mL for 48 h	95–100%	Embregts et al. (2017)
	Common carp (2–4 g)	pcDNA3-SVCV-G (Alginate encapsulation for oral)	Glycoprotein (G) gene of SVCV	Oral gavage IM (Prime only)	3 (oral) 0.1 (IM)	Immersion (75 dpv)	8×10^6 PFU/mL for 48 h	Oral: 5–20% Injection: 92%	Embregts et al. (2019a)
	Common carp (5 g)	BG/pEG-G (in *E.coli* ghost cell)	Glycoprotein (G) gene of SVCV	IM (Prime only)	1	IP (28 dpv)	1×10^5 TCID$_{50}$/fish	81.48%	Zhang et al. (2019a)
	Koi carp (1.5–4.3 g)	pSGnc	Glycoprotein (G) gene of SVCV	IM (Prime only)	10	IP (28 dpv)	1×10^6 PFU/fish	50–88%	Emmenegger and Kurath (2008)
Orthomyxovirus: (−) *ssRNA virus*									
Infectious salmon anemia virus (ISAV)	Atlantic salmon (10 and 100 g)	pEGFP-HE	ISAV hemagglutinin esterase	IM (Prime only)	0.5, 5, 10	IP (21 dpv)	5×10^3 TCID$_{50}$/fish	39.5–60.5%	Mikalsen et al. (2005)
	Atlantic salmon (30–45 g)	pHE + pIFNa/ pIFNb / pIFNc (salmon IFNs gene as immune adjuvant)	ISAV hemagglutinin esterase	IM (Prime only)	15	IP (56 dpv)	10^4 TCID$_{50}$/fish	pHE + pIFNa: 94% pHE + pIFNb: 76% pHE + pIFNc: 90%	Chang et al. (2015)
Betanodavirus: (+) *ssRNA virus*									
Nervous necrosis virus (NNV)	Asian sea bass (10–15 g)	pFNCPE42	RNA2 Capsid protein	IM (Prime only)	20	IM (21 dpv)	$1 \times 10^{6.5}$ TCID$_{50}$/fish	77.33%	Vimal et al. (2014a)
	Asian sea bass (10–15 g)	pFNCPE42 (Chitosan (CS/TPP) NPs encapsulation)	RNA2 Capsid protein	Oral (via feed for 1 day) (Prime only)	100	IM (21 dpv)	$1 \times 10^{6.5}$ TCID$_{50}$/fish	60%	Vimal et al. (2014b)

European sea bass (6 g)	pNNV (Chitosan encapsulation)	RNA2 Capsid protein	Oral (via feed for 2 days) (Prime only)	10	IM (90 dpv)	1×10^6 TCID$_{50}$/fish	45%	Valero et al. (2016)
Orange-spotted grouper (1.2 g)	pcMGNNV2	NNV capsid protein	IM (Prime only)	1 (μg/g fish)	IM (14 dpv)	2×10^7 TCID$_{50}$/fish	71%	Chen et al. (2015)
Atlantic halibut nodavirus (AHNV)	Turbot (3.6 g)	pVHSV-G	Glycoprotein (G) gene of VHSV from trout	IM (Prime only)	5	IM (8 and 35 dpv)	AHNV $1 \times 10^{6.3}$ TCID$_{50}$/fish	8 dpv: 100% 35 dpv: 63%	Sommerset al. (2003)

Alphavirus: (+) ssRNA virus

Salmonid alphavirus 3 (SAV-3)	Atlantic salmon (46 g)	pmaxFP-E1+ pmaxFP-E2 (E1 + E2 (50μg) protein for boost)	Spike protein (E1 and E2) genes of SAV-3	IM (Prime: pDNA— Boost: Subunit; 6 weeks interval)	20	IM (21 dpv)	SAV-3 2×10^6 TCID$_{50}$/fish	79%	Xu et al. (2012)
	Atlantic salmon (30 g)	pCSP	SAV-3 complete structural polyprotein, C-E3-E2-6K-E1	IM (Prime only)	15	IP (70 dpv)	SAV-3 5000 PFU/fish	Pancreas disease (PD) lesion: 0%	Chang et al. (2017)

Reovirus: dsRNA virus

Grass carp reovirus (GCRV)	Grass carp (1.5–1.8 g)	pcDNAvp7 (single-walled carbon nanotubes, SWCNTs)	Major capsid protein (vp7) gene of GCRV	IM (Prime only)	1, 5, 10	IP (28 dpv)	1×10^5 TCID$_{50}$/fish	72.5–100%	Zhu et al. (2015)
	Grass carp (1.1 g)	pEGFP-vp5 (single-walled carbon nanotubes, SWCNTs)	Outer capsid protein (vp5) gene of GCRV	Immersion (3 h) IM (Prime only)	Imm.: 20 mg/L IM: 5μg/fish	IP (28 dpv)	1×10^4 TCID$_{50}$/fish	Immersion: 96.7% IM: 100%	Wang et al. (2015)

(continued)

Table 19.1 (continued)

Virus	Vaccination					Challenge		RPS	Reference
	Fish (size in g)	Vaccine	Antigenic gene	Modality	Dose (µg/fish)	Mode (dpv)	Dose		
	Grass carp (1.5–1.8 g)	pcDNAvp7 (naked) DH5α-BG/pcDNA-vp7 (in ghost bacteria)	Major capsid protein (vp7) gene of GCRV	IM (Prime only)	1, 2.5, 5	IP (21 dpv)	1×10^5 TCID$_{50}$/fish	pcDNAvp7: 5.56%, 23.33%, 42.22% DH5α-BG/ pcDNA-vp7: 53.33%, 72.22%, 90%	Hao et al. (2017)
	Grass carp (15–20 g)	pC-S6 pC-S10	S6: Capsid protein (vp4); S10: Non-structural protein (NS38) of GCRV	IM (Prime-Boost, 4 weeks interval)	10	IP (42 dpv)	1×10^5 TCID$_{50}$/fish	pC-S6: 59.9% PC-S10: 23.1	Chen et al. (2018)
	Grass carp (12 g)	pcDNA3.1-s11	Capsid protein (vp35) gene of GCRV	IM (Prime only)	10	IP (21 dpv)	1×10^5 TCID$_{50}$/fish	60%	Gao et al. (2018)
Aquabirnavirus: dsRNA virus									
Infectious pancreatic necrosis virus (IPNV)	Brown trout Rainbow trout (1–1.5 g)	pcDNA-VP2 (Alginate encapsulation for oral)	Capsid protein (VP2) gene of IPNV	Oral gavage (Prime only)	10	Immersion (15 and 30 dpv)	2×10^6 TCID$_{50}$/mL in 1 L water for 4 h	Brown trout: 15 dpv: 83%; 30 dpv: 84% Rainbow trout: 15 dpv: 78–82%; 30 dpv: 67–83%	Ana et al. (2010)
	Rainbow trout (1.5 g)	pcDNA-VP2 (Alginate encapsulation for oral)	Capsid protein (VP2) gene of IPNV	Oral (via feed for 3 days) (Prime only)	30 (10µg / fish/day)	Immersion (15 and 30 dpv)	1×10^6 TCID$_{50}$/mL in 4 L water for 2 h	15 dpv: 85.9% 30 dpv: 78.2%	Ballesteros et al. (2014)

	Rainbow trout (1.5 g)	pcDNA-VP2 (Alginate encapsulation for oral)	Capsid protein (VP2) gene of IPNV	Oral gavage (Prime only)	10	Immersion (15 dpv)	5×10^5 TCID$_{50}$/mL in 4 L water for 2 h	60%	Ballesteros et al. (2015b)
	Rainbow trout (3 g)	pcDNA3.1-VP2 (Alginate (Alg)/ Chitosan (CS-TPP) encapsulation)	Capsid protein (VP2) gene (PTA motif) of IPNV	Oral (via feed) (Prime-Boost, 15 dpv)	10, 25	IP (30 dpv)	2×10^6 TCID$_{50}$/fish	Alg-pcDNAVP2: 59–82% CS-TPP-pcDNAVP2: 47–70%	Ahmadivand et al. (2017)
	Rainbow trout (4–05 g)	pcDNA3.1-VP2	Capsid protein (VP2) gene (PTA motif) of IPNV	IM (Prime only)	2, 5, 10	IP (30 dpv)	5×10^6 TCID$_{50}$/fish	76.45–88.23%	Ahmadivand et al. (2018)
	Atlantic salmon (30 g)	pDNA-TA	Capsid protein (VP2) gene of IPNV	IM (Prime only)	20	Co-habitation (28 dpv)	IP - 1×10^6 TCID$_{50}$ shedder fish put at 12% total biomass	27.2%	Munang'andu et al. (2012)
	Atlantic salmon (30–40 g)	pVP2 Liposome DOTAP (1,2-dioleoyl-3-trimethylammonium-propane) encapsulation	Capsid protein (VP2) gene of IPNV	Oral (via feed for 7 consecutive days) (Prime only)	Single dose: 1 mg$_{pDNA}$/kg$_{fish}$·d Double dose: 2 mg$_{pDNA}$/kg$_{fish}$·d	IP (30 dpv)	2×10^1 TCID$_{50}$/fish	Single dose: 58.2% Double dose: 66.7%	Reyes et al. (2017)
IHNV/IPNV	Rainbow trout (5 g)	pCh-IHN/IPN	Glycoprotein (G) gene of Chinese IHNV (Sn1203)/ Capsid protein (VP2–VP3) gene of Chinese IPNV (ChRtm213)	IM (Prime only)	1	IP (30 and 60 dpv)	IHNV Sn1203: 10^2 PFU/fish IPNV ChRtm213: 10^6 PFU/fish	IHNV Sn1203: 30 dpv— 93.3% 60 dpv— 89.4% IPNV ChRtm213: 30 dpv— 86.7% 60 dpv— 92.3%	Xu et al. (2017b)

19.3.1.2 Orthomyxovirus

Infectious Salmon Anemia Virus (ISAV) is the single representative of orthomyxovirus infecting salmonid fish. It is a negative sense single stranded RNA virus. Two experimental DNA vaccines, encoding the viral Hemagglutinin Esterase (HE) protein, have been reported so far (Mikalsen et al. 2005; Chang et al. 2015). It was observed that the virus protein suppresses the host immune response and hinders successful vaccine protection (<60% RPS) against IP challenge with ISAV in vaccinated Atlantic salmon. However, the same DNA vaccine (pHE), when co-administered with immune adjuvant plasmids expressing type I IFN subtypes (IFN a, b, c), conferred higher protection (76–94%) as against pHE (14–60% RPS) or pIFNs (10–20% RPS) (Chang et al. 2015). This illustrates that some DNA vaccines may need immune adjuvants to generate a strong protective immune response.

19.3.1.3 Betanodavirus

Nervous Necrosis Virus (NNV) is a fish betanodavirus having positive sense single stranded RNA genome. Four studies on DNA vaccine encoding capsid protein of NNV have been reported to provide moderate protection in different fish, viz., 77.33% RPS (IM vaccinated) and 60% RPS (oral vaccinated with chitosan encapsulation) in Asian seabass (Vimal et al. 2014a, b), 45% RPS in European seabass (oral vaccinated with chitosan encapsulation) (Valero et al. 2016), and 71% RPS in orange-spotted grouper (IM vaccination) (Chen et al. 2015). In case of turbot, DNA vaccine encoding NNV capsid protein reported to provide no protection (Sommerset et al. 2005) but when DNA vaccine encoding VHSV-G protein was used, high protection (63–100% RPS) was observed against NNV challenge (Sommerset et al. 2003). In addition, all the studies showed higher protection when challenged after 2–3 weeks post vaccination. Thus, the above two phenomenon, viz., VHSV-G cross-protection and short-duration protection, illustrated that the conferred protection from NNV-DNA vaccines are mainly due to innate immune mechanisms.

19.3.1.4 Alphavirus

Salmonid Alphavirus 3 (SAV-3), representing alphavirus (+ssRNA virus), is an important pathogen causing Pancreas Disease (PD) in Atlantic salmon. Apart from the commercialized DNA vaccine (CLYNAV), as mentioned above, two studies are reported for experimental DNA vaccine against SAV-3. Chang et al. (2017) reported a DNA vaccine, encoding the entire structural polyproteins of SAV-3, resulted in reduction in virus replication and lesion development after SAV-3 challenge in vaccinated fish, whereas Xu et al. (2012) showed 79% RPS using a DNA vaccine, containing only the spike proteins (E1 and E2) genes.

19.3.1.5 Reovirus

Grass Carp Reovirus (GCRV), a double stranded RNA virus, is a representative fish reovirus. Five studies, involving experimental DNA vaccines against GCRV, are reported in which different viral proteins encoding genes are used, viz., VP7 (Zhu et al. 2015; Hao et al. 2017), VP5 (Wang et al. 2015), VP4 and NS38 (Chen et al. 2018), and VP35 (Gao et al. 2018). Among them, VP5 encoding DNA vaccine showed the best protection (96.7–100% RPS). DNA vaccine encoding VP7 when encapsulated in single-walled carbon nanotubes (SWCNTs) reported 72–100% RPS (Zhu et al. 2015) but the naked pDNA only provides 42% protection (Hao et al. 2017). Limited protection (59.9–60% RPS) was observed for DNA vaccines encoding VP4 and VP35, whereas the DNA vaccine encoding non-structural protein (NS38) showed very low protection (23% RPS).

19.3.1.6 Aquabirnavirus

Infectious Pancreatic Necrosis Virus (IPNV), belonging to aquabirnavirus (dsRNA virus), is an important pathogen causing pancreatic diseases in salmonid fish. Oral delivery of alginate or chitosan encapsulated DNA vaccines, encoding VP2 gene of IPNV, have been reported to show higher protection (>80% RPS) with small size fish (1–2 mg) and higher vaccine dose (10–25µg/fish) by Ana et al. (2010); Ballesteros et al. (2014, 2015b) and Ahmadivand et al. (2017) in trout. In case of Atlantic salmon, varying level of protection was observed depending on the modality of vaccination, viz., 27.2% RPS by IM vaccination (Munang'andu et al. 2012) and 58.2–66.7% RPS by oral (via feed) (Reyes et al. 2017).

19.3.2 DNA Vaccine Against DNA Viruses (Table 19.2)

19.3.2.1 Herpesvirus

Cyprinid herpes virus and Ictalurid herpes virus are important fish DNA viruses, belonging to alloherpesviridae family, infecting cyprinid fish (common carp/koi carp) and channel catfish, respectively. A number of experimental DNA vaccines have been reported for Cyprinid herpes virus-3 (CyHV-3) or Koi herpes virus (KHV), using viral G protein gene (ORF 25) as the antigenic gene. High protection (77.7–96.7% RPS) was recorded in common carp or koi carp by Nuryati et al. (2010), Zhou et al. (2014b), Liu et al. (2018), and Embregts et al. (2019b), whereas moderate protection (63.37% RPS) was reported by Aonullah et al. (2017). Other ORFs were also tested for DNA vaccine construction, viz., ORF 81 by Zhou et al. (2014a) and ORF 149 by Hu et al. (2020) wherein they reported 85% and 81.9% RPS, respectively. In addition, one recent study involving DNA vaccine against Cyprinid herpes virus-2 (CyHV-2) in Prussian carp was also reported where 70% RPS was found (Yuan et al. 2020). For DNA vaccine against Channel catfish virus (CCV) or Ictalurid herpes virus-1 (IHV-1), 2 different ORFs (59 and 6) were tested, and both recorded complete protection (100% RPS) in channel catfish (Nusbaum et al. 2002).

19.3.2.2 Iridovirus

Besides herpes virus, iridovirus constitutes a wide variety of fish DNA viruses. Fish iridoviruses mainly belong to 2 genera, *Lymphocystivirus* and *Megalocytivirus*. Lymphocystis disease virus (LCDV) represents the genera *Lymphocystivirus*

Table 19.2 Status report of experimental DNA vaccine against DNA viruses

Virus	Fish (size in g)	Vaccination Vaccine	Antigenic gene	Modality	Dose (µg/fish)	Challenge Mode (dpv)	Dose	RPS	Reference
Herpesvirus: dsDNA virus									
Koi herpes virus (KHV) or Cyprinid herpes virus 3 (CyHV-3)	Common carp (10–15 g)	pAct-GP	Glycoprotein-25 of KHV	IM (Prime only)	12.5	IP (30 dpv)	KHV 1×10^6 PFU/fish	96.7%	Nuryati et al. (2010)
	Common carp (200–300 g)	pIRES-ORF25	Envelope protein (ORF 25) of CyHV-3	IM (Prime-2 Boost, 3 weeks interval)	1, 10, 50	IP (56 dpv)	KHV-CJ $10^{6.75}$ TCID$_{50}$/fish	77.7–86.1%	Zhou et al. (2014b)
	Common carp (246 g)	pIRES-ORF81	Structural protein (ORF 81) of KHV	IM (Prime-2 Boost, 2 weeks interval)	1, 10, 50	IP (42 dpv)	KHV-CJ $10^{6.75}$ TCID$_{50}$/fish	82.5–85%	Zhou et al. (2014a)
	Common carp (0.2 g)	pGP25 in _E. coli_	Glycoprotein-25 of KHV	Immersion (30 min, 10 mL of heat-killed _E. coli_-pGP25) (Prime only)	1.3×10^8 CFU/mL V^8: 800 fish/L V^{12}: 1200 fish/L	IM (30 dpv)	KHV filtrate 10^{-3}/fish	V^8: $63.37 \pm 14.37\%$ V^{12}: $9.89 \pm 4.76\%$	Aonullah et al. (2017)
	Common carp (200–300 g)	pIRES-ORF25+pIL-1β (carp IL-1β gene as immune adjuvant)	Envelope protein (ORF 25) of CyHV-3	IM (Prime-2 Boost, 2 weeks interval)	40	IP (42 dpv)	KHV-CJ $10^{6.78}$ TCID$_{50}$/fish	85.2%	Liu et al. (2018)
	Common carp (25–30 g)	pcDNA3-ORF25	Envelope protein (ORF 25) of CyHV-3	IM (Prime only)	0.5 (µg/g fish)	Immersion/Co-habitation (90 dpv)	Immersion: KHV 0.7×10^4 PFU/mL for 2 h Co-habitation: 1:5 ratio of shedders (IP injected with 5000PFU of KHV) to vaccinated carp	84–89%	Embregts et al. 2019b)

Virus	Fish	Construct	Antigen	Route	Dose	Challenge	Pathogen	Efficacy	Reference
	Koi carp (200–300 g)	pcDNA-ORF149 (Single-walled carbon nanotubes, SWCNTs)	Envelope protein (ORF 149) of CyHV-3	IM (Prime only)	1, 5, 10	IP (28 dpv)	KHV 2.04×10^7 TCID$_{50}$/fish	61.1–81.9%	Hu et al. (2020)
Cyprinid herpesvirus 2 (CyHV-2)	Prussian carp (20 g)	pEGFP-N1-ORF25 (Liposome encapsulation Lipofectamine™ 2000)	Envelope protein (ORF 25) of CyHV-2	IM (Prime only)	10	IP (21 dpv)	CyHV-2 10^4 copies/fish	70%	Yuan et al. (2020)
Channel catfish virus (CCV or IHV-1)	Channel catfish (4–8 cm)	pCMV-ORF 59/ pCMV-ORF 6	Envelope glycoprotein (ORF 59)/ Membrane protein (ORF 6) gene of CCV	IM (Prime only)	50	Immersion (42 dpv)	CCV 1×10^5 TCID$_{50}$/mL in 4 L water for 1 h	pORF59-100% pORF6-100%	Nusbaum et al. (2002)
Iridovirus: dsDNA virus									
Lymphocystis disease virus (LCDV)	Flounder (60–80 g)	pEGFP-N2-LCDV 0.6 kb	LCDV 0.6 kb ORF 0147 L gene	IM (Prime only)	5, 15	Co-habitation (21 dpv)	1:1 ratio LCDV infected fish to vaccinated fish	Tumor growth: 0%	Zheng et al. (2006)
	Flounder (50–100 g)	pEGFP-N2-LCDV 0.6 kb (PLGA NPs encapsulation)	LCDV 0.6 kb ORF 0147 L gene	Oral gavage (Prime only)	30	IM (28 dpv)	LCDV 4×10^3 TCID$_{50}$/fish	Nodule growth: Vaccinated:16.7% (vs. Control: 100%)	Tian and Yu (2011)
Infectious spleen and kidney necrosis virus (ISKNV)	Chinese perch (50 g)	pcMCP (QCDC adjuvant)	Major capsid protein gene of ISKNV	IM (Prime only)	20	IP (28 dpv)	ISKNV 10 TCID$_{50}$/fish	80%	Fu et al. (2014)
	Chinese perch (40–50 g)	pcDNA-093 pcDNA-MCP	Transmembrane protein, ORF093 (093)/Major capsid protein (MCP) gene of ISKNV	IM (Prime only)	25	IP (28 dpv)	ISKNV 10 TCID$_{50}$/fish	pcDNA-093: 50% pcDNA-MCP: 63%	Li et al. (2015)

(continued)

Table 19.2 (continued)

Virus	Fish (size in g)	Vaccination				Challenge		RPS	Reference
		Vaccine	Antigenic gene	Modality	Dose (µg/fish)	Mode (dpv)	Dose		
	Mandarin (50 g)	pcDNA086 (QCDC adjuvant)	Early protein helicase, ORF086 gene of ISKNV	IM (Prime only)	25	IP (21 dpv)	ISKNV 10 $TCID_{50}$/fish	63%	Fu et al. (2015)
Singapore grouper iridovirus (SGIV)	Grouper (8 g)	pcDNA-36 pcDNA-39 pcDNA-72	CaaX motif (ORF 36)/Protein kinase domain (ORF 39), Major capsid protein (MCP) gene of SGIV	IM (Prime only)	30	IP (21 dpv)	SGIV 7×10^4 $TCID_{50}$/fish	pcDNA-36: 58.3% pcDNA-39: 66.7% pcDNA-72: 66.7%	Ou-Yang et al. (2012)
	Orange-spotted grouper (116 g)	pcDNA3.1–19R	ORF19R gene of SGIV	IM (Prime only)	45, 90	IP (20 dpv)	SGIV 5×10^6 $TCID_{50}$/fish	45µg: 49.9% 90µg: 75%	Yu et al. (2019)
Largemouth bass virus (LMBV)	Largemouth bass (130 g)	pCDNA3.1(+)-MCP-Flag	Major capsid protein (MCP) gene of LMBV	IM (Prime only)	40	IM (30 dpv)	LMBV 4.8×10^7 $TCID_{50}$/fish	63%	Yi et al. (2020)
Turbot reddish body iridovirus (TRBIV)	Turbot (70 g)	pVAX1-TRBIV-MCP	Major capsid protein (MCP) gene of TRBIV	IM (Prime only)	5	IP (20 dpv)	TRBIV 7×10^4 $TCID_{50}$/fish	65.9%	Zheng et al. (2017)
	Turbot (18 g)	pVAX1-TRBIV-MCP (Chitosan NPs encapsulation)	Major capsid protein (MCP) gene of TRBIV	Oral gavage	10	IP (20 dpv)	TRBIV 7×10^4 $TCID_{50}$/fish	68%	Zheng et al. (2016)
Red seabream iridovirus (RSIV)	Red seabream (5–10 g)	pCMV-MCP pCMV-569	Major capsid protein (MCP)/ Transmembrane protein, ORF569 (569) gene of RSIV	IM (Prime only)	25	IP (30 dpv)	RSIV 1×10^3 $TCID_{50}$/fish	pMCP: 42.8–68.7% p569: 67.7–71.4% pMCP + p569: 67.7–71.4%	Caipang et al. (2006)

| Rock bream iridovirus (RBIV) | pCN86 | Turbot (10.2 g) | P86 gene of RBIV-C1 isolate from China | IM (Prime only) | 20 | IP (30 dpv) | RBIV-C1 2×10^5 copies/fish | 72% | Zhang et al. (2012) |
| | pMMP | Rock bream (27.2 g) | ORF008L Myristoylated membrane protein gene of RBIV | IM (Prime only) | 45, 90 | IP (28 and 56 dpv) | RBIV 6.7×10^5 $TCID_{50}$/fish | 28 dpv: 73.36% 56 dpv: 46.72% | Jung et al. (2018) |

which causes disease in flounder, whereas Megalocytiviruses are emerging pathogens that are widespread in a diverse range of environments and hosts. DNA vaccine, expressing LCDV 0.6 kb ORF 0147L, showed significant reduction of nodule/tumor growth in vaccinated flounder (Zheng et al. 2006; Tian and Yu 2011). Tian et al. (2008a, b, c) also reported high antibody titer, following oral vaccination with chitosan/alginate/PLGA encapsulated LCDV-pDNA. Several studies on DNA vaccines against different fish megalocytiviruses are reported but all of them showed limited protection, viz., pDNA against Infectious spleen and kidney necrosis virus (ISKNV) recorded 50–80% RPS in Chinese perch/mandarin (Fu et al. 2014, 2015; Li et al. 2015), pDNA against Singapore grouper iridovirus (SGIV) recorded 49.9–75% RPS in grouper (Ou-Yang et al. 2012; Yu et al. 2019), pDNA against Largemouth bass virus (LMBV) recorded 63% RPS in largemouth bass (Yi et al. 2020), pDNA against Turbot reddish body iridovirus (TRBIV) recorded 65.9–68% RPS in turbot (Zheng et al. 2016), pDNA against Red seabream iridovirus (RSIV) recorded 42.8–71.4% RPS in red seabream (Caipang et al. 2006), and pDNA against Rock bream iridovirus (RBIV) recorded 72% RPS in turbot (Zhang et al. 2012) and 73.36% RPS in rock bream (Jung et al. 2018). Most of these studies used major capsid protein (MCP) as antigenic gene but other viral ORFs were also tested.

19.3.3 DNA Vaccine Against Gram Negative Bacteria (Table 19.3)

19.3.3.1 Enterobacteriaceae

Edwardsiella tarda is the enterobacterium responsible for edwardsiellosis diseases in a wide number of economically important fish species. Several DNA vaccines against *E. tarda* have been reported, wherein different antigenic genes of *E. tarda* are tested, viz., Eta6 and FliC (Jiao et al. 2009), Eta6 and N163 (Jiao et al. 2010), Esa1 (Sun et al. 2011b), Eta2 (Sun et al. 2011a), GroEL (Liu et al. 2016a), OmpC (Liu et al. 2017), GAPDH (Kumari et al. 2018; Kole et al. 2018), and FlgD (Liu et al. 2016c). Most of the studies involve IM vaccination. However, oral and immersion routes were also tested by encapsulating the pDNA with chitosan or PLGA-chitosan nanoparticles (Kole et al. 2018; Leya et al. 2020). Irrespective of the antigenic genes used as DNA vaccines, moderate RPS (55–81%) were recorded in flounder and rohu, whereas higher protection of 87.3% RPS was reported in zebrafish by Liu et al. (2016c).

19.3.3.2 Aeromonadaceae

Motile *Aeromonas* species are important fish pathogens which causes bacterial hemorrhagic septicemia in species of cultured and wild freshwater fish. Among them, *Aeromonas hydrophila* is the predominant pathogen but other species like *A. veronii* and *A. sorbia* also exist. Two DNA vaccines against *A. hydrophila* are reported so far, viz., pDNA encoding aerA gene of *A. hydrophila* provides moderate to high protection (57.3–83.7% RPS) in grass carp (Liu et al. 2016b) and another one used yeast vector with OmpG antigenic gene for crucian carp, wherein only 46.7% RPS was recorded (Han et al. 2019). Apart from these, DNA vaccines against

Table 19.3 Status report of experimental DNA vaccine against Gram negative bacteria

Bacteria	Fish (size in g)	Vaccination Vaccine	Antigenic gene	Modality	Dose (µg/fish)	Challenge Mode (dpv)	Dose	RPS	Reference
Enterobacteriaceae									
Edwardsiella tarda	Flounder (10 g)	pCE6	*E. tarda* antigen, Eta6 and flagellin, FliC	IM (Prime only)	20	IP (28 dpv)	8×10^6 CFU/fish	72%	Jiao et al. (2009)
	Flounder (9.6 g)	pN163E6	*E. tarda* antigen, Eta6 and N-terminal 163 residues of flagellin, N163	IM (Prime only)	20	IP (28 dpv)	8×10^6 CFU/fish	70.2%	Jiao et al. (2010)
	Flounder (9.3 g)	pCEsa1	D15-like surface antigen, Esa1 of *E. tarda*	IM (Prime only)	25	IP (30 and 60 dpv)	1×10^6 CFU/fish	30 dpv: 75% 60 dpv: 71%	Sun et al. (2011b)
	Flounder (9.6 g)	pCEta2	*E. tarda* antigen, Eta2	IM (Prime only)	20	IP (28 and 56 dpv)	2×10^6 CFU/fish	28 dpv: 67% 56 dpv: 68%	Sun et al. (2011a)
	Flounder (30 g)	pCG-GroEL (rIL–2 of flounder as adjuvant)	*E.tarda* antigen, GroEL	IM (Prime only)	20	IP (35 dpv)	1×10^6 CFU/fish	60%	Liu et al. (2016a)
	Flounder (30 g)	pCG-OmpC	*E. tarda* antigen, OmpC	IM (Prime only)	20	IP (42 dpv)	1×10^7 CFU/fish	55%	Liu et al. (2017)
	Rohu (IMC) (20 g)	pGPD + IFN (Rohu IFN-γ gene as immune adjuvant)	GAPDH gene of *E. tarda*	IM (Prime-Boost, 21 days interval)	10	IP (35 dpv)	1×10^5 CFU/fish	63.16%	Kumari et al. (2018)
					10	IP (35 dpv)			

(continued)

Table 19.3 (continued)

Bacteria	Fish (size in g)	Vaccination				Challenge		RPS	Reference
		Vaccine	Antigenic gene	Modality	Dose (µg/fish)	Mode (dpv)	Dose		
	Rohu (IMC) (25 g)	pGPD + IFN (Rohu IFN-γ gene as immune adjuvant) (Chitosan NPs conjugation for oral and immersion)	GAPDH gene of E. tarda	Oral (via feed for 14 days, Prime-Boost, Immersion (2 h); IM (Prime-Boost, 14 days interval)			8.7×10^4 CFU/ fish	Oral: 81.82% Immersion: 72.73% Injection: 63.62%	Kole et al. (2018)
	Rohu (IMC) (15 g)	pGPD + IFN (Rohu IFN-γ gene as immune adjuvant) (PLGA-Chitosan NPs / PLGA NPs encapsulation)	GAPDH gene of E. tarda	Immersion (20 s hyperosmotic, 4.5% NaCl dip followed by vaccine dip for 1 h) (Prime only)	1 mg/L	IP (30 dpv)	1×10^5 CFU/ fish	PLGA-Chitosan-NPs-p (GPD + IFN): 64.7% PLGA-NPs-p (GPD + IFN): 52.94%	Leya et al. (2020)
	Zebra fish (0.2–0.3 g)	pCL-flgD-I-C5a-W (Zebra fish C5a gene as immune adjuvant)	Antigenic flagellar protein, FlgD gene of E. tarda	IM (Prime only)	0.1	IM (30 dpv)	2×10^4 CFU/ fish	87.3%	Liu et al. (2016c)
Aeromonadaceae									
Aeromonas hydrophila	Grass carp (6 g)	pEGFP-aerA (Single-walled carbon nanotubes, SWCNTs)	Antigenic protein, aerA gene of A. hydrophila	IM (Prime only)	1, 5, 10	IP (21 dpv)	1.5×10^5 CFU/ fish	57.3–83.7%	Liu et al. (2016b)
						IP (30 dpv)		46.7%	

Pathogen	Fish (weight)	Construct	Antigen/gene	Route	Dose	HK yeast	Immunization	CFU	RPS	Reference
	Crucian carp (6 g)	OVA-ompG (In yeast vector JMB84) (Heat-killed yeast cells)	Antigenic protein, ompG gene of A. hydrophila	Oral (via feed for 4 weeks at 3% body weight of fish)		1.5×10^8 HK yeast cells/g meal powder		1.5×10^5 CFU/fish		Han et al. (2019)
Aeromonas veronii	Spotted sand bass (10 g)	pOMP38P pOMP48P	Omp38/Omp48 gene of A. veronii	IM (Prime only)	20		IP (28 dpv)	5×10^6 CFU/fish	pOMP38P: 53.8% pOMP48P: 60.2% pOMP38P + pOMP48P: 61.6%	Vazquez-Juarez et al. (2005)
Vibrionaceae										
Vibrio anguillarum	Asian sea bass (10 g)	pVAOMP38	Omp38 porin protein (38-kDa) gene of V. anguillarum	IM (Prime only)	20		IP (35 dpv)	1.2×10^6 CFU/fish	55.6%	Kumar et al. (2007)
	Asian sea bass (10 g)	pVAOMP38 (Chitosan NPs encapsulation)	Omp38 porin protein (38-kDa) gene of V. anguillarum	Oral (via feed) (Prime only)	50		IP (21 dpv)	1.2×10^6 CFU/fish	46%	Kumar et al. (2008)
	Flounder (9.38 g)	pEGFP-N1/m-empA7	Extracellular zinc metalloprotease gene of V. anguillarum	IM (Prime only)	5, 20, 20		IM (28 dpv)	3.5×10^6 CFU/fish	5µg—57.5% 20µg—71.4% 50µg—85.7%	Yang et al. (2009)
	Flounder (11.3 g)	pSiVa1 (Sia10 of S. iniae as co-antigen)	OmpU gene of V. anguillarum	IM (Prime only)	15		IP (30 and 60 dpv)	1×10^5 CFU/fish	30 dpv: 81% 60 dpv: 80%	Sun et al. (2012)
	Flounder (35 g)	pOmpK	OmpK gene of V. anguillarum	IM (Prime only)	20		IP (42 dpv)	1×10^6 CFU/fish	50%	Xu et al. (2019c)
	Flounder (35 g)	pVAA	VAA gene of V. anguillarum	IM (Prime only)	20		IP (42 dpv)	1×10^6 CFU/fish	50%	Xing et al. (2019)
	Flounder (35 g)	pVAA-IRES-IL2 (Flounder IL-2 gene as immune adjuvant)	VAA gene of V. anguillarum	IM (Prime only)	20		IP (49 dpv)	1×10^6 CFU/fish	64.1%	Xu et al. (2019a)

(continued)

Table 19.3 (continued)

Bacteria	Fish (size in g)	Vaccination Vaccine	Antigenic gene	Modality	Dose (µg/fish)	Challenge Mode (dpv)	Dose	RPS	Reference
	Flounder (35 g)	pHsp33	Hsp33 gene of *V. anguillarum*	IM (Prime only)	20	IP (42 dpv)	1×10^6 CFU/fish	42.86%	Xu et al. (2019b)
	Flounder (35 g)	pCCL3-VAA pCCL4-VAA pCCL19-VAA pCCL21-VAA (Flounder CXCL3, CXCL4, CXCL19, CXCL21 genes as immune adjuvant)	VAA gene of *V. anguillarum*	IM (Prime only)	20	IP (49 dpv)	1×10^6 CFU/fish	pCCL3-VAA: 62.16% pCCL4-VAA: 83.78% pCCL19-VAA: 78.38% pCCL21-VAA: 72.97%	Xu et al. (2020b)
Vibrio harveyi	Grouper (150–200 g)	pcFlaA	Antigenic FlaA gene of *V. harveyi*	IM (Prime only)	50	IP (28 dpv)	1×10^7 CFU/fish	64.7%	Qin et al. (2009)
	Flounder (10.3 g)	pDegQ pVhp1 pDV (DegQ +Vhp1)	Antigenic DegQ and Vhp1 genes of *V. harveyi*	IM (Prime only)	15	IP (30 dpv)	1×10^7 CFU/fish	pDegQ: 64.1% pVhp1: 56.5% pDV: 84.6%	Hu and Sun (2011)
	Orange-spotted grouper (10 g)	pcDNA-GPx	Glutathione peroxidase gene of *V. harveyi*	IM (Prime only)	20	IM (35 dpv)	1×10^7 CFU/fish	77.5%	Wang et al. (2017)
	Hump head snapper (40–50 g)	pcDNA-Ls-IL6-OmpW (Snapper IL-6 gene as immune adjuvant)	Antigenic OmpW gene of *V. harveyi*	IM (Prime only)	20	IP (35 dpv)	1×10^7 CFU/fish	76%	Huang et al. (2019)

Pathogen	Fish (weight)	Construct	Gene	Delivery	Dose (µg)	Challenge route (time)	Challenge dose	RPS	Reference
Vibrio alginolyticus	Golden pompano (18.5 g)	pCTssJ	TssJ from the T6SS gene of *V. harveyi*	IM (Prime only)	20	IP (30 dpv)	1×10^5 CFU/fish	69.11%	Sun et al. (2020)
	Red snapper (20 g)	pcDNA-flaA	Flagellin flaA gene of *V. alginolyticus*	IM (Prime only)	8	IP (28 dpv)	1×10^7 CFU/fish	88%	Liang et al. (2011)
	Crimson snapper (50 g)	pcDNA-OmpW	Antigenic OmpW gene of *V. alginolyticus*	IM (Prime only)	10	IM (49 dpv)	1×10^6 CFU/fish	92.53%	Cai et al. (2013)
	Orange-spotted grouper (30 g)	pcacfA + pcMyD88 (Grouper MyD88 gene as immune adjuvant)	Antigenic AcfA gene of *V. alginolyticus*	IM (Prime only)	25	IP (70 dpv)	2×10^6 CFU/fish	83.3%	Huang et al. (2017)
Vibrio mimicus	Grass carp (50 g)	pcDNA3.1-IcIp-OVepis (in DH5α ghost bacteria) (Invariant chain-like protein (IcIp) gene of grass carp as immune adjuvant)	Antigenic OmpU and VMH gene of *V. mimicus*	Oral gavage (Prime-Boost, 14 days interval)	10µg (DH5α-BGs-pcDNA3.1-IcIp-OVepis)	IP (30 dpv)	2×10^7 CFU/fish	81.11%	Cao et al. (2019)
Vibrio parahemolyticus	Turbot (16.2 g)	pEGFP-NI/m-vps	Mutated serine protease (Ser[318]-Pro) gene of *V. parahemolyticus*	IM (Prime only)	10, 50	IP (28 dpv)	7.5×10^8 CFU/fish	10µg—80.55% 50µg—96.11%	Liu et al. (2011)
	Black sea bream (80 g)	pEGFP-N2-OMPK (Chitosan NPs encapsulation)	Antigenic OmpK gene of *V. parahemolyticus*	Oral (via feed) (Prime only)	50	IP (21 dpv)	6.4×10^8 CFU/fish	72.3%	Li et al. (2013)
Photobacterium damselae subsp. piscicida	Flounder (10 g)	pPPA1	7 kDa lipoprotein gene of *P. damselae*	IM (Prime only)	10	Immersion (30 dpv)	1×10^5 CFU/ mL for 1 h	69.2–90.9%	Kato et al. (2015)

A. veronii, expressing Omp38 and Omp48 genes, were also reported for spotted sand bass providing 53–62% protection (Vazquez-Juarez et al. 2005).

19.3.3.3 Vibrionaceae

Vibrio species are member of the Vibrionaceae which causes classical vibriosis in salmonid and non-salmonid fish worldwide. A number of experimental DNA vaccines against different *Vibrio* species have been reported till date, viz., 9 against *V. anguillarum,* 5 against *V. harveyi,* 3 against *V. alginolyticus,* 2 against *V. parahemolyticus,* and 1 against *V. mimicus.* For DNA vaccine against *V. anguillarum,* various antigenic genes were tested, viz., Omp38 encoding porin protein (Kumar et al. 2007, 2008), extracellular zinc metalloprotease gene (Yang et al. 2009); OmpU (Sun et al. 2012), OmpK (Xu et al. 2019c), VAA gene (Xing et al. 2019; Xu et al. 2019a, 2020b), Hsp33 (Xu et al. 2019b). Most of the study reported limited protection (42–74% RPS). However, more than 80% RPS were also reported in flounder at varying condition, viz., at high pDNA dose (50μg/fish) (Yang et al. 2009), employing co-antigen (Sia10) of *Streptococcus iniae* (Sun et al. 2012), adding immune adjuvant (flounder CXCL4) (Xu et al. 2020b). Antigenic genes like FlaA, DegQ, Vhp1, GPx, OmpW, and TssJ were used for construction of DNA vaccines against *V. harveyi* yielding 64–84% RPS in different fish species (Qin et al. 2009; Hu and Sun 2011; Wang et al. 2017; Huang et al. 2019; Sun et al. 2020), whereas for DNA vaccine against *V. alginolyticus* encoding antigenic genes flaA, OmpW and AcfA conferred high protection, 88% in red snapper (Liang et al. 2011), 92.53% in crimson snapper (Cai et al. 2013) and 83.3% in orange-spotted grouper (Huang et al. 2017), respectively. Likewise, mutated serine protease (Ser^{318}-Pro) and OmpK antigenic genes act as effective DNA vaccine against *V. parahemolyticus* in turbot (Liu et al. 2011) and black sea bream (Li et al. 2013), respectively. In case of *V. mimicus,* a DNA vaccine encoding OmpU and VMH genes were tested in grass carp (Cao et al. 2019). Besides, DNA vaccine against *Photobacterium damselae subsp. piscicida* of Vibrionaceae was also reported for flounder encoding antigenic lipoprotein gene, wherein upto 90.9% RPS was recorded (Kato et al. 2015).

19.3.4 DNA Vaccine Against Gram Positive Bacteria (Table 19.4)

19.3.4.1 Streptococcaceae

Among Gram positive bacterial diseases, streptococcal infection of fish are the most important affecting a variety of wild and cultured fish. At least 6 defined species are considered of significance as fish pathogens: *Lactococcus garvieae, Lactococcus piscium, Streptococcus iniae, Streptococcus agalactiae, Streptococcus parauberis,* and *Vagococcus salmoninarum.* However, reports of experimental DNA vaccine have been recorded only against *S. iniae* and *S. agalactiae.* The DNA vaccine against *S. iniae* tested as many as 3 antigenic genes Sia10, SagH, and α-enolase. The Sia10 encoding vaccines yielded high protection of 92.3% in turbot (Sun et al. 2010) and 80–92% in flounder (Sun et al. 2012; Hu et al. 2012). The SagH encoding pDNA also recorded high protection (80–92%) in flounder (Liu et al. 2019), whereas

Table 19.4 Status report of experimental DNA vaccine against Gram positive bacteria

Bacteria	Fish (size in g)	Vaccination				Challenge		RPS	Reference
		Vaccine	Antigenic gene	Modality	Dose (µg/fish)	Mode (dpv)	Dose		
Streptococcaceae									
Streptococcus iniae	Turbot (9 g)	pSia10	Antigenic Sia10 gene of *S. iniae*	IM (Prime only)	25	IP (30 dpv)	High dose: 4×10^7 CFU/ fish Low dose: 5×10^6 CFU/ fish	High dose: 73.9% Low dose: 92.3%	Sun et al. (2010)
	Flounder (11.3 g)	pSiVa1 (OmpU of *V. anguillarum* as co-antigen)	Antigenic Sia10 gene of *S. iniae*	IM (Prime only)	15	IP (30 and 60 dpv)	1×10^6 CFU/ fish	30 dpv: 81% 60 dpv: 80%	Sun et al. (2012)
	Flounder (10.2 g)	pSia10Hsp70 (Flounder Hsp70 gene as immune adjuvant)	Antigenic Sia10 gene of *S. iniae*	IM (Prime only)	15	IP (30 dpv)	1×10^7 CFU/ fish	92%	Sun et al. (2012)
	Flounder (13.2 g)	pSagH	Antigenic SagH gene of *S. iniae*	IM (Prime only)	15	IP (30 and 60 dpv)	1×10^6 CFU/ fish	30 dpv: 83.01– 92.62% 60 dpv: 80.65– 90.58%	Liu et al. (2019)
	Channel catfish (50 g)	pcENO + pcIL-8 (Channel catfish IL-8 gene as immune adjuvant)	Antigenic α-enolase gene of *S. iniae*	IM (Prime only)	25	IP (30 and 60 dpv)	1.2×10^7 CFU/ fish	30 dpv: 80% 60 dpv: 73.33%	Wang et al. (2016)

(continued)

Table 19.4 (continued)

Bacteria	Fish (size in g)	Vaccination				Challenge		RPS	Reference
		Vaccine	Antigenic gene	Modality	Dose (μg/fish)	Mode (dpv)	Dose		
Streptococcus agalactiae	Nile tilapia (25 g)	pEno	Antigenic α-enolase gene of *S. iniae*	IM (Prime only)	15	IP (30 dpv)	2.7×10^7 CFU/ fish	30 dpv: 63%	Kayansamruaj et al. (2017)
	Nile tilapia (100 g)	pVAX1-sip (Electroporated in attenuated *Salmonella typhimurium* SL7207 aroA mutant)	Antigenic Sip gene of *S. agalactiae*	Oral Gavage (Prime-2 Boost, 1 week interval)	$10^7, 10^8, 10^9$ CFU/fish	IP (30 dpv)	1×10^8 CFU/ fish	10^7: 47% 10^8: 53% 10^9: 57%	Huang et al. (2014)
	Nile tilapia (50 g)	pcSip (cationic-PLGA encapsulation)	Antigenic Sip gene of *S. agalactiae*	Oral Gavage (Prime only)	20, 50, 100	IP (28 dpv)	4×10^7 CFU/ fish	41.67– 95.83%	Ma et al. (2017)
	Nile tilapia (100 g)	SL7207-pVAX1-sip (attenuated *Salmonella typhimurium* SL7207 aroA mutant)	Antigenic Sip gene of *S. agalactiae*	Oral: Gavage and Feed pellet (Prime-2 Boost, 1 week interval)	2×10^9 CFU/ fish	IP (30 dpv)	1×10^5 CFU/ fish	Gavage: 57% Feed pellet: 63%	Zhu et al. (2017)
	Nile tilapia (60 g)	p45F2 p42E2	Antigenic 45F2 and 42E2 genes of *S. agalactiae*	IM (Prime only)	10	IP (28 dpv)	1×10^7 CFU/ fish	p45F2: 76.19% p42E2: 76.19%	Pumchan et al. (2020)

Mycobacteriaceae

Mycobacterium marinum	Striped bass (40–50 g)	pCMV-85A	Antigenic Ag85A gene of *M. marinum*	IM (Prime only)	25, 50	IP (90 dpv)	8×10^5 CFU/fish	25 µg: 80% 50 µg: 80%	Pasnik and Smith (2005)
Nocardiaceae									
Nocardia seriolae	Amberjack (56–64 g)	pAg85L	Antigenic Ag85L gene of *N. seriolae*	IM (Prime only)	10	IP (28 dpv)	3×10^5 CFU/fish	73.1–94.9%	Kato et al. (2014)
	Hybrid snakehead (30 g)	pcDNA-FHA	Antigenic Forkhead-associated domain protein (FHA) gene of *N. seriolae*	IM (Prime only)	25	IP (35 dpv)	3.31×10^5 CFU/fish	62.64%	Chen et al. (2019a)
	Hybrid snakehead (30 g)	pcDNA-DnaK pcDNA-GroEL	Antigenic DnaK and GroEL genes of *N. seriolae*	IM (Prime only)	25	IP (35 dpv)	3.31×10^5 CFU/fish	pcDnaK: 74% pcGroEL: 89.33%	Chen et al. (2019b)
	Hybrid snakehead (30 g)	pcDNA-RpsA pcDNA-RplL	Antigenic 30S ribosomal protein S1 (RpsA) and 50S ribosomal protein L7/L12 (RplL) genes of *N. seriolae*	IM (Prime only)	25	IP (35 dpv)	3.16×10^6 CFU/fish	pcRpsA: 71.08% pcRpl1L: 78.31%	Chen et al. (2020)

the α-enolase encoding pDNA conferred moderate protection, 73–80% in channel catfish (Wang et al. 2016), and 63% in Nile tilapia (Kayansamruaj et al. 2017).

Regarding DNA vaccine against *S. agalactiae*, 3 studies using antigenic Sip gene and one using antigenic 45F2 and 42E2 genes have been reported for vaccinating Nile tilapia. The Sip encoding pDNA were delivered by oral route using ghost bacteria (Huang et al. 2014; Zhu et al. 2017) or PLGA encapsulation (Ma et al. 2017), wherein PLGA encapsulation resulted in protection up to 96% but ghost bacteria delivery yielded limited protection (47–63%). Whereas, IM vaccination of pDNA encoding 45F2 and 42E2 genes recorded 76.19% RPS in both the cases (Pumchan et al. 2020).

19.3.4.2 Mycobacteriaceae

Mycobacterium marinum is considered the primary causative agent of fish mycobacteriosis or fish tuberculosis, a subacute to chronic wasting disease, affecting a wide variety of fish species in both freshwater and saltwater. An experimental DNA vaccine encoding antigenic Ag85A gene of *M. marinum* was reported by Pasnik and Smith (2005), where 80% protection was observed in striped bass.

19.3.4.3 Nocardiaceae

Nocardia seriolae is an important pathogen affecting several fish species like yellowtail, amberjack, hybrid snakehead, pompano, four-finger threadfin, big-eye trevally, snapper, and grouper. Four studies on DNA vaccine against *N. seriolae* have been reported till date, 1 for amberjack and 3 for hybrid snakehead. Antigenic Ag85L gene was used for vaccination in amberjack, resulting 80% protection (Kato et al. 2014), whereas for snakehead 5 different antigenic genes, viz., Forkhead-associated domain protein (FHA), DnaK, GroEL 30S ribosomal protein S1 (RpsA), and 50S ribosomal protein L7/L12 (RplL) genes of *N. seriolae* were tested, among which GroEL recorded the highest RPS of 89.33% (Chen et al. 2019a, b).

19.3.4.4 DNA Vaccine Against Parasites (Table 19.5)

Unlike virus and bacteria, there are very limited studies on DNA vaccine against fish parasites. Two studies on DNA vaccine against *Ichthyophthirius multifiliis* in channel catfish are reported by Xu et al. (2019d, 2020a) where Ich immobilization antigen gene was used as pDNA with or without QCDC adjuvant. The RPS was low in both the cases (20% without adjuvant and 48.9% with adjuvant). Similarly, an experimental DNA vaccine, containing codon changed immobilization antigen (iAg), was reported against *Cryptocaryon irritans* for grouper which yielded 46% protection (Priya et al. 2012). Also, a DNA vaccine (encoding antigenic metalloprotease gene) against *Cryptobia salmositica* for rainbow trout was reported to lower the parasitemia indicators in vaccinated fish which resulted in their faster recovery (Tan et al. 2008).

Table 19.5 Status report of experimental DNA vaccine against parasites

Parasite	Fish (size in g)	Vaccination				Challenge		RPS	Reference
		Vaccine	Antigenic gene	Modality	Dose (µg/fish)	Mode (dpv)	Dose		
Ichthyophthirius multifiliis	Channel catfish (13 g)	pcDNA3.1-IAg52b	Ich immobilization antigen gene	IM (Prime only)	10	Immersion (21 dpv)	20,000 theronts/fish in 10 L for 1 h	20%	Xu et al. (2019d)
	Channel catfish (13 g)	pcDNA3.1-IAg52b (QCDC adjuvant)	Ich immobilization antigen gene	IM (Prime only)	10, 20	Immersion (21 dpv)	5000, 10,000, 15,000 theronts/ fish in 10 L for 1 h	15–48.9%	Xu et al. (2020a)
Cryptocaryon irritans	Grouper (40–50 g)	pcDNA3.1-optiAg	Codon changed immobilization antigen (iAg) of *C. irritans*	IM (Prime-Boost, 2 days interval)	1 (µg/ g fish)	Immersion (7 dpv)	1.8×10^4 theronts/L in 5 L for 10–20 min	46%	Priya et al. (2012)
Cryptobia salmositica	Rainbow trout (51.5 g)	pEGFP-N1-MP	Metalloprotease gene of *C. salmositica*	IM (Prime only)	50	IP (49 dpv)	20,000 parasite/fish	Lower parasitemia at peak: pMP: 3.56×10^6 parasites/mL blood; Control: 5.6×10^6 parasites/mL blood Delayed peak: pMP: 4 wpc; Control: 3 wpc Faster recovery: pMP: 8.27×10^5 parasites/mL blood; Control: 1.82×10^6 parasites/mL blood (at 8 wpc)	Tan et al. (2008)

19.4   Delivery Route of DNA Vaccine in Fish

Contrary to the intra-peritoneal (IP) injection route, which is used for conventional whole inactivated vaccines, intramuscular (IM) injection is the most common and efficient delivery method of DNA vaccines in fish. IM injection of purified plasmid has often proved as an effective route to elicit protective immune response in fish. The enhanced efficacy of the IM route for DNA vaccines depends on efficient uptake of the plasmids and the associated immune responses in the muscle cells. The myofibrils act as the most efficient cell type for plasmid uptake and protein expression (Lu et al. 2003). In addition, the membrane invaginations and T tubules present on the muscle cells (Wolff et al. 1992) as well as the increased hydrostatic pressure in the muscle tissue at the time and site of injection (Ahlén et al. 2016) also facilitate efficient cellular intake of the plasmid. In terms of immune response, several authors have suggested that the injection site damage to the muscle tissue/cells generates a pro-inflammatory response, which in turn promotes the migration of leukocytes, including APCs to the site, resulting into antigen processing, presentation and subsequent orchestration of cell-mediated and humoral immune responses (Lorenzen et al. 2005; Embregts et al. 2017; Collins et al. 2019). Although the efficiency of IM injection method is well perceived, its labor-intensiveness and non-feasibility for mass vaccination have limited its applicability.

To address these issues, oral or immersion delivery routes for DNA vaccination in fish are now being investigated with varying results (Kumar et al. 2008; Adomako et al. 2012; Li et al. 2013; Ballesteros et al. 2015a; Vimal et al. 2014b; Ahmadivand et al. 2017; Reyes et al. 2017; Zhu et al. 2017; Ma et al. 2017; Kole et al. 2018; Cao et al. 2019; Leya et al. 2020). For facilitation of the mucosal delivery of DNA vaccine in fish, incorporation of plasmid DNA in different nano- or microparticles, viz., chitosan, alginate, PLGA, liposome, carbon nanotubes, are being tested, showing some good promising results. Unlike the IM injection of naked pDNA method, where myocytes play a key role in plasmid uptake and antigen presentation, for mucosal delivery of nanoparticles encapsulated pDNA, the particle size influences the delivery of plasmid to a particular type of immune cells ensuring optimal antigen presentation (Fifis et al. 2004). For example, nanoparticles are preferentially phagocytosed by dendritic cells which results in antigen presentation via MHC class I and activation of cytotoxic T-cell response, whereas microparticles are more likely phagocytosed by macrophages which in turn results in antigen presentation via MHC class II, followed by a humoral response (Slütter and Jiskoot 2016).

19.5   Correlates of Protection of DNA Vaccination

Correlates of protection (CoPs) are important immune markers which can be statistically correlated with vaccine efficacy. CoPs are of two types, viz., mechanistic CoPs (mCoP) which are directly related to protection like specific antibodies and non-mechanistic CoPs (nCoP) which are indirect immune signatures for disease prevention like innate immune genes (Plotkin and Gilbert 2012). For human or veterinary vaccines, CoPs are well defined as the threshold units of specific antibodies for protection. However, for fish vaccines a systematic analysis method of CoPs is lacking even though it is widely perceived that both induction of innate immunity and antibody response are vital in disease prevention in fish (Anderson et al. 1996; Lorenzen et al. 1998; McLauchlan et al. 2003; Long et al. 2017). Moreover, it has also been shown that the CoPs for fish vaccines are highly dependent on the mode of vaccination (immunogen, dose, formulation, and prime-boost regime), tissue-specific or compartmentalization of immune response to infection and vaccination, pathogen types, mode of challenge, and also the host fish (Plotkin 2013; Munang'andu et al. 2013; Salinas 2015; Magadan et al. 2015; Dubey et al. 2016). In line with the other types of fish vaccine, DNA vaccine also needs established surrogate markers that can be used to reliably predict the vaccine efficacy. As DNA vaccination induces early innate immune responses as well as late systemic and memory responses in the host, both these immune signatures can act as the CoPs for DNA vaccines. Although, there have not been any efforts to statistically correlate these immune signatures with the protection, but numerous studies on specific antibodies and gene expression quantification, following DNA vaccination in fish, are reported.

Based on these studies, several authors showed relatively higher neutralizing (for viral vaccine) or bactericidal (for bacterial vaccine) antibody titers in DNA vaccinated fish. However, they do not establish the protection threshold units of the antibody titer which can confer protection against respective pathogen challenge (LaPatra et al. 2000; Vazquez-Juarez et al. 2005; Kumar et al. 2007; Yang et al. 2009; Jiao et al. 2010; Liu et al. 2011, 2019; Hu et al. 2012; Zhou et al. 2014a, b; Vimal et al. 2014a, b; Ballesteros et al. 2014; Chang et al. 2015; Zheng et al. 2016; Wang et al. 2016; Robertsen et al. 2016; Lazarte et al. 2017; Ahmadivand et al. 2017; Huang et al. 2017; Kayansamruaj et al. 2017; Wang et al. 2017; Zhang et al. 2017; Lim et al. 2019; Xu et al. 2019a, b, c, 2020b; Chen et al. 2019a, b, 2020; Cao et al. 2019; Yi et al. 2020). However, various other studies on DNA vaccination in fish showed low to moderate antibody titer but relatively better protection and vice versa. For example, Corbeil et al. (2000a) and LaPatra et al. (2000) reported 100% protection against IHNV challenge in DNA vaccinated rainbow trout, but the respective neutralizing antibody titer (Nabs) value was significantly different, viz., <20 (Corbeil et al. 2000a) and 320–1280 (LaPatra et al. 2000).

Likewise, several studies also showed upregulation of different immune genes, following DNA vaccination in fish. Among them, the genes which were consistently upregulated by DNA vaccination were type I IFNs, Mx, and interferon-stimulated genes (ISGs) in case of viral vaccine (Yasuike et al. 2007; Ana et al. 2010; Ballesteros et al. 2014, 2015a, b; Fu et al. 2014; Wang et al. 2015; Valero et al. 2016; Robertsen et al. 2016; Zheng et al. 2016; Lazarte et al. 2017; Embregts et al. 2017; Xu et al. 2017a, b; Zhang et al. 2018, 2019a, b; Ahmadivand et al. 2018; Chen et al. 2018; Embregts et al. 2019a, b; Hu et al. 2020) and cytokines like IL-1β, TNF-α for bacterial vaccine (Liu et al. 2011, 2016a, b, c, 2017, 2019; Sun et al. 2010, 2011a, b, 2012; Jiao et al. 2010; Huang et al. 2014, 2017; Zhu et al. 2017; Ma et al.

2017; Kayansamruaj et al. 2017; Chen et al. 2019a, b, 2020). However, in most studies, careful assessments have not been performed to statistically correlate the level of signature molecule(s) with protection. Other immune genes such as MHC class I, MHC class II, CD4, CD8, TLRs, and IgM also showed upregulation in many instances but in inconsistent manner to act as CoPs. In addition, the downregulated genes, representing central "checkpoints" or "controlling units" during an inflammatory response, are seldomly reported as immune markers in validating the CoPs of DNA vaccine. Thus, it can be inferred that comprehensive studies need to be performed to find out the exact correlates or surrogates of protection of fish DNA vaccine.

19.6 Safety Aspects of Fish DNA Vaccines

Besides the potency of DNA vaccines in inducing various innate and adaptive immune responses and providing protection to fish, several other advantages of DNA vaccine are also widely acknowledged, viz., relatively inexpensive to produce, easy and identical method of production processes for all DNA vaccines, and the ease of modifications (Hansen et al. 1991). However, some potential risks are also associated with DNA vaccination, e.g., risk of autoimmunity, immune tolerance against the expressed antigen, too high CTL response resulting in myositis, chromosomal integration and disruption of biological processes, injection site inflammation, and tissue destruction (Evensen and Leong 2013; Myhr and Dalmo 2005; Myhr 2017). Among these, the issue related to the plasmid persistence and chromosomal integration is the major constraint in the success of DNA vaccines in terms of safety and policy (Lorenzen and LaPatra 2005; Gillund et al. 2008). Briefly, the DNA vaccinated fish when consumed, residual plasmid containing elements such as human viral promoter regions (such as the CMV promoter) or antibiotic resistance genes, may result into harmful consequences if integrating into the consumers' genome or taken up by their gut microflora (Hølvold et al. 2014; Collins et al. 2019). However, as the plasmid degrades rapidly after vaccination (Tonheim et al. 2008) and considerably long time interval is there between the vaccination and harvest in a normal fish culture production cycle, the chance of consuming residual plasmid by consumer through fish fillet is extremely low. Likewise, the digestion process reduces the chances of antibiotic resistance genes reaching the gut flora to negligible level (Schalk et al. 2006). Apart from consumer, there is also a risk of release of plasmid into the environment in two ways, viz., plasmid leakage (on a local scale) during the intramuscular injection process (Rambabu et al. 2005) and plasmid shedding from the vaccinated fish (Reyes et al. 2017). The main environmental risk is related to the recombination of the antibiotic resistance genes with the environmental bacteria flora and their subsequent propagation (Berglund 2015). But rapid degradation of the released plasmid in the crude environment and the ability of the engineered plasmids to propagate in very limited bacterial hosts reduce the risk of uptake of the plasmid by environmental bacteria (Glenting and Wessels 2005; Schalk et al. 2006). Although, the safety issues relating to consumer and environmental risks are considered to be negligible (Gillund et al. 2008; Myhr 2017), studies are being conducted to test alternative *E. coli* strains which would allow the removal of antibiotic resistant genes from DNA vaccines (Cranenburgh et al. 2001; Peubez et al. 2010). In addition, there has been attempt to replace the CMV promoter by fish promoter (Sepúlveda et al. 2019).

19.7 Regulatory Aspects of Fish DNA Vaccines

With respect to the abovementioned safety issues, different countries have adopted different regulatory mechanisms for DNA vaccines for fish. In Asian and African countries, application of DNA vaccines in aquaculture is strictly restricted till date. In the USA and Canada, their application is permitted with no requirements for labelling of DNA vaccinated fish as GMOs, whereas in Europe due to the uncertainties with regard to the persistence of a DNA vaccine, the vaccinated fish are required to be labelled as a GMO.

In addition, for licensing of DNA vaccines for aquaculture, various aspects need to be considered: (1) public perception and acceptance, (2) regulatory and environmental concerns, (3) risk-benefit, (4) feasibility of producing the vaccine at a scale and cost appropriate for the fish industry, and (5) intellectual property issues (Nielsen et al. 2011).

Authorities like The Canadian Food Inspection Agency (CFIA), US Food and Drug Administration (FDA), Norwegian Medicines Agency (NoMA), and European Medicines Agency (EMA) are assigned the role to frame guidelines for fish DNA vaccines as well as to conduct risk assessment of DNA vaccines in Canada, USA, Norway, and European Union (EU), respectively (EMEA 2012; FDA 2007).

19.8 Concluding Remarks and Future Perspectives

DNA vaccination is an appropriate and simple concept, that are being developed for fish vaccination for the last two decades. They are most often been evaluated in relation to fighting against persistent and hard-to-combat viral infections in fish, but many promising results have also been obtained against bacterial infection. To increase the efficacy of DNA vaccines, various strategies are continuously being employed, viz., improvements in vaccine carriers to increase the uptake in APCs and enhanced presentation of transgene peptides/antigens, application of nano/microparticles to increase the level of cross presentation by APCs which may result in enhanced antibody response as well as cell-mediated immunity, employment of additional molecular adjuvants such as cytokines, IFN molecules to boost the immune response. However, despite the substantial efforts made since the late 1990s, only two DNA vaccines for aquaculture have reached to the commercial stage. Nevertheless, the recent approval of the first DNA vaccine (CLYNAV) for the aquaculture sector in Europe is believed to test the public acceptance of such a vaccine. Regarding future perspectives, the safety and regulatory uncertainties related to the consumer and environmental risks of plasmid persistence and chromosomal integration need to be studied and addressed in a much wider way to wane out the general mistrust among the scientists and public. In addition, detailed studies on the mechanisms of pDNA uptake, from the moment of administration until the stage of transcription and translation in the nucleus, along with the associated immune responses and their potential immune markers also need to be investigated in order to improve the vaccine potency.

References

Adomako M, St-Hilaire S, Zheng Y, Eley J, Marcum RD, Sealey W, Donahower BC, LaPatra S, Sheridan PP (2012) Oral DNA vaccination of rainbow trout, Oncorhynchus mykiss (Walbaum), against infectious haematopoietic necrosis virus using PLGA [Poly (D, L-Lactic-Co-Glycolic Acid)] nanoparticles. J Fish Dis 35(3):203–214

Ahlén G, Frelin L, Höolmstrm F, Smetham G, Augustyn S, Sällberg M (2016) A targeted controlled force injection of genetic material in vivo. Mol Ther Meth Clin Dev 3:16016

Ahmadivand S, Soltani M, Behdani M, Evensen Ø, Alirahimi E, Hassanzadeh R, Soltani E (2017) Oral DNA vaccines based on CS-TPP nanoparticles and alginate microparticles confer high protection against infectious pancreatic necrosis virus (IPNV) infection in trout. Dev Comp Immunol 74:178–189

Ahmadivand S, Soltani M, Behdani M, Evensen Ø, Alirahimi E, Soltani E, Hassanzadeh R, Ashrafi-Helan J (2018) VP2 (PTA motif) encoding DNA vaccine confers protection against lethal challenge with infectious pancreatic necrosis virus (IPNV) in trout. Mol Immunol 94:61–67

Alonso M, Chiou PP, Leong JA (2011) Development of a suicidal DNA vaccine for infectious hematopoietic necrosis virus (IHNV). Fish Shellfish Immunol 30(3):815–823

Ana I, Saint-Jean SR, Pérez-Prieto SI (2010) Immunogenic and protective effects of an oral DNA vaccine against infectious pancreatic necrosis virus in fish. Fish Shellfish Immunol 28 (4):562–570

Anderson ED, Mourich DV, Fahrenkrug SC, LaPatra S, Shepherd J, Leong JAC (1996) Genetic immunization of rainbow trout (Oncorhynchus mykiss) against infectious hematopoietic necrosis virus. Mol Mar Biol Biotechnol 5(2):114–122

Aonullah AA, Nuryati S, Murtini S (2017) Efficacy of koi herpesvirus DNA vaccine administration by immersion method on Cyprinus carpio field scale culture. Aquacult Res 48(6):2655–2662

Ballesteros NA, Saint-Jean SR, Perez-Prieto SI (2014) Food pellets as an effective delivery method for a DNA vaccine against infectious pancreatic necrosis virus in rainbow trout (Oncorhynchus mykiss, Walbaum). Fish Shellfish Immunol 37(2):220–228

Ballesteros NA, Alonso M, Saint-Jean SR, Perez-Prieto SI (2015a) An oral DNA vaccine against infectious haematopoietic necrosis virus (IHNV) encapsulated in alginate microspheres induces dose-dependent immune responses and significant protection in rainbow trout (Oncorhynchus mykiss). Fish Shellfish Immunol 45(2):877–888

Ballesteros NA, Saint-Jean SR, Perez-Prieto SI (2015b) Immune responses to oral pcDNA-VP2 vaccine in relation to infectious pancreatic necrosis virus carrier state in rainbow trout Oncorhynchus mykiss. Vet Immunol Immunopathol 165(3–4):127–137

Berglund B (2015) Environmental dissemination of antibiotic resistance genes and correlation to anthropogenic contamination with antibiotics. Infect Ecol Epidemiol 5(1):28564

Bootz A, Karbach A, Spindler J, Kropff B, Reuter N, Sticht H, Winkler TH, Britt WJ, Mach M (2017) Protective capacity of neutralizing and non-neutralizing antibodies against glycoprotein B of cytomegalovirus. PLoS Pathog 13(8):e1006601

Boudinot P, Blanco M, de Kinkelin P, Benmansour A (1998) Combined DNA immunization with the glycoprotein gene of viral hemorrhagic septicemia virus and infectious hematopoietic necrosis virus induces double-specific protective immunity and nonspecific response in rainbow trout. Virology 249(2):297–306

Byon JY, Ohira T, Hirono I, Aoki T (2005) Use of a cDNA microarray to study immunity against viral hemorrhagic septicemia (VHS) in Japanese flounder (Paralichthys olivaceus) following DNA vaccination. Fish Shellfish Immunol 18(2):135–147

Byon JY, Ohira T, Hirono I, Aoki T (2006) Comparative immune responses in Japanese flounder, Paralichthys olivaceus after vaccination with viral hemorrhagic septicemia virus (VHSV) recombinant glycoprotein and DNA vaccine using a microarray analysis. Vaccine 24 (7):921–930

Cai SH, Lu YS, Jian JC, Wang B, Huang YC, Tang JF, Ding Y, Wu ZH (2013) Protection against Vibrio alginolyticus in crimson snapper Lutjanus erythropterus immunized with a DNA vaccine containing the ompW gene. Dis Aquat Organ 106(1):39–47

Caipang CM, Takano T, Hirono I, Aoki T (2006) Genetic vaccines protect red seabream, Pagrus major, upon challenge with red seabream iridovirus (RSIV). Fish Shellfish Immunol 21 (2):130–138

Cao J, Zhu XC, Liu XY, Yuan K, Zhang JJ, Gao HH, Li JN (2019) An oral double-targeted DNA vaccine induces systemic and intestinal mucosal immune responses and confers high protection against Vibrio mimicus in grass carps. Aquaculture 504:248–259

Chang CJ, Sun B, Robertsen B (2015) Adjuvant activity of fish type I interferon shown in a virus DNA vaccination model. Vaccine 33(21):2442–2448

Chang CJ, Gu J, Robertsen B (2017) Protective effect and antibody response of DNA vaccine against salmonid alphavirus 3 (SAV 3) in Atlantic salmon. J Fish Dis 40(12):1775–1781

Chen SP, Peng RH, Chiou PP (2015) Modulatory effect of CpG oligodeoxynucleotide on a DNA vaccine against nervous necrosis virus in orange-spotted grouper (Epinephelus coioides). Fish Shellfish Immunol 45(2):919–926

Chen DD, Yao YY, Cui ZW, Zhang XY, Peng KS, Guo X, Wang B, Zhou YY, Li S, Wu N, Zhang YA (2018) Comparative study of the immunoprotective effect of two DNA vaccines against grass carp reovirus. Fish Shellfish Immunol 75:66–73

Chen J, Tan W, Wang W, Hou S, Chen G, Xia L, Lu Y (2019a) Identification of common antigens of three pathogenic Nocardia species and development of DNA vaccine against fish nocardiosis. Fish Shellfish Immunol 95:357–367

Chen J, Wang W, Hou S, Fu W, Cai J, Xia L, Lu Y (2019b) Comparison of protective efficacy between two DNA vaccines encoding DnaK and GroEL against fish nocardiosis. Fish Shellfish Immunol 95:128–139

Chen J, Chen Z, Wang W, Hou S, Cai J, Xia L, Lu Y (2020) Development of DNA vaccines encoding ribosomal proteins (RplL and RpsA) against Nocardia seriolae infection in fish. Fish Shellfish Immunol 96:201–212

Collins C, Lorenzen N, Collet B (2019) DNA vaccination for finfish aquaculture. Fish Shellfish Immunol 85:106–125

Corbeil S, Kurath G, LaPatra SE (2000a) Fish DNA vaccine against infectious hematopoietic necrosis virus: efficacy of various routes of immunisation. Fish Shellfish Immunol 10 (8):711–723

Corbeil S, LaPatra SE, Anderson ED, Kurath G (2000b) Nanogram quantities of a DNA vaccine protect rainbow trout fry against heterologous strains of infectious hematopoietic necrosis virus. Vaccine 18(25):2817–2824

Cranenburgh RM, Hanak JA, Williams SG, Sherratt DJ (2001) Escherichia coli strains that allow antibiotic-free plasmid selection and maintenance by repressor titration. Nucleic Acids Res 29 (5):e26–e26

Dalmo RA (2018) DNA vaccines for fish: review and perspectives on correlates of protection. J Fish Dis 41(1):1–9

Dubey S, Avadhani K, Mutalik S, Sivadasan SM, Maiti B, Paul J, Girisha SK, Venugopal MN, Mutoloki S, Evensen Ø, Karunasagar I (2016) Aeromonas hydrophila OmpW PLGA nanoparticle oral vaccine shows a dose-dependent protective immunity in rohu (Labeo rohita). Vaccine 4(2):21

Einer-Jensen K, Delgado L, Lorenzen E, Bovo G, Evensen Ø, LaPatra S, Lorenzen N (2009) Dual DNA vaccination of rainbow trout (Oncorhynchus mykiss) against two different rhabdoviruses, VHSV and IHNV, induces specific divalent protection. Vaccine 27(8):1248–1253

Embregts CW, Rigaudeau D, Vesely T, Pokorová D, Lorenzen N, Petit J, Houel A, Dauber M, Schütze H, Boudinot P, Wiegertjes GF (2017) Intramuscular DNA vaccination of juvenile carp against spring viremia of carp virus induces full protection and establishes a virus-specific B and T cell response. Front Immunol 8:1340

Embregts CW, Rigaudeau D, Tacchi L, Pijlman GP, Kampers L, Veselý T, Pokorová D, Boudinot P, Wiegertjes GF, Forlenza M (2019a) Vaccination of carp against SVCV with an oral DNA vaccine or an insect cells-based subunit vaccine. Fish Shellfish Immunol 85:66–77

Embregts CW, Tadmor-Levi R, Veselý T, Pokorová D, David L, Wiegertjes GF, Forlenza M (2019b) Intra-muscular and oral vaccination using a Koi Herpesvirus ORF25 DNA vaccine does not confer protection in common carp (Cyprinus carpio L.). Fish Shellfish Immunol 85:90–98

EMEA (2012) Committee for the medicinal products for human use: concept paper on guidance for DNA vaccines. http://www.ema.europa.eu/docs/en_GB/document library/Scientific_guideline/2012/03/WC500124898.pdf

Emmenegger EJ, Kurath G (2008) DNA vaccine protects ornamental koi (Cyprinus carpio koi) against North American spring viremia of carp virus. Vaccine 26(50):6415–6421

Evensen Ø, Leong JA (2013) DNA vaccines against viral diseases of farmed fish. Fish Shellfish Immunol 35:1751–1758

Fifis T, Gamvrellis A, Crimeen-Irwin B, Pietersz GA, Li J, Mottram PL, McKenzie IF, Plebanski M (2004) Size-dependent immunogenicity: therapeutic and protective properties of nano-vaccines against tumors. J Immunol 173(5):3148–3154

Food and Drug Administration (FDA) (2007) Center for Biologics Evaluation and Research: guidance for industry. Considerations for plasmid DNA vaccines for infectious disease indications. http://www.fda.gov/downloads/biologicsbloodvaccines/guidancecomplianceregulatoryinformation/guidances/vaccines/ucm091968.pdf

Fu X, Li N, Lin Q, Guo H, Zhang D, Liu L, Wu S (2014) Protective immunity against infectious spleen and kidney necrosis virus induced by immunization with DNA plasmid containing mcp gene in Chinese perch Siniperca chuatsi. Fish Shellfish Immunol 40(1):259–266

Fu X, Li N, Lin Q, Guo H, Liu L, Huang Z, Wu S (2015) Early protein ORF086 is an effective vaccine candidate for infectious spleen and kidney necrosis virus in mandarin fish Siniperca chuatsi. Fish Shellfish Immunol 46(2):200–205

Gao Y, Pei C, Sun X, Zhang C, Li L, Kong X (2018) Plasmid pcDNA3.1-s11 constructed based on the S11 segment of grass carp reovirus as DNA vaccine provides immune protection. Vaccine 36(25):3613–3621

Garver KA, LaPatra SE, Kurath G (2005) Efficacy of an infectious hematopoietic necrosis (IHN) virus DNA vaccine in Chinook Oncorhynchus tshawytscha and sockeye O. nerka salmon. Dis Aquat Organ 64(1):13–22

Gillund F, Tonheim T, Seternes T, Dalmo RA, Myhr AI (2008) DNA vaccination in aquaculture—expert judgements of impact on environment and fish health. Aquaculture 284:25–34

Glenting J, Wessels S (2005) Ensuring safety of DNA vaccines. Microb Cell Fact 4(1):1–5

Han B, Xu K, Liu Z, Ge W, Shao S, Li P, Yan N, Li X, Zhang Z (2019) Oral yeast-based DNA vaccine confers effective protection from Aeromonas hydrophila infection on Carassius auratus. Fish Shellfish Immunol 84:948–954

Hansen E, Fernandes K, Goldspink G, Butterworth P, Umeda PK, Chang KC (1991) Strong expression of foreign genes following direct injection into fish muscle. FEBS Lett 290 (1–2):73–76

Hao K, Chen XH, Qi XZ, Yu XB, Du EQ, Ling F, Zhu B, Wang GX (2017) Protective immunity of grass carp induced by DNA vaccine encoding capsid protein gene (vp7) of grass carp reovirus using bacterial ghost as delivery vehicles. Fish Shellfish Immunol 64:414–425

Hart LM, Lorenzen N, Einer-Jensen K, Purcell MK, Hershberger PK (2017) Influence of temperature on the efficacy of homologous and heterologous DNA vaccines against viral hemorrhagic septicemia in pacific herring. J Aquat Anim Health 29(3):121–128

Heppell J, Davis HL (2000) Application of DNA vaccine technology to aquaculture. Adv Drug Deliv Rev 43(1):29–43

Hølvold LB, Myhr AI, Dalmo RA (2014) Strategies and hurdles using DNA vaccines to fish. Vet Res 45(1):21

Hu YH, Sun L (2011) A bivalent Vibrio harveyi DNA vaccine induces strong protection in Japanese flounder (Paralichthys olivaceus). Vaccine 29(26):4328–4333

Hu YH, Dang W, Zhang M, Sun L (2012) Japanese flounder (Paralichthys olivaceus) Hsp70: adjuvant effect and its dependence on the intrinsic ATPase activity. Fish Shellfish Immunol 33 (4):829–834

Hu F, Li Y, Wang Q, Wang G, Zhu B, Wang Y, Zeng W, Yin J, Liu C, Bergmann SM, Shi C (2020) Carbon nanotube-based DNA vaccine against koi herpesvirus given by intramuscular injection. Fish Shellfish Immunol 98:810–818

Huang LY, Wang KY, Xiao D, Chen DF, Geng Y, Wang J, He Y, Wang EL, Huang JL, Xiao GY (2014) Safety and immunogenicity of an oral DNA vaccine encoding Sip of Streptococcus agalactiae from Nile tilapia Oreochromis niloticus delivered by live attenuated Salmonella typhimurium. Fish Shellfish Immunol 38(1):34–41

Huang Y, Cai S, Pang H, Jian J, Wu Z (2017) Immunogenicity and efficacy of DNA vaccine encoding antigenic AcfA via addition of the molecular adjuvant Myd88 against Vibrio alginolyticus in Epinephelus coioides. Fish Shellfish Immunol 66:71–77

Huang P, Cai J, Yu D, Tang J, Lu Y, Wu Z, Huang Y, Jian J (2019) An IL-6 gene in humphead snapper (Lutjanus sanguineus): Identification, expression analysis and its adjuvant effects on Vibrio harveyi OmpW DNA vaccine. Fish Shellfish Immunol 95:546–555

Jiao XD, Zhang M, Hu YH, Sun L (2009) Construction and evaluation of DNA vaccines encoding Edwardsiella tarda antigens. Vaccine 27(38):5195–5202

Jiao XD, Hu YH, Sun L (2010) Dissection and localization of the immunostimulating domain of Edwardsiella tarda FliC. Vaccine 28(34):5635–5640

Jones SR (2001) Plasmids in DNA vaccination. In: Schleef M (ed) Plasmids for therapy and vaccination. Wiley-VCH Verlag GmbH, Weinheim, pp 169–191

Jung MH, Nikapitiya C, Jung SJ (2018) DNA vaccine encoding myristoylated membrane protein (MMP) of rock bream iridovirus (RBIV) induces protective immunity in rock bream (Oplegnathus fasciatus). Vaccine 36(6):802–810

Kaattari SL, Zhang HL, Khor W, Kaattari IM, Shapiro DA (2002) Affinity maturation in trout: clonal dominance of high affinity antibodies late in the immune response. Dev Comp Immunol 26(2):191–200

Kanellos T, Sylvester ID, D'Mello F, Howard CR, Mackie A, Dixon PF, Chang KC, Ramstad A, Midtlyng PJ, Russell PH (2006) DNA vaccination can protect Cyprinus Carpio against spring viraemia of carp virus. Vaccine 24(23):4927–4933

Kato G, Kato K, Jirapongpairoj W, Kondo H, Hirono I (2014) Development of DNA vaccines against Nocardia seriolae infection in fish. Fish Pathol 49(4):165–172

Kato G, Yamashita K, Kondo H, Hirono I (2015) Protective efficacy and immune responses induced by a DNA vaccine encoding codon-optimized PPA1 against Photobacterium damselae subsp. piscicida in Japanese flounder. Vaccine 33(8):1040–1045

Kayansamruaj P, Dong HT, Pirarat N, Nilubol D, Rodkhum C (2017) Efficacy of α-enolase-based DNA vaccine against pathogenic Streptococcus iniae in Nile tilapia (Oreochromis niloticus). Aquaculture 468:102–106

Kim CH, Johnson MC, Drennan JD, Simon BE, Thomann E, Leong JA (2000) DNA vaccines encoding viral glycoproteins induce nonspecific immunity and Mx protein synthesis in fish. J Virol 74(15):7048–7054

Kole S, Kumari R, Anand D, Kumar S, Sharma R, Tripathi G, Makesh M, Rajendran KV, Bedekar MK (2018) Nanoconjugation of bicistronic DNA vaccine against Edwardsiella tarda using chitosan nanoparticles: evaluation of its protective efficacy and immune modulatory effects in Labeo rohita vaccinated by different delivery routes. Vaccine 36(16):2155–2165

Kumar SR, Parameswaran V, Ahmed VI, Musthaq SS, Hameed AS (2007) Protective efficiency of DNA vaccination in Asian seabass (Lates calcarifer) against Vibrio anguillarum. Fish Shellfish Immunol 23(2):316–326

Kumar SR, Ahmed VI, Parameswaran V, Sudhakaran R, Babu VS, Hameed AS (2008) Potential use of chitosan nanoparticles for oral delivery of DNA vaccine in Asian sea bass (Lates calcarifer) to protect from Vibrio (Listonella) anguillarum. Fish Shellfish Immunol 25 (1–2):47–56

Kumari R, Kole S, Soman P, Rathore G, Tripathi G, Makesh M, Rajendran KV, Bedekar MK (2018) Bicistronic DNA vaccine against Edwardsiella tarda infection in Labeo rohita: construction and comparative evaluation of its protective efficacy against monocistronic DNA vaccine. Aquaculture 485:201–209

Kurath G (2008) Biotechnology and DNA vaccines for aquatic animals. Rev Sci Tech 27(1):175

Kurath G, Garver KA, Corbeil S, Elliott DG, Anderson ED, LaPatra SE (2006) Protective immunity and lack of histopathological damage two years after DNA vaccination against infectious hematopoietic necrosis virus in trout. Vaccine 24(3):345–354

LaPatra SE, Corbeil S, Jones GR, Shewmaker WD, Kurath G (2000) The dose-dependent effect on protection and humoral response to a DNA vaccine against infectious hematopoietic necrosis (IHN) virus in subyearling rainbow trout. J Aquat Anim Health 12(3):181–188

LaPatra SE, Corbeil S, Jones GR, Shewmaker WD, Lorenzen N, Anderson ED, Kurath G (2001) Protection of rainbow trout against infectious hematopoietic necrosis virus four days after specific or semi-specific DNA vaccination. Vaccine 19(28–29):4011–4019

Lazarte JM, Kim YR, Lee JS, Im SP, Kim SW, Jung JW, Kim J, Lee WJ, Jung TS (2017) Enhancement of glycoprotein-based DNA vaccine for viral hemorrhagic septicemia virus (VHSV) via addition of the molecular adjuvant, DDX41. Fish Shellfish Immunol 62:356–365

Leya T, Ahmad I, Sharma R, Tripathi G, Kurcheti PP, Rajendran KV, Bedekar MK (2020) Bicistronic DNA vaccine macromolecule complexed with poly lactic-co-glycolic acid-chitosan nanoparticles enhanced the mucosal immunity of Labeo rohita against Edwardsiella tarda infection. Int J Biol Macromol 156:928–937

Li L, Lin SL, Deng L, Liu ZG (2013) Potential use of chitosan nanoparticles for oral delivery of DNA vaccine in black seabream Acanthopagrus schlegelii Bleeker to protect from Vibrio parahaemolyticus. J Fish Dis 36(12):987–995

Li N, Fu X, Guo H, Lin Q, Liu L, Zhang D, Fang X, Wu S (2015) Protein encoded by ORF093 is an effective vaccine candidate for infectious spleen and kidney necrosis virus in Chinese perch Siniperca chuatsi. Fish Shellfish Immunol 42(1):88–90

Liang HY, Wu ZH, Jian JC, Huang YC (2011) Protection of red snapper (Lutjanus sanguineus) against Vibrio alginolyticus with a DNA vaccine containing flagellin flaA gene. Lett Appl Microbiol 52(2):156–161

Lim HJ, Abdellaoui N, Kim KH (2019) Effect of miR-155 as a molecular adjuvant of DNA vaccine against VHSV in olive flounder (Paralichthys olivaceus). Fish Shellfish Immunol 88:225–230

Liu MA (2003) DNA vaccines: a review. J Intern Med 253(4):402–410

Liu MA (2011) DNA vaccines: an historical perspective and view to the future. Immunol Rev 239 (1):62–84

Liu R, Chen J, Li K, Zhang X (2011) Identification and evaluation as a DNA vaccine candidate of a virulence-associated serine protease from a pathogenic Vibrio parahaemolyticus isolate. Fish Shellfish Immunol 30(6):1241–1248

Liu F, Tang X, Sheng X, Xing J, Zhan W (2016a) DNA vaccine encoding molecular chaperone GroEL of Edwardsiella tarda confers protective efficacy against edwardsiellosis. Mol Immunol 79:55–65

Liu L, Gong YX, Liu GL, Zhu B, Wang GX (2016b) Protective immunity of grass carp immunized with DNA vaccine against Aeromonas hydrophila by using carbon nanotubes as a carrier molecule. Fish Shellfish Immunol 55:516–522

Liu X, Xu J, Zhang H, Liu Q, Xiao J, Zhang Y (2016c) Design and evaluation of an Edwardsiella tarda DNA vaccine co-encoding antigenic and adjuvant peptide. Fish Shellfish Immunol 59:189–195

Liu F, Tang X, Sheng X, Xing J, Zhan W (2017) Construction and evaluation of an Edwardsiella tarda DNA vaccine encoding outer membrane protein C. Microb Pathog 104:238–247

Liu L, Gao S, Luan W, Zhou J, Wang H (2018) Generation and functional evaluation of a DNA vaccine co-expressing Cyprinid herpesvirus-3 envelope protein and carp interleukin-1 beta. Fish Shellfish Immunol 80:223–231

Liu C, Hu X, Cao Z, Sun Y, Chen X, Zhang Z (2019) Construction and characterization of a DNA vaccine encoding the SagH against Streptococcus iniae. Fish Shellfish Immunol 89:71–75

Long A, Richard J, Hawley L, LaPatra SE, Garver KA (2017) Transmission potential of infectious hematopoietic necrosis virus in APEX-IHN (R)-vaccinated Atlantic salmon. Dis Aquat Organ 122:213–221

Lorenzen N, LaPatra SE (1999) Immunity to rhabdoviruses in rainbow trout: the antibody response. Fish Shellfish Immunol 9(4):345–360

Lorenzen N, LaPatra SE (2005) DNA vaccines for aquacultured fish. Rev Sci Tech 24(1):201–213

Lorenzen N, Lorenzen E, Einer-Jensen K, Heppell J, Wu T, Davis H (1998) Protective immunity to VHS in rainbow trout (Oncorhynchus mykiss, Walbaum) following DNA vaccination. Fish Shellfish Immunol 8(4):261–270

Lorenzen E, Einer-Jensen K, Martinussen T, LaPatra SE, Lorenzen N (2000) DNA vaccination of rainbow trout against viral hemorrhagic septicemia virus: a dose–response and time–course study. J Aquat Anim Health 12(3):167–180

Lorenzen N, Lorenzen E, Einer-Jensen K, LaPatra SE (2002) Immunity induced shortly after DNA vaccination of rainbow trout against rhabdoviruses protects against heterologous virus but not against bacterial pathogens. Dev Comp Immunol 26(2):173–179

Lorenzen E, Lorenzen N, Einer-Jensen K, Brudeseth B, Evensen Ø (2005) Time course study of in situ expression of antigens following DNA-vaccination against VHS in rainbow trout (Oncorhynchus mykiss Walbaum) fry. Fish Shellfish Immunol 19(1):27–41

Lorenzen E, Einer-Jensen K, Rasmussen JS, Kjaer TE, Collet B, Secombes CJ, Lorenzen N (2009) The protective mechanisms induced by a fish rhabdovirus DNA vaccine depend on temperature. Vaccine 27(29):3870–3880

Lu QL, Bou-Gharios G, Partridge TA (2003) Non-viral gene delivery in skeletal muscle: a protein factory. Gene Ther 10(2):131–142

Ma YP, Ke H, Liang ZL, Ma JY, Hao L, Liu ZX (2017) Protective efficacy of cationic-PLGA microspheres loaded with DNA vaccine encoding the sip gene of Streptococcus agalactiae in tilapia. Fish Shellfish Immunol 66:345–353

Magadan S, Sunyer OJ, Boudinot P (2015) Unique features of fish immune repertoires: Particularities of adaptive immunity within the largest group of vertebrates. Results Probl Cell Differ 57:235–264

McLauchlan PE, Collet B, Ingerslev E, Secombes CJ, Lorenzen N, Ellis AE (2003) DNA vaccination against viral haemorrhagic septicaemia (VHS) in rainbow trout: size, dose, route of injection and duration of protection—early protection correlates with Mx expression. Fish Shellfish Immunol 15(1):39–50

Mikalsen AB, Sindre H, Torgersen J, Rimstad E (2005) Protective effects of a DNA vaccine expressing the infectious salmon anemia virus hemagglutinin-esterase in Atlantic salmon. Vaccine 23(41):4895–4905

Millard EV, Bourke AM, LaPatra SE, Brenden TO, Fitzgerald SD, Faisal M (2017) DNA vaccination partially protects muskellunge against viral hemorrhagic septicemia virus (VHSV-IVb). J Aquat Anim Health 29(1):50–56

Munang'andu HM, Evensen Ø (2015) A review of intra-and extracellular antigen delivery systems for virus vaccines of finfish. J Immunol Res 2015:960859

Munang'andu HM, Fredriksen BN, Mutoloki S, Brudeseth B, Kuo TY, Marjara IS, Dalmo RA, Evensen Ø (2012) Comparison of vaccine efficacy for different antigen delivery systems for infectious pancreatic necrosis virus vaccines in Atlantic salmon (Salmo salar L.) in a cohabitation challenge model. Vaccine 30(27):4007–4016

Munang'andu HM, Fredriksen BN, Mutoloki S, Dalmo RA, Evensen Ø (2013) Antigen dose and humoral immune response correspond with protection for inactivated infectious pancreatic necrosis virus vaccines in Atlantic salmon (Salmo salar L). Vet Res 44(1):7

Myhr AI (2017) DNA vaccines: regulatory considerations and safety aspects. Curr Issues Mol Biol 22:79–88

Myhr A, Dalmo RA (2005) Introduction of genetic engineering in aquaculture: ecological and ethical implications for science and governance. Aquaculture 250:542–554

Nielsen KN, Fredriksen BN, Myhr AI (2011) Mapping uncertainties in the upstream: the case of PLGA nanoparticles in salmon vaccines. NanoEthics 5:57–71

Nuryati S, Alimuddin A, Sukenda S, Soejoedono RD, Santika A, Pasaribu FH, Sumantadinata K (2010) Construction of a DNA vaccine using glycoprotein gene and its expression towards increasing survival rate of KHV-infected common carp (Cyprinus carpio). J Nat Indones 13 (1):47–52

Nusbaum KE, Smith BF, DeInnocentes P, Bird RC (2002) Protective immunity induced by DNA vaccination of channel catfish with early and late transcripts of the channel catfish herpesvirus (IHV-1). Vet Immunol Immunopathol 84(3–4):151–168

Ou-Yang Z, Wang P, Huang Y, Huang X, Wan Q, Zhou S, Wei J, Zhou Y, Qin Q (2012) Selection and identification of Singapore grouper iridovirus vaccine candidate antigens using bioinformatics and DNA vaccination. Vet Immunol Immunopathol 149(1–2):38–45

Pasnik DJ, Smith SA (2005) Immunogenic and protective effects of a DNA vaccine for Mycobacterium marinum in fish. Vet Immunol Immunopathol 103(3–4):195–206

Peñaranda MM, LaPatra SE, Kurath G (2011) Specificity of DNA vaccines against the U and M genogroups of infectious hematopoietic necrosis virus (IHNV) in rainbow trout (Oncorhynchus mykiss). Fish Shellfish Immunol 31(1):43–51

Peubez I, Chaudet N, Mignon C, Hild G, Husson S, Courtois V, De Luca K, Speck D, Sodoyer R (2010) Antibiotic-free selection in E. coli: new considerations for optimal design and improved production. Microb Cell Fact 9(1):65

Plotkin SA (2013) Complex correlates of protection after vaccination. Clin Infect Dis 56:1458–1465

Plotkin SA, Gilbert PB (2012) Nomenclature for immune correlates of protection after vaccination. Clin Infect Dis 54:1615–1617

Priya TJ, Lin YH, Wang YC, Yang CS, Chang PS, Song YL (2012) Codon changed immobilization antigen (iAg), a potent DNA vaccine in fish against Cryptocaryon irritans infection. Vaccine 30 (5):893–903

Pumchan A, Krobthong S, Roytrakul S, Sawatdichaikul O, Kondo H, Hirono I, Areechon N, Unajak S (2020) Novel chimeric Multiepitope vaccine for Streptococcosis disease in nile tilapia (Oreochromis niloticus Linn.). Sci Rep 10(1):1–3

Qin Y, Su Y, Wang S, Yan Q (2009) Immunogenicity and protective efficacy of Vibrio harveyi pcFlaA DNA vaccine in Epinephelus awoara. Chinese J Oceanol Limnol 27(4):769

Rambabu KM, Rao SH, Rao NM (2005) Efficient expression of transgenes in adult zebrafish by electroporation. BMC Biotechnol 5(1):29

Reyes M, Ramírez C, Ñancucheo I, Villegas R, Schaffeld G, Kriman L, Gonzalez J, Oyarzun P (2017) A novel "in-feed" delivery platform applied for oral DNA vaccination against IPNV enables high protection in Atlantic salmon (Salmon salar). Vaccine 35(4):626–632

Robertsen B, Chang CJ, Bratland L (2016) IFN-adjuvanted DNA vaccine against infectious salmon anemia virus: Antibody kinetics and longevity of IFN expression. Fish Shellfish Immunol 54:328-32

Salinas I (2015) The mucosal immune system of teleost fish. Biology 4:525–539

Schalk JA, Mooi FR, Berbers GA, Aerts LA, Ovelgönne H, Kimman TG (2006) Preclinical and clinical safety studies on DNA vaccines. Hum Vaccin 2(2):45–53

Seo JY, Kim KH, Kim SG, Oh MJ, Nam SW, Kim YT, Choi TJ (2006) Protection of flounder against hirame rhabdovirus (HIRRV) with a DNA vaccine containing the glycoprotein gene. Vaccine 24(7):1009–1015

Sepúlveda D, Lorenzen E, Rasmussen JS, Einer-Jensen K, Collet B, Secombes CJ, Lorenzen N (2019) Time-course study of the protection induced by an interferon-inducible DNA vaccine against viral haemorrhagic septicaemia in rainbow trout. Fish Shellfish Immunol 85:99–105

Shirota H, Petrenko L, Hong C, Klinman DM (2007) Potential of transfected muscle cells to contribute to DNA vaccine immunogenicity. J Immunol 179(1):329–336

Slütter B, Jiskoot W (2016) Sizing the optimal dimensions of a vaccine delivery system: a particulate matter. Expert Opin Drug Deliv 13(2):167–170

Sommerset I, Lorenzen E, Lorenzen N, Bleie H, Nerland AH (2003) A DNA vaccine directed against a rainbow trout rhabdovirus induces early protection against a nodavirus challenge in turbot. Vaccine 21(32):4661–4667

Sommerset I, Skern R, Biering E, Bleie H, Fiksdal IU, Grove S, Nerland AH (2005) Protection against Atlantic halibut nodavirus in turbot is induced by recombinant capsid protein vaccination but not following DNA vaccination. Fish Shellfish Immunol 18(1):13–29

Sun Y, Hu YH, Liu CS, Sun L (2010) Construction and analysis of an experimental Streptococcus iniae DNA vaccine. Vaccine 28(23):3905–3912

Sun Y, Liu CS, Sun L (2011a) Comparative study of the immune effect of an Edwardsiella tarda antigen in two forms: subunit vaccine vs DNA vaccine. Vaccine 29(11):2051–2057

Sun Y, Liu CS, Sun L (2011b) Construction and analysis of the immune effect of an Edwardsiella tarda DNA vaccine encoding a D15-like surface antigen. Fish Shellfish Immunol 30(1):273–279

Sun Y, Zhang M, Liu CS, Qiu R, Sun L (2012) A divalent DNA vaccine based on Sia10 and OmpU induces cross protection against Streptococcus iniae and Vibrio anguillarum in Japanese flounder. Fish Shellfish Immunol 32(6):1216–1222

Sun Y, Ding S, He M, Liu A, Long H, Guo W, Cao Z, Xie Z, Zhou Y (2020) Construction and analysis of the immune effect of Vibrio harveyi subunit vaccine and DNA vaccine encoding TssJ antigen. Fish Shellfish Immunol 98:45–51

Takano T, Iwahori A, Hirono I, Aoki T (2004) Development of a DNA vaccine against hirame rhabdovirus and analysis of the expression of immune-related genes after vaccination. Fish Shellfish Immunol 17(4):367–374

Tan CW, Jesudhasan P, Woo PT (2008) Towards a metalloprotease-DNA vaccine against piscine cryptobiosis caused by Cryptobia salmositica. Parasitol Res 102(2):265–275

Tian J, Yu J (2011) Poly (lactic-co-glycolic acid) nanoparticles as candidate DNA vaccine carrier for oral immunization of Japanese flounder (Paralichthys olivaceus) against lymphocystis disease virus. Fish Shellfish Immunol 30(1):109–117

Tian J, Sun X, Chen X, Yu J, Qu L, Wang L (2008a) The formulation and immunisation of oral poly (DL-lactide-co-glycolide) microcapsules containing a plasmid vaccine against lymphocystis disease virus in Japanese flounder (Paralichthys olivaceus). Int Immunopharmacol 8 (6):900–908

Tian J, Yu J, Sun X (2008b) Chitosan microspheres as candidate plasmid vaccine carrier for oral immunisation of Japanese flounder (Paralichthys olivaceus). Vet Immunol Immunopathol 126 (3–4):220–229

Tian JY, Sun XQ, Chen XG (2008c) Formation and oral administration of alginate microspheres loaded with pDNA coding for lymphocystis disease virus (LCDV) to Japanese flounder. Fish Shellfish Immunol 24(5):592–599

Tonheim TC, Bøgwald J, Dalmo RA (2008) What happens to the DNA vaccine in fish? A review of current knowledge. Fish Shellfish Immunol 25(1–2):1–18

Traxler GS, Anderson E, LaPatra SE, Richard J, Shewmaker B, Kurath G (1999) Naked DNA vaccination of Atlantic salmon Salmo salar against IHNV. Dis Aquat Organ 38(3):183–190

Utke K, Kock H, Schuetze H, Bergmann SM, Lorenzen N, Einer-Jensen K, Köllner B, Dalmo RA, Vesely T, Ototake M, Fischer U (2008) Cell-mediated immune responses in rainbow trout after DNA immunization against the viral hemorrhagic septicemia virus. Dev Comp Immunol 32 (3):239–252

Valero Y, Awad E, Buonocore F, Arizcun M, Esteban MÁ, Meseguer J, Chaves-Pozo E, Cuesta A (2016) An oral chitosan DNA vaccine against nodavirus improves transcription of cell-mediated cytotoxicity and interferon genes in the European sea bass juveniles gut and survival upon infection. Dev Comp Immunol 65:64–72

Vazquez-Juarez RC, Gomez-Chiarri M, Barrera-Saldaña H, Hernandez-Saavedra N, Dumas S, Ascencio F (2005) Evaluation of DNA vaccination of spotted sand bass (Paralabrax maculatofasciatus) with two major outer-membrane protein-encoding genes from Aeromonas veronii. Fish Shellfish Immunol 19(2):153–163

Vimal S, Farook MA, Madan N, Majeed SA, Nambi KS, Taju G, Sundarraj N, Venu S, Subburaj R, Thirunavukkarasu AR, Hameed AS (2014a) Development, distribution and expression of a DNA vaccine against nodavirus in Asian Seabass, Lates calcarifer (Bloch, 1790). Aquacult Res 47(4):1209–1220

Vimal S, Majeed SA, Nambi KS, Madan N, Farook MA, Venkatesan C, Taju G, Venu S, Subburaj R, Thirunavukkarasu AR, Hameed AS (2014b) Delivery of DNA vaccine using chitosan–tripolyphosphate (CS/TPP) nanoparticles in Asian sea bass, Lates calcarifer (Bloch, 1790) for protection against nodavirus infection. Aquaculture 420:240–246

Wang Y, Liu GL, Li DL, Ling F, Zhu B, Wang GX (2015) The protective immunity against grass carp reovirus in grass carp induced by a DNA vaccination using single-walled carbon nanotubes as delivery vehicles. Fish Shellfish Immunol 47(2):732–742

Wang E, Long B, Wang K, Wang J, He Y, Wang X, Yang Q, Liu T, Chen D, Geng Y, Huang X (2016) Interleukin-8 holds promise to serve as a molecular adjuvant in DNA vaccination model against Streptococcus iniae infection in fish. Oncotarget 7(51):83938

Wang H, Zhu F, Huang Y, Ding Y, Jian J, Wu Z (2017) Construction of glutathione peroxidase (GPx) DNA vaccine and its protective efficiency on the orange-spotted grouper (Epinephelus coioides) challenged with Vibrio harveyi. Fish Shellfish Immunol 60:529–536

Wolff JA, Malone RW, Williams P, Chong W, Acsadi G, Jani A, Felgner PL (1990) Direct gene transfer into mouse muscle in vivo. Science 247(4949):1465–1468

Wolff JA, Dowty ME, Jiao SH, Repetto GA, Berg RK, Ludtke JJ, Williams PH, Slautterback DB (1992) Expression of naked plasmids by cultured myotubes and entry of plasmids into T tubules and caveolae of mammalian skeletal muscle. J Cell Sci 103(4):1249–1259

Xing J, Xu H, Tang X, Sheng X, Zhan W (2019) A DNA vaccine encoding the VAA gene of Vibrio anguillarum induces a protective immune response in flounder. Front Immunol 10:499

Xu C, Mutoloki S, Evensen Ø (2012) Superior protection conferred by inactivated whole virus vaccine over subunit and DNA vaccines against salmonid alphavirus infection in Atlantic salmon (Salmo salar L.). Vaccine 30(26):3918–3928

Xu L, Zhao J, Liu M, Kurath G, Ren G, LaPatra SE, Yin J, Liu H, Feng J, Lu T (2017a) A effective DNA vaccine against diverse genotype J infectious hematopoietic necrosis virus strains prevalent in China. Vaccine 35(18):2420–2426

Xu L, Zhao J, Liu M, Ren G, Jian F, Yin J, Feng J, Liu H, Lu T (2017b) Bivalent DNA vaccine induces significant immune responses against infectious hematopoietic necrosis virus and infectious pancreatic necrosis virus in rainbow trout. Sci Rep 7(1):1–1

Xu H, Xing J, Tang X, Sheng X, Zhan W (2019a) Generation and functional evaluation of a DNA vaccine co-expressing Vibrio anguillarum VAA protein and flounder interleukin-2. Fish Shellfish Immunol 93:1018–1027

Xu H, Xing J, Tang X, Sheng X, Zhan W (2019b) Immune response and protective effect against Vibrio anguillarum induced by DNA vaccine encoding Hsp33 protein. Microb Pathog 137:103729

Xu H, Xing J, Tang X, Sheng X, Zhan W (2019c) Intramuscular administration of a DNA vaccine encoding OmpK antigen induces humoral and cellular immune responses in flounder

(Paralichthys olivaceus) and improves protection against Vibrio anguillarum. Fish Shellfish Immunol 86:618–626

Xu DH, Zhang D, Shoemaker C, Beck B (2019d) Immune response of channel catfish (Ictalurus punctatus) against Ichthyophthirius multifiliis post vaccination using DNA vaccines encoding immobilization antigens. Fish Shellfish Immunol 94:308–317

Xu DH, Zhang D, Shoemaker C, Beck B (2020a) Dose effects of a DNA vaccine encoding immobilization antigen on immune response of channel catfish against Ichthyophthirius multifiliis. Fish Shellfish Immunol 106:1031–1041

Xu H, Xing J, Tang X, Sheng X, Zhan W (2020b) The effects of CCL3, CCL4, CCL19 and CCL21 as molecular adjuvants on the immune response to VAA DNA vaccine in flounder (Paralichthys olivaceus). Dev Comp Immunol 103:103492

Yang H, Chen J, Yang G, Zhang XH, Liu R, Xue X (2009) Protection of Japanese flounder (Paralichthys olivaceus) against Vibrio anguillarum with a DNA vaccine containing the mutated zinc-metalloprotease gene. Vaccine 27(15):2150–2155

Yasuike M, Kondo H, Hirono I, Aoki T (2007) Difference in Japanese flounder, Paralichthys olivaceus gene expression profile following hirame rhabdovirus (HIRRV) G and N protein DNA vaccination. Fish Shellfish Immunol 23(3):531–541

Ye J, Kaattari IM, Kaattari SL (2011) The differential dynamics of antibody subpopulation expression during affinity maturation in a teleost. Fish Shellfish Immunol 30(1):372–377

Yi W, Zhang X, Zeng K, Xie D, Song C, Tam K, Liu Z, Zhou T, Li W (2020) Construction of a DNA vaccine and its protective effect on largemouth bass (Micropterus salmoides) challenged with largemouth bass virus (LMBV). Fish Shellfish Immunol 106:103–109

Yu NT, Zheng XB, Liu ZX (2019) Protective immunity induced by DNA vaccine encoding viral membrane protein against SGIV infection in grouper. Fish Shellfish Immunol 92:649–654

Yuan X, Shen J, Pan X, Yao J, Lyu S, Liu L, Zhang H (2020) Screening for protective antigens of Cyprinid herpesvirus 2 and construction of DNA vaccines. J Virol Methods 280:113877

Zhang M, Hu YH, Xiao ZZ, Sun Y, Sun L (2012) Construction and analysis of experimental DNA vaccines against megalocytivirus. Fish Shellfish Immunol 33(5):1192–1198

Zhang C, Zhao Z, Zha JW, Wang GX, Zhu B (2017) Single-walled carbon nanotubes as delivery vehicles enhance the immunoprotective effect of a DNA vaccine against spring viremia of carp virus in common carp. Fish Shellfish Immunol 71:191–201

Zhang C, Zhao Z, Liu GY, Li J, Wang GX, Zhu B (2018) Immune response and protective effect against spring viremia of carp virus induced by intramuscular vaccination with a SWCNTs-DNA vaccine encoding matrix protein. Fish Shellfish Immunol 79:256–264

Zhang C, Zhao Z, Li J, Song KG, Hao K, Wang J, Wang GX, Zhu B (2019a) Bacterial ghost as delivery vehicles loaded with DNA vaccine induce significant and specific immune responses in common carp against spring viremia of carp virus. Aquaculture 504:361–368

Zhang C, Zheng YY, Gong YM, Zhao Z, Guo ZR, Jia YJ, Wang GX, Zhu B (2019b) Evaluation of immune response and protection against spring viremia of carp virus induced by a single-walled carbon nanotubes-based immersion DNA vaccine. Virology 537:216–225

Zheng FR, Sun XQ, Liu HZ, Zhang JX (2006) Study on the distribution and expression of a DNA vaccine against lymphocystis disease virus in Japanese flounder (Paralichthys olivaceus). Aquaculture 261(4):1128–1134

Zheng F, Liu H, Sun X, Zhang Y, Zhang B, Teng Z, Hou Y, Wang B (2016) Development of oral DNA vaccine based on chitosan nanoparticles for the immunization against reddish body iridovirus in turbots (Scophthalmus maximus). Aquaculture 452:263–271

Zheng F, Liu H, Sun X, Qin X, Xu Z, Wang B (2017) Construction and expression of DNA vaccine against reddish body iridovirus and evaluation of immune efficacy in turbot (Scophthalmus maximus). Aquacult Res 48(8):4174–4183

Zhou J, Xue J, Wang Q, Zhu X, Li X, Lv W, Zhang D (2014a) Vaccination of plasmid DNA encoding ORF81 gene of CJ strains of KHV provides protection to immunized carp. In Vitro Cell Dev Biol Anim 50(6):489–495

Zhou JX, Wang H, Li XW, Zhu X, Lu WL, Zhang DM (2014b) Construction of KHV-CJ ORF 25 DNA vaccine and immune challenge test. J Fish Dis 37(4):319–325

Zhu B, Liu GL, Gong YX, Ling F, Wang GX (2015) Protective immunity of grass carp immunized with DNA vaccine encoding the vp7 gene of grass carp reovirus using carbon nanotubes as a carrier molecule. Fish Shellfish Immunol 42(2):325–334

Zhu L, Yang Q, Huang L, Wang K, Wang X, Chen D, Geng Y, Huang X, Ouyang P, Lai W (2017) Effectivity of oral recombinant DNA vaccine against Streptococcus agalactiae in Nile tilapia. Dev Comp Immunol 77:77–87

Robertsen B, Chang CJ, Bratland L (2016) IFN-adjuvanted DNA vaccine against infectious salmon anemia virus: Antibody kinetics and longevity of IFN expression. Fish Shellfish Immunol 54:328-32

Sepúlveda D, Lorenzen E, Rasmussen JS, Einer-Jensen K, Collet B, Secombes CJ, Lorenzen N (2019) Time-course study of the protection induced by an interferon-inducible DNA vaccine against viral haemorrhagic septicaemia in rainbow trout. Fish Shellfish Immunol 85:99-105

Bacteriophage Therapy in Aquaculture: An Overview

20

Md. Idrish Raja Khan and Tanmoy Gon Choudhury

Abstract

In the present scenario, the development of drug-resistant bacteria poses a global threat to all living kinds including aquatic animals. The phenomenon calls for prompt action, through development and timely adoption of alternative strategies in order to sustain the quality as well as to ensure safety of the aquatic produce. In view of antimicrobial resistance especially antibiotic abuse, efforts made towards the advancement of the biological control approaches such as probiotic, symbiotic, and bacteriophage have been accelerated. In recent times, the employment of the biocontrol approach through the applications of lytic bacteriophages for therapy of bacterial infection have leaped over other bioagents. Bacteriophages are bacteria-specific viruses that precisely infect host bacteria and ultimately kill them. Ever since their discovery in the early nineteenth century, the phage therapy enjoyed fleeting popularity in western countries owing to exploratory researches and scientific explanation with regard to their successful clinical trials. In the post antibiotic discovery era, the significance of the phage was ignored. However, after the emergence of antimicrobial resistance, a new craze for therapy was appeared either as prophylactic or therapeutic approach including the aquaculture industry. Most of the therapy in aquaculture is still in the laboratory stage, and is limited to in vitro characterisation and lab-based efficacy which have emerged as the major obstacle in its adoption at the farm level. In this chapter, an effort has been made to draw a connecting line between the current state of information about bacteriophages and what could be the possible strategies for the development of field-based therapy towards the sustenance of aquaculture.

M. I. R. Khan · T. G. Choudhury (✉)
Department of Aquatic Health and Environment, College of Fisheries, Central Agricultural University (Imphal), Lembucherra, Tripura, India

© The Author(s), under exclusive license to Springer Nature Singapore Pte Ltd. 2021
P. K. Pandey, J. Parhi (eds.), *Advances in Fisheries Biotechnology*,
https://doi.org/10.1007/978-981-16-3215-0_20

Keywords

Antimicrobial resistance · Phage therapy · Biocontrol · Prophylactic or therapeutic

20.1 Introduction

Over the past few decades, the aquaculture sector has served the nutritional needs of the people throughout the globe. The contribution from Asian subcontinent was maximum, i.e. 89% of total volume and 79% of the total value of fish production globally (Bostock et al. 2010). However, there are several factors which continue to play a crucial role in limiting the aquaculture production such as infectious diseases, especially those of bacterial origin. As per an assessment of Lafferty et al. (2014), the bacterial infection alone accounts for about 34% of total outbreaks encountered in the aquaculture system. Additionally, the indiscriminate use of chemotherapeutics to mitigate the disease problem has caused the rise in antimicrobial resistance (AMR) strain and the situation can exaggerate by the emergence of superbugs. According to Van Boeckel et al. (2019), the application of chemical therapeutics, especially antibiotics, for rearing of the farm animals including aquatic animals, accounts for about 73% of all antibiotic usage throughout the globe. In the recent past, various chemical agents have been used either as a prophylactic treatment or as growth enhancers. This would have paved the way that, due to the emergence of drug-resistant aetiological agents, the pathological condition that was resolved easily earlier is becoming a major setback to aquaculture production (Gelband et al. 2015). Consequently, researchers all over the world have been engaged with the development of alternative treatment approaches. In light of the investigation for substitute, the biocontrol strategy via bacteriophages could be considered as a sustainable option. The phage therapy, however, is an aged approach but the latest developments in the identification of potential isolates and their multidimensional application strategies have also fuelled the investigations towards the use of bacteriophages as a biological tool for health management in aquaculture.

20.2 Brief About Bacteriophages

Bacteriophages are the viruses which are obligate intracellular parasites of bacteria; they ultimately kill or lyse the host cell and release new progenies (Al-Sum and Al-Dhabi 2014). Bacteriophages are informally called phages, which is derived from a Greek word "phagein" meaning "to devour". They utilize the bio-machinery of the bacterial host for all kinds of metabolic support in order to survive (Al-Sum and Al-Dhabi 2014). As the natural environment is replete with loads of bacterial host, the occurrence of phages is natural and can flourish in soil up to 10^{7-8} virions g^{-1} and in water approximately 10^7 virions mL^{-1} either in fresh or saline environment (Ninawe et al. 2020; Park et al. 2020). According to Abedon et al. (2011), the total count of bacteriophages on the earth is about 10 times the total bacterial host thriving in different environments, which accounts for about 10^{30-31}. The International Committee on the Taxonomy of Viruses (ICTV) is responsible for the typing of phages and they have classified bacteriophages into 19 families, among which a few are well characterized including *Microviridae, Myoviridae, Inoviridae, Podoviridae* and *Siphoviridae* (Simmonds et al. 2017; Adriaenssens et al. 2018; Walker et al. 2019). The vast abundance and diversity of phages in the biosphere provides an already equipped resource to mine for the potential phages for a variety of purposes (Nikolich and Filippov 2020). Employment of precise killing capability of phages to control lethal bacterial pathogens is called as phage therapy or phagotherapy. The putative phages are composed of proteinous outer shell/capsid measuring about 24 to 200 nm in size, which contains proteins and nucleic acids (either DNA or RNA) ranging 17 and 700 kb in length (Ackermann 2003; Sharma et al. 2017). The majority of phages possess a tail (variable in size) in their structure with tail fibres on it which helps in the precise identification and adherence to the bacterial host (Kowalska et al. 2020).

The life cycle of bacteriophages can be categorized into two stages, first is lytic (virulent) and second, temperate. In the first lytic cycle, the phages adhere themselves to bacterial host followed by taking control of the host's bio-molecular machinery to proliferate and ultimately kill the host bacteria, concurrently releasing its progeny phages. The lytic phages are responsible for the production of two specific proteins to kill the host, "holins and endo-lysins". The protein, holins work in synergy with the endo-lysins and are responsible for the perforation on the bacterial cell followed by the destruction of cell wall after phage multiplication (Cisek et al. 2017). In the second temperate lysogenic stage, after the infection of bacterial host the phage genome shifts to dormant stage "prophage" which can exist within the host in the form of a plasmid and can last for many generations and can make its genes (including virulent genes) functional for the host bacterium. However, any sudden exposure or any triggering factor such as DNA damage, UV exposure and antibiotic treatment might lead the conversion of lysogenic phage to lytic stage (Letchumanan et al. 2016; Kowalska et al. 2020). Temperate phages are favourable to bacteria because they might encode for antibiotic resistance gene or some other potent genes; additionally, these lethal genes can be horizontally transferred to another bacterium in the residing environment (Lin et al. 2017). On the contrary, virulent lytic phages kill the bacterial cells directly where the possibility of any genes transfer is limited, which make lytic phages a desirable candidate for therapeutic bacteriophage therapy (Jassim and Limoges 2014; Letchumanan et al. 2016). However, according to the report of Freifelder (1987), the prevalence of lysogenic phage compared to lytic phages is as more as 90% in nature, which makes phage isolation a crucial state in development of phage therapy. There are few literature who vote for another third phage variant, a carrier state of the lysogenic stage termed as pseudolysogenic cycle, where the phage genetic material does not replicate but instead remains inactivated within the host till the occurrence of favourable condition (such as nutrient availability which hinders the bacteriophage gene expression). Once the favourable situation prevails, carrier state might be initiated with either the lytic cycle or the commencement of true lysogeny (Sieiro et al. 2020).

20.3   History of Bacteriophage Researches

Ernst Hankin in 1896 was the first one to demonstrate the presence of certain unidentified antimicrobial compounds against *Vibrio cholera* which are heat labile, filterable and transmissible, from the waters sample of the Ganges river system of India (Hankin 1896); however, he was not able to come to a conclusion regarding the reason behind anti-bacterial activity (Twort 1915; D'Hérelle 1917; Summers 2005). Later, in 1915, Frederick Twort, a British pathologist, was the first to demonstrate the presence of an "ultra-microscopic virus" that could affect bacteria; however, he also failed to explain the phenomenon, including the existence of virus (Summers 2005). Two years later in the year 1917, a French-Canadian microbiologist Felix d'Herelle observed a similar clear zone phenomenon in stool samples of bacillary dysentery patients. Unlike Twort, this time, d'Herelle was able to explain the presence of "invisible microbe", a virus which he termed as "Bacteriophage" (Brunoghe and Maisin 1921). Later, during the 1920s, various clinical trials on phagotherapy were carried out in Eastern Europe and the Soviet Union, where therapy was used for the treatment of variety of diseases including bubonic plague and cholera in India (Nikolich and Filippov 2020). Despite encouraging initial success of the page therapy, their application as antimicrobial approach was declined because of the discovery of antibiotics in the mid-nineteenth century.

20.4   Bacterial Diseases in Aquaculture and Its Control Measures

Despite the fact that aquaculture is one of the fastest rising food-production sectors in the world, it is currently plagued by frequent and severe outbreaks of diseases. The sector is under threat from several groups of pathogen such as bacteria, fungi, viruses, and parasites. Among all these concerns, the bacterial pathogens can endure well in both fresh water and marine water aquatic ecosystem without their host; and the attribute favours them as major impediments to the aquaculture industry. The situation is further exaggerated by the adopted intensive culture practices and human anthropogenic activities which has led the foundation for the adulteration in the optimal physico-chemical quality of the aquatic environment (Pridgeon and Klesius 2012). Till now, about 13 bacterial genera have been identified as pathogenic to aquatic organisms including fish, which comprises both gram-negative pathogens (*Edwardsiella, Aeromonas, Vibrio, Flavobacterium, Pseudomonas, Yersinia, Francisella, Piscirickettsia, Photobacterium* and *Tenacibaculum*) and gram-positive (*Renibacterium, Lactococcus* and *Streptococcus*) (Pridgeon and Klesius 2012; Gui and Zhang 2018).

To control bacterial disease outbreak in an aquatic system, feeding fishes with drug-medicated feed, especially antibiotics, is a general practice. At present, the addition of various kinds of nutraceuticals or functional food is very well accepted to remediate the situation either as a prophylactic or therapeutic agent (Pridgeon and Klesius 2012). However, the approach is usually expensive and maybe ineffective for therapeutic purposes as infection-weaken fish do not accept any kind of feed especially medicated feed. Additionally, frequent and sub-therapeutic level of chemical additives or drugs over an extended period led the base for the development of AMR among pathogens (Cunha 2009). Substitutes for antimicrobial agents with similar or enhanced protection are therefore urgently needed to provide robust protection against variety of bacterial aetiological agents in target organisms. At present, the application of various kinds of vaccines, immunostimulant of natural or chemical origin is very well accepted in commercial aquaculture farms, along with several biocontrol strategies such as application of probiotic, bacteriophages and symbiotic. Among these alternative strategies, phagotherapy emerges as a sustainable substitute to chemical therapeutics, since phage application has the potential to not only eliminate the virulent pathogens precisely but can also to help in the creation of homeostasis in aquatic environment by minimizing the application of chemicals and other remedial drugs to achieve the goals of "One Health" approach of WHO.

20.5 Research on Bacteriophage Therapy in Aquaculture

Although bacteriophages were discovered way back at the beginning of the nineteenth century, however, the focus of research on its therapeutic potential against bacterial diseases was limited to a certain part of the world because of the poor understanding of phage life cycle and bacteria-phage interactions (Almeida et al. 2009). Furthermore, with the discovery of antibiotics, the application of phages remains underexplored. However, in some places such as Eastern Europe and in the Soviet Union, they successfully demonstrated several clinical trials on human patients which laid the foundation to the future work (Park et al. 2020). Moreover, the emergence of multi-drug resistant bacteria has substantially encouraged researchers to explore the potential of phagetherapy; because, phages can be employed as bioagents against wide range of bacterial pathogens. Owing to the specificity of phages to their host, the probability of disrupting natural microflora of aquatic environment or host inhabiting beneficial bacteria will be null which is very unlikely with the administration of common broad-spectrum antibiotics (Fortuna et al. 2008). The very first attempt to employ phage therapy in aquaculture was made in the year 1981 in Taiwan against *Aeromonas hydrophila* in loach (*Misgurnus anguillicaudatus*) (Wu et al. 1981). Nowadays, work associated with the phagotherapy against bacterial pathogens in aquaculture has been accepted worldwide and encouraging researchers to explore the application and efficacy of phage therapy in different circumstances under various culture conditions (Table 20.1).

Table 20.1 Isolation and application of bacteriophage in aquaculture

Bacteriophage application in finfish

Pathogen	Disease (lesion)	Organism	Bacteriophage	Phage administration	Treatment	Reference
Aeromonas salmonicida	Furunculosis	Brook trout (*Salvelinus fontinalis*)	HER 110	Immersion	The treatment at MOI 100 not only delayed the onset of infection by 7 days; additionally, bacteriophage reduced the total mortality from 100% to 10%	Imbeault et al. (2006)
		Atlantic salmon (*Salmo salar*) and Rainbow trout (*Oncorhynchus mykiss*)	O, R and B	Intraperitoneal injection, oral feeding and immersion	No adverse effect was observed. However, using a combination of all three phages by injection only delayed the death, but didn't affect the result as none of the treatments was able to provide protection against infection	Verner-Jeffreys et al. (2007)
		Rainbow trout (*Oncorhynchus mykiss*)	PAS-1	Intramuscular injection	Fish treated with MOI of 10,000 showed a significant survival rate of 26.7%. The surviving fish did not show ulcerative lesions and remained healthy until 14 days post administration	Kim et al. (2015)

		Senegalese sole (*Solea senegalensis*)	AS-A	Immersion	After 72 h of infection, fish juveniles treated with phages at MOI of 100 showed no mortality contrary to 36% mortality in the untreated control group	Silva et al. (2016)
Edwardsiella ictaluri	Edwardsiellosis or enteric septicaemia	Japanese eel (*Anguilla Japonica*)	Phages ET-1	–	Phages were very effective with lysing capacity of 92.6% against 27 bacterial hosts. Additionally, at MOI 0.08 phages were able to reduce down the bacterial count by 99.9% in water	Wu (1982)
		Channel catfish (*Ictalurus punctatus*)	ΦeiDWF, ΦeiAU and ΦeiMSLS (*Siphoviridae*)	–	The in vitro analysis reveals the lysing capacity of phages, which can be used for therapeutic application	Carrias et al. (2011)
		Ayu (*Plecoglossus altivelis*)	–	Intraperitoneal injection	Higher protection was observed in fish that were first injected with phages and then 1 h later injected with the pathogen, whereas the fish that was first injected with the pathogen and then the phages only showed	Mahmoud and Nakai (2012)

(continued)

Table 20.1 (continued)

Pathogen	Disease (lesion)	Organism	Bacteriophage	Phage administration	Treatment	Reference
					delayed mortality compared with the control	
E. tarda	Edwardsiellosis or Edwardsiella septicaemia	Zebrafish (Danio rerio)	ETP-1 (Podoviridae)	Immersion	The fish were bath exposed to phages for 12 days and concurrently infected with E. tarda, the result revealed the elevated survival in treatment in comparison to control until 4 days post challenge	Nikapitiya et al. (2020)
E. tarda and A. hydrophila	Hemorrhagic septicaemia and Edwardsiellosis	Japanese eel (A. japonica)	Different bacteriophages combination	Immersion	At MOI of 11.5 the bacterial count was reduced 3 times within 2 h of exposure. Whereas in pond water, 250-folds reduction at MOI of 0.23 in 8 h. Additionally, the count of E. tarda was dropped by 85% even in the absence of phage in the pond water after 48 h of exposure	Hsu et al. (2000)
A. hydrophila	Haemorrhagic septicaemia or		pAh1-C and pAh6-C		Both of the intraperitoneal and oral	Jun et al. (2013)

Motile Aeromonas Septicaemia (MAS)	Cyprinid loach (*Misgurnus anguillicaudatus*)		Intraperitoneal injection and oral feeding	administration improved the survival	Le et al. (2018)
	Striped catfish (*Pangasianodon hypophthalmus*)	*A. hydrophila* Φ2 and *A. hydrophila* Φ-5	Intraperitoneal injection	The survival rate of catfish at MOI 100 was 100%, compared to the 18.3% survival in the control devoid of phage treatment	
	Loach (*Misgurnus anguillicaudatus*)	Akh-2 (*Siphoviridae*)	Immersion	Mortality rates were 16%, 53%, 57% and 56.67% after 24, 48, 72 and 96 h, respectively when compared to the control group with 100% mortality; most of the surviving fish showed no disease symptoms	Akmal et al. (2020)
A. hydrophila and *Pseudomonas fluorescens*	Rainbow trout (*Oncorhynchus mykiss*)	–	Bacteriophage cocktail BAFADOR®, containing 3 bacteriophages against *A. hydrophila* and 4 against *P. fluorescens* was used for immersion or feeding of fish	Stimulation of non-specific immune system and reduction of mortality	Schulz et al. (2019a)

(continued)

Table 20.1 (continued)

Pathogen	Disease (lesion)	Organism	Bacteriophage	Phage administration	Treatment	Reference
		European eels (*Anguilla anguilla*)	–	Fish were fed with bacteriophage cocktail BAFADOR® containing 3 bacteriophages against *A. hydrophila* and 4 against *P. fluorescens*	Stimulation of cellular and humoral immunity and reduction in mortality	Schulz et al. (2019b)
Flavobacterium columnare	Columnaris disease	Catfish (*Clarias batrachus*)	FCP1–FCP9 FCP1 (*Podoviridae*)	Intramuscular injection, bath and oral feeding	Phage treatment led to the disappearance of gross clinical signs, negative bacteriological test, detectable phage and 100% survival	Prasad et al. (2011)
		Rainbow trout (*Oncorhynchus mykiss*) and zebrafish (*Danio rerio*)	FCL-2	–	Reduced mortality	Laanto et al. (2015)
F. psychrophilum	Systemic bacterial coldwater disease (CWD)	Rainbow trout (*Oncorhynchus mykiss*) and other species of trouts	FpV-1 to FpV-22	–	Significant lytic capacity against with broad host range	Stenholm et al. (2008)
		Ayu fish (*Plecoglossus altivelis*)	PFpW-3, PFpC-Y (*Myoviridae*) PFpW-6, PFpW-7 (*Podoviridae*) PFpW-8 (*Siphoviridae*)	–	PFpW-3 displayed significant lytic capacity	Kim et al. (2010)

		Host	Phage	Treatment	Outcome	Reference
		Atlantic salmon (*Salmo salar*) and rainbow trout (*Oncorhynchus mykiss*)	–	Intraperitoneal injection	Mortality decreased in the range of 16% to 100%	Castillo et al. (2012)
Lactococcus garvieae	Lactococcosis	Yellowtail (*Seriola quinqueradiata*)	PlgY, PLgY-16, PLgY-30, PLgW-1 (*Siphoviridae*)	Intraperitoneal injection and oral feeding	Both administered phage prevented fish from experimental *L. garvieae* infection. Mortality drops from 90% to 45% (for injection), whereas for oral mortality drop from 65% to 10%.	Nakai et al. (1999)
P. aeruginosa	Ulcerative lesions	Catfish (*Clarias gariepinus*)	–	On-spot treatment	The therapy efficiently cured the infected fish within 8 to 10 days with a sevenfold reduction of the lesion with untreated infection control	Khairnar et al. (2013)
P. plecoglossicida	Bacterial haemorrhagic ascites disease	Ayu (*Plecoglossus altivelis*)	PPpW-3 (*Podoviridae*) PPpW-4 (*Myoviridae*) and a combination of both PPpW-3/PPpW-4	Oral feeding	At MOI 1, mortality drop from 65% to 22%	Park et al. (2000)
		Ayu (*Plecoglossus altivelis*)	PPpW-3, PPpW-4	Oral	Phage-receiving fish showed high protection against infection and	Park and Nakai (2003)

(continued)

Table 20.1 (continued)

Pathogen	Disease (lesion)	Organism	Bacteriophage	Phage administration	Treatment	Reference
Streptococcus iniae	Streptococcosis	Japanese flounder (*Paralichthys olivaceus*)	PSiJ31, PSiJ32, PSiJ4, and PSiJ42	Intraperitoneal injection	mortality drop from 90 to 26% Mortalities of fish receiving phages were significantly lower than the control, ranging from 80% to 0%	Matsuoka et al. (2007)
Streptococcus agalactiae	–	Nile tilapia (*Oreochromis niloticus*)	–	Immersion	Treated fish had survival rates of 60% with a delayed mean death time of about 3 days in comparison to control	Jun et al. (2017)
Vibrio anguillarum	Vibriosis	Atlantic salmon (*S. salar*)	ALMED, CHOED, ALME, CHOD, CHOB	Immersion	At MOI of 1 and 20, the treatment increased the survival of fish up to 100%. Mortality drop from 95 to 30% at MOI 1 and at MOI 20 from 95% to 0%	Higuera et al. (2013)
		Atlantic cod (*G. morhua*) and turbot (*Scophthalmus maximus*) larvae	KVP40	Immersion	The maximum reduction in mortality varied from 29% to 92% for turbot and from 49% to 86%; notably, reduction in mortality	Rørbo et al. (2018)

					was not significant in the majority of cases	
Bacteriophage application in shellfish						
V. alginolyticus	Skin ulceration and viscera ejection	Sea cucumber (*Apostichopus japonicus*)	–	Immersion	Increased survival in a range of 73, 50 and 47% at MOI of 10, 1 and 0.1, respectively, whereas the no phage treatment group only had 3% of survival rate	Zhang et al. (2015)
		Live prey (*Artemia salina*)	jSt2 and jGrm1	Immersion	At MOI 100, 93% reduction of presumptive *Vibrio* concentration after 4 h of treatment	Kalatzis et al. (2016)
V. harveyi	Luminous vibriosis	Larvae of *Penaeus monodon*	VHLM (*Myoviridae*)	Immersion	The laboratory trial showed that survival was enhanced up to 80% with two doses of bacteriophage, whereas survival rate in control was only 25%	Vinod et al. (2006)
		Larvae of *P. monodon*	Viha8, Viha10 (*Siphoviridae*) Viha9, Viha11	Immersion	Mortality drops from 88% to 32% compared to antibiotic treatment	Karunasagar et al. (2005, 2007)
		Penaeid shrimp	Viha1 to Viha7 (six from *Siphoviridae* and one Viha4 from *Myoviridae*)	–	All the phages were found to be highly lytic with different lytic spectrum. Three of the phages (Viha1, Viha3 and Viha7) caused 65%	Shivu et al. (2007)

(continued)

Table 20.1 (continued)

Pathogen	Disease (lesion)	Organism	Bacteriophage	Phage administration	Treatment	Reference
					of the strains to lyse while Viha2, Viha4 and Viha6 caused 40% of the host strains to lyse. Viha5 had a narrow spectrum (14%)	
		Tropical rock lobster (*Panulirus ornatus*)	VhCCS-06 (*Siphoviridae*)	–	Phages were able to eliminate the host bacterial count up to 1.2×10^7 CFU mL^{-1} compared to control 9.3×10^7 CFU mL^{-1}	Stomps et al. (2010)
		Shrimp larvae (*P. monodon*)	Bacteriophages VHM1, VHM2 and VHS1	Immersion	The phages were applied alone and in different cocktail combinations. Larval survival was in a range of 60%–88.3% after 96 h in the phage treatment group, compared to 26.6% to 35% survival in the control treatments without phage	Stalin and Srinivasan (2017)
		Abalone (*Haliotis laevigata*)	vB_VhaS-a, vB_VhaS (*Siphoviridae*)	Immersion	The treatment was revealed survival of about 70%	Wang et al. (2017)
			VHP6b	Immersion	After 10 days, mortality in the treated group was	Patil et al. (2014)

		Host	Phage	Route	Findings	Reference
		Black tiger shrimp (*P. monodon*)		Immersion	20% when compared to >70% of control treatment	Choudhury et al. (2012, 2019)
		Black tiger shrimp (*P. monodon*)	Phage V		Optimum activity of *V. harveyi* phage was observed at salinity of 25 ppt, pH of 7, TDS of 11.25 mg mL^{-1} and temperature of 30 °C. Combination of recombinant shrimp lysozyme and *V. harveyi* phage significantly improved the phage activity	
V. parahaemolyticus	Vibriosis	Brine shrimp (*Artemia franciscana*)	–	–	Single dose was efficient enough to eliminate the pathogens. However, when the phage treatment was delayed, it was ineffective to control the mortality	Martinez-Diaz and Hipólito-Morales (2013)
		Whiteleg shrimp (*Litopenaeus vannamei*) larvae	A3S and Vpms1	Immersion	At MOI of 0.1, the infection was counteracted and an early application (at 6 h post-infection) was	Lomelí-Ortega and Martinez-Díaz (2014).

(continued)

Table 20.1 (continued)

Pathogen	Disease (lesion)	Organism	Bacteriophage	Phage administration	Treatment	Reference
		Shrimp (*Penaeus vannamei*)	–	Oral diet and immersion	effective to avoid mortality	
					Mortality in groups treated 1 h after bacterial infection was 100%, whereas prophylactic use of phages resulted in mortality varied from 25% to 50%	Luo et al. (2018)
		Blue mussels (*Mytilus edulis*)	–	Immersion	Phage cocktail was effective in significantly reducing *V. parahaemolyticus* to undetectable numbers in mussels	Onarinde and Dixon (2018)
		Oysters	*Siphoviridae* pVp-1	Immersion and surface application	After 72 h of phage application with bath immersion, bacterial growth was reduced up to 1.4×10 CFU mL^{-1} in the treatment group as compared to control (8.9×10^6 CFU mL^{-1}). Whereas, after 12 h of phage surface application, the bacterial growth was	Jun et al. (2014)

Bacteria	Disease	Host	Phage	Application	Results	Reference
Vibrio sp. VA-F3	—	Shrimp (L. vannamei)	ValLY-3, VspDsh-1, VspSw-1, VpaJT-1 and ValSw4-1 (Siphoviridae)	—	inhibited by 1.94 $CFU\ mL^{-1}$ of the treatment group to 1.44×10^6 $CFU\ mL^{-1}$ in the control group. Survival rate assessed after 7 days of cultivation reached 91.4% when compared to 20% rate in the untreated control group	Chen et al. (2019)
V. splendidus	Severe epizootics Skin Ulceration Syndrome (SUS)	Sea cucumber (Apostichopus japonicus)	vB_VspS_VS-ABTNL-1 (PVS-1), vB_VspS_VS-ABTNL-2 (PVS-2) and vB_VspS_VS-ABTNL-3 (PVS-3)	Oral feeding	Survival rate during the next 10 days was 18% for the control group, whereas 82% for the phage cocktail, and 65%, 58% and 50% for the three phages applied alone	Li et al. (2016a, b)
V. cyclitrophicus	—	Sea cucumbers (A. japonicus)	vB_VcyS_Vc1	Oral feeding	Reduced mortality	Li et al. (2016a, b)
V. coralliilyticus	Massive mortality of Pacific oyster larvae	Pacific oyster larvae (Crassostrea gigas)	pVco-14 (Siphoviridae)	—	Significantly higher survival rate in treatments compared to the untreated control	Kim et al. (2019)

20.6 Phage-Based Products for Therapy in Aquaculture

The potential and efficacy of phages have encouraged some private companies/ institutes to develop phage-based product for commercial application to treat bacterial diseases in aquaculture which is tabulated below (Table 20.2).

20.7 Strategic Guideline for the Development of Phage Therapy in Aquaculture

For the development of bacteriophages therapy in aquaculture, a set of standard protocols need to be followed (Nakai and Park 2002; Choudhury et al. 2017) (Fig. 20.1). This includes isolation and characterization of phage (Fig. 20.2), in vivo and in vitro therapeutic potentiality testing, safety testing and regulatory approval, etc.

20.8 Dose and Mode of Application for Phage Therapy

There are several modes of application of phage therapy reported by many researchers since its discovery. However, the application of phage in the aquaculture system includes direct release of phages in the culture system, injection through intramuscular or intraperitoneal mode, immersion, oral administration through feed, anal intubation, etc. Among all these reported modes, release of phages directly into the culture system is the most preferred method (Shivu et al. 2007; Choudhury et al.

Table 20.2 Phage-based products for therapy in aquaculture

Name of the Company/ Institute	Product description	References
Intralytix	Phage therapy (as cocktail of phage) to control *Vibrio tubiashii* and *V. coralliitycis* infections in oyster	Intralytix I (2018)
Phage Biotech Ltd	Phage therapy to treat *V. harveyi* infections in shrimp	Phage Biotech (2017)
Mangalore Biotech Laboratory	Phage formulation (LUMI-NIL MBL) to control luminous vibriosis in shrimp	Mangalore Biotech Laboratory (2019)
Fixed Phage Ltd	Binds the phages in feed pallets for phage therapy aquaculture.	Mattey (2020)
ACD Pharma	Phage-based solutions against Yersiniosis in Atlantic salmon	ACD Pharma (2017)
Proteon pharmaceutical	Phage-based product BAFADOR® to targets aquaculture pathogens *Pseudomonas* spp. and *Aeromonas* spp. via immersion	Grzelak (2017)
ICAR-CIBA	LUMI^{PHAGE} for biocontrol of luminous bacteria in shrimp larvae	ICAR-CIBA (2017)

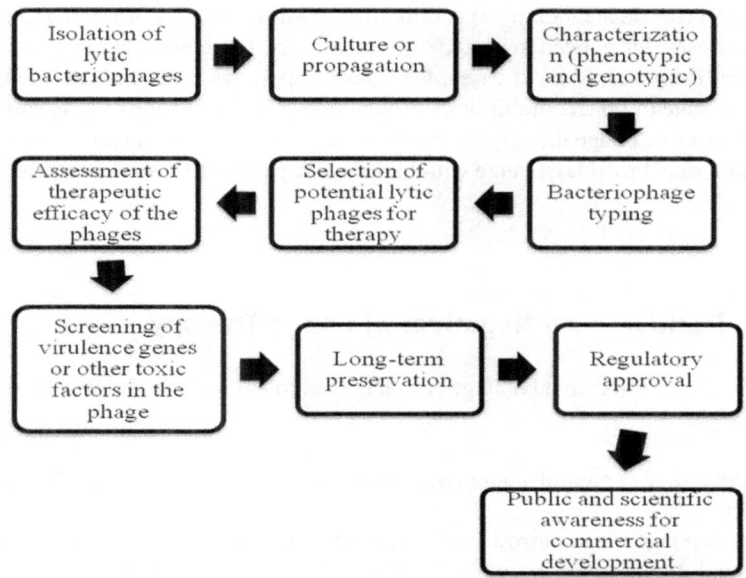

Fig. 20.1 Strategies for bacteriophage therapy in aquaculture

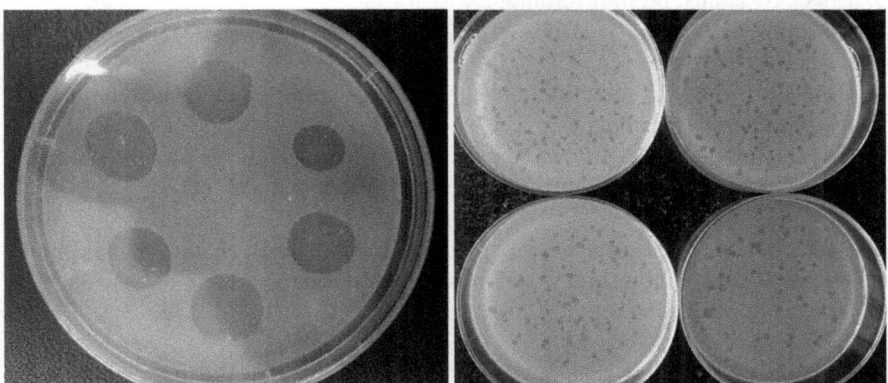

Fig. 20.2 Zone of lysis by bacteriophages and plaques formed by bacteriophage

2017; Silva et al. 2016). In recent times, various combination of phage as "cocktail" has gained a lot of interest among researchers as futuristic bacteriophage approach. Cocktail of diverse combinations such as phage-phage, phage-probiotic, phage-immunostimulant and phage-antibiotic are demonstrated in the literature (Fischetti et al. 2006; Chan et al. 2013; Choudhury et al. 2019). There are advantages and disadvantages to each mode of application; which often depends on the nature of the bacterial pathogen (Martinez-Diaz and Hipólito-Morales 2013; Richards 2014).

For effective phage therapy, it is important to know the exact dose of application. Various doses have been reported by researchers for both laboratory and field condition. However, in most cases, the dose of application depends on the type of pathogen, state of phage, multiplicity of infection (MOI) of phage or lytic capability, etc. For effective phage therapy, researchers may attempt to isolate phage with a high replication rate, broad host range with high lytic capacity at lower doses (Choudhury et al. 2017).

20.9 Positives and Negatives of Phage Therapy

Several well-established advantages of phage treatment include (Barrow et al. 1998; Nakai 2010):

1. Because of the natural abundance, phage isolation is comparatively easy and cheap.
2. Bacteriophages have narrow host range indicating that phages are very specific to host and do not harm the endemic intestinal or environmental microflora.
3. No inherent toxicity and environment friendly.
4. Self-replicating capability eliminates the necessity of multiple administrations.
5. Effective against biofilm-forming bacteria.
6. Bacteriolytic capability of phages allows them to eliminate MDR (multi-drug resistant) bacteria.
7. Because of the high specificity, phages do not contribute to the development of resistance among pathogens.
8. Administration of phages can be very feasible because of the multimodal application such as oral, aerosols, immersion, injection, and topical.

Bacteriophage application has an immense potential but even then, the feasibility, accessibility and field efficacy still remains a concern, which roots to several drawbacks in phage therapy:

1. Because of the high specificity of phages, the pathogenic bacteria must be identified before therapy, which may prove to be a realistic and practical challenge in the field condition.
2. Difficult to extrapolate in vivo efficacy in comparison to in vitro results.
3. Temperate phages can transfer lethal or toxic genes to harmless bacteria.
4. Because of the robust nature of the host bacteria phage resistance can be developed by bacteria.
5. Contradictory opinion on interaction with the immune responses of fish/shellfish.
6. There might be practical difficulties, e.g. injecting large numbers of animals, acceptance of phage mediated feed to diseased fish.
7. Conversion of lytic phage to lysogenic state is still a mystery among phage experts and may be a concern prior to application.

20.10 Conclusion

Bacteriophage therapy has been reintroduced in the system after the rise of drug-resistant bacteria and to cater the necessity of finding an alternative to chemotherapeutic application. Owing to the host specificity of phage and lytic capability, it can prove to be an attractive approach in that it provides a ray of hope against AMR. At present, the potential phagotherapy has established its efficacy in preventing or controlling the bacterial infections in both freshwater and marine water in various target species of fish and shellfish origin. Bacteriophage therapy has been intensively researched and developed against various clinical conditions in the area of biomedical application. However, in aquaculture, the therapy is not yet fully investigated. The lack of in vitro and in vivo research on optimization and efficacy in different culture condition existing in diverse aquatic environments has led to the challenge we are facing today, with the development of effective field-based formulation. It is high time that attempts are made to address the concerns that have arisen over time, and research efforts should therefore be conceptualized and aimed at establishing sustainable phage therapy.

References

Abedon ST, Kuhl JS, Blasdel BG, Kutter EM (2011) Phage treatment of human infections. Bacteriophage 1:66–85

ACD Pharma (2017) Bacteriophage therapy—ACD Pharma. https://acdpharma.com/2021/nyhet/acd-pharmas-work-with-bacteriophages/. Accessed 9 Jul 2021

Ackermann HW (2003) Bacteriophage observations and evolution. Res Microbiol 154(4):245–251

Adriaenssens EM, Wittmann J, Kuhn JH, Dann Turner D, Sullivan MB, Dutilh BE, Jang HB, Zyl LJV, Klumpp J, Lobocka M, Switt AIM (2018) Taxonomy of prokaryotic viruses: 2017 update from the ICTV bacterial and archaeal viruses subcommittee. Arch Virol 166:1125–1129

Akmal M, Rahimi-Midani A, Hafeez-Ur-Rehman M, Hussain A, Choi TJ (2020) Isolation, characterization, and application of a bacteriophage infecting the fish pathogen *Aeromonas hydrophila*. Pathogens 9(3):215

Almeida A, Cunha A, Gomes NCM, Alves E, Costa L, Faustino MAF (2009) Phage therapy and photodynamic therapy: low environmental impact approaches to inactivate microorganisms in fish farming plants. Mar Drugs 7:268–313

Al-Sum AB, Al-Dhabi NA (2014) Isolation of bacteriophage from Mentha species in Riyadh, Saudi Arabia. J Pure Appl Microbiol 8(2):945–949

Barrow P, Lovell M, Berchieri A (1998) Use of lytic bacteriophage for control of experimental *Escherichia coli* septicemia and meningitis in chickens and calves. Clin Diagn Lab Immunol 5:294–298

Biotech P (2017) Developments. Phage Biotech, Tel Aviv

Bostock J, McAndrew B, Richards R, Jauncey K, Telfer T, Lorenzen K, Little D, Ross L, Handisyde N, Gatward I, Corner R (2010) Aquaculture: global status and trends. Philos Trans R Soc Lond B Biol Sci 365(1554):2897–2912

Brunoghe R, Maisin J (1921) Essais de therapeutique au moyen du bacteriophage du staphylocoque. C R Soc Biol 85:1020–1021

Carrias A, Welch TJ, Waldbieser GC, Mead DA, Terhune JS, Liles MR (2011) Comparative genomic analysis of bacteriophages specific to the channel catfish pathogen, *Edwardsiella ictaluri*. Virol J 8:6

Castillo D, Higuera G, Villa M, Middelboe M, Dalsgaard I, Madsen L, Espejo RT (2012) Diversity of *Flavobacterium psychrophilum* and the potential use of its phages for protection against bacterial cold water disease in salmonids. J Fish Dis 35(3):193–201

Chan BK, Abedon ST, Loc-Carrillo C (2013) Phage cocktails and the future of phage therapy. Future Microbiol 8:769–783

Chen L, Fan J, Yan T, Liu Q, Yuan S, Zhang H, Yang J, Deng D, Huang S, Ma Y (2019) Isolation and characterization of specific phages to prepare a cocktail preventing *Vibrio* sp. Va-F3 infections in shrimp (*Litopenaeus vannamei*). Front Microbiol 10:2337

Choudhury TG, Maiti B, Venugopal MN, Karunasagar I (2012) Effect of Total Dissolved Solids and Temperature on Bacteriophage Therapy against Luminous vibriosis in Shrimp. Bamidgeh 64:761

Choudhury TG, Tharabenahalli Nagaraju V, Gita S, Paria A, Parhi J (2017) Advances in bacteriophage research for bacterial disease control in aquaculture. Rev Fish Sci Aquac 25(2):113–125

Choudhury TG, Maiti B, Venugopal MN, Karunasagar I (2019) Influence of some environmental variables and addition of r-lysozyme on efficacy of *Vibrio harveyi* phage for therapy. J Biosci 44 (1):8

Cisek AA, Dąbrowska I, Gregorczyk KP, Wyżewski Z (2017) Phage therapy in bacterial infections treatment: one hundred years after the discovery of bacteriophages. Curr Microbiol 74 (2):277–283

Cunha BA (2009) Antibiotic resistance. Med Clin N Am 84:1407–1429

D'Hérelle F (1917) Sur un microbe invisible antagoniste des bacilles dysentériques. C R Acad Sci 165:373–375

Fischetti VA, Nelson D, Schuch R (2006) Reinventing phage therapy: are the parts greater than the sum? Nat Biotechnol 24(12):1508–1511

Fortuna W, Miedzybrodzki R, Weber-Dabrowska B, Gorski A (2008) Bacteriophage therapy in children: facts and prospects. Med Sci Monit 14(8):RA126–RA132

Freifelder DM (1987) Microbial genetics. Jones and Bartlett, Portolla Valley, CA

Gelband H, Miller-Petrie M, Pant S, Gandra S, Levinson J, Barter D, White A, Laxminarayan R (2015) The state of the World's antibiotics 2015. Wound Heal S Afr 8:30–34

Grzelak J (2017) BAFADOR® presented at the international bacteriophage conference in Tbilisi. https://www.proteonpharma.com/bafador-at-the-international-bacteriophageconference-in-tibilisi/. Accessed 9 Jul 2021

Gui L, Zhang QY (2018) Disease prevention and control. In: Aquaculture in China: success stories and modern trends, pp 577–598

Hankin EH (1896) An outbreak of cholera in an officers' mess. Br Med J 2:1817–1819

Higuera G, Bastías R, Tsertsvadze G, Romero J, Espejo RT (2013) Recently discovered *Vibrio anguillarum* phages can protect against experimentally induced vibriosis in Atlantic salmon, *Salmo salar*. Aquaculture 392:128–133

Hsu CH, Lo CY, Liu JK, Lin C (2000) Control of the eel (Anguilla japonica) pathogens, *Aeromonas hydrophila* and *Edwardsiella tarda*, by bacteriophages. J Fish Soc Taiwan 27(1):21–31

ICAR-CIBA (2017). http://www.ciba.res.in/images/aquaticdiv/adv/EHP,WSSV,%20Lumiphage%202.pdf

Imbeault S, Parent S, Lagacé M, Uhland CF, Blais JF (2006) Using bacteriophages to prevent furunculosis caused by *Aeromonas salmonicida* in farmed brook trout. J Aquat Anim Health 18 (3):203–214

Intralytix I (2018) Intralytix, Inc. http://www.intralytix.com/

Jassim SAA, Limoges RG (2014) Natural solution to antibiotic resistance: bacteriophages "the living drugs". World J Microbiol Biotechnol 30:2153–2170

Jun JW, Kim JH, Shin SP, Han JE, Chai JY, Park SC (2013) Protective effects of the *Aeromonas* phages pAh1-C and pAh6-C against mass mortality of the cyprinid loach (*Misgurnus anguillicaudatus*) caused by *Aeromonas hydrophila*. Aquaculture 416–417:289–295

Jun JW, Kim HJ, Kil Yun S, Chai JY, Park SC (2014) Eating oysters without risk of vibriosis: application of a bacteriophage against *Vibrio parahaemolyticus* in oysters. Int J Food Microbiol 188:31–35

Jun JW, Han JE, Giri SS, Tang KF, Zhou X, Aranguren LF, Kim HJ, Yun S, Chi C, Park SC (2017) Phage application for the protection from acute Hepatopancreatic necrosis disease (AHPND) in *Penaeus vannamei*. Indian J Microbiol 58:114–117

Kalatzis PG, Bastias R, Kokkari C, Katharios P (2016) Isolation and characterization of two lytic bacteriophages, φSt2 and φGrn1; phage therapy application for biological control of *Vibrio alginolyticus* in aquaculture live feeds. PLoS One 11(3):e0151101

Karunasagar I, Vinod MG, Kennedy B, Vijay A, Deepanjali A, Umesh K, Karunasagar I (2005) Biocontrol of bacterial pathogens in aquaculture with emphasis on phage therapy. In: Diseases in Asian Aquaculture V, Fish Health Section, Asian Fisheries Society, Proceedings of the Fifth Symposium on Diseases in Asian Aquaculture

Karunasagar I, Shivu MM, Girisha SK, Krohne G, Karunasagar I (2007) Biocontrol of pathogens in shrimp hatcheries using bacteriophages. Aquaculture 268(1–4):288–292

Khairnar K, Raut MP, Chandekar RH, Sanmukh SG, Paunikar WN (2013) Novel bacteriophage therapy for controlling metallo-beta-lactamase producing *Pseudomonas aeruginosa* infection in catfish. BMC Vet Res 9(1):264

Kim JH, Gomez DK, Nakai T, Park SC (2010) Isolation and identification of bacteriophages infecting ayu *Plecoglossus altivelis* altivelis specific *Flavobacterium psychrophilum*. Vet Microbiol 140:109–115

Kim JH, Choresca CH, Shin SP, Han JE, Jun JW, Park SC (2015) Biological control of *Aeromonas salmonicida* subsp. Salmonicida infection in rainbow trout (*Oncorhynchus mykiss*) using Aeromonas phage PAS-1. Transbound Emerg Dis 62:81–86

Kim HJ, Jun JW, Giri SS, Chi C, Yun S, Kim SG, Kim SW, Kang JW, Han SJ, Kwon J, Oh WT (2019) Application of the bacteriophage pVco-14 to prevent *Vibrio coralliilyticus* infection in Pacific oyster (*Crassostrea gigas*) larvae. J Invertebr Pathol 167:107244

Kowalska JD, Kazimierczak J, Sowińska PM, Wójcik EA, Siwicki AK, Dastych J (2020) Growing trend of fighting infections in aquaculture environment—opportunities and challenges of phage therapy. Antibiotics 9(6):301

Laanto E, Bamford JKH, Ravantti JJ, Sundberg LR (2015) The use of phage FCL-2 as an alternative to chemotherapy against columnaris disease in aquaculture. Front Microbiol 6:1–9

Lafferty KD, Harvell CD, Conrad JM, Friedman CS, Kent ML, Kuris AM, Powell EN, Rondeau D, Saksida SM (2014) Infectious diseases affect marine fisheries and aquaculture economics. Ann Rev Mar Sci 7:471–496

Le TS, Nguyen TH, Vo HP, Doan VC, Nguyen HL, Tran MT, Tran TT, Southgate PC, Kurtböke Dİ (2018) Protective effects of bacteriophages against *Aeromonas hydrophila* causing motile Aeromonas septicemia (MAS) in striped catfish. Antibiotics 7(1):16

Letchumanan V, Chan KG, Pusparajah P, Saokaew S, Duangjai A, Goh BH, Ab Mutalib NS, Lee LH (2016) Insights into bacteriophage application in controlling vibrio species. Front Microbiol 7:01114

Li Z, Li X, Zhang J, Wang X, Wang L, Cao Z, Xu Y (2016a) Use of phages to control *Vibrio splendidus* infection in the juvenile sea cucumber *Apostichopus japonicus*. Fish Shellfish Immunol 54:302–311

Li Z, Zhang J, Li X, Wang X, Cao Z, Wang L, Xu Y (2016b) Efficiency of a bacteriophage in controlling Vibrio infection in the juvenile sea cucumber *Apostichopus japonicus*. Aquaculture 451:345–352

Lin DM, Koskella B, Lin HC (2017) Phage therapy: an alternative to antibiotics in the age of multi-drug resistance. World J Gastrointest Pharmacol Ther 8(3):162–173

Lomelí-Ortega CO, Martínez-Díaz SF (2014) Phage therapy against *Vibrio parahaemolyticus* infection in the whiteleg shrimp (*Litopenaeus vannamei*) larvae. Aquaculture 434:208–211

Luo X, Liao G, Liu C, Jiang X, Lin M, Zhao C, Tao J, Huang Z (2018) Characterization of bacteriophage HN 48 and its protective effects in Nile tilapia *Oreochromis niloticus* against *Streptococcus agalactiae* infections. J Fish Dis 41(10):1477–1484

Mahmoud M, Nakai T (2012) Bacteriophage therapy of *Edwardsiella ictaluri* infection in ayu *Plecoglossus altivelis*. In: Proceedings of the 5th global fisheries and aquaculture research conference, Faculty of Agriculture, Cairo University, Giza, Egypt, 1-3 October 2012

Mangalore Biotech Laboratory (2019) MANGALORE BIOTECH LAB: Products. http://mangalorebiotech.com/products.html

Martinez-Diaz SF, Hipólito-Morales A (2013) Efficacy of phage therapy to prevent mortality during the vibriosis of brine shrimp. Aquaculture 400:120–124

Matsuoka S, Hashizume T, Kanzaki H, Iwamoto E, Park SC, Yoshida T, Nakai T (2007) Phage therapy against β-hemolytic streptococcicosis of Japanese flounder *Paralichthys olivaceus*. Fish Pathol 42:181–189

Mattey M (2020) Fixed Phage Ltd. Treatment of bacterial infections in aquaculture. U.S. Patent 10,849,942

Nakai T (2010) Application of bacteriophages for control of infectious diseases in aquaculture. In: Bacteriophages in the control of food-and waterborne pathogens. American Society of Microbiology

Nakai T, Park SC (2002) Bacteriophage therapy of infectious diseases in aquaculture. Res Microbiol 153:13–18

Nakai T, Sugimoto R, Park KH, Matsuoka S, Mori K, Nishioka T, Maruyama K (1999) Protective effects of bacteriophage on experimental *Lactococcus garvieae* infection in yellowtail. Dis Aquat Organ 37:33–41

Nikapitiya C, Chandrarathna HPSU, Dananjaya SHS, De Zoysa M, Lee J (2020) Isolation and characterization of phage (ETP-1) specific to multidrug resistant pathogenic *Edwardsiella tarda* and its in vivo biocontrol efficacy in zebrafish (*Danio rerio*). Biologicals 63:14–23

Nikolich MP, Filippov AA (2020) Bacteriophage therapy: developments and directions. Antibiotics 9(3):135

Ninawe AS, Sivasankari S, Ramasamy P, Kiran GS, Selvin J (2020) Bacteriophages for aquaculture disease control. Aquac Int 28(5):1925–1938

Onarinde BA, Dixon RA (2018) Prospects for biocontrol of *Vibrio parahaemolyticus* contamination in blue mussels (*Mytilus edulus*)—a year-long study. Front Microbiol 9:1043

Park SC, Nakai T (2003) Bacteriophage control of *Pseudomonas plecoglossicida* infection in ayu *Plecoglossus altivelis*. Dis Aquat Organ 53(1):33–39

Park SC, Shimamura I, Fukunaga M, Mori KI, Nakai T (2000) Isolation of bacteriophages specific to a fish pathogen, *Pseudomonas plecoglossicida*, as a candidate for disease control. Appl Environ Microbiol 66(4):1416–1422

Park SY, Han JE, Kwon H, Park SC, Kim JH (2020) Recent insights into *Aeromonas salmonicida* and its bacteriophages in aquaculture: a comprehensive review. J Microbiol Biotechnol 30 (10):1443–1457

Patil JR, Desai SN, Roy P, Durgaiah M, Saravanan RS, Vipra A (2014) Simulated hatchery system to assess bacteriophage efficacy against *Vibrio harveyi*. Dis Aquat Organ 112:113–119

Prasad Y, Kumar AD, Sharma AK (2011) Lytic bacteriophages specific to *Flavobacterium columnare* rescue catfish, *Clarias batrachus* (Linn.) from columnaris disease. J Environ Biol 32(2):161–168

Pridgeon JW, Klesius PH (2012) Major bacterial diseases in aquaculture and their vaccine development. Anim Sci Rev 7:1–16

Richards GP (2014) Bacteriophage remediation of bacterial pathogens in aquaculture: a review of the technology. Bacteriophage 4(4):e975540

Rørbo N, Rønneseth A, Kalatzis PG, Rasmussen BB, Engell-Sørensen K, Kleppen HP, Wergeland HI, Gram L, Middelboe M (2018) Exploring the effect of phage therapy in preventing *Vibrio anguillarum* infections in cod and turbot larvae. Antibiotics 7(2):42

Schulz P, Pajdak-Czaus J, Robak S, Dastych J, Siwicki AK (2019a) Bacteriophage-based cocktail modulates selected immunological parameters and post-challenge survival of rainbow trout (*Oncorhynchus mykiss*). J Fish Dis 42:1151–1160

Schulz P, Robak S, Dastych J, Siwicki AK (2019b) Influence of bacteriophages cocktail on European eel (*Anguilla anguilla*) immunity and survival after experimental challenge. Fish Shellfish Immunol 84:28–37

Sharma S, Chatterjee S, Datta S, Prasad R, Dubey D, Prasad RK, Vairale MG (2017) Bacteriophages and its applications: an overview. Folia Microbiol 62(1):17–55

Shivu MM, Rajeeva BC, Girisha SK, Karunasagar I, Krohne G, Karunasagar I (2007) Molecular characterization of *Vibrio harveyi* bacteriophages isolated from aquaculture environments along the coast of India. Environ Microbiol 9:322–331

Sieiro C, Areal-Hermida L, Pichardo-Gallardo Á, Almuiña-González R, de Miguel T, Sánchez S, Sánchez-Pérez Á, Villa TG (2020) A hundred years of bacteriophages: can phages replace antibiotics in agriculture and aquaculture? Antibiotics 9(8):493

Silva YJ, Moreirinha C, Pereira C, Costa L, Rocha RJM, Cunha Â, Gomes N, Calado R, Almeida MA (2016) Biological control of *Aeromonas salmonicida* infection in juvenile Senegalese sole (*Solea senegalensis*) with phage AS-A. Aquaculture 450:225–233

Simmonds P, Adams MJ, Benko M, Breitbart M, Brister JR, Carstens EB, Davison AJ, Delwart E, Gorbalenya AE, Harrach B, Hull R (2017) Consensus statement: virus taxonomy in the age of metagenomics. Nat Rev Microbiol 15:161–168

Stalin N, Srinivasan P (2017) Efficacy of potential phage cocktails against *Vibrio harveyi* and closely related Vibrio species isolated from shrimp aquaculture environment in the south east coast of India. Vet Microbiol 207:83–96

Stenholm AR, Dalsgaard I, Middelboe M (2008) Isolation and characterization of bacteriophages infecting the fish pathogen *Flavobacterium psychrophilum*. Appl Environ Microbiol 74:4070–4078

Stomps CC, Hoj L, Bourne DG, Hall MR, Owens L (2010) Isolation of lytic bacteriophage against *Vibrio harveyi*. J Appl Microbiol 108:1744–1750

Summers WC (2005) Bacteriophage research: early history. In: Kutter E, Sulakvelidze A (eds) Bacteriophages: biology and applications. CRC, Boca Raton, FL

Twort F (1915) An investigation on the nature of ultra-microscopic viruses. Lancet 186:1241–1243

Van Boeckel TP, Pires J, Silvester R, Zhao C, Song J, Criscuolo NG, Gilbert M, Bonhoeffer S, Laxminarayan R (2019) Global trends in antimicrobial resistance in animals in low-and middle-income countries. Science 365(6459):eaaw1944

Verner-Jeffreys DW, Algoet M, Pond MJ, Virdee HK, Bagwell NJ, Roberts EG (2007) Furunculosis in Atlantic salmon (*Salmo salar* L.) is not readily controllable by bacteriophage therapy. Aquaculture 270(1–4):475–484

Vinod MG, Shivu MM, Umesha KR, Rajeeva BC, Krohne G, Karunasagar I, Karunasagar I (2006) Isolation of *Vibrio harveyi* bacteriophage with a potential for biocontrol of luminous vibriosis in hatchery environments. Aquaculture 255(1–4):117–124

Walker PJ, Siddell SG, Lekowitz EJ, Mushegian AR, Dempsey DM, Dutilh BE, Harrach B, Harrision RL, Hendrickson RC, Junglen S, Knowles NJ (2019) Changes to virus taxonomy and international code of virus classification and nomenclature ratified by the international committee on taxonomy of viruses. Arch Virol 164:2417–2429

Wang Y, Barton M, Elliott L, Li X, Abraham S, O'Dea M, Munro J (2017) Bacteriophage therapy for the control of *Vibrio harveyi* in greenlip abalone (*Haliotis laevigata*). Aquaculture 473:251–258

Wu JL (1982) Isolation and application of a new bacteriophage, ET-1, which infect *Edwardsiella tarda*, the pathogen of edwardsiellosis. Rep Fish Dis Res (Taiwan) 4:8–17

Wu JL, Lin HM, Jan L, Hsu YL, Chang LH (1981) Biological control of fish bacterial pathogen, *Aeromonas hydrophila*, by bacteriophage AH 1. Fish Pathol 15(3–4):271–276

Zhang J, Cao Z, Li Z, Wang L, Li H, Wu F, Jin L, Li X, Li S, Xu Y (2015) Effect of bacteriophages on *Vibrio alginolyticus* infection in the sea cucumber, *Apostichopus japonicus* (Selenka). J World Aquac Soc 46(2):149–158

Disease Diagnostic Tools for Health Management in Aquaculture

21

Vikash Kumar, Suvra Roy, B. K. Behera, and Basanta Kumar Das

Abstract

Disease diagnosis is an essential part of fish disease management and prevention. Starting from microscopic characterization and morphological descriptions, the diagnostic methods for aquatic animal pathogens have advanced to molecular characterization and probe-based diagnosis. Apart from microscopic, histological, microbiological and immunological techniques, the molecular tools ensure early detection of infectious disease and avoid severe mortalities and production losses to aquaculture. Additionally, the molecular sequencing techniques presently allow great advancement in biology and are improving diagnosis and control of pathogens by studying the complete genomes of pathogens. With the advancement in diagnostic tools, the specificity, sensitivity and speed of diagnosis have improved significantly in recent years which aids in epidemiological studies as well as in identification of causes of disease outbreaks or the presence of pathogens to prevent further spread of infectious fish diseases in aquaculture.

Keywords

Aquaculture · Disease diagnosis · Diagnostic techniques · Fish health management

V. Kumar (✉) · S. Roy · B. K. Behera · B. K. Das
Aquatic Environmental Biotechnology and Nanotechnology (AEBN) Division, ICAR-Central Inland Fisheries Research Institute (CIFRI), Barrackpore, West Bengal, India

© The Author(s), under exclusive license to Springer Nature Singapore Pte Ltd. 2021
P. K. Pandey, J. Parhi (eds.), *Advances in Fisheries Biotechnology*,
https://doi.org/10.1007/978-981-16-3215-0_21

21.1 Introduction

Aquaculture, the farming of aquatic animals and plants, continues to dominate the food producing sector in the world. The total global aquaculture production in 2017 was 111.9 million tonnes (an increase of 4.9% compared to 2016), with over 91% of global production currently being produced within the Asian region (Dong et al. 2021; Kumar et al. 2020). The contribution of aquaculture to the total production of aquatic animals from capture and aquaculture combined has risen steadily from 25.7% in 2000 to 46.4% in 2017, with an average annual growth rate of 4.8% during 2011–2017, eclipsing the rapid growth rates of the pig (<2.5%) and poultry (~3%) sectors (FAO 2019; Tacon 2020).With total aquaculture production valued at over US$ 250 billion, the aquaculture sector represents a very diverse group of different animal species and aquatic plants. In 2017, 328 species ranging from the production of unicellular *Chlorella* algae within the indoor bioreactor to the production of Atlantic salmon (*Salmo salar*) in outdoor floating net cages were reported to contribute in world aquaculture production (FAO 2019; Kumar 2020). As with capture fisheries, fish still represent the largest major species group of aquaculture production by weight (53.4 MT or 47.7% total aquaculture), followed by aquatic plants (31.8 MT or 28.4% total), molluscs (17.4 MT or 15.4% total), crustaceans (8.4 MT or 7.5% total), amphibians and reptiles (471,758 tonnes) and miscellaneous invertebrate animals (422,124 tonnes) (FAO 2019). Moreover, as the global human population continues to expand at a high rate and is expected to reach over nine billion by 2030, the aquaculture will play an important role in providing global food and nutritional security to people in both developed and developing countries, and also support the livelihood and provide jobs to the global populations (FAO 2019; Verdegem 2013).

However, due to the global demand, the pressure for intensification and expansion of aquaculture systems has made most aquaculture businesses fragile. Intensive and high density land-based culture results in suboptimal oxygen level, accumulation of metabolic waste products and leftover feed, injuries and wounds due to animal interactions. Together, this can create stressful environmental conditions that can significantly affect the health status of the cultured animals (Woo and Gregory 2014). Also, the degraded aquatic environment leads to decreased growth performance and reduced immune response. In turn, this is an ideal ground for disease outbreaks, having a devastating impact on socio-economic development worldwide (Herbeck et al. 2013; Kalaimani et al. 2013). In the aquaculture industry, economic losses from disease outbreak have been estimated by the FAO to be over US $9 billion per year, which is approximately 15% of the value of world farmed fish and shellfish production. The bacterial disease problem in shrimp aquaculture has escalated since the late 2010, when the industry collapsed in South-Asian countries. Therefore, there is an urgent need to thoroughly understand the diagnostic methods in the aquaculture system; it will be crucially important to ensure early detection of infectious diseases and avoid severe mortalities and production losses in aquaculture.

21.2 Diagnostic Tools Used in Fish Disease

The term diagnosis is derived from the latin word '*diagnōsis*' meaning to distinguish. Diagnosis in medicine refers to the process of determining or identifying a disease by examining the nature and circumstances including sign and symptoms of disease condition. In aquaculture, disease diagnosis is an integral part of aquatic animal health management. In fact, during diagnosis for particular disease conditions, various techniques and procedures are used in order to identify the nature of disease and precisely distinguish primary and secondary causative agent or pathogen involved (Noga 2010). Apart from searching the causative agent for disease, some diagnostic techniques are also used to screen the health of aquatic animals to certify that they do not have any sub-clinical infections. In general, proper diagnosis helps to adopt accurate therapies and avoid indiscriminate use of chemotherapies, that might be associated with development of antibiotic resistance in bacterial pathogens and alteration of host gut microbiota and immunity (Wang and Li 2009; Woo and Bruno 2011). Moreover, the new diagnostic tools that aim to reduce the burden of infectious diseases and most often have life-saving impacts are, in general, not well understood and require more complex analysis.

Investigations of potential disease condition are initiated by observation of case history, for instance, feeding and swimming behaviour, followed by sampling and analysis. Since the initial symptoms of several diseases are common, it is necessary to carry out extensive laboratory analysis, which may include traditional microscopic, histological and microbiological diagnostic techniques (Schmitt and Henderson 2005). Moreover, in several instances the above-mentioned techniques are not sufficient to detect the aetiological agent of a disease and hence, more advanced immunological (ELISA) or molecular biological (PCR, sequencing) methods are used for confirmatory diagnosis (Fig. 21.1).

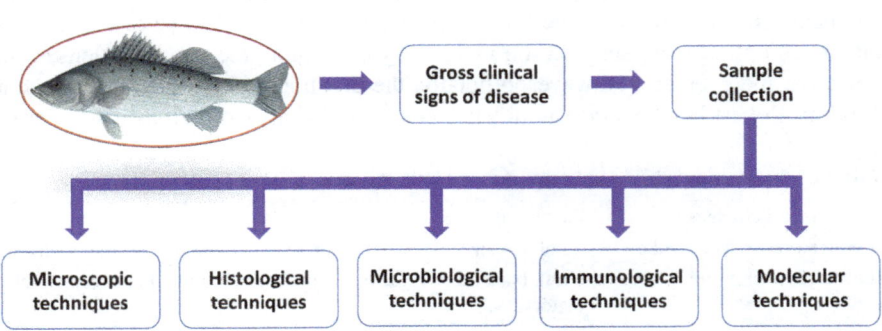

Fig. 21.1 Stages in the diagnosis of fish disease

21.2.1 Case History of Disease: An Important Prerequisite of Disease Diagnosis

During disease diagnosis, three pieces of data are generally collected: the case history, physical examination findings and diagnostic test results. Among these three components, the case history of the infection from a farm or diseased sample provides important information that helps to determine the circumstances under which the disease has developed. In addition, it gives presumptive ideas about the nature of problem that might have to deal with disease conditions (Tesser and Norman 2016). Although the clinical signs may not directly indicate the aetiology of disease, these combined with other diagnostic techniques would aid in confirmatory diagnosis of disease condition in aquaculture. For instance, as indicated in Table 21.1, the fish can have very similar behavioural signs during disease, caused by biotic (viral, bacterial, parasite) or abiotic factors (environmental, toxicants).

Moreover, the mortality pattern can give a fairly good idea whether the problem is due to environmental factors, microbial agents or pollutants (Love and Lewbart 1997). For instance, during infectious disease, a gradual acceleration in morbidity and mortality pattern are observed, except in case of virulent viral (e.g. white spot syndrome virus) or bacterial (e.g. acute hepatopancreatic necrosis disease) infections where 90–100% mortality is recorded. Moreover, if overnight mass mortality (up to 100%) is observed, it can be due to oxygen depletion or any other environmental factors like chemical pollutants or toxicants. In case of nutritional deficiency, the clinical signs are more protracted, and mortality is seldom observed (Noga 2010; Woo and Bruno 2011). The information, collected from the case history, might start from recording the behavioural changes, and morphological abnormalities to stocking information. The details of information normally collected are summarized in Fig. 21.2.

21.2.2 Microscopy Techniques in Disease Diagnosis

The term microscope comes from two Greek words 'mikros' meaning small and 'skopein' meaning to look. The first microscope was believed to be discovered by the Dutch optician Janssen and son in 1595 by combining the two glass lenses, an objective and eyepiece. However, officially, the first microscope was discovered in 1667 by Robert Hooke and Antoni van Leeuwenhoek, also known as the father of

Table 21.1 The fish behavioural signs during biotic and abiotic factors

Behavioural changes	Factors involved
Reduced or no feed intake	Viral, bacterial or parasitic infection and environmental factors
Lethargic swimming	Viral, bacterial, fungal or parasitic infection and environmental factors
Spinning and erratic swimming	Viral, parasitic and environmental pollutants

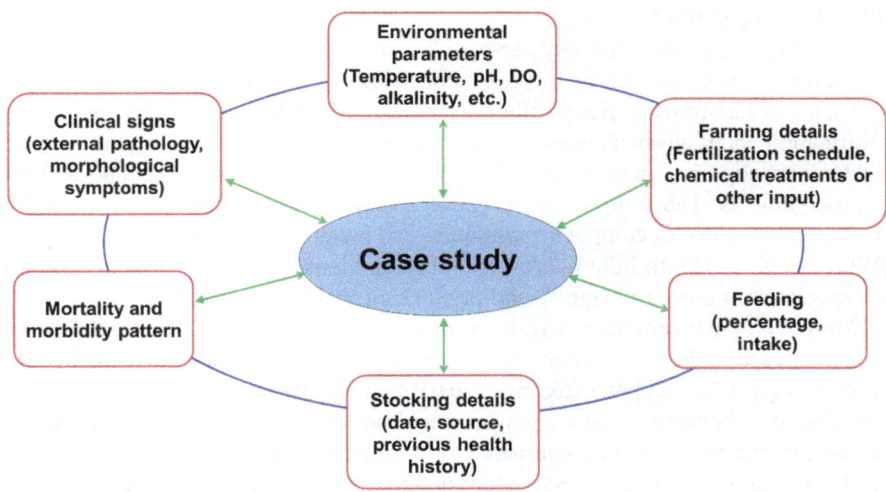

Fig. 21.2 Case history information collected and carefully examined in order to potentially diagnose a disease with increased specificity

microscopy (Harris 2019). Soon after discovery, Antoni van Leeuwenhoek observed several elements, including bacteria and named it 'animalcules'. Two centuries later, in 1882, Louis Pasteur demonstrated that 'animalcules' are pathogenic bacteria, *Bacillus tuberculosis* (first identified pathogenic bacteria), responsible for causing disease in humans. Afterwards, for many years microscopy has been the only tool available for the detection of microbes through inspection of blood smears or tissue specimens (Jorink 2018). It is also an important diagnostic tool for rapid and inexpensive (e.g. light microscopy) analysis of clinical samples for infectious diseases. However, sample preparation for direct observation is often time-consuming, labour intensive, and proper diagnosis depends on qualified laboratory technicians. Nevertheless, in remote areas, where resources are limited, this technique (e.g. light microscopy) has proven to be an important tool in providing vital information on nature of infectious agents, whether the problem is due to bacterium, parasite or fungus (Groisberg et al. 2017).

In general, microscopes are considered as the most basic tool for fish bacterial and parasitic disease diagnosis. However, with the advancement in microscopy techniques, e.g. electron microscopes, the microscopy tools can be used for viral disease diagnosis in fish (Lakshman 2019). The tissue or samples are often treated with different colour stains, causing them to stand out from the background, although wet mounts of unstained samples can be used to detect fungi, parasites (including helminth eggs and larvae) and motile organisms. Moreover, numerous colour stains have been developed to identify microbes, cellular structures, or indicators of infection in tissue samples, under the microscope. However, no stain is 100% specific for a particular microbial pathogen. The microscopy techniques are generally divided into two parts, (a) light microscopy and (b) electron microscopy.

21.2.2.1 Light Microscopy

Light microscopes are extremely versatile instruments that can be adapted to examine sample of any size, whole or sectioned, living or dead, wet or dry, hot or cold, and static or fast-moving. They offer a wide range of contrast techniques, providing information about physical, chemical, and biological attributes of the specimen. The light microscope is generally used for interpreting histological slides and stained microorganisms. The samples can be observed with the eye directly, recorded by photographic, video or computer techniques, and image components can be analysed (Witten et al. 2017). In light microscope, different lenses are used to focus light on the specimen in order to magnify and produce an image.

Moreover, the magnification varies and greatly depends on the types and number of lenses, used in the microscope. Depending on the number of lenses, there are two types of light microscopes (Lakshman 2019): simple light microscope (it has low magnification because it uses a single lens) and the compound light microscope (it has a higher magnification compared to the simple microscope because it uses at least two sets of lenses, an objective lens and an eyepiece). The magnification range of compound microscope extends from $4\times$, $10\times$ or $40\times$ objective lenses and multiplied by the power of the eyepiece which is typically $10\times$.

21.2.2.2 Electron Microscopy

An electron microscope is an advanced form of microscope that uses a beam of accelerated electrons as a source of illumination instead of light. It was first built by Ernst Ruska in 1931 (a German engineer and academic professor), and to this day the same principles are used behind all the prototype of modern electron microscopes. The electron microscope was designed with an objective to achieve high resolution of images and magnify the objects in nanometres, which are done by the use of electron beam in vacuum and images are captured on a phosphorescent screen (Lakshman 2019). Based on operating styles, there are two types of electron microscopes.

Transmission electron microscope (TEM) is generally used to examine thin slices or sections of cells or tissues. These microscopes have a maximum magnification of around $\times 1,000,000$, but images can be enlarged photographically beyond that. TEM facilitates study of the inner structure and analysis of the cellular features on an atomic scale (in the range of a few nanometers), the structures in cells that are not visible with the light microscope (El-Naggar et al. 2016). TEM can be used, among other things, to visualize the interior of cells (in thin sections), the organization of molecules in viruses and cytoskeletal filaments (prepared by the negative staining technique), the structure of protein molecules (contrasted by metal shadowing) and the arrangement of protein molecules in cell membranes (by freeze-fracture).

Scanning electron microscope (SEM) has been widely used in disease diagnosis to characterize the surface structure of biomaterials and to measure cell attachment and changes in morphology of microbes. In addition, SEM is also useful for determining the number and distribution of microorganisms that adhere to surfaces. SEM uses finely focused electron beams to reveal the detailed surface characteristics of a specimen and provide information relating to its three-dimensional structure (Adel et al. 2019). As the name suggests, in SEM, the images are formed by scanning the surface of specimen in a raster pattern from a focused electron beam. The SEMs are often used at lower magnifications (up to $\times 30,000$). As compared to TEM, the resolution limit of a SEM is lower than that of a TEM (approximately 50 nm).

21.2.3 Histological Techniques in Disease Diagnosis

Histological methods are becoming quite common in diagnostic methodology in aquatic animals for quite some time. The term is derived from the Greek 'histos' meaning tissue, refers to the 'science of tissues'. Histology involves the microscopic examination of thin, stained tissue sections in order to study their structure, the cellular organization of body tissues and organs (Kumar et al. 2014a, b). On the other hand, in histopathology the investigation is done to determine changes which occur in tissues due to pathogens and diseases. Hence, histopathology is used as a promising diagnostic tool to identify the pathological conditions associated with diseases. The light microscopy techniques are widely used for histological tissue analysis. With the advancement of microscopy techniques, e.g. electron microscope, the visualization of subcellular structure to a greater extent is becoming possible. In general, the fine cellular structure with a magnification of $1000\times$ or less can be distinguished at the level of light microscopy. While, in electron microscope, the magnification of above $1000\times$ up to as high as $10,000,000\times$, with a resolution of 50 picometers (0.05 nm) is employed to visualize the detailed ultrastructure of the cell cytoplasm, organelles and membranes. Hence, the histological techniques are now embracing the study of the structures of both tissue and cells, and the relationship between these structures and physiological function (Mishra et al. 2020).

The first step of any histologic analysis is sample collection. The fish tissues are put in aqueous fixative, which preserves the morphology of tissues and cells, so that they withstand further preparatory steps without change. Tissues should be fixed immediately after death of fish to avoid disintegration of tissues or cells by the action of their own enzymes. After fixation, tissues are gradually dehydrated using a graded series of alcohol. The tissues are then subjected to clearing by a liquid that mixes completely with both alcohol and wax, the embedding agent. The tissues are then embedded in molten paraffin wax and the blocks are cooled to harden the wax blocks which are subjected to sectioning or cutting using a microtome (Sirri et al. 2020). The thin sections (3–5μm) are made and then spread on warm water bath and mounted onto glass microscope slides. The wax is removed from the sections before staining.

Moreover, during examination the tissue sections are prepared and stained with suitable dyes and later examined under light microscope for changes, resulting from infectious or non-infectious disease. Haematoxylin and eosin (H&E) are the most widely used histological stains because of their ability to reveal a wide range of

Table 21.2 Different colour stains used to distinguish fish cellular microstructures

Cellular structure	Staining methods		
	H & E (Haematoxylin and Eosin)	PAS/MY (Periodic acid-Schiff's reagent/ Metalin yellow)	Thionin
Nucleus (non-ovarian)	Deep blue-violet	Deep blue-violet	Deep blue with distinct chromatin
Nucleolus (non-ovarian)	Orange-pink to red-violet depending on cell type	Blue-grey to yellow-grey	–
Chromatin (non-ovarian)	Blue-black	Blue-black	Deep blue
Cytoplasm (non-ovarian)	Varying shades of pink and orange	Shades of yellow or yellow-tan	Light blue
Collagen	Pale pink	Magenta	–
Muscle	Deep pink	Bright yellow	–
Nuclei of primary oocytes	Deep blue-violet	Pale yellow-grey	–
Chromatin of primary oocytes	Blue-black	Pale yellow-grey	–
Cytoplasm of pre-vitellogenic oocytes	Varying shades of pink and orange	Deep blue-violet	–
Erythrocytes (red blood cells)	Deep blue-violet (nucleus) and varying shades of pink and orange (cytoplasm)	–	Green
Mucus	–	–	Red-purple to violet
Cartilage matrix	–	–	Red-purple to violet
Mast cell granules	–	–	Deep violet

different tissue components. In brief; the stains differentiate cellular structure in tissue samples which have been summarized in Table 21.2.

21.2.3.1 Histochemistry

Histochemistry comprises of two words histo and chemistry, which means the chemistry of tissues. Histochemistry has become one of the most widely used techniques in disease diagnosis to localize and visualize the cellular components, tissues and other living structures. This technique uses different stains and indicators, which reacts with the cellular components, to develop tiny coloured structures that could be easily observed under a microscope. Moreover, the immunohistochemistry (IHC) introduced in 1970 is a technique used to assess the presence of a specific protein or antigen in cells by use of a specific antibody that binds to it. The antibody allows visualization of the protein under a microscope. Immunocytochemistry is a valuable tool to study the presence and subcellular localization of proteins (Ekman and Norrgren 2003; Kumar et al. 2019). In addition, IHC plays an important role for the detailed study of undiagnosed diseases. Although IHC is an evolving technique of fish disease diagnosis, it is developed for the diagnosis of viruses such as infectious pancreatic necrosis virus (IPNV), infectious salmon anaemia virus (ISAV) and noda virus in paraffin-embedded tissue sections. The viral antigen is localized by an antibody, raised against the virus and subsequent detection steps result in a coloured product that is later visualized by light microscopy.

The histological techniques provide detailed description of tissue pathology and highlight the sequence of cellular changes and their progression, caused by infectious and non-infectious diseases. By examining tissue sections, viruses, bacteria, fungi and parasites can be identified and using immune histochemical techniques, the confirmatory diagnosis of the important infectious agents is possible.

21.2.4 Microbiological Techniques in Disease Diagnosis

The term microbiology, derived from Greek word mĩkros- 'small'; bios- 'life' and logia, is the study of microorganisms. In microbiological techniques, we try to identify pathogenic unicellular or multicellular organisms belonging to bacteria, virus, yeast or fungi from diseased fish samples. For microbiological analysis, the samples must be collected aseptically, prior to any treatment and possibly from site of infection or lesions. For instance (a) skin wounds, samples should be taken at the margin of an acute lesion, (b) if enteritis is suspected, sample from the gut and (c) in case of septicaemia, the best site for sampling is the posterior kidney (moreover, posterior kidney is the target organ for many diseases and harbours more than 70% of bacteria because of extensive blood supply and large number of phagocytes lining the renal sinusoids) (Austin and Austin 1989). Moreover, when diseased tissues are exposed to the surrounding environment, such as with surface ulcers and/or gills, there is an inherent risk of contamination and difficulties in discerning the relevance of resulting bacterial growth, i.e. which is the pathogen, a secondary invader of already dead or diseased tissue, or a chance contaminant. The internal tissues are less likely to be plagued by contaminants provided they have been derived from the freshly dead or moribund fish rather than specimens that have been deceased for numerous hours. The microbiological tools used for effective diagnosis of microbial pathogens are summarized below:

21.2.4.1 Staining Methods
Prior to microbiological analysis (e.g. culture and biochemical assays), various staining methods are used to confirm that the disease associated changes are possibly due to microbial infection or abiotic factors. For instance, for bacterial disease diagnosis, the specimens are treated with Gram stain to visualize Gram-positive and Gram-negative bacteria (Austin and Austin 2012). In addition, if bacterial pathogens are not visible in Gram stains, other stains or other identification methods

Table 21.3 Staining methods used to identify various bacterial pathogens

Staining methods	Microorganism
Gram stain (crystal violet, Gram's iodine, ethanol, safranin)	1. Gram-positive bacteria—purple or blue; 2. Gram-negative bacteria—pink or red
Acid-fast stain (carbolfuchsin and methylene blue)	To distinguish two types of gram-positive bacteria, based on the presence or absence of waxy mycolic acids in their cell walls Acid-fast bacteria (mycolic acids)—red or pink colour against blue background
Capsule staining (negative staining technique, staining around the cells as capsules do not absorb basic dyes—India ink or nigrosine)	1. The capsule visible as halos surrounding the yeast cells—India ink or nigrosine 2. Colour the bacterial cell (crystal violet and copper sulphate, positive stain); Colours the background but not the capsule, leaving halo around each bacterial cell (India ink or nigrosine—negative stain)
Endospore staining (malachite green and safranin)	1. Bacterial endospore—green 2. Bacterial vegetative cells—pink 3. Endospore within vegetative cells—green with surrounding pink
Flagella staining (tannic acid, pararosaniline or basic fuchsin)	Bacterial flagella are visible (pink-red colour), if present
Fluorescent stains	1. Bacteria and fungi—Acridine orange 2. Mycobacteria—auramine-rhodamine and auramine O 3. Fungi, especially dermatophytes—calcofluor white
Romanowsky stains (Wright and Giemsa stain) (methylene blue, azure B and eosin Y)	Detection of parasites in blood, intracellular inclusion bodies formed by viruses and intracellular bacteria 1. Cell cytoplasm—orange to pink 2. Nucleus—blue to purple
Wheatley Gomori Trichrome stain	Detection of yeast cells and intestinal protozoa (e.g. microsporidia)

are used, e.g. acid-fast stain if mycobacteria are suspected. The details of the stains used in microbiological diagnosis are summarized in Table 21.3.

21.2.4.2 Culture of Microbial Pathogens in Selective Medium

After preliminary identification of infectious pathogens, an attempt is made to culture the pathogen from pathological material, followed by identification using phenotypic data in comparison to the published diagnostic schemes, such as those in 'Bergey's Manual of Systematic Bacteriology'. Having acquired diseased tissue, a comparatively narrow range of selective media is used for culture of microbial pathogens. Generally, the selective media lack imagination in terms of the characteristics of the host and thus, the likely nutritional needs of the pathogen, i.e. the nutrients likely to be available in situ to the pathogen (Austin 2011). For instance, Zobell's marine 2216E agar and/or thiosulphate citrate bile salt agar

Table 21.4 Culture media used for the growth of microbial pathogens

Non-selective medium	
Brain heart infusion agar + 3% (w/v) NaCl	*Shewanella putrefaciens*
Cytophaga agar prepared in seawater	*Tenacibaculum* spp.
Flexibacter maritimus medium (FMM)	*Tenacibaculum* spp.
Marine 2216E agar	*Photobacterium damselae, Vibrio spp.*
Tryptone soya agar + 1–2% (w/v) NaCl	*Aeromonas salmonicida, Aliivibrio salmonicida, V. harveyi, V. splendidus, V. vulnificus*
Non-selective medium-enriched	
AUSTRAL-Salmonid Rickettsial Septicaemia (SRS) broth + L-cysteine	*Piscirickettsia salmonis*
Blood agar + 0.5–1.5% (w/v) NaCl	*Aliivibriowodanis, Moritella marina, Moritella viscosa, Photobacterium damselae subsp. piscicida, V. splendidus, V. tapetis*
Cystine heart agar supplemented with 1% (w/v) haemoglobin	*Francisella* spp.
Glucose asparagine agar	*Nocardia* spp.
Kidney disease medium 2 (KDM2)	*R. salmoninarum*
Löwenstein–Jensen medium/dorset egg medium	*Mycobacterium* spp., *Nocardia* spp.
Middlebrook 7H10 medium	*Mycobacterium* spp.
Ogawa egg medium	*Mycobacterium gordonae*
Selective medium	
Coomassie brilliant blue agar	*Aeromonas salmonicida*
Selective kidney disease medium (SKDM)	*R. salmoninarum*
Thiosulphate citrate bile salt sucrose agar (TCBS)	*Photobacterium damselae, Vibrio spp.*

(TCBS; if vibrios were suspected) are often used with incubation at 15–37 °C for 1–7 days for diseased marine fish (Table 21.4). For fastidious pathogens, such as *Mycobacterium* and *Renibacterium salmoninarum*, specialized media are used for the recovery. Moreover, the desired outcome after incubation is the presence of dense bacterial growth; scant growth of colonies being indicative only of the presence of contaminants (Austin and Austin 2012). Conventionally, pure dense growth from diseased tissues was considered to be indicative of the recovery of the actual microbial pathogen. Nevertheless, the immediate goal of disease diagnosis is recovery of pure cultures, although the reasons for the choices of which colonies/growth to use from the isolation plates may well be highly subjective. The most common method to isolate individual microbial cells and produce a pure culture is to prepare a streak plate. The streak plate method is a way to physically separate the microbial population and is generally done by spreading the inoculum back and forth with an inoculating loop over the solid agar plate. Upon incubation, colonies grow, and single colony forming unit is isolated from the biomass.

21.2.4.3 Biochemical Test

Biochemical tests are most frequently used microbiological technique for disease diagnosis. They specifically test the metabolic and enzymatic products, produced by infectious agent. Biochemical tests are also used for fermentation products, acids, alcohol or gases that may be products of metabolic pathways. Few biochemical tests used in identification of microbial pathogens are discussed below:

The **indole test** is performed to demonstrate the ability of certain bacteria to produce tryptophanase, an enzyme that cleaves amino acid tryptophan to indole, which accumulates in the medium (Huang et al. 2020). Indole production test is important in the identification of Enterobacteria. Most strains of *Escherichia coli, Proteus vulgaris, P. rettgeri, Morganella morganii* and *Providencia* species break down the amino acid tryptophan with the release of indole. The test organism is inoculated into tryptone broth, a rich source of the amino acid tryptophan.

The **Catalase test** demonstrates the presence of catalase, an enzyme that catalyses the release of oxygen from hydrogen peroxide (H_2O_2) (Dias et al. 2019). It is used to differentiate those bacteria that produce an enzyme catalase, such as staphylococci, from non-catalase producing bacteria such as streptococci. In general, catalase-positive bacteria include strict aerobes as well as facultative anaerobes, although they all have the ability to respire using oxygen as a terminal electron acceptor. Moreover, catalase-negative bacteria may be anaerobes, or they may be facultative anaerobes that only ferment and do not respire using oxygen as a terminal electron acceptor (i.e. Streptococci).

The objective of **lipase test** is to identify bacteria, capable of producing the exoenzyme lipase. The lipase test is used to detect and enumerate lipolytic bacteria, e.g. Enterobacteriaceae, *Clostridium, Staphylococcus* and *Neisseria* from diseased fish samples. Tributyrin agar is used as a differential medium that tests the ability of an organism to produce a lipase exoenzyme, that hydrolyzes tributyrin oil (Patchimpet et al. 2019). A lipase positive microbial pathogen produces lipase and breaks down the tributyrin and can be visualized as a clear halo around the areas where the lipase-producing organism has grown in agar plate.

21.2.5 Immunological Techniques in Disease Diagnosis

The immunological techniques, based on antigen-antibody reactions, are highly specific and sensitive. These techniques opened the possibility of rapid diagnoses directly from infected fish tissues, even on the fish farm. The immunological tools are used for the qualitative and quantitative estimation of the pathogens and/or the protective antibody, even without the aid of the instruments (Vadstein et al. 2013; Kumar et al. 2018).This method has the advantage over other traditional method. It can detect sub-clinical/latent/carrier sate of infection and can also discriminate the antigenic differences. For detecting the presence of pathogens, array of polyclonal and monoclonal antibody-based diagnostic kits are available for various aquatic animal pathogens. However, availability of highly specific monoclonal antibodies reduced the chance of misidentification as there was always the risk of cross-reactions with resultant erroneous conclusions (Lulijwa et al. 2019). Additionally, specificity of the antibodies also limits the usefulness of immunological techniques because major antigens are not conserved among life stages of certain pathogens. The most commonly used immunological techniques are described below:

21.2.5.1 Agglutination Test

Agglutination tests detect antibody or antigen, involved in agglutination of bacteria, red cells, or antigen- or antibody-coated latex particles. They rely on the bivalent nature of antibodies, which can form cross-link with particulate antigens. In agglutination tests, an antigen reacts with its corresponding antibody, resulting in formation of visible clumping of bacterial cells (Toranzo et al. 1987). Agglutination tests are frequently used for initial confirmation of specific pathogens. Since antibodies to the target organism may cross-react with other organisms and auto-agglutination may occur, these must be considered as screening tests and further confirmation will usually be necessary.

21.2.5.2 Agar Gel Immuno-Diffusion Assay

This technique is used to detect antibody produced in serum (the fluid, non-cellular part of blood) in response to infection (Sherman et al. 1984). In this test, antibody is placed in a centre well of the agar plate and antigens (specific or nonspecific) are placed in surrounding wells. These two test components passively diffuse out of the well into the agar. If the serum sample contains antibodies to antigens at the proper concentrations they bind, forming an interlaced antigen-antibody complex that precipitates in the agar. The precipitation is visible to the unaided eyes as a thin white line; also called precipitin line is formed. The test is best interpreted by using a known positive control serum sample (provided in several commercial kits) in the assay for comparison. However, this technique has largely been replaced with more sensitive ELISA technology.

21.2.5.3 ELISA

ELISA (enzyme-linked immunosorbent assay), both specific and sensitive, is a plate-based assay technique, designed for detecting and quantifying peptides, proteins, antibodies and hormones. Originally described by Engvall and Perlmann (1971), the method enables analysis of protein samples, immobilized in microplate wells, using specific antibodies. In ELISA, an antigen (target macromolecules) is immobilized to a solid surface and then complexed with an antibody that is linked to a reporter enzyme (typically, alkaline phosphatase or horseradish peroxidase). Detection is accomplished by assessing the antibody-enzyme conjugated activity via incubation with a substrate (o-phenylenediamine for alkaline phosphatase) to produce a measurable product. A positive reaction in terms of a colour change would develop within 60 min. The most crucial element of the detection strategy is a highly specific antibody-antigen interaction (Adams et al. 1995).

ELISAs are typically performed in 96-well (or 384-well) polystyrene plates, which will passively bind antibodies and proteins. The binding and immobilization of reagents make ELISAs simple to design and perform. Having the reactants of the ELISA immobilized to the microplate surface enables easy separation of bound from non-bound material during the assay. This ability to wash away non-specifically bound materials makes the ELISA a powerful tool for measuring specific analytes within a crude preparation.

21.2.5.4 Fluorescent Antibody Technique

Fluorescent antibody technique was originally developed by Coons et al. (1942) and established by his group in 1950. As the principle of the technique, proteins such as serum antibodies were labelled by chemical combination with fluorescent dyes without their effect on the biological or immunological properties of the proteins. The serum antibodies labelled with fluorescent dyes (fluorescein isothiocyanate) which are called fluorescent antibody can be used for detection of specific antigen in the preparation for examination by the fluorescent microscopy which is illuminated by ultraviolet light source and equipped with some combined filters. This method offers straight-forward detection of antigens, using fluorescently labelled antigen-specific antibodies. Presence of fluorescing cells, which were observed with the fluorescence microscope, demonstrates a positive reaction (Ainsworth et al. 1986). However, fluorescence fads after 20–30 min, and hence, the observation must be made immediately after adding the antibody.

Although the use of immunological techniques for disease diagnosis have declined and largely have been replaced with molecular tools, ELISA and few other immunological techniques are still routinely used for the identification of cultures in the laboratory and disease diagnosis. For instance, the ICAR-Central Institute of Freshwater Aquaculture, Bhubaneswar has developed two agglutination and ELISA kits for use by fish farmers in the field conditions.

21.2.6 Molecular Techniques in Disease Diagnosis

Although the previously described techniques provide useful information on causative agent of fish disease, they are often time-consuming and less useful in detection of microbes from asymptotic fish. On the other hand, the molecular tools are potentially faster and more sensitive, and pathogen can be detected during early stage of infection to avoid disease outbreak and production losses (Kumar et al. 2014a, b; Clinton et al. 2020). In recent years, a series of advanced molecular diagnostic techniques have been introduced and incorporated in the workflow of several fish disease diagnostic laboratories. These techniques have improved the efficacy, accuracy and speed of detection and help in identification and characterization of disease-causing pathogens (Cunningham 2002). The molecular techniques have extremely high levels of sensitivity and can even detect a single pathogenic cell well below the level associated with occurrences of clinical disease. This raises the issue about the significance of positive results. For instance, if overt disease signs have been observed, then positivity with molecular tools provides strong indication of the identity of the pathogen. However, if clinical signs are absent, then positivity may suggest the presence of asymptomatic or carrier fish or background populations of the pathogen present in the aquatic environment. Hence, the interpretation of results, observed from molecular techniques, is very crucial. The common molecular techniques include polymerase chain reaction, northern blotting, restriction length polymorphism, flow cytometry, southern blotting, fluorescent in situ hybridization, agarose gel electrophoresis and sequencing that are based on either the properties of nucleic acids (deoxyribonucleic acid, DNA and ribonucleic acid, RNA) or proteins of the target agents. Few selected molecular techniques are summarized below:

21.2.6.1 Polymerase Chain Reaction (PCR)

PCR is one of the most important scientific advances in molecular biology, developed by Kary Mullis in 1983. PCR is a common fish diagnosis laboratory technique used to make millions or billions of copies of a particular region of DNA (Mullis et al. 1986; Steffan and Atlas 1991). In PCR, there are three major steps, which are repeated for 30–40 cycles. This is done by an automated cycler, which can heat and cool the tubes with the reaction mixture in a very short time. At first, high temperature is applied to the original double-stranded DNA molecule (the sample DNA that contains target sequence, also known as template DNA) to separate the strands and generate single-stranded DNA molecules. The short pieces of single-stranded DNA molecules (known as forward and reverse primers) that are complementary to the template DNA binds with target DNA molecules. The DNA polymerase (generally TaqDNA polymerase from *Thermus aquaticus* is used) begins synthesizing new DNA from both forward and reverse end of primer with the help of dNTP or deoxynucleotide triphosphates (provide A, T, G and C nucleotide bases, essential building blocks of new DNA strand) and generates new strands of DNA (Fig. 21.3).

The regions of DNA template amplified usually vary between 150 and 3000 base pairs (bp) in length. After PCR, million-fold of a specific section of a DNA molecule are generated that can be subsequently detected by gel electrophoresis. The PCR product, following gel electrophoresis, can be used for sequencing or digested with restriction enzymes and cloned into a plasmid (Altinok and Kurt 2003; Roy et al. 2019). In fish disease diagnosis, PCR can also be used for detection of bacterium or DNA virus: if the pathogen is present, it may be possible to amplify regions of its DNA from a blood or tissue sample.

21.2.6.2 Sequencing Techniques

The identification of infectious agents by characterizing the DNA or RNA genome makes sequencing methods an attractive approach for pathogen detection. The sequencing methods enable diagnostic laboratories to diagnose, interrogate, and track infectious diseases in fish. The sequencing methods are used to determine the primary structure of biomacromolecules such as nucleic acids, proteins and polysaccharides. Nucleic acid sequencing, including DNA and RNA sequencing, is the most commonly used sequencing technique which determines the order of nucleotides in nucleic acid sequences (Gu et al. 2019).With technological advancements, the cost of high-throughput or next-generation sequencing has been significantly reduced by several orders of magnitude, because of that it has emerged

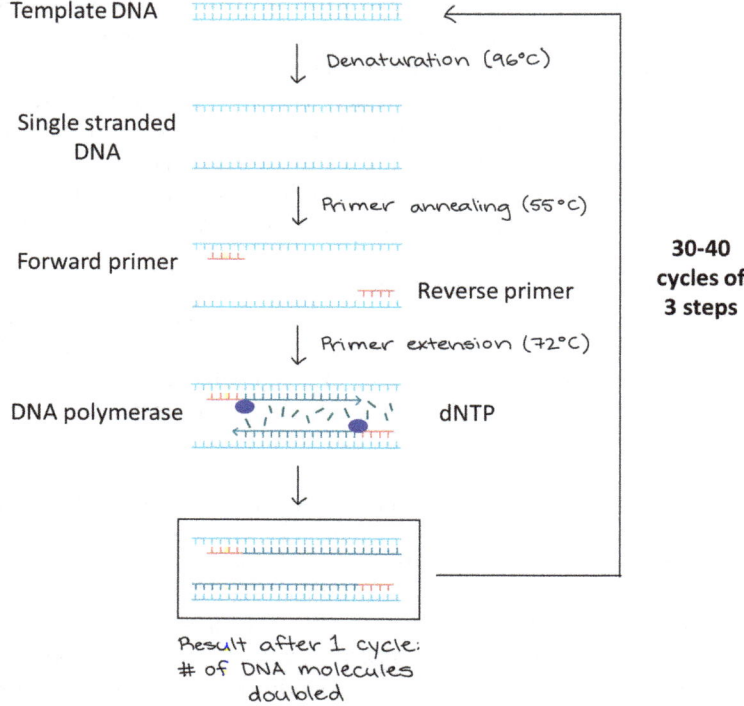

Fig. 21.3 Schematic representation of PCR cycles

as an enabling technological platform for the detection and taxonomic characterization of microorganisms from infected or diseased samples in fish.

The sequencing technology was started in mid to late 1970s. Maxam and Gilbert established the Maxam–Gilbert sequencing based on chemical fracture in 1977 (Maxam and Gilbert 1977). In the same year, Sanger proposed the dideoxy chain termination method for sequencing (Sanger et al. 1977). Later, Smith et al. (1986) developed a semi-automated method for DNA sequence analysis based on the principles of Sanger sequencing and fluorescence detection. These sequencing methods including Maxam–Gilbert sequencing, Sanger dideoxy sequencing, and fluorescence automated sequencing and hybrid sequencing are collectively known as first-generation DNA sequencing technology. Moreover, in the mid-1990s, next-generation sequencing (NGS) has emerged as important advancement in sequencing method. Next-generation sequencing has the potential to accelerate biological and biomedical research, increase throughput and reduce production scale and labour costs (Shendure and Ji 2008). It is suitable for not only genome sequencing and genome re-sequencing but also transcriptome analysis (RNA-Seq), the characterization of DNA-protein interaction (ChIP-sequencing) and epigenomic studies (de Magalhaes et al. 2010; Yu et al. 2019).

21.3   Conclusion and Future Perspective

The aquaculture environment is a highly dynamic system and the parameters of abiotic (mainly water temperature and oxygen concentration) and biotic (microbial load) can change substantially during the culture period. This can often result in the development of stressful environmental condition, making the commercial aquaculture species more susceptible to microbial pathogen, resulting in mass mortality and huge economic losses. Therefore, diagnosis of the nature and cause of mortality and identification of causative agents is necessary for effective treatments to avoid disease outbreak. Starting from study of case history of disease to observation of clinical signs, the diagnostic techniques have evolved over time and several advanced techniques like microscopic, histological, microbiological, immunological and molecular tools are used for rapid detection and identification of disease causative agents with high specificity and sensitivity. Moreover, the molecular techniques have improved the accuracy of microbial identification and progressed from use only in specialized laboratories to those involved with routine diagnostics. It is prudent to remember that immunological techniques currently also offer the possibility of rapid, field-based diagnoses, as exemplified by ELISA kits. However, because of high sensitivity there are always concerns about positive results in both immunological and molecular techniques. If it recognizes the microbial pathogen from natural environmental pathogen populations (false positive) that are not necessarily involved in fish disease, then the therapeutic strategies and management measures used may not be useful and can generate unwanted economic pressure to farmers. Additionally, if the diagnostic methods are not used efficiently there are always concerns about false results, which could elicit mass mortality and put economic pressure on the farmers. Hence, for the effective diagnosis, all available information including the clinical signs and symptoms of disease along with results of microscopic, histological, microbiological, immunological and molecular analysis must be considered before making firm conclusions. As it is commonly said, diagnoses need to help, not hinder the management of fish diseases.

Acknowledgments Authors are thankful to ICAR-Central Inland Fisheries research Institute (ICAR-CIFRI) for ample help and support.

References

Adams A, Thompson KD, Morris D, Farias C, Chen SC (1995) Development and use of monoclonal antibody probes for immunohistochemistry, ELISA and IFAT to detect bacterial and parasitic fish pathogens. Fish Shellfish Immunol 5(8):537–547

Adel A, El-Ganainy S, Ahmed M, Morsy K, Mostafa N (2019) Light and scanning Electron microscopy on Cucullanus aliyaii Akhtar and Mujib (2012) (Nematoda: Cuculanidae) from the Rabbitfish Siganus canaliculatus of the Red Sea, Egypt. Egyptian academic journal of biological sciences. E Med Entomol Parasitol 11(2):95–103

Ainsworth AJ, Capley G, Waterstreet P, Munson D (1986) Use of monoclonal antibodies in the indirect fluorescent antibody technique (IFA) for the diagnosis of Edwardsiellaictaluri. J Fish Dis 9(5):439–444

Altinok İ, Kurt İ (2003) Molecular diagnosis of fish diseases: a review. Turk J Fish Aquat Sci 3 (2):131–138

Austin B (2011) Taxonomy of bacterial fish pathogens. Vet Res 42(1):20

Austin B, Austin DA (eds) (1989) Methods for the microbiological examination of fish and shellfish. Ellis Horwood, Chichester, pp 317–327

Austin B, Austin DA (2012) Bacterial fish pathogens, vol 481. Springer, Dordrecht, p 482

Clinton M, Kintner AH, Delannoy CM, Brierley AS, Ferrier DE (2020) Molecular identification of potential aquaculture pathogens adherent to cnidarian zooplankton. Aquaculture 518:734801

Coons AH, Creech HJ, Jones RN, Berliner E (1942) The demonstration of pneumococcal antigen in tissues by the use of fluorescent antibody. J Immunol 45(3):159–170

Cunningham CO (2002) Molecular diagnosis of fish and shellfish diseases: present status and potential use in disease control. Aquaculture 206(1–2):19–55

de Magalhaes JP, Finch CE, Janssens G (2010) Next-generation sequencing in aging research: emerging applications, problems, pitfalls and possible solutions. Ageing Res Rev 9:315–323

Dias JAR, Abe HA, Sousa NC, Silva RDF, Cordeiro CAM, Gomes GFE, Maria AN et al (2019) Enterococcus faecium as potential probiotic for ornamental neotropical cichlid fish, Pterophyllum scalare (Schultze, 1823). Aquac Int 27(2):463–474

Dong H, Roy S, Zheng X, Kumar V, Das BK, Duan Y, Sun Y, Zhang J (2021) Dietary teprenone enhances non-specific immunity, antioxidative response and resistance to hypoxia induced oxidative stress in Lateolabrax maculatus. Aquaculture 533:736126

Ekman E, Norrgren L (2003) Pathology and immunohistochemistry in three species of salmonids after experimental infection with Flavobacterium psychrophilum. J Fish Dis 26(9):529–538

El-Naggar MM, Cable J, ZakyArafa S, El-Abbassy SA, Kearn GC (2016) Scanning and transmission electron microscopy of the histopathological impact of Macrogyrodactylus clarii (Monogenea: Gyrodactylidae) on the gills of catfish, Clariasgariepinus. Folia Parasitol 63:017

Engvall E, Perlmann P (1971) Enzyme-linked immunosorbent assay (ELISA) quantitative assay of immunoglobulin G. Immunochemistry 8(9):871–874

FAO (2019) State of fisheries and aquaculture in the world. FAO, Rome

Groisberg R, Roszik J, Conley A, Patel SR, Subbiah V (2017) The role of next-generation sequencing in sarcomas: evolution from light microscope to molecular microscope. Curr Oncol Rep 19(12):78

Gu W, Miller S, Chiu CY (2019) Clinical metagenomic next-generation sequencing for pathogen detection. Annu Rev Pathol Mech Dis 14:319–338

Harris PJ (2019) Microscopy and literature. Endeavour 43(3):100695

Herbeck LS, Unger D, Wu Y, Jennerjahn TC (2013) Effluent, nutrient and organic matter export from shrimp and fishponds causing eutrophication in coastal and back-reef waters of NE hainan, tropical China. Cont Shelf Res 57:92–104

Huang L, Qi W, Zuo Y, Alias SA, Xu W (2020) The immune response of a warm water fish orange-spotted grouper (Epinephelus coioides) infected with a typical cold water bacterial pathogen Aeromonas salmonicida is AhR dependent. Dev Comp Immunol 113:103779

Jorink E (2018) 8 insects, philosophy and the microscope. In: Worlds of natural history. Cambridge University Press, Cambridge, p 131

Kalaimani N, Ravisankar T, Chakravarthy N, Raja S, Santiago TC, Ponniah AG (2013) Economic losses due to disease incidences in shrimp farms of India. Fish Technol 50:80–86

Kumar V (2020) Acute hepatopancreatic necrosis disease (AHPND) in shrimp: virulence, pathogenesis and mitigation strategies. University of Ghent. Faculty of Bioscience Engineering

Kumar V, Kumar K, Raman RP, Prasad KP, Roy R, Kumar K, Kumar K (2014a) Haematological and histopathological changes during carrageenan induced acute inflammatory response in Labeorohita (Hamilton, 1822) fingerlings. Int J Curr Microbiol App Sci 3(7):794–802

Kumar V, Roy S, Barman D, Kumar A (2014b) Immunoserological and molecular techniques used in fish disease diagnosis: a mini review. Int J Fish Aquat 1(3):111–117

Kumar V, Kumar K, Raman RP, Prasad KP, Kumar N, Kumar S, Roy S (2018) Evaluation of cellular induction, soluble components of proteins and expression of pro-inflammatory genes in Labeorohita fingerlings. J Environ Biol 39(4):486–492

Kumar V, Bels LD, Couck L, Baruah K, Bossier P, Broeck WVD (2019) PirABVP toxin binds to epithelial cells of the digestive tract and produce pathognomonic AHPND lesions in germ-free brine shrimp. Toxins 11(12):717

Kumar V, Roy S, Baruah K, Van Haver D, Impens F, Bossier P (2020) Environmental conditions steer phenotypic switching in acute hepatopancreatic necrosis disease-causing Vibrio parahaemolyticus, affecting PirAVP/PirBVP toxins production. Environ Microbiol 22 (10):4212–4230

Lakshman M (2019) Application of conventional electron microscopy in aquatic animal disease diagnosis: a review. J Entomol Zool Stud 7:470–475

Love NE, Lewbart GA (1997) Pet fish radiography: technique and case history reports. Vet Radiol Ultrasound 38(1):24–29

Lulijwa R, Alfaro AC, Merien F, Meyer J, Young T (2019) Advances in salmonid fish immunology: a review of methods and techniques for lymphoid tissue and peripheral blood leucocyte isolation and application. Fish Shellfish Immunol 95:44–80

Maxam AM, Gilbert W (1977) A new method for sequencing DNA. Proc Natl Acad Sci U S A 74:560–564

Mishra SS, Das R, Sahoo SN, Swain P (2020) Biotechnological tools in diagnosis and control of emerging fish and shellfish diseases. In: Genomics and biotechnological advances in veterinary, poultry, and fisheries. Academic Press, Cambridge, MA, pp 311–360

Mullis K, Faloona F, Scharf S, Saiki RK, Horn GT, Erlich H (1986) Specific enzymatic amplification of DNA in vitro: the polymerase chain reaction. In: Cold Spring Harbor symposia on quantitative biology, vol 51. Cold Spring Harbor Laboratory Press, Cold Spring Harbor, NY, pp 263–273

Noga EJ (2010) Fish disease: diagnosis and treatment. Wiley, Hoboken, NJ

Patchimpet J, Sangkharak K, Klomklao S (2019) Lipolytic activity of viscera extract from three freshwater fish species in Phatthalung, Thailand: comparative studies and potential use as dishwashing detergent additive. Biocatal Agric Biotechnol 19:101143

Roy S, Kumar V, Bossier P, Norouzitallab P, Vanrompay D (2019) Phloroglucinol treatment induces transgenerational epigenetic inherited resistance against vibrio infections and thermal stress in a brine shrimp (Artemia franciscana) model. Front Immunol 10:2745

Sanger F, Nicklen S, Coulson AR (1977) DNA sequencing with chain-terminating inhibitors. Proc Natl Acad Sci U S A 74:5463–5467

Schmitt B, Henderson L (2005) Diagnostic tools for animal diseases. Revue scientifiqueet technique-Office international des épizooties 24(1):243

Shendure J, Ji H (2008) Next-generation DNA sequencing. Nat Biotechnol 26:1135–1145

Sherman DM, Markham RJ, Bates F (1984) Agar gel immunodiffusion test for diagnosis of clinical paratuberculosis in cattle. J Am Vet Med Assoc 185(2):179–182

Sirri R, Tura G, Budai J, Beraldo P, Fiorentino M, Barbé T, Galeotti M, Sarli G, Mandrioli L (2020) Histological and immunohistochemical characterization of 17 gonadal tumours in koi carp (Cyprinus carpio koi). J Fish Dis 44(3):273–285

Smith LM, Sanders JZ, Kaiser RJ et al (1986) Fluorescence detection in automated DNA sequence analysis. Nature 321:674–679

Steffan RJ, Atlas R (1991) Polymerase chain reaction: applications in environmental microbiology. Annu Rev Microbiol 45(1):137–161

Tacon AGJ (2020) Trends in global aquaculture and Aquafeed production: 2000–2017. Rev Fish Sci Aquac 28:43–56

Tesser CD, Norman AH (2016) Differentiating clinical care from disease prevention: a prerequisite for practicing quaternary prevention. Cadernos de saudepublica 32:e00012316

Toranzo AE, Baya AM, Roberson BS, Barja JL, Grimes DJ, Hetrick FM (1987) Specificity of slide agglutination test for detecting bacterial fish pathogens. Aquaculture 61(2):81–97

Vadstein O, Bergh Ø, Gatesoupe FJ, Galindo-Villegas J, Mulero V, Picchietti S, Scapigliati G, Makridis P, Olsen Y, Dierckens K, Defoirdt T (2013) Microbiology and immunology of fish larvae. Rev Aquacult 5:S1–S25

Verdegem MCJ (2013) Nutrient discharge from aquaculture operations in function of system design and production environment. Rev Aquac 5:158–171

Wang G, Li D (2009) A fish disease diagnosis expert system using short message service. In: 2009 WRI International conference on communications and mobile computing, vol 3. IEEE, Piscataway, NJ, pp 299–303

Witten PE, Harris MP, Huysseune A, Winkler C (2017) Small teleost fish provide new insights into human skeletal diseases. In: Methods in cell biology, vol 138. Academic Press, Cambridge, MA, pp 321–346

Woo PT, Bruno DW (2011) Fish diseases and disorders. Viral, bacterial and fungal infections, 2nd edn. CABI, Wallingford

Woo PT, Gregory DWB (eds) (2014) Diseases and disorders of finfish in cage culture. CABI, Wallingford

Yu X, Jiang W, Shi Y, Ye H, Lin J (2019) Applications of sequencing technology in clinical microbial infection. J Cell Mol Med 23(11):7143–7150

Prospect and Challenges of Biofloc Technology for Sustainable Aquaculture Development

22

Sudhansu Shekhar Mahanand and Pramod Kumar Pandey

Abstract

Aquaculture is considered the sunrise sector in terms of growth worldwide. There is an urgent need to ensure that this growth continues in a sustainable manner in order to protect the environment and preserve the natural resources of the world. The fundamental concept of biofloc technology (BFT) is premised on the principle that waste is processed into biofloc which serves as a natural food within the culture system. Biofloc Technology reduces toxic metabolites and helps to convert nitrogenous waste into fish or shrimp biomass through microbial mass protein processing. Along with other vital minerals, biofloc contains 12–50% crude protein. Biofloc also includes essential fatty acids and amino acids, which are sufficient to meet the nutritional needs of fish and shrimp. The 3-week biofloc image, collected by the scanning electron microscope, showed the presence of various species of bacteria, fungi, protozoa, rotifers, etc. in varying shapes and sizes ranging from 10 to 100μm. It is shown that biofloc microorganisms can partially replace the protein content of a diet and therefore reduce the requirement of fishmeal for the preparation of fish feeds.

Keywords

Biofloc technology · Microorganisms · Nitrogenous waste

S. S. Mahanand (✉)
College of Fisheries (CAU), Lembucherra, Tripura (West), India

P. K. Pandey
ICAR-Directorate of Coldwater Fisheries Resources, Bhimtal, Uttarakhand, India

© The Author(s), under exclusive license to Springer Nature Singapore Pte Ltd. 2021
P. K. Pandey, J. Parhi (eds.), *Advances in Fisheries Biotechnology*,
https://doi.org/10.1007/978-981-16-3215-0_22

22.1   Introduction

The global population is increasing at a rapid pace; it is estimated that the global population will hit nine billion by 2050. There is increasing evidence of persistent hunger, malnutrition and under-nutrition adversely affecting a large portion of the population as it continues to grow exponentially. The primary explanation to such adverse change is the paradigm of eating from a healthy diet to insufficient and deficient diets. The United Nations Food and Agricultural Organisation has estimated that more than 500 million people are undernourished worldwide. The two sources of protein viz. plant and animal are to ensure the health of the world population by providing nutritionally balanced food, especially in terms of protein. Fish is a rich source of protein that is easily digestible which also provides nutrition through polyunsaturated fatty acids, vitamins and minerals. As the population grows exponentially, the gap between demand and supply of fish is expected to increase. In order to satisfy the demand, intensification of aquaculture practices is inevitable that its impact on the environment becomes heavier as the aquaculture industry continues to grow intensively to meet the insatiable demands. Water quality management remains a major challenge in such intensive aquaculture systems due to the huge amount of waste generated during the culture cycle in the form of unused feed and the production of faecal matter. The collective ecosystem is deteriorating because of the adverse impacts of aquaculture discharges. In addition, the demands for fishmeal and fish oil for the production of fish feed are also growing. The periodic exchange and substitution of pond water is one of the popular methods employed to eliminate unnecessary nitrogen. However, due to environmental legislation, this approach is practised in a constrained manner. Another method is based on means of promoting and improving the comparatively inert nitrate species' nitrification of ammonium and nitrites. This is also accomplished by the use of biofilters, which are essentially immobile surfaces that serve as substrates for nitrifying bacteria. In a regulated setting, a high surface area with immobilised nitrifying biomass allows for a high nitrifying ability. However, high costs are an issue associated with biofiltration. The elimination of ammonia from the water by its assimilation into microbial proteins through the incorporation of carbonaceous materials to the environment is an additional technique that is becoming popular and attracting attention. Added carbohydrates will theoretically solve the issue of inorganic aggregation of nitrogen if properly modified. The possible use of microbial protein as a source of feed protein for fish or shrimp is another essential part of this process. This mechanism is generally referred to as biofloc technology (BFT). In intensive culture activities, the latest strategy for preserving the nitrogenous waste is to exchange and refresh the pond water regularly. This technique, however, may lead to extreme environmental destruction and high pumping costs (Avnimelech 1999). For the stabilisation of the aquaculture environment, two techniques are gaining worldwide popularity: (1) water reuse and (2) biofloc. Due to the high initial capital costs needed by the recirculating aquaculture method (RAS), the reuse of water has become a common practice in small-capacity systems (Schneider et al. 2006). In order to offset invest-ment costs that may also contribute to the accumulation of minerals, medication residues, dangerous feed compounds and metabolites affecting the hygiene, quality and protection of the cultivated species, high stocking densities and production are needed (Martins et al. 2009a, b). BFT has recently gained a lot of popularity worldwide, especially for tilapia and shrimp cultivation.

Bioflocs, which are conglomerates of bacteria, algae, protozoa and other dead organic particles, along with detritus, are the essential elements of the BFT system. Biofloc is a special ecosystem of rich, energetic particles that are suspended in comparatively low water. The fundamental concept of biofloc (BFT) technology is that waste is stored and processed into biofloc as a natural food within the culture system. This can be accomplished by continuous water column aeration and agitation and the addition of carbon sources as substrates of organic matter to allow aerobic decomposition and sustain high suspended microbial floc levels in fed and/or fertilised ponds. Biofloc technology is a technique focused on in-situ development of microorganisms that is environmentally sustainable. Algae and bacteria have a number of stimulatory or inhibitory effects on each other, resulting in water quality mechanisms becoming complex (Cole 1982). Via microbial assimilation by the inclusion of extra carbonaceous materials, inorganic nitrogenous waste is minimised (Avnimelech 1999, 2007). This methodology is applicable to both comprehensive and intensive aquaculture systems. In addition, it is believed that the heterotrophic microbial biomass has a limiting effect on pathogenic bacteria (Michaud et al. 2006). Due to its bottom dwelling habit and resistance to environmental changes, biofloc technology is applied in shrimp farming. In order to determine the larval development and reproductive success of shrimps and Nile tilapia, experiments were carried out. In contrast to standard culture activities, better breeding efficiency was observed in shrimp reared in the biofloc system. Similarly, increased production in larval growth was also observed.

Several experiments have been carried out using BFT for tilapia and shrimp culture (Azim and Little 2008; Asaduzzaman et al. 2009, 2010; Kuhn et al. 2008, 2010; Ray et al. 2011; Emerenciano et al. 2012a, b). Kuhn et al. (2009) assessed the development of *Litopenaeus vannamei* using microbial floc meal as a supplement for fish meal and soybean protein and concluded that without any discrepancies in survival, microbial floc meal outperformed the control diets in terms of weight gain per week. Two forms of bioflocs collected from the biological treatment of fish effluent as feed ingredients for *Litopenaeus vannamei* were further analysed by Kuhn et al. (2010). The results showed that bioflocs developed using either sequential batch reactor (SBR) or membrane biological reactor (MBR) systems could substitute fish meal and soybean protein.

BFT is typically sufficient for the cultivation of both herbivorous and omnivorous marine plants. Various researchers have stated (Avnimelech 1999; Fontenot et al. 2007) that biofloc formation is maximised at C:N = 10. Natural carbon is typically applied to water to preserve the above-mentioned C:N ratio. The amount of biofloc emitted per unit mass of organic carbon added was not clearly stated. Different scientists have explored the feasibility of using biofloc as a shrimp feed (Kuhn et al. 2008, 2009, 2010; Emerenciano et al. 2012a, b). The findings of all such experiments reveal that bioflocs can be a replacement for shrimp for fishmeal and soybean protein. In several experiments, biofloc was produced by incorporating carbon sources as organic matter to preserve the correct C:N ratio in the culture ecosystem itself (Azim and Little 2008; Azim et al. 2008). The water quality was observed to be extremely variable in such situations and the main water quality parameters such as pH, NH_4^+-N, NO_2--N, etc. exceeded their allowable limit in several instances. The idea of externally adding biofloc to preserve the right C:N ratio has not yet been investigated. In the context of aquaculture development, the above-mentioned issues are yet to be discussed.

In order to increase environmental control over the growth and development of aquatic animals, the biofloc method was designed. In aquaculture, feed costs (accounting for 60% of the overall production cost) are powerful influencing factors and the supply of water/land is the most limiting factor affecting development. Waste water treatment is required for the high stocking density and rearing of aquatic animals. As an alternative to aquaculture, the biofloc method is a waste water technology that has acquired great significance over the years. BFT is adopted in the new blue revolution because nutrients can be constantly recycled and reused in the culture form, benefiting from minimal and zero exchange of water.

22.2 History of BFT

Steve Surfling is known to be one of the founders of BFT. He founded solar aquafarms in California, where he was farming shrimp and fish based on the advancement of the principle of active microbial suspension ('microbial soup'). This eventually led to the advancement of aqua farming based on 'biofloc'. Heterotrophic bacteria are produced and these bacteria, called flocs, consume nitrogen waste 10 to 100 times more effectively than algae and transform it into high-protein feed. In 1980, Ifremer initiated the research programme 'Ecotron' to learn more about BFT. Israel and the USA (Waddell Mariculture Center) began R&D with Tilapia and White Shrimp in BFT during the same time. In the farming of Tilapia, *Penaeus monodon*, Pacific white shrimp *Litopenaeus vannamei*, giant freshwater prawn *Macrobrachium rosenbergii*, *Fenneropenaeus merguiensis* and *Litopenaeus stylirostris* (Rosenberry 2010) family, it has become a commonly used technology.

In 1988, Sopomer farm in Tahiti (French Polynesia) began the commercial application of BFT using 1000 m concrete tanks and small water exchange, setting a world production record (20–25 ton/ha/year with two crops). Central America's aquaculture farm in Belize produces roughly 26 tonnes/ha/cycle using 1.6 ha of living ponds. Green house BFT farms are actively run in Maryland, USA. In Latin and Central America, and in small scale in the USA, South Korea, Brazil, China, Italy, Indonesia, Australia and India, the effective introduction of this technology has been extended. Prof. Yoram Avnimelech and his team from Israel are currently engaged in devising strategies to modify and support this technology, thereby making significant contribution towards its utilisation in aquaculture. Because of its many benefits, the methodology developed in Israel and later spread to many other nations.

Several nations are at present pursuing R&D on this technology to improve the production of fish. Other BFT concerns such as environmental effects, food efficiency, energy kinetics, imaging and bacterial detection and economics, low cost production, etc. have also been highlighted by researchers for further application of BFT in various fish species.

22.3   Biofloc Production

The theory behind this approach entails that the nitrogen cycle is created by maintaining a higher C:N ratio by stimulating heterotrophic microbial growth, which assimilates the nitrogen waste that can be used as feed by the cultivated spices. BFT is not only effective in handling waste but also provides marine animals with nutrients.

In practice, there are two forms of biofloc processing: in-situ and ex-situ. In the in-situ biofloc cell, in order to maintain the C:N ratio above 10 for the optimal growth of heterotrophic microbial biomass, carbon sources are introduced to the culture system. With floc aggregation, a transformation in the colour from green to brown is observed. This approach involves a long process to transform autotrophs to heterotrophic structures, which ideally requires 7 to 10 days. Heterotrophic bacterial colony is generated in a well-managed environment, which in turn serves as a feed and also assimilates the wastes provided by the cultured organism. For holding the floc under suspension, continuous aeration is necessary in this method. With the help of a biofloc reactor, ex-situ floc is generated where the C:N ratio is adjusted, floc production is externally monitored and the floc is applied to the culture device at regular intervals. The consistency of the floc can be manually controlled in this method and the dried floc may be used as an additional source of protein in the organism's diet. By using locally available sources of carbohydrates such as maize, cassava, tapioca, wheat and cellulose, several researchers have developed biofloc in the culture method.

Increased C:N ratio by carbon addition potentially increases the conversion of harmful inorganic nitrogen species to microbial biomass which is usable as fuel for cultivated animals. In an aquaculture method, the optimal C:N ratio of 10 can be maintained by adding various locally available, cheap sources of carbon and/or reducing the protein content of feed (Avnimelech 1999). Microbial organic matter breakdown contributes to new bacteria being formed, which is responsible for 40 to 60% of the organic matter metabolised. The quantity of organic matter needed for an intensive pond can be determined technically on the basis of the volume of nitrogen excreted by the aquaculture species (Fig. 22.1). Many researchers have developed biofloc from various sources of carbon for the culture of different organisms (Table 22.1).

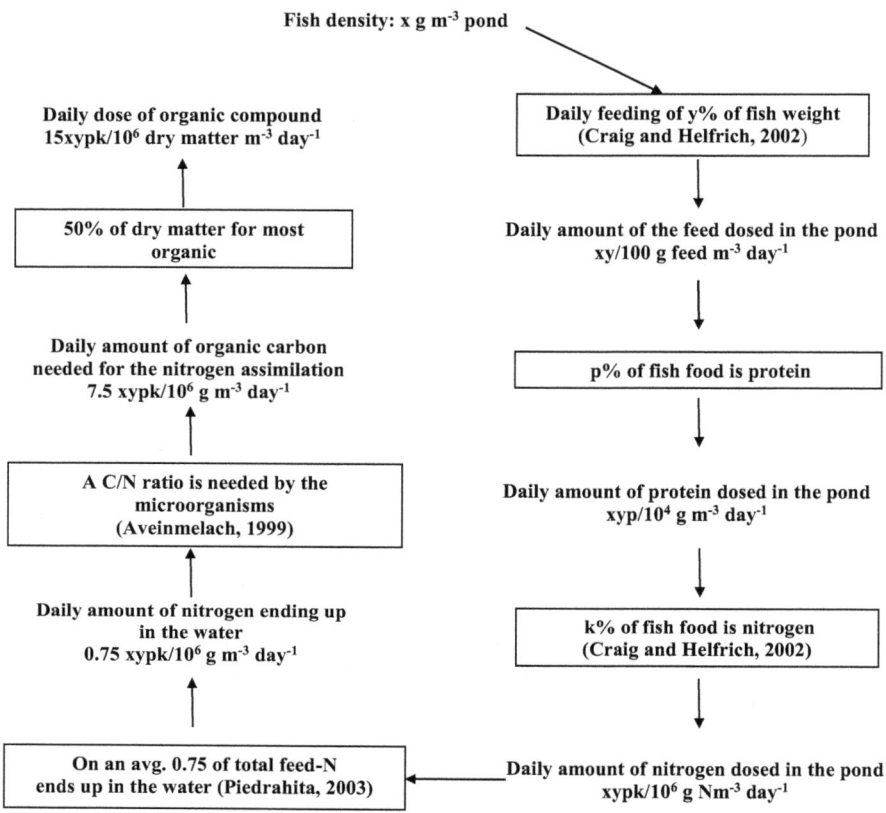

Fig. 22.1 Schematic calculation of daily amount of organic carbon needed by bioflocs for removal of inorganic nitrogen

22.4   Nutritional Value of Biofloc as Fish Feed

Biofloc's nutritional composition varies depending on environmental situation, carbon source added, TSS amount, salinity, stocking depth, light strength, population and ratio of phytoplankton and bacteria, etc. Table 22.2 depicts the proximate composition derived from various analyses. Biofloc is reported to have high nutritional value. Dry protein weight ranges from 12 to 50%, lipid ranges from 0.5 to 12.5% and ash content varies from 13 to 50%. The dry matter-based biofloc contains 38% protein, 3% lipid, 6% carbohydrate, 12% ash and 19 kJ g^{-1} energy in terms of consistency (Azim and Little 2008). Tacon et al. (2002) and Wasielesky et al. (2006) reported a consistent amount of crude protein of 35% in biofloc, but less than the 43% crude protein, described by Jory (2001). The nutritional content of

Table 22.1 Different carbon sources applied for biofloc production system for various species

Carbon sources	Cultured species	References
Cassava meal	*Penaeus monodon*	Avnimelech and Mokady (1988)
Sorghum meal	*Oreochromis niloticus*	Avnimelech et al. (1989)
Molasses	*Litopenaeus vannamei* and *Penaeus monodon*	Burford et al. (2004)
Tapioca	*Litopenaeus vannamei*	Hari et al. (2004)
Starch	*Oreochromis mossambicus*	Avnimelech (2007)
Wheat flour	*Oreochromis niloticus*	Azim and Little (2008)
Tapioca	*Macrobrachium rosenbergii*	Asaduzzaman et al. (2008)
Cellulose	*Oreochromis niloticus*	Avnimelech and Kochba (2009)
Dextrose	*Litopenaeus vannamei*	Suita (2009)
Acetate and glucose	*Macrobrachium rosenbergii*	Crab et al. (2010)
Wheat bran + molasses	*Farfantepenaeus paulensis*	Emerenciano et al. (2011)
Wheat flour	*Labeo rohita*	Mahanand et al. (2012)
Wheat bran + molasses	*Farfantepenaeus brasiliensis*	Emerenciano et al. (2012a, b)
Molasses + dextrose + rice flour	*Litopenaeus vannamei*	Serra et al. (2015)
Molasses + wheat flour + starch	*Litopenaeus vannamei*	Khanjani et al. (2017)
Wheat flour + molasses	*Oreochromis niloticus*	Mirzakhani et al. (2019)
Molasses + palm sap	*Litopenaeus vannamei*	Abbaszadeh et al. (2019a, b)

biofloc was researched by Emerenciano et al. (2012a, b) who reported 30% crude protein. Ballester et al. (2010) found that bioflocs contained 30.4% crude protein, 4.7% crude fat, 8.3% fibre, 39.2% ash and 29.1% dry matter-based nitrogen-free extract. It is a healthy source of (Table 22.3) vitamins and minerals. Mahanand et al. (2013) reported that biofloc contained essential fatty acids derived from wheat flour (Table 22.4). In the biofloc study, which is important for fish growth, there were 21.11% polyunsaturated, 35.63% monounsaturated and 43.1% saturated fatty acids. Tacon (1990) stated that for common carp, about 1% of dietary essential fatty acids 18:2 ω6 and 18:3 ω3 are needed. These values were 12% and 4.4% in the biofloc tests. Therefore, the established biofloc appears to be sufficient enough for successful use as fish feed, particularly for herbivorous and omnivorous species. Ekasari et al. (2010) reported that there was higher overall n-6 PUFAs in bioflocs with glycerol as the carbon source than in those with glucose. This study suggests that the nutritional content of bioflocs is likely to be compromised by microbiota.

Amino acid is also important for aquatic organisms such as fish and shrimp to develop. The usage of protein in fish and shrimp relies on the nutritional supply of essential amino acids. Table 22.5 displays the amino acid composition (g/100 g dry floc) of the biofloc produced with wheat flour and molasses. Several studies have

Table 22.2 Proximate composition of biofloc obtained from different studies

Crude protein (%)	Carbohydrate (%)	Lipid (%)	Crude fibre (%)	Ash (%)	References
43	–	12.5	–	26.5	McIntosh et al. (2000)
31.2	–	2.6	–	28.2	Tacon et al. (2002)
12.0–42.0	–	2.0–8.0	–	22.0–46.0	Soares et al. (2004)
31.1	23.6	0.50	–	44.8	Wasielesky et al. (2006)
26.0–41.9	–	1.2–2.3	–	18.3–40.7	Ju et al. (2008)
49	36.4	1.13	12.6	13.4	Kuhn et al. (2009)
38.8	25.3	>0.1	16.2	24.7	Kuhn et al. (2010)
35.40	–	1.1	15.03	15.38	Mahanand et al. (2012)
28.8–43.1	–	2.1–3.6	8.7–10.4	21.1–42.9	Maicá et al. (2012)
30.4	29.1	0.5	0.8	39.2	Emerenciano et al. (2012a, b)
18.2–29.3	22.8–22.9	0.4–0.7	1.5–3.5	43.7–51.8	Emerenciano et al. (2012a, b)
18.4–26.3	20.2–35.7	0.3–0.7	2.1–3.4	34.5–41.54	Emerenciano et al. (2012a, b)
28–30.4	18.1–22.7	0.5–0.6	3.1–3.2	35.8–39.6	Emerenciano et al. (2012a, b)
27.43	–	0.86	–	39.8	Khanjani et al. (2017)
23.8	–	1.14	–	21.81	Khanjani et al. (2017)
30.73	–	2.18	–	29.97	Khanjani et al. (2017)
–	–	0.5–0.8	6.8–8.9	7.5–9.3	Abbaszadeh et al. (2019a, b)

Table 22.3 Mean (±SD) Energy content, mineral content and C:N ratio of biofloc produce from wheat flour as a carbohydrate source

Composition (%DM)	Biofloc
Energy (kJ g^{-1})	18.78 ± 0.19
C:N ratio	7 ± 0.48
Phosphorous	1.40 ± 0.59
Potassium	0.69 ± 0.21
Calcium	1.78 ± 0.90
Magnesium	0.29 ± 0.09
Sodium	2.51 ± 1.28
Manganese (mg kg^{-1})	28.12 ± 12.01
Iron (mg kg^{-1})	295.158 ± 68.58
Copper (mg kg^{-1})	26.84 ± 25.14
Zinc (mg kg^{-1})	310.50 ± 188.06
Boron (mg kg^{-1})	27.73 ± 13.26

Table 22.4 Fatty acid composition (% lipid) of biofloc produce from wheat flour as a carbohydrate source

Components[a]	Biofloc
12:0	0.3
13:0	0.2
13:1	0.3
14:0	2.9
14:1	4.5
15:0	2.3
15:1	0.9
16:0	29.3
16:1	7.9
16:2	2.0
17:0	0.2
17:1	0.3
18:0	4.6
18:1ω9	20.8
18:2ω6	12.7
18:3ω6	0.6
18:3ω3	4.4
20:0	0.2
20:1ω9	0.5
20:3ω6	0.3
20:4ω6	0.1
22:0	1.9
20:4ω3	–
22:1ω11	0.4
20:5ω3	0.4
21:5ω3	0.3
22:5ω6	0.01
24:0	1.2
24:1	0.03
22:6ω3	0.3
Total −ω3	5.4
Total −ω6	13.71
Total PUFA	21.11
Total saturates	43.1
Total monoenes	35.63
Total lipid %	4.22

The ω values represent the methyl end chain from the centre of double bond farthest from the carboxyl end
[a]First and second figures represent carbon chain length, number of double bonds

documented that biofloc contains vital nutrients such as proteins, lipids, vital fatty acids, minerals, vitamins, carotenoids and digestive enzymes in order to promote absorption and thereby enhance nutritional status. It may be used partly as a feed and

Table 22.5 Amino acid compositions (g/100 g dry floc) of the biofloc produced with wheat flour and molasses

Amino acids	Wheat flour	Molasses
Alanine	2.85	3.14
Arginine	1.75	1.36
Aspartic acid	4.65	5.10
Glutamic acid	6.87	7.45
Glycine	1.23	1.87
Histidine	2.15	2.42
Isoleucine	1.35	1.42
Leucine	2.48	2.56
Methionine	0.91	1.06
Cysteine	0.18	0.31
Phenylalanine	1.65	1.75
Proline	1.45	1.83
Serine	2.12	2.65
Taurine	0.15	0.18
Theanine	1.76	1.88
Tryptophan	0.28	0.34
Tyrosine	1.04	1.12
Valine	1.38	1.42

also in the pelletised feed of brood stock. The cost-effectiveness of commercial-scale processing and drying of biofloc solids is a problem. More studies need to be undertaken in this area.

22.5 Water Quality Management in Biofloc System

Maintenance and control of water quality are important activities for the aquaculture sustainability of the biofloc system. It is important to continuously track selected physico-chemical parameters of water, viz. temperature, pH and dissolved oxygen (DO), total ammonia nitrogen (TAN), nitrite nitrogen, nitrate nitrogen and total suspended solids (TSS), floc size, alkalinity and orthophosphate.

22.5.1 Temperature

For microbial metabolism, the temperature plays vital role. For tropical fish, the optimal temperature is 28 to 30 °C. A range of 8–30 °C can be provided by nitrifying bacteria, but the performance at 16 °C is decreased by 50% and at 10 °C by 80% (Saraswathy et al. 2019). Microbial growth might be influenced by low temperatures (<20 °C). The temperature is closely related to the quantity of dissolved oxygen in water (Boyd 1998).

22.5.2 Dissolved Oxygen (DO)

DO concentrations above 4 mg L^{-1} are considered important for the survival of fish and other aquatic life. Since the culture animals and biofloc are in the same water, in-situ biofloc systems have a very high oxygen requirement. Therefore, dissolving oxygen should be retained above 7 mg L^{-1} to ensure the proper functioning of the system.

22.5.3 pH

pH is a measure of the environment's acid base status and acts as a production metric. 6.8 to 8 is the ideal pH in biofloc system (Emerenciano et al. 2017). In BFT, pH below 7.0 is common but it might influence the process of nitrification. The growth output of shrimp in the biofloc system may be impaired by pH below 7 for extended periods of time (Furtado et al. 2011).

22.5.4 Alkalinity

Alkalinity is water's ability to buffer or resist changes in pH in response to acid or base additions. Higher alkalinity levels can aid the assimilation of nitrogen by heterotrophic bacteria and chemoautotrophic bacteria in the process of nitrification. The alkalinity in the biofloc system should be above 100 ppm.

22.5.5 Total Ammonia Nitrogen (TAN)

Ammonia is harmful to aquatic organisms and is known to be a significant limiting factor for fish abundance and stocking density. TAN concentraion of less than 1.0 mg L^{-1} is advisable for healthy growth of all aquatic species in intensive culture systems. TAN concentrations should be kept below 1.0 mg L^{-1} in the biofloc system. Ammonia nitrogen concentrations in *Litopenaeus vannamei* may be less than 1.2 and 6.5 mg L^{-1} in post-larvae and juveniles (Frías-Espericueta et al. 2000).

22.5.6 Nitrite Nitrogen

Nitrite-N is an intermediary result of bacterial nitrification which, even at low concentrations, is considered toxic to marine organisms. Nitrite toxicity affects oxygen delivery, essential compound oxidation and tissue damage. The concentration of nitrite nitrogen should be kept under 1.0 mg L^{-1}.

22.5.7 Nitrate Nitrogen

Nitrate, being the stable form, is readily taken by algae from the various sources of nitrogen and hence helps to increase primary development (Goldman and Horne 1983). Via the process of nitrification, ammonia is oxidised to nitrate in a two-step process by two classes of aerobic chemoautotrophic bacteria that extract energy through the oxidation of ammonia or nitrite for carbon dioxide fixation. Nitrate has been confirmed to be less toxic to fish than nitrite or TAN (Meade 1985; Lyssenko and Wheaton 2006). Nitrate concentration in biofloc culture does not reach 20 mg L^{-1}.

22.5.8 Orthophosphate

In biofloc system, 0.5 to 20 mg L^{-1} is the optimal range of orthophosphate. Within this optimal range there is no harmful to culture animals.

22.5.9 Total Suspended Solids (TSS)

TSS consists of planktons, food grains that are not digested, suspended clay particles, etc. In an intensive system, the suspended clay particles are highly unwanted (Boyd 1982). In shrimp and fish ponds, the TSS should be limited to around 200 and 400 mg L^{-1}, respectively. An accumulation of TSS inhibits the respiratory mechanism of animals, leading to discomfort, or leading to death by clogging gills in extreme situations. The amount of TSS in biofloc culture should be held below 500 mg L^{-1}.

22.5.10 Settling Solids (SS)

Large suspended solids amount is calculated by imhoff cone. The ideal range of concentrations of settleable solids for shrimp is 10 to 15 mL/L and 25 to 50 mL/L for tilapia under biofloc system.

22.6 Placement of Aerators in Biofloc Ponds

The aerators used in tanks and biofloc ponds are equivalent to those used in the aquaculture system. In the aquaculture pond, all specific types of mechanical aerators are used, but in the biofloc pond, vertical pumps, pump sprayers, propeller-aspirator-pumps, paddle wheels and diffuse-air systems are most popular. Gravity aerators, nozzle aerators and pure oxygen touch devices are also used sometimes.

Proper positioning of aerators plays an essential part in the successful filtering of water through wetlands. The optimal location to put the aerator in a rectangular pond is at the centre of one of the long sides of the pond that channels water parallel to the short side of the pond. The worse positioning of the aerator would undoubtedly be the location in the corner of the pond to direct water diagonally around the pond. For fast water circulation, the long arm paddle wheel aerators that are usually positioned perpendicular to the pond dykes must be turned and guided towards the middle at about 30°. The positioning methods of the aerators aimed at ensuring the optimum water distribution can differ based on the size and form of the bath. In rupturing the stratification of water parameters such as O_2, temperature and salinity, water circulation in the pond is critical. The location of the aerator on the pond should be determined taking into account the predominant wind speed in order to efficiently disperse the oxygen-saturated water in the pond and facilitate the movement of water in it.

Installing a secondary device targeted at sludge aggregation sites to encourage sludge particle resuspension is a very effective approach; it is often feasible in comparatively small ponds or in ponds fitted with high aeration power. In limited ($50–1000$ m^2) wetlands, such a structure has been used (Crab et al. 2009). In these ponds, the main aeration system is based on a number of paddle wheel aerators which produce a radial flow pattern. In the middle, sludge can be collected. Above this area, an air lift tube may be mounted, causing an upward flow and resuspending the fine particles of sludge back into the aerated water body. The placement of aerators is done in a manner that minimises sludge pile formation. In lieu of paddle wheels, the air lift or aspirator aerator does not mix the sludge.

22.7   Identification of Biofloc Using Scanning Electron Microscope (SEM)

The size of the floc varies from 10 to 100µm. Rod-shaped bacteria (bacilli) (3.962µm), spherical algae (1.408 nm), algae moss clumps (5.06µm), micro-tubular flagella spirilla form (0.5µm), peanut-shaped diatom (2.7µm), algae mats (3.613µm), died diatom cells, protozoa, rotifer and other microbes were found in the floc. A 3-week biofloc image with 6000 magnifications is shown in Fig. 22.2. Related microorganism forms with varying shapes and sizes detected by a confocal microscope at magnification of 10 were identified by Azim and Little (2008), Azim et al. (2008) and Emerenciano et al. (2013).

22.8   Prospective of Biofloc Technology

BFT was designed in order to promote intensive culture, while relying on lower investment, maintenance costs and integrating feed recycling capacity. The development of this technology is largely focused on a set of motives, values and necessary operating technologies. Higher efficiency and higher biosecurity can be achieved due to this technology. It decreases the contamination of water and the possibility of bacteria being released and propagated. It eliminates the use of

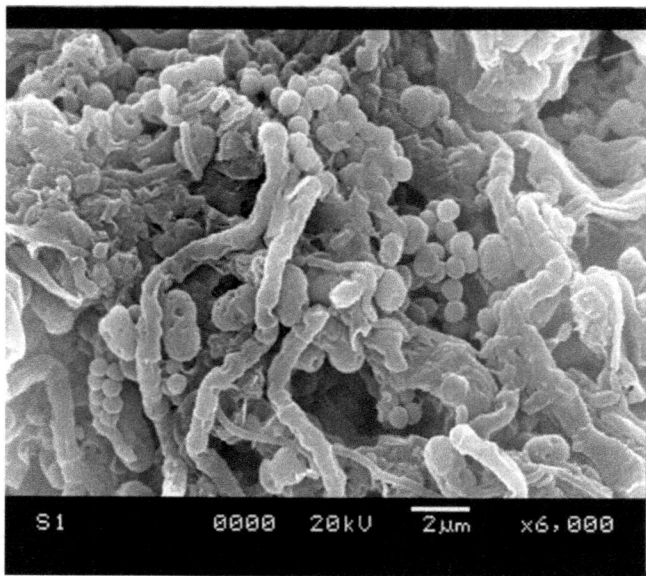

Fig. 22.2 SEM image of 3 weeks' biofloc at 6000 magnification

protein-rich feed and the expense of regular feed, thus minimising the burden on fisheries to catch, i.e. the use of inexpensive food fish and garbage fish for the formulation of fish feed. Microorganism intake in BFT reduces FCR and, consequently, feed prices. In addition, the microbial population is able to easily use and transform dissolved nitrogen leached from shrimp faeces and unconsumed food into microbial protein. Such characteristics make the BFT minimum-exchange scheme an alternative to comprehensive aquaculture.

22.9   Conclusion and Challenges of BFT

BFT provides aquaculture with a sustainable tool to resolve its environmental, social and economic problems which seems to be on the rise. This technology would allow for the advancement of aquaculture into an environmentally sustainable approach and provide biosecurity. Biofloc technology plays a crucial role in the control of nutrients and thus avoids the accumulation of toxic nutrients that under the zero water exchange system affect the growth efficiency and survival of cultured species. With a very high ammonia load, BFT can be used to process waste water. As the use of biofloc grows in quality, it promises to offer more environmentally sustainable and less costly solutions to tackle the issues of food security in future. However, this technology faces many issues which poses as threats to its survival. The physical, chemical and biological interactions in the biofloc structures are very complex as a matter of fact. Further experiments on correct placement of aerators, incorporation of the current system of RAS, raceways and polyculture systems have also been questioned by researchers. Further research is required for its further development, fine-tuning and application to make this technique a keystone of sustainable aquaculture in the future.

References

Abbaszadeh A, Keyvanshokooh S, Yavari V, Naderi M (2019a) Proteome modifications of Pacific white shrimp *(Litopenaeus vannamei)* muscle under biofloc system. Aquacult Nutr 25:358–366

Abbaszadeh A, Yavari V, Hoseini SJ, Nafisi M, Mozanzadeh MT (2019b) Effects of different carbon sources and dietary protein levels in a biofloc system on growth performance, immune response against white spot syndrome virus infection and cathepsin L gene expression of *Litopenaeus vannamei*. Aquacult Res 50:1162–1176

Asaduzzaman A, Wahab MA, Verdegm MCJ, Huque S, Salam MA, Azim ME (2008) C/N ratio control and substrate addition for periphyton development jointly enhance freshwater prawn *Macrobrachium rosenbergii* production in ponds. Aquaculture 280(1–4):117–123

Asaduzzaman M, Wahab MA, Verdegem MCJ, Benerje S, Akter T, Hasan MM, Azim ME (2009) Effects of addition of tilapia *Oreochromis niloticus* and substrates for periphyton developments on pond ecology and production in C/N-controlled freshwater prawn *Macrobrachium rosenbergii* farming systems. Aquaculture 287(3–4):371–380

Asaduzzaman M, Rahman MM, Azim ME, Islam MA, Wahab MA, Verdegem MCJ, Verreth JAJ (2010) Effects of C/N ratio and substrate addition on natural food communities in freshwater prawn monoculture ponds. Aquaculture 306(1–4):127–136

Avnimelech Y (1999) Carbon/nitrogen ratio as a control element in aquaculture systems. Aquaculture 176(3–4):227–235

Avnimelech Y (2007) Feeding with microbial flocs by tilapia in minimal discharge bioflocs technology ponds. Aquaculture 264(1–4):140–147

Avnimelech Y, Kochba M (2009) Evaluation of nitrogen uptake and excretion by tilapia in bio floc tanks, using 15N tracing. Aquaculture 287:163–168

Avnimelech Y, Mokady S (1988) Protein biosynthesis in circulated ponds. In: Pullin RSV, Bhukaswan T, Tonguthai K, Maclean JL (eds) Proceedings of second international symposium on Tilapia in aquaculture. Department of Fisheries of Thailand and ICLARM, Manila, pp 301–309

Avnimelech Y, Mokady S, Schoroder GL (1989) Circulated ponds as efficient bioreactors for single cell protein production. Bamdigeh 41:58–66

Azim ME, Little DC (2008) The biofloc technology (BFT) in indoor tanks: water quality, biofloc composition, and growth and welfare of Nile tilapia *(Oreochromis niloticus)*. Aquaculture 283 (1–4):29–35

Azim ME, Little DC, Bron JE (2008) Microbial protein production in activated suspension tanks manipulating C: N ratio in feed and the implications for fish culture. Bioresour Technol 99 (9):3590–3599

Ballester E, Abreu P, Cavalli R, Emerenciano M, De Abreu L, Wasielesky W Jr (2010) Effect of practical diets with different protein levels on the performance of *Farfantepenaeus paulensis* juveniles nursed in a zero exchange suspended microbial flocs intensive system. Aquacult Nutr 16:163–172

Boyd CE (1982) Water quality management for pond fish culture. Auburn University, Alabama, p 318

Boyd CE (1998) Pond water aeration systems. Aquacult Eng 18(1):9–40

Burford MA, Thompson PJ, McIntosh RP, Bauman RH, Pearson DC (2004) The contribution of flocculated material to shrimp *(Litopenaeus vannamei)* nutrition in a high-intensity, zero-exchange system. Aquaculture 232(1–4):525–537

Cole JJ (1982) Interactions between bacteria and algae in aquatic ecosystems. Annu Rev Ecol Syst 13:291–314

Crab R, Kochba M, Verstraete W, Avnimelech Y (2009) Biofloc technology application in over-wintering of tilapia. Aquacult Eng 40:105–112

Crab R, Chielens B, Wille M, Bossier P, Verstraete W (2010) The effect of different carbon sources on the nutritional value of bioflocs, a feed for *Macrobrachium rosenbergii* postlarvae. Aquacult Res 41(4):559–567

Ekasari J, Crab R, Verstraete W (2010) Primary nutritional content of bio-flocs cultured with different organic carbon sources and salinity. Hayati J Biosci 17:125–130

Emerenciano M, Ballester ELC, Cavalli RO, Wasielesky W (2011) Effect of biofloc technology (BFT) on the early postlarval stage of pink shrimp *Farfantepenaeus paulensis*: growth performance, floc composition and salinity stress tolerance. Aquac Int 19:891–901

Emerenciano M, Ballester ELC, Cavalli RO, Wasielesky W (2012a) Biofloc technology application as a food source in a limited water exchange nursery system for pink shrimp *Farfantepenaeus brasiliensis* (Latreille, 1817). Aquac Res 43(3):447–457

Emerenciano M, Cuzon G, Goguenheim J, Gaxiola G, Aquacop et al (2012b) Floc contribution on spawning performance of blue shrimp Litopenaeus stylirostris. Aquacult Res 44:75–85

Emerenciano M, Gaxiola G, Cuzon G (2013) Biofloc technology (BFT): a review for aquaculture application and animal food industry. In: Biomass Now-cultivation and utilization, pp 301–328

Emerenciano MGC, Martinez-Cordova LR, Martinez-Porchas M, Miranda-Baeza A (2017) Biofloc technology (BFT): a tool for water quality management in aquaculture. Water quality. In Tech, London, pp 91–109

Fontenot Q, Bonvillain C, Kilgen M, Boopathy R (2007) Effects of temperature, salinity, and carbon:nitrogen ratio on sequencing batch reactor treating shrimp aquaculture wastewater. Bioresour Technol 98(9):1700–1703

Frías-Espericueta MG, Harfush-Meléndez M, Páez-Osuna F (2000) Effects of ammonia on mortality and feeding of postlarvae shrimp Litopenaeus vannamei. Bull Environ Contam Toxicol 65:98–103

Furtado PS, Poersch LH, Wasielesky W (2011) Effect of calcium hydroxide, carbonate and sodium bicarbonate on water quality and zoo technical performance of shrimp Litopenaeus vannamei reared in bio-flocs technology (BFT) systems. Aquaculture 321:130–135

Goldman C, Horne AJ (1983) Limnology, 1st edn. McGraw Hill, Tokyo, p 464

Hari B, Madhusoodana K, Varghese JT, Schrama JW, Verdegem MCJ et al (2004) Effects of carbohydrate addition on production in extensive shrimp culture systems. Aquaculture 24 (1–4):179–194

Jory DE (2001) Feed management practices for a healthy pond environment. In: Browdy CL, Jory CL (eds) The new wave, proceedings of the special session on sustainable shrimp culture. The world Aquac Soc, Baton Rouge, LA, pp 118–143

Ju ZY, Forster I, Conquest L, Dominy W, Kuo WC, Horgen FD (2008) Determination of microbial community structures of shrimp floc cultures by biomarkers and analysis of floc amino acid profiles. Aquacult Res 39:118–133

Khanjani MH, Sajjadi MM, Alizadeh M, Sourinejad I (2017) Nursery performance of Pacific white shrimp (*Litopenaeus vannamei* Boone, 1931) cultivated in a biofloc system: the effect of adding different carbon sources. Aquacult Res 48:1491–1501

Kuhn DD, Boardman D, Craig SR, Flick GJ, Mclean E (2008) Use of microbial floc generated from tilapia effluent as a nutritional supplement for shrimp, *Litopenaeus Vannamei* in recirculating aquaculture system. J World Aquac Soc 39(1):72–78

Kuhn DD, Boardman GD, Lawrence AL, Marsh L, Flick GJ Jr (2009) Microbial floc meal as a replacement ingredient for fish meal and soybean protein in shrimp feed. Aquaculture 296 (1–2):51–57

Kuhn DD, Lawrence AL, Boardman GD, Patnaik S, Marsh L, Flick GJ Jr (2010) Evaluation of two types of bioflocs derived from biological treatment of fish effluent as feed ingredients for Pacific white shrimp, *Litopenaeus vannamei*. Aquaculture 303(1–4):28–33

Lyssenko C, Wheaton F (2006) Impact of positive ramp short-term operating disturbances on ammonia removal by trickling and submerged-upflow biofilters for intensive recirculating aquaculture. Aquacult Eng 35(1):26–37

Mahanand SS, Moulick S, Rao PS (2012) Optimum formulation of feed for rohu, *Labeo rohita* (Hamilton), with biofloc as a component. Aquac Int 21(2):347–360

Mahanand SS, Moulick S, Rao PS (2013) Water quality and growth of rohu, Labeo rohita, in a biofloc system. J Appl Aquac 25(2):121–131

Maicá PF, Borba MR, Wasielesky W (2012) Effect of low salinity on microbial floc composition and performance of *Litopenaeus vannamei* (Boone) juveniles reared in a zero-water-exchange super-intensive system. Aquacult Res 43:361–370

Martins CIM, Pistrin MG, Ende SSW, Eding EH, Verreth JAJ (2009a) The accumulation of substances in Recirculating Aquaculture Systems (RAS) affects embryonic and larval development in common carp Cyprinus carpio. Aquaculture 291(1–2):65–73

Martins CIM, Ochola D, Ende SSW, Eding EH, Verreth JAJ (2009b) Is growth retardation present in Nile tilapia Oreochromis niloticus cultured in low water exchange recirculating aquaculture systems? Aquaculture 298(1–2):43–50

McIntosh D, Samocha TM, Jones ER, Lawrence AL, McKee DA, Horowitz S, Horowitz A (2000) The effect of a bacterial supplement on the high-density culturing of *Litopenaeus vannamei* with low-protein diet in outdoor tank system and no water exchange. Aquacult Eng 21:215–227

Meade WJ (1985) Allowable ammonia for fish culture. Prog Fish Cult 47(3):135–145

Michaud L, Blancheton JP, Bruni V, Piedrahita R (2006) Effect of particulate organic carbon on heterotrophic bacterial populations and nitrification efficiency in biological filter. Aquacult Eng 34(3):224–233

Mirzakhani N, Ebrahimi E, Jalali SAH, Ekasari J (2019) Growth performance, intestinal morphology and non specific immunity response of Nile tilapia (Oreochromis niloticus) fry culture in biofloc systems with different carbon sources and input C:N ratios. Aquaculture 512:734235

Ray AJ, Dillon KS, Lotz JM (2011) Water quality dynamics and shrimp (Litopenaeus vannamei) production in intensive, mesohaline culture systems with two levels of biofloc management. Aquac Eng 45(3):127–136

Rosenberry B (2010) Controlling pH in biofloc ponds. The shrimp news international. http://www.shrimpnews.com/FreeReportsFolder/phContronBioflocPonds.html. Accessed 15 Jul 2020

Saraswathy R, Muralidhar M, Panigrahi A, Lalitha N, Kumararaja P (2019) Biofloc technology for water quality management in aquaculture. In: Training manual on Biofloc technology for nursery and growout aquaculture, CIBA TM series 2019, vol 15, p 172

Serra FP, Gaona CA, Furtado PS, Poersch LH, Wasielesky W (2015) Use of different carbon sources for the biofloc system adopted during the nursery and grow-out culture of *Litopenaeus vannamei*. Aquac Int 23:1325–1339

Schneider O, Blancheton JP, Varadi L, Eding EH, Verreth JAJ (2006) Cost price and production strategies in European recirculation system. Linking tradition & technology highest quality for the consumer. WIAS, Firenze

Soares R, Jackson C, Coman F, Preston N et al (2004) Nutritional composition of flocculated material in experimental zero-exchange system for *Penaeus monodon*. In: Proceedings of Australian Aquaculture, 2004. WAS, Sydney, p 89

Suita SM (2009) The use of Dextrose as a carbon source in the development of bio-flakes and performance of white shrimp (Litopenaeus vannamei) cultivated in a system without water renewal. Master's dissertation, Federal University of Rio Grandae, Brazil

Tacon AGJ (1990) In standard methods for the nutrition and feeding of farmed fish and shrimp. Argent Laboratories Press, Washington, DC, p 454

Tacon AGJ, Cody JJ, Conquest LD, Divakaran S, Forster IP, Decamp OE (2002) Effect of culture system on the nutrition and growth performance of Pacific white shrimp *Litopenaeus vannamei* (Boone) fed different diets. Aquacult Nutr 8:121–137

Wasielesky W, Atwood H Jr, Stokes A, Browdy CL (2006) Effect of natural production in a zero exchange suspended microbial floc based super-intensive culture system for white shrimp *Litopenaeus vannamei*. Aquaculture 258(1–4):396–403

Biofilm in Aquaculture Production

23

P. K. Pandey and V. Santhana Kumar

Abstract

In intensive aquaculture practices, lesser biomass of fish is produced in comparison to the feed they consume. The wastewater discharges from the aquaculture system deteriorates the aquatic ecosystem. Hence, it warrants an effective technology to manage waste nutrient generation and also to reduce the use of fish meal as feed. Biofilm-based aquaculture was found to be a promising technology for extensive aquaculture practices which reduces both feed use and waste nutrient generation. Besides, biofilm acts as a quality protein source, habitat for shrimps during moulting process, immunostimulant and improves fish production, yield, survival and health and also effectively converts waste nitrogen into quality feed for fish/shrimps. Recently, substrate installation along with proper C:N was found to increase the fish/shrimp growth, making biofilm-based aquaculture possible for intensive culture practices. It has been found that biofilm plays a major role starting from nursery rearing to post-stocking management of table size fish in making the intensive aquaculture practices more sustainable.

Keywords

Biofilm · Aquaculture · Fish · Growth · Sustainability · Nutrient management

P. K. Pandey (✉)
ICAR-Directorate of Coldwater Fisheries Research, Bhimtal, Uttarakhand, India

V. S. Kumar
ICAR-Central Inland Fisheries Research Institute, Kolkata, West Bengal, India

© The Author(s), under exclusive license to Springer Nature Singapore Pte Ltd. 2021

P. K. Pandey, J. Parhi (eds.), *Advances in Fisheries Biotechnology*,
https://doi.org/10.1007/978-981-16-3215-0_23

401

23.1   Introduction

Aquaculture is the most important food producing sector and it meets the protein requirement of 40% of world's population. Aquaculture contributes about 46% of the global fish production of 179 million tonnes and 52% of fish requirements for human consumption. Capture fish production has diminished steeply in the last few decades and lot of marine fish stocks are already in the declining phase, which makes aquaculture an alternative option to meet the growing protein demand, especially in developing countries (FAO 2020). Intensive aquaculture practices are the need of the hour to meet the growing protein demands considering available land and water resources. However, intensification has associated shrinking problems such as pollution, disease and ecological impacts (Beveridge et al. 1997; Priedahitra 2003). Therefore, one needs a sustainable aquaculture technology to increase the fish production without affecting the environment. At the same time, the technology should be economically viable, especially one that reduces feed usage. Feed accounts for a major part of the operational cost in intensive aquaculture and excess feeding leads to increased pollution, disease incidence and economic loss to the industry (Priedahitra 2003).

In recent years, biofilm-based aquaculture has been gaining momentum, which is applicable to both freshwater and brackish water aquaculture. It has proved to be an effective nitrogen management technique for sustainable aquaculture (Crab et al. 2007). It effectively recycles the waste nutrients into fish feed and reduces both feed cost and nutrient pollution both in-situ and ex-situ (Kumar et al. 2017). The maintenance of optimum water quality is very essential for well-being of the cultured animal. Biofilm is a microbial consortium associated with the matrix of extracellular polymeric substances and bound to the submerged substrates, actively involved in nutrient recycling of the aquatic ecosystem. The bacteria present in the biofilm were found to be effective in recycling the nutrients and removal of unwanted nutrients from the system (Meyer-Reil 1994). It also reduces the toxic inorganic nitrogen, present in the culture system via nitrification and through heterotrophic bacteria (Avnimelech 2007). Installation of aquamat/substrate for the biofilm growth also improves the growth and survival of fish/shrimps. It also acts as excellent hideouts for freshwater prawns and shrimps during the time of moulting. The presence of microalgae in the biofilm serves as an excellent feed for the cultured animals, thereby the FC is reduced (Anand et al. 2013; Kumar et al. 2017). The economic analysis of the biofilm-based aquaculture showed increase in the profit margin when compared to the conventional aquaculture system (Kumar et al. 2019). Besides, manipulation of C:N ratio and the biofilm growth in substrate also improve the growth of both heterotrophic and autotrophic community in the biofilm in a well-balanced manner (Azim and Asaeda 2005). Despite the benefits of biofilm-based aquaculture, commercialization of this technology is absent with a few exceptions. Therefore, the present chapter aims to consolidate the research work carried out during the last decades on biofilm-based aquaculture (both freshwater and brackish water) in India and other parts of the world, suitable candidate species and the role of biofilm in sustainable aquaculture production.

23.2 Biofilm

There are so many terminologies used to refer the attached organisms to the submerged substrates. Biofilm is defined as the organisms found attached to the submerged substrates which can be natural or artificial materials. The organisms include the aggregate of algae, heterotrophic and autotrophic bacteria and other flora and fauna attached to the substrates. It plays a major role in nutrient cycling, energy flow and trophic transfer in aquatic ecosystems (Azim et al. 2005). Other terms are "Aufwuchs" and "Periphyton". The only difference in aufwuchs from periphyton/ biofilm is that the former also includes the non-attached fauna associated with the biofilm (Van Dam et al. 2002). In the limnological term, periphyton refers to the microfloral community associated with the submerged substrates in water (Wetzel 1983). A German word "Aufwuchs" refers to all organisms that are attached or move upon the submerged substrates in water. It includes algae, fungi, bacteria, protozoan and other associated animal components (Azim et al. 2005). The biofilm community comprises of attached algae, bacteria, protozoa, zooplankton and other invertebrates. The term biofilm is also used in other fields such as medical, food processing, wastewater and drinking water technology. The following steps were involved in biofilm formation: Initially, there will be a deposition of dissolved organic molecules like mucopolysaccharides on the surface of the substrates due to electrostatic forces. Following that, bacteria start adsorbing onto the substrate and attach by using mucus layers, aggregates together, cell division occurs, and formation of colonies through-out the substrate. After a day to week, algae start developing over the bacterial layer. The algae includes 4 different groups like Bacillariophyceae (14 genera), Chlorophyceae (12 genera), Cyanophyceae (10 genera) and Euglenophyceae (3 genera). Zooplankton, after algal settlement, will find their way and hides inside the substrate for consuming the algal population. In periphyton-based aquaculture, the dominant zooplanktons are rotifers, copepods, ciliates, daphnia and moina.

23.2.1 Organisms Associated with Biofilm

The taxonomic composition of the biofilm mostly depends upon the factors such as grazing pressure, water quality and the substrate types (Azim and Asaeda 2005). In biofilm-based aquaculture, most of the studies documented the algal composition of biofilm, the heterotrophic organisms have been neglected in the analysis. The taxonomic composition has been studied in various aquaculture systems, installed with different types of substrate. Biofilm (aquamat™ as a substrate) in *L. vannamei* culture consists of both algae and zooplankton. Among the algae, the Bacillariophyceae (14 genera), Chlorophyceae (12 genera), Cyanophyceae (10 genera) and Euglenophyceae (3 genera) were found. Cyanophyceae were the most dominant groups of periphytic algae associated with the aquamat™. The identified algal genera were *Amphora, Coscinodiscus, Cyclotella Cymbella, Diatoma, Fragellaria, Melosira, Navicula, Nitzschia, Pinnularia, Pleurosigma, Skeletonema, Surirella, Synedra, Anabaena, Anacystis, Aphanothece, Gloeocapsa,*

Gomphosphaeria, Lyngbya, Microcystis, Nostoc, Oscillatoria, Phormidium, Spirulina, Chaetophora, Chlamydomonas, Chlorella, Closterium, Enteromorpha, Gonatozygon, Mougeotia, Oedogonium, Scenedesmus, Sphaerocystis, Volvox, Ulothrix, Euglena and Gyrodinium. Zooplankton community includes rotifers, copepods, ciliates, daphnia and moina. Similar kinds of observation were made by Anand et al. (2019) in the biofilm (bamboo as a substrate) based *P. monodon* culture. Similarly, in mullet culture, Bacillariophyceae was the most abundant groups, present in the substrate. Audelo-Naranjo et al. (2011) used artificial substrates (Aquamats™) for the intensive farming of shrimp at two densities and found that the periphyton was composed of primary producers such as diatoms and cyanobacteria; primary consumers such as rhizopods, heliozoans, ciliates, flagellates, foraminifera, copepods, rotifers and gastrotriches, and some detritivorous metazoans such as nematodes and amphipods. Taxonomic composition of biofilm present in the pink shrimp nursery also dominated by pennate diatoms (Bacillaria paradoxa, Surirella sp. and Pleurosigma sp.), centric diatoms (Melosira dubia), filamentous cyanobacteria (*Oscillatoria* sp.), ciliates (mainly from vorticellid and tintinnid groups) and nematodes (Ballester et al. 2007). During the sea cucumber culture in substrate installed system, similar types of observation have been made. Besides diatoms, several propagules of macroalgae, including *Gracilaria* sp., *Sargassum* sp., three species of sea slugs, *Stylocheilus striatus, S. longicauda* and *Bursatella leachii* and crustose coralline algae were also observed (Gorospe et al. 2019). In freshwater aquaculture ponds, the biofilm communities include Bacillariophyceae, Chlorophyceae, Cyanophyceae and Euglenophyceae, and two groups of zooplankton: Crustacea and Rotifera. But, chlorophyceae dominated the other algal groups (Azim et al. 2002a). In bamboo installed polyculture system (*Macrobrachium rosenbergii + Oreochromis niloticus*), Haque et al. (2015) noticed that chlorophyceae was the most dominant group in biofilm, followed by Bacillariophyceae. Among the chlorophyceae, *Chlorella, Pediastrum, Scenedesmus* and *Ceratium* were the dominant algae. Besides algae, other benthic macroinvertebrates such as chironomids, oligochaetes and mollusks were also recorded from the biofilm. The earlier studies clearly indicated the dominance of algal groups in the biofilm depending on the nutrient quality of the culture water and also it varied based on the salinity of the water.

23.2.2 Biochemical Composition of Biofilm

Biofilm consists of both cellular biomass and extracellular polymeric substances. The major proportions of both the components are organic matter (10–90%) and other molecules such as protein, nucleic acids and heteropolymers (Nielsen et al. 1997). Nutritive quality of biofilm mostly depends on its taxonomic composition, habitat, water quality, grazing pressure and the substrates used for their growth. The quality of biofilm grown in closed systems was found to be better than their quality in open water ecosystems. There exists high variation in nutrient composition of the biofilm while using different types of substrates (Azim et al. 2005). The proximate composition of biofilm grown on different substrates in different habitats is provided in Table 23.1. The nutritional composition was well documented by various authors and also found appropriate for the fish dietary requirement (Azim et al. 2005). Biofilm not only meets the whole dietary requirement of fish/shrimp, it provides quality protein source to the cultured animals. The experimental diet prepared from biofilm showed improved growth and survival of shrimp when compared to the conventional feed (Anand et al. 2013). Apart from the nutrient source, biofilm is a rich source of growth promoters (Kuhn et al. 2010), immune stimulants (Supamattaya et al. 2005), bioactive compounds (Ju et al. 2008) and essential fatty acids. The above said properties of the biofilm were mainly attributed to the presence of bacteria and phytoplankton associated with the biofilm. It has been found that bacteria present in the biofilm are an important food source for penaeid shrimp which promoted its growth and survival in pond aquaculture systems (Keshavanath and Gangadhar 2005). In addition, other biofilm associated organisms such as protozoans, nematodes and polychaetes also serve as a quality nutrient source for the fish due to their high protein to energy conversion ratio (Ballester et al. 2007).

23.3 Factors Influencing Biofilm Formation

23.3.1 Nutrients

Nutrients have a profound effect on biofilm productivity in aquaculture system. In aquaculture, feed and input water are the major sources of nutrients. Increased nutrient availability in the system increases the biofilm mat density and thickness, and also the dominance of cyanobacteria has been noticed. The dominance of diatoms in the biofilm community depends upon the availability of silica in the water. High Si:P and N:P ratios favoured diatoms in the biofilm community, grown in reservoirs (Baffico and Pedrozo 1996) and the Baltic Sea (Sommer 1996). Inorganic carbon is another important nutrient which influences the biofilm productivity. The 2 mM concentration of inorganic carbon increased the productivity of biofilm fourfold when compared to the treatment without carbon addition (Jones et al. 2002). Phosphorus was found to be the limiting nutrient in freshwater ecosystem and affects the biofilm growth (Vymazal et al. 1994). The impact of different nutrients on biofilm development has been well documented in streams (Azim et al. 2005). The increase in production was mainly attributed to the increase in phytoplankton growth on the biofilm. Nitrogen is the limiting nutrient in marine waters and in shrimp aquaculture; it is a major burden for the shrimp growth and survival. Biofilm was found to utilize the waste nitrogen present in aquaculture system and convert it into useful biomass with the manipulation of carbon (Kumar et al. 2017). Asaduzzaman et al. (2010) conducted the experiment on *M. rosenbergii* by using substrate installed ponds manipulated with carbon. Anand et al. (2012) conducted an outdoor trial with *P. monodon* by using bamboo as a substrate along with the addition of rice flour as carbon source. The C:N ratio of 10:1 and 20:1 was maintained to support the growth of microbial community in biofilm. They have

Table 23.1 The biochemical composition of biofilm grown in different substrates installed in aquaculture system

Substrate type	Habitat/culture system	Protein (% DM)	Lipid (% DM)	Carbohydrate (% DM)	Ash (% DM)	Energy (kJ/g DM)	References
Sugarcane bagasse	Mud bottom cemented tank	9.4	0.33	38	23	–	Mridula et al. (2003)
Bamboo	Earthen pond/carp culture	23–32	2–5	33	15–29	14–20	Azim et al. (2002a, b, c)
Hizol branch	Earthen pond/carp culture	15	6	38	41	12	Azim et al. (2002b)
Jute stick	Earthen pond/carp culture	13	1.5	–	31	14	Azim et al. (2002c)
Bamboo	*Penaeus monodon* culture/FRP tank	25.96	2.65	67.25	32.75	298.58	Anand et al. (2013)
Aquamat ™	*L. vannamei*/HDPE lined pond; earthen bottom	24.97	2.67		37.5	278.11	Kumar et al. (2017)
Nylon mesh	Earthen pond/Nile tilapia	25.75	4.54		11.4–12.69	10.51	Tammam et al. (2020)

found that the addition of carbon in the ratio of 20:1 in the substrate installed pond reduced the inorganic nitrogen NH3–N by 48.2%, NO_3–N by 41.6%, NO_2–N by 42.7% compared with other treatments. Kumar et al. (2017) cultured the *L. vannamei* in the aquamat™ installed pond with the C:N ratio of 15:1 which significantly reduced the total nitrogen (ammonia, nitrite and nitrate) by 25% when compared with the control pond. In feed-based aquaculture, uneaten feed and excreted faeces are the vital sources of the nutrient waste accumulation inside the culture system. It has been noted that between 20 and 40% of the feed nutrient is incorporated into the fish body and the remaining is excreted (Funge-Smith and Briggs 1998). Therefore, the main strategies to reduce the nutrient waste in effluents should be to reduce its generation or recycling it into harvestable biomass. Waste generation can be reduced by improving the nutrient retention ability of fish (Amirkolaie 2011). Periphyton has the ability to activate digestive enzymes such as amylase, cellulase, protease, lipase, trypsin and chymotrypsin in shrimp (Anand et al. 2013), which in turn improves the feed digestibility and reduces feed wastage. Periphyton has the ability to utilize waste nitrogen and convert it into harvestable biomass (Audelo-naranjo et al. 2010; Kumar et al. 2017). Further, application of carbon source in periphyton system augments the conversion of toxic nitrogen into less toxic forms and improves the pond water quality. Immobilization of toxic nitrogen species occurs much more rapidly by heterotrophic assimilation than by nitrifying bacteria. These carbon source applications in the substrate (Aquamat™) installed pond provides an environment with high concentrations of organic matter, supports both heterotrophs and autotrophs in the aphotic and photic zone, respectively, removing nitrogen compounds in the culture water (Suryakumar and Avnimelech 2017; Kumar et al. 2017). The adsorption of organic matter over the substrate also reduces sediment accumulation in the pond bottom. Azim et al. (2003) showed that addition of substrate for biofilm adherence to fertilized Tilapia culture ponds, significantly improved growth and doubled the nitrogen retention when compared to the control ponds. It has been noticed that the installation of aquamat in vertical position in the shrimp culture system reduced the nitrogen content (25%) in earthen pond (Azim et al. 2003; Kumar et al. 2017) and in experimental tanks (Audelo-naranjo et al. 2010) as it was consumed by both periphyton and shrimp biomass in the system. This clearly indicates that the culture practices integrated with substrate installation, reduces the nutrient waste generation in-situ, acts as a feed source for the cultured animals and excludes the usage of separate effluent treatment system (ETS) in aquaculture farm.

23.3.2 Grazing

Biofilm community is grazed by most of the aquatic animals existing in the ecosystem. Invertebrate grazers such as gastropods, crustaceans (freshwater prawns, *P. monodon*, *L. vannamei*) and vertebrates such as fishes (rohu, mullets, rabbit fish, Nile tilapia) effectively graze the biofilm community but not all components of biofilm community will be grazed equally. Invertebrate snails, generally, graze upon the algal component of the biofilm. The biofilm grazing rate by the animals mostly depends on the size of the grazers, animal density and the availability of other food sources. The biofilm grazing rate was estimated to be in the range of 0.03–0.9 mg AFDW/individual/day (Vermaat 2005). In aquaculture system, Huchette and Beveridge (2005) cultured the Nile tilapia in cages containing biofilm and studied their feeding preference on biofilm community. The study revealed that Nile tilapia grazed upon large diatoms, filamentous algae, present in the biofilm community. In periphytic substrate, rohu is found to graze upon the plankton, present in the subperiphytic layer of the substrate and is regarded as a suitable candidate species for biofilm-based aquaculture (Majumder and Saikia 2020). *L. vannamei* has also proven to be an effective grazer of biofilm community in aquaculture (Kumar et al. 2017). Biswas et al. (2017) also noticed that mullet has gained weight and survival due to their effective grazing upon biofilm. Grazing also positively influences the biofilm growth by renewing the algal cells via removing the dead and senescent cells, and keeps the algal assemblage in a productive, early-successional stage. Studies have also observed a fourfold increase in periphyton specific productivity in grazed periphyton compared to ungrazed controls (Norberg 1999).

23.3.3 Substrate

In biofilm-based aquaculture, a variety of synthetic materials can be used to support the biofilm growth viz. PVC pipes (Keshavanath et al. 2001), plastic sheets (Shresta and Knud-Hansen 1994; Tidwell et al. 1998) and custom-designed materials such as Aquamats™ (Bratvold and Browdy 2001). Organic substrates such as bamboo, hizol, coconut husks, jute sticks, sugarcane bagasses, paddy stem and other tree branches have been successfully utilized for the growth of biofilm. The nature of the substrate highly influences the biofilm production, growth, composition, nutrient quality and water quality. The protein content of the biofilm, grown in bamboo tree, was 50% higher than the other substrates studied, but there was no change in the community of biofilm grown in different substrates. Besides, another study compared both synthetic (PVC) and natural substrate (sugarcane bagasse) for biofilm production. It has been found that sugarcane bagasse is a best substrate to support the biofilm growth and improved culture water quality when compared to PVC substrates. Another study by Khatoon et al. (2007) showed that bamboo produced more biofilm biomass when compared to other synthetic substrates such as PVC, ceramic tiles, plastic sheets and fibrous scrubbers. The change in the biomass production of biofilm on different substrates highly depends on the biodegradability of the substrates, texture, nutrient leaching and the presence of toxic substances on the substrates. In case of shrimp culture, artificial substrate such as Aquamats™ installed vertically inside the pond improved the growth, health and survival of the shrimp. This is mainly due to their role as hideouts during the time of moulting (Kumar et al. 2017). Therefore, the choice of substrates will be based upon the types of species cultured, cost of substrates and availability of raw material for the substrates.

23.3.4 Light and Temperature

Sunlight is a major source of illumination in the aquaculture system for the growth of plankton communities in biofilm. The biofilm population mostly comprises of autotrophic and heterotrophic organisms and their dominance mostly depends on the quantity and quality of available light. The change in biofilm assemblages on different depth of water bodies has revealed the importance of light on the biofilm productivity. Moreover, algal components play a major role in supporting the non-algal components of biofilm by providing nutrient exudates and luring more biofilm assemblage organism (Azim et al. 2005). It shows that algae is an integral part of biofilm and determines the biofilm productivity. Hence, irradiance plays a vital role in improving the biofilm community. In a recent study in *L. vannamei* culture systems, it was observed that there was dominance of algal community in upper part of the substrate, exposed to sunlight and heterotrophic dominance was noticed in lower part of the substrate (Kumar et al. 2017). Similarly, shifting of dominance from diatom to anaerobic photosynthetic green and purple bacteria has been noticed in an aquatic ecosystem (Goldsborough 1993).

Temperature also largely determines the composition of the periphyton community. Earlier studies have found that at high temperatures, *Scenedesmus* dominated whereas low temperature favoured the growth of *Navicula* (Vermaat and Hootsmans 1994; Azim et al. 2005). The temperature-influenced succession of periphyton communities was also noticed earlier in temperate countries (i.e. diatom dominance in spring season to green algae/cyanobacteria dominance during summer). Very few studies have been conducted which noticed the effect of temperature and light incidence on biofilm growth and composition. More studies need to be undertaken to determine the optimum temperature and light required for the growth of biofilm and their influence on biofilm quality.

23.4 Role of Biofilm in Aquaculture

23.4.1 Biofilm in Nursery System

Nursery rearing of finfish involves the rearing of spawn to fingerling stage. For shrimps, the nursery phase includes raising of wild-caught or hatchery produced post-larvae up to the stage it could be stocked in grow out ponds. Utilization of biofilm in both shrimp and fish nursery are very well documented. The growth, survival and the weight of *Labeo fimbriatus* fry were greatly enhanced by the addition of sugarcane bagasse as a substrate for biofilm development and it was also found that the stocking density could be doubled (Gangadhar et al. 2015). The improved growth was mainly due to influence of the biofilm on digestive enzyme activity in fry (Gangadhar et al. 2016). Sakr et al. (2015) evaluated the effect of lowering protein diets in substrates installed Nile tilapia juvenile rearing system. The result revealed that the diet containing 15% protein, in the substrate installed ponds, significantly improved the growth of juveniles when compared with high protein diet treatments without substrates. Gorospe et al. (2019) found that biofilm development was directly correlated with the development of sea cucumber juveniles in an ocean nursery system. The installation of substrate (HDPE sheet) in the nursery rearing of *L. vannamei* in RAS-based culture system doubled the shrimp biomass in comparison to the treatment without the installation of substrates (Tierney et al. 2020). Raising the mullet fry in bamboo substrates (10% of the total pond surface area) significantly improved their growth and survival (95%). In contrast, the introduction of substrates (meshed *Happa*) in marine cages did not show significant changes in the growth and survival of Mullet stocked at the density of 1 m^2/cage. Similarly, biofilm did not influence the growth of sea bream (Richard et al. 2010). Another study by Thompson et al. (2002) found that biofilm alone did not support the growth of shrimp and suggested that feed addition along with biofilm improved the shrimp growth. The contrasting/varied results noticed by earlier studies on growth and production of cultured animals could be attributed to the difference in stocking density, substrate surface area, cultured environment and associated organisms (Tidwell and Bratvold 2005).

23.4.2 Biofilm in Stocking Management

23.4.2.1 As a Feed Ingredient

The presence of high protein content, organic matter and other nutrients in biofilm make it a suitable feed ingredient for the cultured animals. Apart from direct consumption of biofilm by the grazers in the aquaculture system, its usage as a feed ingredient has also been successfully experimented. In the earlier study by Anand et al. (2013), periphyton grown in the natural bamboo substrate was dried and supplemented in shrimp basal diet. It has been found that periphyton supplementation at 6% in basal diet significantly improved the digestive enzyme activities in shrimp and also enhanced the growth and production. The weight of the individuals also greatly increased in periphyton supplemented diet. The study suggests that inclusion of periphyton in diet at an optimum level would improve the production of the cultured animal by significantly enhancing the digestion capability of the animals. In another study on *Farfantepenaeus paulensis*, an increased growth was noticed due to the consumption of biofilm grown on the substrate (Abreu et al. 2007). Besides, the algal species to the biofilm was isolated and provided as a feed supplement in *P. monodon* nursery system. The result revealed that the marine periphytic diatoms significantly improved the growth and survival of the *P. monodon* post-larvae (Khatoon et al. 2009).

23.4.2.2 As an Immunostimulant/Vaccine

Fish immune system is comprised of both innate and adaptive immunity. Shrimps have only non-specific immune response. Biofilm has immunostimulant property and greatly improves both fish and shellfish immunity. Biofilm usage, as both feed supplement and direct exposure to the cultured fishes, improves their immune responses. It has been reported that biofilm improves non-specific immune response in *L. vannamei* and *P. monodon* (Anand et al. 2013, 2015; Zhang et al. 2010). Other study by Tammam et al. (2020) showed that biofilm had the potential to elucidate both specific and non-specific immune response in Nile tilapia. The biofilm developed over substrate is capable to elicit immune response in fish and shrimps (Verdegem et al. 2005). Usage of biofilm as a dietary ingredient also improves the immune response and disease resistance of shrimp (Anand et al. 2015). In addition, development of vaccine using biofilm is a new area for improving the health of aquaculture animals. Azad et al. (1997) developed biofilm-based *Aeromonas hydrophila* as oral vaccine and the efficacy was evaluated in Indian major carps (IMCs). The antibody titre and immune protection were higher in biofilm vaccinated IMCs when compared to free cell vaccine. Further, higher retention period of vaccine was reported (Azad et al. 2000).

23.4.2.3 In-Situ Water Quality Management by Biofilm

Water quality management is very important for sustainable fish production. Numerous studies have earlier reported the improvement of water quality through biofilm-based aquaculture system. Nitrification is the major process in the aquaculture system which determines the water quality of the system. In aquaculture, feed is the major source of nitrogen which disturbs the natural nitrogen cycle in case of intensive or semi-intensive aquaculture (Kumar et al. 2017). The excess feed based nitrogen is converted into inorganic nitrogen species such as ammonia which in turn affects the health of the cultured animal. But the nitrifying bacteria present in the biofilm significantly reduce ammonia by converting it into nitrate via nitrite. Besides, other heterotrophic community, associated with the biofilm, also directly consumes inorganic nitrogen for their growth. The balanced C:N ratio in substrate installed treatment significantly improves the water quality of the system and reduces the ammonia nitrogen in the system (Azim et al. 2002a; Kumar et al. 2017). Biofilm application in *Catla catla* fingerling rearing showed improvement in water quality (Pradeep et al. 2003). In freshwater ponds, biodegradable substrate installation made the water quality favourable for the growth of fishes. The level of ammonia is significantly low in the biofilm treatment when compared to the normal aquaculture system (Keshavanath et al. 2012). In brackish water ponds, non-biodegradable substrates installation also reduces the ammonia nitrogen and nitrite nitrogen and favours the growth of *L. vannamei* (Kumar et al. 2017). The reduction of toxic nitrogen in the water could be possible due to the presence of both nitrifying bacteria and microalgae. Microalgae also use this element for their growth (Thompson et al. 2002).

Alkalinity and pH are two other, important water quality parameters which need to be maintained for welfare of culture animals. It has been observed that substrate installation in *L. vannamei* culture significantly improves the pH (7.4–8.4) and alkalinity (88–230 mg/L as $CaCO_3$) of the culture system throughout the culture period. The photosynthesis of autotrophic community in the biofilm increases pH and alkalinity of the culture system (Kumar et al. 2017). Numerous studies have reported the improvement of water quality in biofilm-based eco system, using different types of substrates (Anand et al. 2013; Asaduzzaman et al. 2010). The turbidity of the water is also reduced by the biofilm; this is due to the trapping of organic matter or the accumulation of suspended matters over the submerged substrate (Van Dam et al. 2002). Reduction in dissolved oxygen (DO) level of the substrate installed tanks has been noticed by various researchers which is due to the shading effects of substrates and also due to the increased biochemical oxygen demand (BOD) (Anand et al. 2019).

23.4.3 Biofilm in Post-stocking Management

23.4.3.1 Biofilm in Aquaculture Effluent Treatment

During the last few decades, researchers utilized microalgae for the treatment of aquaculture effluents (Oswald 2003; Levy et al. 2017). The cost of harvesting the algae from the system impedes its usage further. Periphyton grown in vertical plastic net substrates and algal turf scrubbers was found to be a feasible option for mitigating nutrient pollution, caused due to aquaculture effluents (Erler et al. 2004; Valeta and Verdegem 2009).

Marine periphyton-based biofilters which allow the growth of marine biofilm on a plastic net based substrate in a bioreactor contain fish mariculture effluents. It has been noted that the periphyton removes total ammonia nitrogen (TAN) and dissolved inorganic nitrogen (DIN) in the range between 0.11 and 1.2 $gNm^{-2} day^{-1}$. Further, they revealed that the vertical integration of plastic net as well as the biomass weight and effluent retention time significantly affected the nutrient removal efficiency of marine periphyton (Levy et al. 2017). Sereti et al. (2004) studied the role of periphyton in maintaining water quality in a recirculating aquaculture system (RAS). It has been revealed that the incorporation of periphyton mat in RAS instead of biological filters improves the growth and nutrient (N, P) retention in Nile tilapia. Scrapping of periphyton from the periphyton mat at regular time intervals and feeding to the culture organisms directly or indirectly improves the nutrient retention capacity. The combination of finfish and periphyton in the effluent treatment process improves the nutrient retention of the farm effluents when compared to utilizing periphyton alone as effluent treatment system (Erler et al. 2004). The fish-periphyton system mesocosm has been successfully used as a tertiary nutrient removal system. It consists of series of the 375-L tank, each containing vertical plastic mesh to support the periphytic algal growth and the algal grazing tilapia fish (*Tilapia mossambica*), into which wastewater has been pumped and tested. This system reduces total phosphorus (TP) and total nitrogen (TN) by 82% and 23%, respectively from the wastewater with an average removal rate of 27 mg of total phosphorus (TP) $m^{-2} d^{-1}$ and 108 mg of total nitrogen (TN) $m^{-2} d^{-1}$. This system is covered by US patent no. 5254252 (Rectenwald and Drenner 2000). Other than artificial substrates, natural substrates such as sugarcane bagasse have also been used for treating the aquaculture effluents. The study utilized bagasse, a natural highly fibrous lignocellulosic byproduct of sugarcane for supporting the growth of periphyton, applied at the dose of 1 to 6 g L^{-1}, significantly reduces the ammonia concentration from the shrimp wastewater within 24 h (Krishnani et al. 2006). It paves the way to research on screening the eco-friendly agriculture byproducts in the effective treatment of aquaculture effluents.

23.4.3.2 Biofilm in Fish Growth

The growth of fishes in aquaculture ponds mainly depends upon the culture environment and the availability of feed in the system. Biofilm development inside the culture system found to improve the water quality and also acts as a feed ingredient, which directly improves the fish growth. In freshwater aquaculture system, natural biodegradable substrates are used for supporting the growth of biofilm and it has been found that differences exist in growth performance of animals grown in different substrates. For example, the growth of rohu and common carp in the treatment, using sugarcane bagasses as a substrate was found to be 2–3 times higher when compared with the treatment using other substrates like bamboo leaf, *Eichhornia*, paddy straw and palm leaf (Ramesh et al. 1999). Another study found that the performance of freshwater fish like *Cirrhinus mrigala* was better in natural biodegradable substrate when compared with the non-biodegradable substrate like plastic sheet and tiles (Pandey et al. 2014). A 20% increase in production of freshwater prawn, grown in biofilm-based pond, was noticed by Tidwell et al. (2011). Similar kind of growth improvement was also noticed in brackish water shrimp farming. The growth and survival of the *L. vannamei* cultured in biofilm pond was very high in comparison to the normal pond. Most of the studies on *L. vannamei* utilized the non-biodegradable substrates for supporting the growth of biofilm (Kumar et al. 2017). Other species such as *Fenneropenaeus paulensis* (Thompson et al. 2002; Ballester et al. 2007), *Penaeus esculentus* (Burford et al. 2003; Arnold et al. 2005) and *P. monodon* (Anand et al. 2013) also showed increased growth, survival and production in substrate installed ponds. The improved growth, survival and production of the cultured animal in biofilm-based culture system is mainly due to the following reasons: (1) increase in food availability, (2) effective nutrient cycling, (3) conversion of toxic nutrient into useful forms, (4) reduction of pathogenic bacteria, (5) improvement of culture water quality, (6) provision of special nutrients (Azim et al. 2005).

23.4.4 Suitable Species for Biofilm-Based Aquaculture

In the aquatic environment, biofilm plays a major role in primary production and supports the food web. It also serves as a food source for fishes present in both natural or culture environment. In freshwater aquaculture, biofilm was highly utilized for polyculture systems by using the substrates such as bamboo, palm leaf, coconut leaf and sugarcane bagasses. The species cultured in the biofilm-based aquaculture system has been listed in Table 23.2. In the biofilm-based polyculture system, freshwater fish species such as *O. niloticus, Catla catla, Cirrhinus mrigala, L. rohita* and *L. calbasu* were grown together and found that *L. rohita* growth was higher in comparison to other species. In another comparative study of both *L. rohita* and *L. gonius*, the former showed 77% more growth when compared to the latter. The increased growth of rohu in the biofilm-based polyculture system is mainly due to the consumption of organisms associated with the biofilm subsurface zone, which may exclude the competition from other planktonic feeders and grazers (Majumder and Saikia 2020). Freshwater prawn has been extensively cultured in the biofilm-based aquaculture systems. There has been an improved productivity, animal welfare and reduced mortality in case of freshwater prawns grown in the biofilm-based aquaculture system (Tidwell et al. 2000). For *L. vannamei* culture, artificial substrate such as aquamat™ has been widely used for biofilm growth in the culture system with a view to improve the biomass, health and moulting behaviour of shrimp. Numerous studies have noticed improved growth and survival of the shrimp, grown in the biofilm-based aquaculture system (Kumar et al. 2017).

23.5   Biofilm Development in Aquaculture Ponds

Aquamats are commercially available black shade netting used for horticulture. They have an area of 20×0.5 m with 1 mm mesh size. It can be installed inside the pond like a vertical gill net, having plastic soft drink bottles as floats and sand-filled bottles as sinkers. A mat is tied on both sides to the pole and is suspended vertically 10 cm below the water surface and 50 cm above the pond bottom. Aquamats are kept parallel to each other, 30 cm apart, to form a battery; they are installed in front of the aerator to provide ample oxygen to attached organisms. In addition, carbon source (molasses) can be added in front of the aerator, in order to use ammonia uptake by heterotrophic organisms in the aquamat (Suryakumar and Avnimelech 2017; Kumar et al. 2017). This depicts that both autotrophic and heterotrophic biomass controls the generation of waste nutrients. Regular scrapping is done to remove excess, suspended solids which settle over the aquamat. Biofilm development starts once the substrate is submerged inside the ponds.

23.6   Economics of Biofilm-Based Aquaculture

Gangadhar and Keshavanath (2002) attempted to calculate the economics of substrate-based polyculture system. The installation of substrate, transport and labour cost warrant increased production cost when compared with the normal fish pond. But the increased fish production and the resultant sale amount compensate the production cost and improve the profit of the polyculture pond. In another study by Azim et al. (2002a), the economics of substrate-based aquaculture has been calculated and the study showed the effect of substrate on the profitability of the culture

Table 23.2 Growth performance of culture animals grown on biofilm-based aquaculture using different substrates

Aquaculture system	Substrate installed	Periphyton productivity (DM basis mg/cm^2)	Culture species	Weight gain/final weight (g)	FCR	Survival (%)	Reference
Freshwater aquaculture	Bamboo+ carbohydrate source addition	2.02–2.56	Rohu+ tilapia	150^R 294^T	–	85.8	Asaduzzaman et al. (2010)
	Sugarcane bagasse	–	*Labeo Fimbriatus* fry	2.2–5.47	–	59.32–88.8	Gangadhar et al. (2015)
	Bamboo side shoots	1.7–2.9	*M. rosenbergii + Oreochromis niloticus*	12.78^M 96.08^T	2.02	66 ^{M} 90.5 ^{T}	Haque et al. (2015)
	Bamboo Bagasse	0.14–0.17^{Bam} 0.11–0.17^{Bag}	*Labeo rohita*	28.63–37.67 ^{Bam} 30.1–34.75^{Bag}		77.78–88.89 ^{Bam} 66.67–100 ^{Bag}	Keshavanath et al. (2012)
	Bamboo branches (*Kanchi*)	1.85	*M. rosenbergii + GIFT tilapia*	40.46 199.35	2.27	78.97 ^{M} 91.57 ^{T}	Haque et al. (2016)
Brackish water aquaculture	Polyethylene screens		*Farfantepenaeus paulensis*	0.6		95	Ballester et al. (2007)
	Bamboo	1.62–2.85	Milkfish (*Chanos chanos*)	13.3–25.5	–	93–96	Garg (2016)
	Bamboo poles	15.06 ± 2.78	Mullet fry	28.39 ± 1.94		94.3	Biswas et al. (2017)
	Aquamat™	4.81	*L. vannamei*	–	1.2–1.5	93	Kumar et al. (2017)
	Bamboo + C: N ratio of 10:1	12.77 ± 1.47 (3.4–22.7)	*Penaeus monodon*	4.85 ± 0.49	2.37	70	Anand et al. (2019)
	High density polyethylene mesh	–	*Litopenaeus vannamei*	1.52 ± 0.19	0.99	50.3	Tierney et al. (2020)
	Nylon mesh	–	*L. vannamei*	20.70 ± 2.18	1.53	92	Sofyani and Sambhu (2020)

system. Azim (2001) compared the economic efficiency of different types of substrates, used for biofilm growth. It has been found that net profit margin of biofilm-based aquaculture is higher in jute stick treatment (46%) and lower in bamboo treatment (14%). Kumar et al. (2019) carried out the economic analysis of substrate installed *L. vannamei* pond aquaculture system. Substrate installation warrants more capital investment relative to the control pond. Higher production in periphyton pond resulted in increased net income generation by 35.4%. Periphyton improves the economic return of the semi-intensive shrimp farming and reduces the breakeven point and feed cost of the culture pond. Though substrate installation warrants more production cost, the increased survival, growth of the cultured animal and reduced commercial feed requirement surge the profitability of the substrate installed aquaculture system. The economic analysis of the substrate installed aquaculture system has been depicted in Table 23.3.

23.7 Knowledge Gaps and Further Recommendation

Biofilm-based aquaculture has been intensively studied for applying in polyculture of carps with *M. rosenbergii* in freshwater culture system. Most of the experimental studies and field trials show improved production and profit of biofilm-based polyculture in earthen ponds, installed with bamboo substrate. In most of the commercial scale aquaculture, farmers do partial harvesting of carps in a timely manner. Hence, it is found difficult to remove and reinstall the substrates more frequently. Also the biofilm productivity and growth increment by considering the partial harvesting have not yet been studied. For high value species like *L. vannamei*, non-biodegradable mats such as polythene mesh, and aquamats™ have been experimented in both experimental and outdoor tanks. Apart from the improved shrimp production, the biofilm-based shrimp culture was also found efficient in treating the effluent nitrogen. More research and effective policies are needed to establish it as an efficient effluent treatment system for intensive farming. The following research gaps also need to be addressed in order to commercialize the biofilm-based aquaculture:

1. Biofilm-based freshwater polyculture and its effectiveness on fish growth and production should be studied by considering the partial harvesting of fishes.
2. Biodegradable substrates were found suitable for biofilm-based freshwater aquaculture, but more research will be needed to find suitable species, substrates and the optimum substrate to biofilm ratio.
3. Field level research on the biofilm-based intensive aquaculture is less in number.
4. Biofilm-based mariculture activities are meagre with a few exceptions on mullet rearing. More marine candidate species could be explored for biofilm-based aquaculture.
5. The intervention of biofilm on integrated multi-trophic aquaculture should be studied further.

Table 23.3 Economics of biofilm-based aquaculture system

Culture system/species	Economic analysis					Culture system information	Reference
	Substrate type	Operational cost (USD $ 1$ = 75.08 rupees; 1 USD = 50 Dkt)	Economic return (USD $)	Net economic return (USD $)	Benefit cost ratio[a]/net profit margin (%)[b]		
Carp culture	Bamboo	15	17	2	14[b]	Pond area: 75 m²; DOC: 135 days	Azim (2001)
	Jutestick	12	17	5	46[b]		
	Kanchi	13	17	4	27[b]		
Polyculture (catla, rohu and common carp)	Sugarcane bagasse	79	864	784	–	Pond area: 1 ha	Gangadhar and Keshavanath (2002)
Nile tilapia culture	Plastic bottles	–	715.39	430.29	2.5[a]	Pond area: 1 ha	Abwao et al. (2014)
Grey mullet fry rearing	Bamboo	5914	11,303	5389	1.91[a]	Pond area: 1 ha; DOC: 120 days	Biswas et al. (2017)
Litopenaeus vannamei culture	Aquamat™	13,979	32,000	18,021	2.3[a]	Pond area: 1 ha; DOC: 120 days	Kumar et al. (2019)

[a]Benefit cost ratio
[b]Net profit margin

23.8 Conclusion

It is very much necessary to develop an aquaculture technology which maximizes the production without affecting the environment. Biofilm is a complex mixture of both autotrophic and heterotrophic organisms, attached in a solid substrate. Utilization of biofilm in aquaculture bears lots of advantages, namely, (1) improvement of water quality, (2) conversion of toxic nitrogen into quality feed, (3) improvement of growth and survival of fish and shrimp, (4) improvement of the health of the animal via extracellular secretions, (5) Acting as excellent hideouts for shrimps/prawns, (6) highly sustainable by removing nutrient wastes from the effluents. Low cost substrates for growing high value species like *L. vannamei* without affecting the productivity need to be explored. Biofilm in marine aquaculture, the candidate species selection in particular, should be studied. In addition, studies addressing the optimum substrate area required for optimum maintenance of biofilm for the cultured aquatic organisms need to be undertaken.

References

Abreu PC, Ballester ELC, Odebrecht C, Wasielesky JRW, Cavalli RO, Graneli W, Anesio AM (2007) Importance of biofilm as food source for shrimp (*Farfantepenaeus paulensis*) evaluated by stable isotopes (δ13C and δ15N). J Exp Mar Biol Ecol 347:88–96

Abwao JO, Boera PN, Munguti JM, Orina PS, Ogello EO (2014) The potential of periphyton based aquaculture for Nile tilapia (*Oreochromis niloticus* L.) production. A review. Int J Fish Aquat Stud 2(1):147–152

Amirkolaie AK (2011) Reduction in the environmental impact of waste discharged by fish farms through feed and feeding. Rev Aquac 3(1):19–26

Anand PSS, Kumar S, Panigrahi A, Ghoshal TK, Dayal JS, Biswas G, Sundaray JK, De D, Raja RA, Deo AD, Pillai SM, Ravichandran P (2012) Effects of C:N ratio and substrate integration on periphyton biomass, microbial dynamics and growth of *Penaeus monodon* juveniles. Aquacult Int 21:511–524. https://doi.org/10.1007/s10499-012-9585-6

Anand PSS, Kohli MPS, Roy SD, Sundaray JK, Kumar S, Sinha A, Pailan GH, Sukham MK (2013) Effect of dietary supplementation of periphyton on growth performance and digestive enzyme activities in *Penaeus monodon*. Aquaculture 392(395):59–68

Anand PS, Kohli MP, Roy SD, Sundaray JK, Kumar S, Sinha A, Pailan GH, Sukham MK (2015) Effect of dietary supplementation of periphyton on growth, immune response and metabolic enzyme activities in *Penaeus monodon*. Aquacult Res 46(9):2277–2288

Anand PS, Balasubramanian CP, Christina L, Kumar S, Biswas G, De D, Ghoshal TK, Vijayan KK (2019) Substrate based black tiger shrimp, *Penaeus monodon* culture: stocking density, aeration and their effect on growth performance, water quality and periphyton development. Aquaculture 507:411–418

Arnold SA, Sellars MJ, Crocos PJ, Coman GJ (2005) Response of juvenile brown tiger shrimp (*Penaeus esculentus*) to intensive culture conditions in a flow through tank system with three dimensional artificial substrate. Aquaculture 246:231–238

Asaduzzaman M, Rahman MM, Azim ME, Islam MA, Wahab MA, Verdegem MCJ, Verreth JAJ (2010) Effects of C/N ratio and substrate addition on natural food communities in freshwater prawn monoculture ponds. Aquaculture 306:127–136

Audelo-Naranjo JM, Martínez-Córdova LR, Voltolina D (2010) Nitrogen budget in intensive cultures of *Litopenaeus vannamei* in mesocosms, with zero water exchange and artificial substrates. Rev Biol Mar Oceanogr 45(3):519–524

Audelo-Naranjo JM, Martínez-Cordova LR, Voltolina D, Gomez-Jimenez S (2011) Water quality, production parameters and nutritional condition of *Litopenaeus vannamei* (Boone, 1931) grown intensively in zero water exchange mesocosms with artificial substrates. Aquacult Res 42:1371–1377

Avnimelech Y (2007) Feeding with microbial flocs by tilapia in minimal discharge bio-flocs technology ponds. Aquaculture 264(1–4):140–147

Azad IS, Shankar KM, Mohan CV, Kalita B (1997) Evaluation of biofilm of *Aeromonas hydrophila* for oral vaccination of carps. Dis Asian Aquacult 9(7):519–528

Azad IS, Shankar KM, Mohan CV, Kalita B (2000) Uptake and processing of biofilm and free-cell vaccines of *Aeromonas hydrophila* in Indian major carps and common carp following oral vaccination antigen localization by a monoclonal antibody. Dis Aquat Organ 43(2):103–108

Azim ME (2001) The potential of periphyton-based aquaculture production systems. Ph.D. Thesis, Wageningen University

Azim ME, Asaeda T (2005) Periphyton structure, diversity and colonization. In: Azim ME, Verdegem MCJ, van Dam AA, Beveridge MCM (eds) Periphyton—ecology, exploitation and management. CABI, Wallingford, pp 15–33

Azim ME, Milstein A, Wahab MA, Verdegam MCJ (2003) Periphyton – water quality relationships in fertilized fish ponds with artificial substrates. Aquaculture 228(1–4):169–187

Azim ME, Verdegem MC, Rahman MM, Wahab MA, Van Dam AA, Beveridge MC (2002a) Evaluation of polyculture of Indian major carps in periphyton-based ponds. Aquaculture 213 (1–4):131–149

Azim ME, Wahab MA, Verdegem MCJ, Van Dam AA, Van Rooij JM, Beveridge MCM (2002b) The effects of artificial substrates on freshwater pond productivity and water quality and the implications for periphyton-based aquaculture. Aquat Living Resour 15:231–241

Azim ME, Verdegem MC, Khatoon H, Wahab MA, Van Dam AA, Beveridge MC (2002c) A comparison of fertilization, feeding and three periphyton substrates for increasing fish production in freshwater pond aquaculture in Bangladesh. Aquaculture 212(1–4):227–243

Azim ME, Verdegem MC, van Dam AA, Beveridge MC (eds) (2005) Periphyton: ecology, exploitation and management. CABI, Wallingford

Baffico GD, Pedrozo FL (1996) Growth factors controlling periphyton production in a temperate reservoir in Patagonia used for fish farming. Lakes Reserv Res Manag 2(3–4):243–249

Ballester ELC, Wasielesky JW, Cavalli RO, Abreau PC (2007) Nursery of the pink shrimp *Farfante penaeus paulensis* in cages with artificial substrates: biofilm composition and shrimp performance. Aquaculture 269:355–362

Beveridge MC, Phillips MJ, Macintosh DJ (1997) Aquaculture and the environment: the supply and demand for environmental goods and services by Asian aquaculture and the implications for sustainability. Aquacult Res 28(10):797–807

Biswas G, Sundaray JK, Bhattacharyyaa SB, Anand PSS, Ghoshala TK, Prem Kumara DD, Sukumaran K, Berac A, Mandal B, Kailasam M (2017) Influence of feeding, periphyton and compost application on the performances of striped grey mullet (*Mugil cephalus* L.) fingerlings in fertilized brackish water ponds. Aquaculture 481:64–71

Bratvold D, Browdy CL (2001) Effect of sand sediment and vertical surfaces (Aquamat TM) on production, water quality and microbial ecology in an intensive Litopenaeus vannamei culture system. Aquaculture 195:81–94

Burford MA, Thompson PJ, McIntosh RP, Bauman RH, Pearson DC (2003) Nutrient and microbial dynamics in high-intensity, zero-exchange shrimp ponds in Belize. Aquaculture 219:393–411

Crab R, Avnimelech Y, Defoirdt T, Bossier P, Verstraete W (2007) Nitrogen removal techniques in aquaculture for a sustainable production. Aquaculture 270:1–14

Erler D, Pollard P, Duncan P, Knibb W (2004) Treatment of shrimp farm effluent with omnivorous finfish and artificial substrates. Aquacult Res 35(9):816–827

FAO (2020) The state of world fisheries and aquaculture 2020. Sustainability in action. FAO, Rome

Funge-Smith SJ, Briggs MR (1998) Nutrient budgets in intensive shrimp ponds: implications for sustainability. Aquaculture 164(1–4):117–133

Gangadhar B, Keshavanath P (2002) Substrates used in aquauculture—economical and ecological implications. Fish Chimes 22(8):14–16

Gangadhar B, Sridhar N, Saurabh S, Raghavendra CH, Raghunath MR, Prasanth H (2015) Influence of periphyton based culture systems on growth performance of fringe-lipped carp *Labeo fimbriatus* (Bloch, 1795) during fry to fingerling rearing. Indian J Fish 62(3):118–123

Gangadhar B, Umalatha NS, Raghavendra CH, Santhosh HJ, Jayasankar P (2016) Growth performance and digestive enzyme activities of fringe-lipped carp *Labeo fimbriatus* (Bloch, 1795) in periphyton based nursery rearing system. Indian J Fish 63(1):125–131

Garg SK (2016) Impacts of grazing by milkfish (Chanos chanos Forsskal) on periphytongrowth and its nutritional quality in inland saline groundwater: fish growth and pond ecology. Ecol Evol Environ Biol 1:41–52

Goldsborough LG (1993) Diatom ecology in the phyllosphere of the common duckweed (Lemna minor L). Hydrobiologia 269/270:463–471

Gorospe JC, Juinio-Menez MA, Southgate PC (2019) Effects of shading on periphyton characteristics and performance of sandfish, *Holothuria scabra* jaeger 1833, juveniles. Aquaculture 512:734307

Haque MR, Islam MA, Rahman MM, Shirin MF, Wahab MA, Azim ME (2015) Effects of C/N ratio and periphyton substrates on pond ecology and production performance in giant freshwater prawn *Macrobrachium rosenbergii* (De man, 1879) and tilapia Oreochromis niloticus (Linnaeus, 1758) polyculture system. Aquacult Res 46(5):1139–1155

Haque MR, Islam MA, Wahab MA, Hoq ME, Rahman MM, Azim ME (2016) Evaluation of production performance and profitability of hybrid red tilapia and genetically improved farmed tilapia (GIFT) strains in the carbon/nitrogen controlled periphyton-based (C/N-CP) on-farm prawn culture system in Bangladesh. Aquacult Rep 4:101–111. https://doi.org/10.4060/ca9229en

Huchette SMH, Beveridge MCM (2005) Periphyton-based cage aquaculture. In: Azim ME, Verdegem MCJ, van Dam AA, Beveridge MCM (eds) Periphyton. Ecology, exploitation and management. CABI Publishing, Wallingford, p 237

Jones PL, Thanuthong T, Kerr P (2002) Preliminary study on the use of synthetic substrate for the juvenile stage production of yabby, *Cherax destructor* (Clark) (Decapoda: Parastacidae). Aquacult Res 33:811–818

Ju ZY, Forster I, Conquest L, Dominy W (2008) Enhanced growth effects on shrimp (*Litopenaeus vannamei*) from inclusion of whole shrimp floc or floc fractions to a formulated diet. Aquacult Nutr 14(6):533–543

Keshavanath P, Gangadhar B (2005) Research on periphyton-based aquaculture in India. In: Periphyton: ecology, exploitation and management. CABI, Wallingford, pp 223–236

Keshavanath P, Gangadhar B, Ramesh TJ, van Rooij JM, Beveridge MCM, Baird DJ, Verdegem MCJ, van Dam AA (2001) Use of artificial substrates to enhance production of freshwater herbivorous fish culture. Aquac Res 32:189–197

Keshavanath P, Manissery JK, Bhat AG, Gangadhara B (2012) Evaluation of four biodegradable substrates for periphyton and fish production. J Appl Aquac 24(1):60–68

Khatoon H, Yusoff F, Banerjee S, Shariff M, Bujang JS (2007) Formtion of periphyton biofilm and subsequent biofouling on different substrates in nutrient enriched brackishwater shrimp ponds. Aquaculture 273(4):470–477

Khatoon H, Banerjee S, Yusoff FM, Shariff M (2009) Evaluation of indigenous marine periphytic Amphora, Navicula and Cymbella grown on substrate as feed supplement in Penaeus monodon postlarval hatchery system. Aquac Nutr 15(2):186–193

Krishnani KK, Parimala V, Gupta BP, Azad IS, Meng X, Abraham M (2006) Bagasse-assisted bioremediation of Ammonia from shrimp farm wastewater. Water Environ Res 78(9):938–950

Kuhn DD, Lawrence AL, Boardman GD, Patnaik S, Marsh L, Flick GJ Jr (2010) Evaluation of two types of bioflocs derived from biological treatment of fish effluent as feed ingredients for Pacific white shrimp, *Litopenaeus vannamei*. Aquaculture 303(1–4):28–33

Kumar VS, Pandey PK, Anand T, Bhuvaneswari R, Kumar S (2017) Effect of periphyton (aquamat) on water quality, nitrogen budget, microbial ecology, and growth parameters of *Litopenaeus vannamei* in a semi-intensive culture system. Aquaculture 479:240–249

Kumar S, Pandey PK, Kumar S, Anand T, Suryakumar B, Bhuvaneswari R (2019) Effect of periphyton (aquamat installation) in the profitability of semi-intensive shrimp culture systems. Indian J Econ Dev 7(1)

Levy A, Milstein A, Neori A, Harpaz S, Shpigel M, Guttman L (2017) Marine periphyton biofilters in mariculture effluents: nutrient uptake and biomass development. Aquaculture 473:513–520

Majumder S, Saikia SK (2020) Ecological intensification for feeding rohu *Labeo rohita* (Hamilton, 1822): a review and proposed steps towards an efficient resource fishery. Aquacult Res 51:3072–3078

Meyer-Reil LA (1994) Microbial life in sedimentary biofilms-the challenge to microbial ecologists. Marine ecology progress series. Oldendorf 112(3):303–311

Mridula RM, Manissery JK, Keshavanath P, Shankar KM, Nandeesha MC, Rajesh KM (2003) Water quality, biofilm production and growth of fringe-lipped carp (*Labeo fimbriatus*) in tanks provided with two solid substrates. Bioresour Technol 87(3):263–267

Nielsen PH, Jahn A, Palmgren R (1997) Conceptual model for production and composition of exopolymers in biofilms. Water Sci Technol 36(1):11–19

Norberg J (1999) Periphyton fouling as a marginal energy source in trophical tilapia cage farming. Aquac Res 30:427–430

Oswald WJ (2003) My sixty years in applied algology. J Appl Phycol 15:99–106

Pandey PK, Bharti V, Kumar K (2014) Biofilm in aquaculture production. Afr J Microbiol Res 8 (13):1434–1443

Pradeep B, Pandey PK, Ayyappan S (2003) Effect of probiotic and antibiotics on water quality and bacterial flora. J Inland Fish Soc India 35(2):68–72

Priedahitra RH (2003) Reducing the potential environmental impacts of tank aquaculture effluents through intensification and recirculation. Aquaculture 226(1–4):35–44

Ramesh MR, Shankar KM, Mohan CV, Varghese TJ (1999) Comparison of three plant substrates for enhancing carp growth through bacterial biofilm. Aquacult Eng 19:119–131

Rectenwald LL, Drenner RW (2000) Nutrient removal from wastewater effluent using an ecological water treatment system. Environ Sci Technol 34(3):522–526

Richard M, Anginot JTMA, Paticat F, Verdegem MCJ, Hussenot JME (2010) Influence of periphyton substrates and rearing density on *Liza aurata* growth and production in marine nursery ponds. Aquaculture 310:106–111

Sakr EM, Shalaby SM, Wassef EA, El-Sayed AM, Moneim AIA (2015) Evaluation of Periphyton as a food source for Nile Tilapia (*Oreochromis niloticus*) juveniles fed reduced protein levels in cages. J Appl Aquac 27:50–60

Sereti V, Verdegem MC, Eding EH, Verreth JA (2004) Periphyton as water treatment in fresh water recirculation system. In: Conference on 'Biotechnologies for quality', Barcelona, Spanje, 20–23 October, pp 741–742

Shresta MK, Knud-Hansen CF (1994) Increasing attached microorganisms biomass as a management strategy for Nile tilapia (Oreochromis niloticus) production. Aquacult Eng 13:101–108

Sofyani A, Sambhu C (2020) Role of artificial substrates on growth performance of pacific white shrimp, Litopenaeus vannamei (Boone, 1931) in semi-arid lands. Indian J Geo-Marine Sci 49(04):576–580

Sommer U (1996) Nutrient competition experiments with periphyton from the Baltic Sea. Mar Ecol Prog Ser 140:161–167

Supamattaya K, Kiriratnikom S, Boonyaratpalin M, Borowitzka L (2005) Effect of a Dunaliella extract on growth performance, health condition, immune response and disease resistance in black tiger shrimp (*Penaeus monodon*). Aquaculture 248:207–216

Suryakumar B, Avnimelech Y (2017) Adapting biofloc technology for use in small-scale ponds with vertical substrate. World Aquacult:54–58

Tammam MS, Wassef EA, Toutou MM, El-Sayed AM (2020) Combined effects of surface area of periphyton substrates and stocking density on growth performance, health status, and immune response of Nile tilapia (*Oreochromis niloticus*) produced in cages. J Appl Phycol. https://doi.org/10.1007/s10811-020-02136-x

Thompson FL, Abreu PC, Wasielesky W (2002) Importance of biofilm for water quality and nourishment in intensive shrimp culture. Aquaculture 203:263–278

Tidwell JH, Bratvolt D (2005) Utility of added substrates in shrimp culture. In: Azim ME, Verdegem MCJ, van Dam AA, Beveridge MCM (eds) Periphyton. Ecology, exploitation and management. CABI Publishing, Wallingford, pp 247–268

Tidwell JH, Coyle SD, Schulmeister G (1998) Effects of added substrate on the production and population characteristics of freshwater prawns (Macrobrachium rosenbergii) to increasing amounts of artificial substrate in ponds. J World Aquacult Soc 31:174–179

Tidwell JH, Coyle S, Van Arnum A, Weibel C (2000) Production response of freshwater prawns Macrobrachium rosenbergii to increasing amounts of artificial substrate in ponds. J World Aquacult Soc 31:452–458

Tierney TW, Fleckenstein LJ, Ray AJ (2020) The effects of density and artificial substrate on intensive shrimp *Litopenaeus vannamei* nursery production. Aquacult Eng 89:102063

Valeta JS, Verdegem MCM (2009) Phosphate removal from aquaculture effluent by algal turf scrubber technology. Malawi J Aquacult Fish 19

Van Dam AA, Beveridge MCM, Azim ME, Verdegem MCJ (2002) The potential of fish production based on periphyton. Rev Fish Biol Fisher 12:1–31

Verdegem MCJ, Eding EH, Sereti V, Munubi RN, Satacruz-Reyes RA, van Dam AA (2005) Similarities between microbial and periphytic biofilms in aquaculture systems. In: Azim ME, Verdegem MCJ, van Dam AA, Berveridge MCM (eds) Periphyton: ecology, exploitation and management. CABI, Wallingford, pp 191–205

Vermaat JE (2005) Periphyton dynamics and influencing factors. In: Azim ME, Verdegem MCJ, van Dam AA, Beveridge MCM (eds) Periphyton. Ecology, exploitation and management. CABI Publishing, Wallingford, pp 35–49

Vermaat JE, Hootsmans MJM (1994) Periphyton dynamics in a temperature-light gradient. In: Van Vierssen W, Hootsmans MJM, Vermaat JE (eds) Lake Veluwe, a Macrophyte-Dominated system under eutrophic stress, Geobotany 21. Kluwer Academic Publishers, The Netherlands, pp 193–212

Vymazal J, Craft CB, Richardson CJ (1994) Periphyton response to nitrogen and phosphorus additions in Florida Everglades. Algol Stud Archiv für Hydrobiologie 7:75–97

Wetzel RG (1983) Attached algal-substrata interactions: fact of myth, and when an-dhow? In: Wetzel RG (ed) Periphyton of freshwater ecosystems. Dr. W. Junk, The Hague, pp 207–215

Zhang B, Lin W, Wang Y, Xu R (2010) Effects of artificial substrates on growth, spatial distribution and non-specific immunity factors of *Litopenaeus vannamei* in the intensive culture condition. Turk J Fish Aquat Sci 10:491–497

Bioremediation of Aquatic Environment

24

P. K. Pandey and K. S. Sukhdhane

Abstract

Indiscriminate release of different pollutants like heavy metals, Polycyclic Aromatic Hydrocarbons (PAHs) and other inorganic pollutants in the aquatic environment have caused worldwide concerns. Remediation of these polluted aquatic environments, following the conventional engineering approaches based on physicochemical methods, is both technically and economically challenging. Remediation of polluted sites using microbial process (bioremediation) has proven effective and reliable due to its eco-friendly features. Bioremediation uses the microorganisms and is the most promising, relatively efficient, and cost-effective technology for decontamination of sites with a wide range of pollutants. The process of bioremediation can either be carried out ex situ or in situ, depending on several factors, which includes the cost, site characteristics, type and concentration of pollutants. Therefore, choosing an appropriate bioremediation technique, which will effectively reduce pollutant concentrations to an innocuous state, is crucial for successful bioremediation. This chapter highlights the understanding of the problems associated with the toxicity of heavy metals and PAHs to the contaminated ecosystems and their viable, sustainable and eco-friendly bioremediation technologies. The chapter also provides insight into the major bioremediation techniques, their principles, advantages, limitations, and prospects.

P. K. Pandey (✉)
College of Fisheries (CAU), Imphal, Lembucherra, Tripura, India

K. S. Sukhdhane
Central Institute of Fisheries Education, Mumbai, Maharashtra, India

© The Author(s), under exclusive license to Springer Nature Singapore Pte Ltd. 2021
P. K. Pandey, J. Parhi (eds.), *Advances in Fisheries Biotechnology*,
https://doi.org/10.1007/978-981-16-3215-0_24

423

Keywords

Bioremediation · Pollution · Environment · Heavy metals · Biodegradation

24.1   Introduction

In the past few decades, environment pollution has been on the rise due to rapid industrialization, increased anthropogenic activities and indiscriminate use of chemical in agriculture practices: Owing to these factors, the society at large has to grapple with issues such as heavy contamination of aquatic environment with heavy metals, Polycyclic Aromatic Hydrocarbons (PAHs), and other inorganic pollution. There are several conventional approaches for the disposal of these pollutants which have their own limitations (Asha and Sandeep 2013). Heavy metals, along with other pollutants, can cause numerous hazards to both human health and environment even at a low concentration because of gradual accumulation (Zeng et al. 2013a, b; Zhang et al. 2014; Nouha et al. 2016). Even though heavy metals are indestructible, they may be turned less toxic to the environment either by chelating via chemical or physical remediation or by shifting the valence by redox reaction (Fan et al. 2008).

Polycyclic aromatic hydrocarbons (PAHs) are ubiquitous in the aquatic environment and are formed by the incomplete combustion or pyrolysis of organic materials such as oil, petroleum gas, coal, or wood. PAHs are considered hazardous because of their effects on cell division, leading to tumors and altered reproduction rates through enzyme induction processes (NRC 1983; Lockhart et al. 1992). PAHs have been recommended as priority pollutants by the United States Environmental Protection Agency (USEPA) and are the target components for study in oil-contaminated sites due to their carcinogenic, mutagenic, and toxic properties (Menzie et al. 1992). Therefore, to effectively restore the polluted aquatic environment in an eco-friendly way, bioremediation technology has been developed by researchers. Bioremediation is an innovative and promising technology which is available for reduction/removal of different pollutants from polluted environments. It is the strategy to utilize the biological activities as much as possible for the quick elimination of pollutants (Kumar et al. 2011). However, due to the nature of pollutants, there is no single way to restore the polluted environment. Verma and Jaiswal (2016) identified different autochthonous microorganisms, present in the polluted environment, as the key to solve most of the challenges associated with environment pollution.

Considering the arguments made above, the aim of the chapter is to document the current trends pertaining to the application of different microorganisms for bioremediation of heavy metals and PAHs which are identified gaps in the thematic area.

24.2 Principle of Bioremediation

"Bioremediation is defined as the process whereby organic wastes are biologically degraded under controlled conditions to an innocuous state, or to levels below concentration limits established by regulatory authorities." This technique is used for converting the pollutants into less harmful or nontoxic substances by removing the toxic elements and degrading organic substances which ultimately lead to mineralization of organic substances into carbon dioxide, water, nitrogen gas etc. It is a technology used for removing the pollutant from the environment, restoring contaminated sites and preventing future pollution. Several members of algae, fungi and bacteria are known to solubilize, transport, and deposit the metals, and detoxify dyes and complex chemicals. Bioremediation technology uses naturally occurring bacteria, fungi, or plants that can degrade or detoxify substances hazardous to mankind and the environment. For degradation, different microorganisms may be isolated from the native or can be brought from the contaminated sites. In the process of degradation, different pollutants get transformed by the living organisms through different reactions that take place as part of the metabolic process.

The effective bioremediation depends upon the activities of different microorganisms that enzymatically attack different pollutants and convert them into harmless or less toxic substance. As bioremediation can be effective only where environmental conditions permit microbial growth and activity, its application often involves the manipulation of environmental parameters to allow microbial growth and degradation to proceed at a faster rate. It usually occurs under the aerobic conditions but in a few cases, anaerobic conditions also allow the degradation processes.

24.3 Bioremediation Strategies

Bioremediation can be broadly classified into ex situ and in situ techniques based on the degree of saturation and aeration of an area. Ex situ technologies are those treatments which involve the physical removal of the contaminated material for treatment process whereas In situ techniques involve treatment of the contaminated material in place.

24.3.1 In Situ Bioremediation

In situ bioremediation is the most desirable approach with lower cost and fewer disturbances which directly involves the contact between microorganisms and dissolved or sorbed pollutants for the biotransformation (Bouwer and Zehnder 1993). In situ allows treatment of the pollutants on site by avoiding the excavation and transport. It is limited by the depth of the soil and also with contact with the pollutants that can be effectively treated. In many cases, in situ treatment happened at the surface, whereas, in few cases, it is up 30 cm depth. The most important land treatments are:

24.3.1.1 Bioventing

It is the method of treating contaminated soil by drawing oxygen through the soil to stimulate microbial activity. It is the most common type of in-situ treatment which involves the supply of air and nutrients through the wells to polluted sites to stimulate indigenous bacteria for biodegradation. It employs low air flow rate for the treatment that is necessary for biodegradation with minimal volatilization and release of pollutants in the atmosphere. It is mainly used for the hydrocarbon degradation in deeper surface. Hohener and Ponsin (2014) opined that the bioventing technique has gained popularity for *in situ* bioremediation, especially for restoring the contaminated sites with the petroleum pollutants.

24.3.1.2 Biosparging

It is the method of treating pollutants by supplying oxygen and nutrients to stimulate naturally occurring indigenous bacteria to degrade the pollutants. In this technique, air is injected in the saturated zones which causes the movement of volatile organic compounds to promote biodegradation. This technique increases the mixing in the saturated zone thereby increasing the contact between soil and groundwater for the treatment. Biosparging allows installation of small diameter air injection points with considerable flexibility for adequate performance. Philp and Atlas (2005) evaluated the effectiveness of biosparging which depends on two important factors i.e., soil permeability and pollutant biodegradability as it determines pollutant bioavailability in microorganisms.

24.3.1.3 Bioaugmentation

It involves the addition of different microorganisms of indigenous or exogenous origin to the contaminated sites. The rationale of this approach is that indigenous microbial populations may not be capable of degrading a wide range of pollutants (Leahy and Colwell 1990). There are certain limitations while using this treatment; for example, nonindigenous organisms rarely compete with the indigenous population to develop and sustain; most soils with long-term exposure to biodegradable waste have indigenous microorganisms that are effective degraders if the land treatment unit is well managed (USEPA 2004; Kumar et al. 2011).

Potential advantage of in situ bioremediation methods includes minimal site disruption, simultaneous treatment of contaminated sites, minimal exposure of public and site personnel, and low cost.

24.3.2 Ex Situ Bioremediation

These techniques involve the excavation or removal of contaminated soil/water from the ground. The method is broadly classified into solid phase system and slurry phase system.

24.3.2.1 Solid Phase Treatment

It is mainly used for organic wastes (leaves, manures, and agricultural wastes) and problematic wastes (Domestic/industrial wastes and sewage sludge/municipal solid wastes). The treatment processes include land farming, composting, biopiles, and hydroponics (Ramos et al. 2011; Kumar et al. 2011; Rayu et al. 2012).

24.3.2.2 Slurry Phase Treatment

The treatment process involves the controlled treatment of excavated soil/water using bioreactor. Slurry contains 10–30% solid in terms of weight.

24.3.2.3 Land Farming

It is a simple bioremediation technique as low cost and less equipment are required for the process. The aim of this technique is to stimulate indigenous microorganisms and facilitate aerobic biodegradation. In this technique, a contaminated soil/water is excavated and spread over a prepared bed and periodically tilled until desired pollutants are degraded. The depth of treatment plays important role in this treatment as it is limited to superficial 10–35 cm of soil. This technique has the potential to reduce the cost of maintenance.

24.3.2.4 Composting

It is a technique that involves combining contaminated soil with nonhazardous organic amendants such as straw, hay or corncobs, or manure to which make it easier to deliver the required air and water to the microorganisms. The presence of these organic amendants enhances the indigenous microorganisms for bioremediation. Elevated temperature is one of the characteristics of composting. Duah-Yentumi and Kuwatsuka (1980) evaluated and found improved degradation of composted material with addition of two herbicides benthiocarb (S-4-chlorobenzyl diethyl-thiocarbamate) and MCPA (4-chloro-2-methylphenoxyacetic acid). Purnomo et al. (2010) was successful to degrade and mineralize the DDT in soil with spent mushroom waste from Pleurotusostreatus.

24.3.2.5 Biopiles

It is a technique which involves ground pilling of the polluted soil, followed by the nutrient amendment and aeration which enhances microbial activities for the bioremediation. This technique is also known as biocells, bioheaps, biomounds, and compost piles and mainly used to bioremediate the petroleum pollutants from contaminated environment. This technique provides favorable environment for the indigenous aerobic and anaerobic microorganisms for biodegradation. Whelan et al. (2015) considered this technique very useful due to constructive features which include cost effectiveness and high biodegradation rate when nutrients, temperature and aeration are adequately controlled. Chemlal et al. (2013) and Akbari and Ghoshal (2014) have evaluated the feasibility of biopiles toward the bioremediation of different soil samples including clay and sandy soil.

24.3.2.6 Bioreactors

Bioreactor is the ex situ method of bioremediation in which a raw material is converted into the specific products, following different biological reactions. It includes different operating modes, based on the market economy and expenditure. The different operating modes are batch, fed-batch, sequencing batch, continuous and multistage methods. The bioreactor supports the natural process of biodegradation by mimicking and maintaining their natural environment to provide optimum growth conditions. The polluted soil (dry matter or slurry) water can be fed into the bioreactor and the mixing increases the rate of degradation of targeted pollutants. The bioreactor has several advantages like controlled bioprocess parameters (temperature, pH, agitation and aeration rates, substrate and inoculum concentrations). Mohan et al. (2004) evaluated the flexible nature of bioreactor design that allows maximum biological degradation while minimizing abiotic losses.

24.3.3 Factors Responsible for Bioremediation

There are several microbiological methods that are currently employed for bioremediation. The monitoring of the efficacy of bioremediation requires careful attention; besides using specific information on abundance of microbial members or genes and microbial processes and different factors of course of attenuation (Table 24.1). These techniques must address multiphasic and heterogenous environment such as different soil and water medium and the atmosphere. El Fantroussi and Agathos (2005) have highlighted the different factors such as chemical nature and concentration of pollutants, physiological characteristics of the environment and their availability to microorganisms. Boopathy (2000) outlined the various factors, including scientific, non-scientific, and regulatory, that limit the use of bioremediation technologies.

24.3.4 Bioremediation of Heavy Metals

Heavy metal pollution due to anthropogenic activities and through industrial discharges is posing significant threats to the aquatic environment. Due to the heavy metals discharges from the different point and non-point sources such as metallurgical ovens (Lee et al. 2006; Govarthanan et al. 2013; Chaturvedi et al. 2015; Zhou et al. 2016), radionuclides (Mclean and Abbe 2008), sewage sludge/wastewater (from industrial, municipal and domestic origin) (Deng et al. 2007; Kumar and Mani 2010; Mapanda et al. 2005; Robinson et al. 2011) and un-hygienic approaches of rapidly growing population, the environment has suffered in a severe way in terms of their detrimental effects. Heavy metals together with other pollutants can cause numerous health hazards to the human health even at low concentration due to their toxicity, non-biodegradability and bioaccumulation (Bahadir et al. 2007; Zeng et al. 2013a, b; Zhang et al. 2014; Nouha et al. 2016). Heavy metals such as cadmium (Cd (II)), lead (Pb (II)), copper (Cu (II)) and chromium (Cr (VI)), which are commonly present in contaminated soils in traces

Table 24.1 Factors responsible for bioremediation

Microbial factors	• Growth until critical biomass is reached • Mutation and horizontal gene transfer • Enzyme induction • Enrichment of the capable microbial populations • Production of toxic metabolites
Environmental factors	• Depletion of preferential substrates • Lack of nutrients • Inhibitory environmental conditions
Substrate	• Too low concentration of contaminants • Chemical structure of contaminants • Toxicity of contaminants • Solubility of contaminants
Biological aerobic vs. anaerobic process	• Oxidation/reduction potential • Availability of electron acceptors • Microbial population present in the site
Growth substrate vs. co-metabolism	• Type of contaminants • Concentration • Alternate carbon source present • Microbial interaction (competition, succession, and predation)
Physicochemical bioavailability of pollutants	• Equilibrium sorption • Irreversible sorption • Incorporation into humic matters
Mass transfer limitations	• Oxygen diffusion and solubility • Diffusion of nutrients • Solubility/miscibility in/with water

but the permanent existence of heavy metals in the polluted aquatic ecosystems can significantly affect the health of human beings (Nogaw and Kido 1996).

There are various physical and chemical methodologies widely used to remove metals from the polluted environment viz. Chemical precipitation, electrochemical treatment, filtration, ion exchange, membrane technology, oxidation/reduction, reverse osmosis, and evaporation (Xiao et al. 2010). However, most of these technologies appear to be expensive, less efficient, labor-intensive, or nonselective in the treating process (Tang et al. 2008). With this in view, it becomes apparent that, the development and application of the more broadly applicable and environment friendly techniques such as bioremediation are very necessary (Banik et al. 2014). It is an ecologically sound and state-of-the-art technique that employs natural biological processes to completely eliminate toxic contaminants.

24.3.5 Sources of Heavy Metal in the Environment

Heavy metals can occur naturally in the environment by the processing of parent material and also through anthropogenic sources. The most significant natural sources are weathering of minerals, erosion and volcanic activity, while the

Table 24.2 Different Sources of heavy metals and its possible impact on the human being

Heavy metals	Anthropogenic sources	Possible impacts
Zinc	• Effluents discharged from industries • Electroplating, manufacture of batteries, galvanization, and metallurgical industries metal processing, leather tanning; chrome plating; metallurgy; wood preservatives; boilers and different cooking system as anti-corrosives	Dizziness, fatigue, etc. (Hess and Schmid 2002)
Copper	• Animal manure • Discharge from industrial factories, such as paint factory, and is disposed by electronic and electrical factories • Discharge from landfalls, mine sites, combustion of fossil fuels, and domestic wastewaters • It can be released by particles from volcanoes, dust, and forest fires into the atmosphere or dissolved compounds in water	Brain and kidney damage, elevated levels result in liver cirrhosis and chronic anemia stomach and intestine irritation (Wuana and Okieimen 2011)
Cadmium	• Production of batteries, alloys, pigments, fertilizers, pesticides; welding; mining; combustion of fossil fuels and municipal wastes; recycling of electronic and cadmium-plated waste	Mutagenic, endocrine; disruptor, lung damage and fragile bones, affects calcium regulation in biological systems (Salem et al. 2000; Degraeve 1981)
Arsenic	• Coal combustion; pesticide formulations; smelting; wood preservative; metal refining; drugs/medicines to treat amoebic dysentery and syphilis; veterinary drugs to treat against parasitic diseases	Affects essential cellular processes, such as oxidative phosphorylation and ATP synthesis (Tripathi et al. 2007)
Mercury	• Chlor-alkali plants, thermal power plants, fluorescent lamps, hospital waste (damaged thermometers, barometers, sphygmomanometers)	Autoimmune diseases, depression, drowsiness, fatigue, hair loss, insomnia (Neustadt and Pieczenik 2007) loss of memory, restlessness disturbance of vision, tremors, temper outbursts, brain damage, lung and kidney (Gulati et al. 2010)
Chromium	• Mining, industrial coolants, chromium salts • Manufacturing, leather tanning	Nose ulcers, runny nose, breathing problems, such as asthma, cough, shortness of breath, or wheezing and hair (Salem et al. 2000)

anthropogenic sources depend upon different human activities. The different anthropogenic sources and their possible impact are mentioned in Table 24.2.

24.3.6 Microbial Bioremediation of Heavy Metals

Microorganisms are ubiquitous and many native microorganisms are present in the aquatic environment which play a very crucial role in elemental biogeochemical cycles of metal transformation. Heavy metals and associated microbes interaction have many beneficial as well as harmful consequences. Many external factors also play a pivotal role in the transformation of heavy metals in the aquatic environment like pH, moisture, temperature, ionic composition, humic compounds, and other biotic organisms. The transformation of heavy metals in an aquatic environment happens due to several of the aforementioned factors. Bioremediation of metals can be carried out by the mobilization or immobilization processes, which can be accomplished by the following mechanisms, i.e., chelation, oxidation-reduction, pH change, biosorption, bioaccumulation, immobilization, biomethylation, or change in organic metallic complex to radionuclides.

Heavy metals bioremediation has been widely studied by many authors (Hassan et al. 2009; Kao et al. 2008; Tuzen et al. 2008). The most common heavy metals which contaminate the soil and water are As, Pb, Hg, and Cr. In the aquatic environment, arsenic is mostly found in two forms: arsenate (As^{5+}) and arsenite (As^{3+}) (Paul et al. 2014). Different microorganism remediates arsenic by using mobilization process. The conversion of As^{3+} to less toxic arsenic, i.e., As^{5+} happens with the help of oxidation and reduction reaction. The common arsenic bio-transforming microorganisms are as follows: *Acinetobacter* sp., *Brevundimonas* sp., *Pseudomonas* sp., *Rhizobium* sp., *Aeromonas* sp. and *Penicillium canescens* (Layton et al. 2014). Another important heavy metal contamination in the aquatic environment is lead. The bioremediation of lead occurs due to various microbial strains and consortia. It occurs through immobilization mechanisms. Jebara et al. (2015) and Joshi et al. (2011) enlisted several indigenous microorganisms that are resistant and withstand the toxic effects. These are *Aspergillus niger, A. terrus, A. versicolor, N. crassa, P. canescens, P. chrysogenum, P. decumbens, P. simplicissimum,* and *Saccharomyces cerevisiae.*

Hg poisoning has a number of side effects on human health. Organic form of mercury, methyl-mercury, is the most toxic form, which gets bioaccumulated and biomagnified in the biota. Methylation occurs via biological pathway and via chemical (methylcobalamin) or photochemical reactions. Jayakumar et al. (2019) isolated 19 mercury-resistant bacteria of *Pseudomonas* and *Bacillus* genera from Thane creek of Maharashtra, India. *Exiguobacterium profundum* 10C, a moderately thermophilic bacterium capable of tolerating 180 mg/L of $HgCl_2$ which was first isolated as bioremediating agents from the said study. The Hg transformation by several indigenous microorganisms as *P. canescens, Rhizopus arrhizus, Shewanella oneidensis, Geobacter sulfurreducens, and G. metallireducens* has been reported (Sinha et al. 2009; Dixit et al. 2015; Wiatrowski et al. 2006).

In the environment, Cr exists in various oxidation states like Cr (0), Cr (III) and the hexavalent Cr (VI) forms (Bhalerao and Sharma 2015). Out of these states, Cr (0) is the most stable state, followed by the Cr (VI) which exists mostly as chromate (CrO_4^{2-}) dichromate ($Cr_2O_4^{2-}$) and chromium trioxide (CrO_3). The CrO_3 is the most toxic form of Cr which has high oxidizing potential. Due to the high solubility and mobility of CrO_3, it affects the cell membranes in living organisms. Chromium can be bio-remediated by several microbial species by using different mechanisms. The major chromium bioremediator are *Pantoea* sp., *Bacillus circulans, B. megaterium, B. coaglans, Zoogloea ramigera, Streptomyces nouresei, Aeromonas caviae, Pseudomonas* sp., and *Staphylococcus xylosus* (Malaviya and Singh 2014; Aryal and Liakopoulou-Kyriakides 2015).

Genetically Modified Organisms (GEMs) are the microbial strains whose genetic makeup is reformed with the help of molecular tools to enhance the biodegradation or biotransformation (Zhang et al. 2015). These kinds of organisms are mainly termed as recombinant DNA technology. The engineered microorganisms have enhanced capabilities to transform the heavy metals to their corresponding less harmful states. Several GEMs have been tried and tested for the removal of heavy metals, i.e., *E. Coli* stain (Yuan et al. 2008); *Sphingomonas desiccabilis* and *B. idriensis* strains (Liu et al. 2011) for Arsenic removal, *Bacillus subtilis* BR151; *Staphylococcus aureus* RN4220 (Bondarenko et al. 2008) for pb removal, *Achromobacter* sp. AO22 (Ng et al. 2009), *Acidithiobacillus ferrooxidans* strain (Sasaki et al. 2005), *Deinococcus geothermalis* (Dixit et al. 2015), *E. coli* JM 109 (Zhao et al. 2005), *E. coli* MC 1061 (Bondarenko et al. 2008), *Pseudomonas* K-62 (Kiyono and Pan-Hou 2006; Kiyono et al. 2009) and *Pseudomonas* sp. for Hg removal. Application of genetically engineered microbial system for the remediation of heavy metal pollution is a cheap and safe alternative due to the selective nature of GE microbes which pose very few health hazards as compared to the other physico-chemical methods.

24.3.7 Bioremediation of Polycyclic Aromatic Hydrocarbons

Polycyclic aromatic hydrocarbons (PAHs) are the group of lipophilic contaminants that are widespread in the aquatic environment. It has aroused great environmental concerns due to their potential hazards on the human and aquatic organisms. PAHs gain entry into the near shore marine environments through spillage of petroleum, industrial discharges and atmospheric fall outs due to shipping activities, storm water drains, and urban run-off (Gevao et al. 1998) and have been identified to pose potential health risks because they are chemically stable and have high recalcitrance to different types of degradation and high toxicity to living organisms (Cheng et al. 2014; Tian et al. 2008). Contamination from PAH is a hazard that is of great concern to aquatic life in sediments, especially in areas close to anthropogenic sources (Liu et al. 2009; Veltman et al. 2012). Because of low water solubility and hydrophobic nature, PAHs in the aquatic environment rapidly gets dissociated with the organic and inorganic particles (Gearing et al. 1980; Chiou et al. 1998) and subsequent decomposition in the aquatic environments. In India, several incidents have happened due to transportation and accidental discharges from the refineries along the coast. During 1970s to 2011, there were about 79 major and minor accidents reported in the Indian waters, spilling large quantities of oil in the environment

(Sukhdhane et al. 2015). PAHs as the major constituent of the crude oil contribute to around 50–98% depending on the origin and carbon items.

The principal process for the successful removal and elimination of PAHs from the aquatic sediments is the bioremediation (Cerniglia 1992; Wilson and Jones 1993; Yuan et al. 2000). The diverse groups of indigenous microorganisms, capable of utilizing and degrading contaminants such as PAHs, might have been present in the aquatic environment (Chaineau et al. 1999).The biodegradation potential of the different strains isolated from the hydrocarbon-contaminated environments was as active as or even higher than those originating from the non-contaminated sediments (Wild and Jones 1986).

24.4 Distribution of Hydrocarbon-Utilizing Microorganisms

Microorganisms are considered to be the major degrader of the natural PAHs contaminants and are of great practical interest for the bioremediation. Hydrocarbon degrading bacteria and fungi are widely distributed in the aquatic environments. Bacteria are the best described hydrocarbons utilizing microorganisms (Atlas et al. 1978; Kiyohara et al. 1982; Wilson and Jones 1993; Olivera et al. 2003). This process is of great interest for the preservation of the aquatic environments by reducing the number of PAHs degrading contaminants. Mulkins-Philips and Stewart (1974) examined the distribution of hydrocarbon-utilizing bacteria, represented by hydrocarbon utilizers which ranged from 10–100% depending on the different studies conducted on PAHs degradation. The variety of hydrocarbons utilizers illustrates their ubiquity and broad enzymatic capacity for the hydrocarbon degradation. Sukhdhane et al. (2019) isolated five different Phenanthrene degrading bacteria belonging to *Bacillus* sp. from PAHs contaminated mangrove sediment of Thane creek, Maharashtra, India which had potential to biodegrade. Braddock et al. (1995) found significantly higher number of hydrocarbon degrading bacteria from polluted environments, affected by the Exxon oil spill at Alaska. Atlas et al. (1978) found that the number of hydrocarbons utilizing microorganisms increased in marine, freshwater and soil biota in acute response to PAHs. Ramsay et al. (2000) reported high number of PAHs degrader in the aquatic ecosystems (10^4 to 10^6 cells g^{-1} sediments) and such indigenous microbial community had considerable potential to degrade the PAHs compounds; moreover the number of PAHs degrader could be increased with increase in PAH compounds. La Rosa et al. (2006) isolated 14 bacterial strains from different PAHs contaminated sites and identified them to the species level, using 16S rRNA full gene sequencing studies. Among different 14 isolates, five were identified as *Achromobacter xylosoxidans*, *Rhodococcus wratislaviensis, Microbacterium paraoxidans,* and *Pseudomonas putida* and three members of *Bacillus* sp. were also identified as the PAHs degrader. Zhang et al. (2004) isolated *Pseudomonas, Microbacterium,* and *Paracoccus* from PAHs contaminated aquatic environments as bioremediation.

24.5   Advantages and Disadvantages of Bioremediation

For successful bioremediation, it is required to have right kind of microorganisms at right place at right time with right environmental factors. The right microorganisms can be bacteria or fungi whose physiological or metabolic responses may cater to the biodegradation process. Bioremediation process has several advantages over conventional techniques like land filling or incineration. The bioremediation technique requires less time and can be performed on site without causing disturbance to the natural process. This technique is eco-friendly and sustainable since it does not use any chemicals for the degradation of the pollutants. It also helps in complete destruction of the pollutants, wherein many hazardous compounds can be transformed to harmless one. Bioremediation has also several disadvantages as all pollutants may not be amenable to the process of biodegradation, especially radio nuclides and some chlorinated compounds. In some cases, microbial metabolism of contaminants may produce toxic metabolites. It is a scientifically intensive procedure which must be tailored to site-specific conditions, which indicates that one has to pursue treatability studies on a small scale before the actual cleanup of the sites.

24.6   Limitation of Bioremediation

The bioremediation process is limited to only those compounds that are biodegradable and not all compounds are susceptible to rapid and complete biodegradation. Some of the products of bioremediation may be more toxic and persistent as compared to parent compounds. The biological processes, involved in bioremediation are very specific and based on the culture medium and it may not be extrapolated from lab to land. Sometime bioremediation process takes long time for the degradation of pollutants. There are several factors responsible for the degradation of pollutants from polluted environment such as depletion of preferential substrate, lack of nutrients, oxidation and reduction potential of different microbial population. The type of microbial transformation depends on whether the compounds serve as primary, secondary or co-metabolic substrate. In the process of bioremediation, there is always the chance of biomagnifications and biotransformation among the different food chains. A continuous search on different indigenous as well as Genetically Modified Organisms is required for better management of aquatic pollution. Therefore, bioremediation is still considered as developing technology to regulate the aquatic pollution.

24.7   Conclusion

Heavy metals and polycyclic aromatic hydrocarbons (PAHs) pollution occur from various anthropogenic and natural sources. Due to their hazardous nature, they should be removed from the environment with cost-effective and eco-friendly techniques such as bioremediation, employing biological agents. Bioremediation is an eco-friendly and sustainable technique for cleaning up of the aquatic environment by enhancing the biodegradation processes that occurs in nature. A bioremediation approach will involve strategic use of all indigenous microorganisms in an engineered way to achieve the detoxification of specific pollutants. Bioremediation techniques for heavy metals and PAHs detoxification have successfully been employed for management of polluted environment and it is gaining importance due to an increased level of understanding about the bioremediation mechanism through continuing research.

References

Akbari A, Ghoshal S (2014) Pilot-scale bioremediation of a petroleum hydrocarbon-contaminated clayey soil from a sub-Arctic site. J Hazard Mater 280:595–602

Aryal M, Liakopoulou-Kyriakides M (2015) Bioremoval of heavy metals by bacterial biomass. Environ Monit Assess 187(4173):1–26

Asha LP, Sandeep RS (2013) Review on bioremediation- potential tool for removing environmental pollution. Int J Basic Appl Chem Sci:2277–2073

Atlas RM, Horowitz A, Busdosh M (1978) Prudhoe crude oil in Arctic marine ice, water, and sediment ecosystems: degradation and interactions with microbial and benthic communities. J Fish Res Board Can 35:585–590

Bahadir T, Bakan G, Altas L, Buykgungar H (2007) The investigation of lead removal by biosorption. An application at storage battery industry wastewaters. Enzyme Microb Technol 41:98–102

Banik S, KC D, MS I (2014) Recent advancements and challenges in microbial bioremediation of heavy metals contamination. JSM Biotech Biomed Eng 2(1):1–9

Bhalerao SA, Sharma AS (2015) Chromium: as an environmental pollutant. Int J Curr Microbiol App Sci 4(4):732–746

Bondarenko O, Rolova T, Kahru A, Ivask A et al (2008) Bioavailability of Cd, Zn and Hg in soil to nine recombinant luminescent metal sensor bacteria. Sensors 8:6899–6923

Boopathy R (2000) Factors limiting bioremediation technologies. Bioresour Technol 74:63–67

Bouwer EJ, Zehnder AJB (1993) Bioremediation of organic compounds: microbial metabolism to work. Trends Biotechnol 11:360–367

Braddock JF, Lindstrom JE, Brown EJ (1995) Distribution of hydrocarbon-degrading microorganisms in sediments from Prince William sound, Alaska, following the Exxon Valdez oil spill. Mar Pollut Bull 30:125–132

Cerniglia CE (1992) Biodegradation of polycyclic aromatic hydrocarbons. Biodegradation 3:351–368

Chaineau CH, Morel J, Dupont J, Bury E et al (1999) Comparison of the fuel oil biodegradation potential of hydrocarbon-assimilating microorganisms isolated from a temperate agriculture soil. Sci Total Environ 227:237–247

Chaturvedi AD, Pal D, Penta S, Kumar A (2015) Ecotoxic heavy metals transformation by bacteria and fungiin aquatic ecosystem. World J Microbiol Biotechnol 31:1595–1603

Chemlal R, Abdi N, Lounici H, Drouiche N et al (2013) Modelling and qualitative study of diesel biodegradation using biopile process in sandy soil. Int Biodeter Biodegr 78:43–48

Cheng JJ, Song J, Ding C, Li X et al (2014) Ecotoxicity of Benz[a]pyrene assessed by soil microbial indicators. Environ Toxicol Chem 33:1930–1936

Chiou CT, McGroddy SE, Kile DE (1998) Partition characteristics of polycyclic aromatic hydrocarbons on soils and sediments. Environ Sci Technol 32:264–269

Degraeve N (1981) Carcinogenic, teratogenic and mutagenic effects of cadmium. Mutat Res 86 (1):115–135

Deng L, Su Y, Su H, Wang X et al (2007) Sorption and desorption of lead (II) from wastewater by green algae *Cladophora fascicularis*. J Hazard Mater 143(1–2):220–225

Duah-Yentumi S, Kuwatsuka S (1980) Effect of organic matter and chemical fertilizers on the degradation of benthiocarb and MCPA herbicides in the soil. Soil Sci Plant Nutr 26(4):541–549

Dixit R, Wasiullah MD, Pandiyan K et al (2015) Bioremediation of heavy metals from soil and aquatic environment: an overview of principles and criteria of fundamental processes. Sustainability 7(2):2189–2212

El Fantroussi S, Agathos SN (2005) Is bioaugmentation a feasible strategy for pollutant removal and site remediation? Curr Opin Microbiol 8:268–275

Fan T, Liu Y, Feng B, Zeng G et al (2008) Biosorption of cadmium (II), zinc (II) and lead (II) by Penicillium simplicissimum: isotherms, kinetics and thermodynamics. J Hazard Mater 160:655–661

Gearing PJ, Gearing JN, Pruell RJ, Wade TL et al (1980) Partitioning of no. 2 fuel oil in controlled estuarine ecosystem, sediments and suspended particulate matter. Environ Sci Technol 14:1129–1135

Gevao B, Hamilton-Taylor J, Jnes KC (1998) PCB and PAH deposition to and exchange at the air–water interface of a small rural lake in Cumbria, UK. Environ Pollut 102:63–75

Govarthanan M, Lee KJ, Cho M, Kim JS et al (2013) Significance of autochthonous Bacillus sp. KK1 on biomineralization of lead in mine tailings. Chemosphere 8:2267–2272

Gulati K, Banerjee B, Lall SB, Ray A (2010) Effects of diesel exhaust, heavy metals and pesticides on various organ systems: possible mechanisms and strategies for prevention and treatment. Indian J Exp Biol 48:710–721

Hassan SH, Kim SJ, Jung AY, Joo JH et al (2009) Biosorptive capacity of cd(II) and cu(II) by lyophilized cells of *Pseudomonas stutzeri*. J Gen Appl Microbiol 55(1):27–34

Hess R, Schmid B (2002) Zinc supplement overdose can have toxic effects. J Pediatr Hematol Oncol 24:582–584

Hohener P, Ponsin V (2014) In situ vadose zone bioremediation. Curr Opin Biotechnol 27:1–7

Jayakumar TK, Purushothaman CS, Sukhdhane KS (2019) Isolation and characterization of mercury resistant bacteria from thane creek, Mumbai. In: One health and ecosystem services 39th annual conference

Jebara SH, Abdelkerim S, Fatnassi IC, Chiboub M et al (2015) Identification of effective Pb resistant bacteria isolated from Lens culinaris growing in lead-contaminated soils. J Basic Microbiol 55:346–353

Joshi PK, Swarup A, Maheshwari S, Kumar R et al (2011) Bioremediation of heavy metals in liquid media through fungi isolated from contaminated sources. Indian J Microbiol 51:482–487

Kao WC, Huang CC, Chang JS (2008) Biosorption of nickel, chromium and zinc by MerP-expressing recombinant Escherichia coli. J Hazard Mater 158(1):100–106

Kiyohara H, Nago K, Yana K (1982) Rapid screen for Bacteria degrading water-insoluble, solid hydrocarbons on agar plates. Appl Environ Microbiol 43:454–457

Kiyono M, Pan-Hou H (2006) Genetic engineering of bacteria for environmental remediation of mercury. J Health Sci 52(3):199–204

Kiyono M, Sone Y, Nakamura R, Pan-Hou H et al (2009) The Mer E protein encoded by transposon Tn21 is a broad mercury transporter in Escherichia coli. FEBS Lett 583:1127–1131

Kumar C, Mani D (2010) Enrichment and management of heavy metals in sewage irrigated soil. Lap Lambert Academic Publishing AG & Co. KG, p 132

Kumar R, Acharya C, Joshi SR (2011) Isolation and analyses of uranium tolerant Serratia marcescens strains and their utilization for aerobic uranium U(VI) bioadsorption. J Microbiol 49(4):568–574

La Rosa G, De Carolisa E, Salia M, Papacchinib M et al (2006) Genetic diversity of bacterial strains isolated from soils, contaminated with polycyclic aromatic hydrocarbons, by 16S rRNA gene sequencing and amplified fragement length polymorphism fingerprinting. Microbiol Res 161:150–157

Layton AC, Chauhan A, Williams DE, Mailloux B et al (2014) Metagenomes of microbial communities in arsenic- and pathogen-contaminated well and surface water from Bangladesh. Genome Announc 2(6):1–2

Leahy JG, Colwell RR (1990) Microbial degradation of hydrocarbons in the environment. Microbiol Rev 54:305–315

Lee CS, Li XD, Shi WZ, Cheung SC et al (2006) Metal contamination in urban, suburban and country park soils of Hong Kong: a study based on GIS and multivariate statistics. Sci Total Environ 356:45–61

Liu Y, Chen L, Huang QH, Li WY et al (2009) Source apportionment of polycyclic aromatic hydrocarbons (PAHs) in surface sediments of the Huangpu River, Shanghai, China. Sci Total Environ 407:2931–2938

Liu S, Zhang F, Chen J, Sun G (2011) Arsenic removal from contaminated soil via biovolatilization by genetically engineered bacteria under laboratory conditions. J Environ Sci 23(9):1544–1550

Lockhart WL, Wagemann R, Tracey B, Sutherland D et al (1992) Presence and implications of chemical contaminants in the freshwaters of the Canadian Arctic. Sci Total Environ 122:165–243

Malaviya P, Singh A (2014) Bioremediation of chromium solutions and chromium-containing wastewaters. Crit Rev Microbiol 42(4):607–633

Mapanda F, Mangwayana EN, Nyamangara J, Giller KE (2005) The effect of long-term irrigation using wastewater on the heavy metal contents of soils under vegetables in Harare, Zimbabwe. Agric Ecosyst Environ 107:151–165

Mclean RI, Abbe GR (2008) Characteristics of Crassostrea ariakensis (Fujita 1913) and Crassostrea virginica (Gmelin 1791) in the discharge area of a nuclear power plant in Central Chesapeake Bay. J Shellfish Res 27(3):517–523

Menzie CA, Potachi BB, Santodonato J (1992) Exposure to carcinogenic PAH in environment. Environ Sci Technol 26:1278–1284

Mohan SV, Sirisha K, Rao NC, Sarma PN et al (2004) Degradation of chlorpyrifos contaminated soil by bioslurry reactor operated in sequencing batch mode: bioprocess monitoring. J Hazard Mater 116:39–48

Mulkins-Philips GJ, Stewart JE (1974) Distrubution of hydrocarbon-utilizing bacteria in northwestern Atlantic waters and coastal sediments. Can J Microbiol 20:955–962

Neustadt J, Pieczenik S (2007) Toxic-metal contamination: mercury. Integr Med 6(2):36–37

Ng SP, Davis B, Polombo EA, Bhave M (2009) Tn5051 like mer containing transposon identified in a heavy metal tolerant strain Achromobacter sp. AO22. BMC Res Notes 7:2–38

Nogaw K, Kido T (1996) Ital–Ital disease and health effects of cadmium. In: Chang LW (ed) Toxicology of metals. CRC Press, Boca Raton, FL, pp 353–369

Nouha K, Kumar RS, Tyagi RD (2016) Heavy metals removal from wastewater using extracellular polymeric substances produced by Cloacibacterium normanense in wastewater sludge supplemented with crude glycerol and study of extracellular polymeric substances extraction by different methods. Bioresour Technol 212:120–129

NRC (National Research Council) (1983) Polycyclic aromatic hydrocarbons: evaluation of sources and effects, committee on pyrene and selected analogues. National Academy Press, Board on Toxicology and Environmental Health Hazard, Commission on Life Sciences, Washington, DC

Olivera NL, Commendatore MG, Delgado O, Esteves JL (2003) Microbial characterization and hydrocarbon biodegradation potential of natural bilge waste microflora. J Ind Microbiol Biotechnol 30:542–548

Paul D, Poddar S, Sar P (2014) Characterization of arsenite-oxidizing bacteria isolated from arsenic-contaminated groundwater of West Bengal. J Environ Sci Health A 49(13):1481–1492

Philp JC, Atlas RM (2005) Bioremediation of contaminated soils and aquifers. In: Atlas RM, Philp JC (eds) Bioremediation: applied microbial solutions for real-world environmental cleanup. American Society for Microbiology (ASM) Press, Washington, DC, pp 139–236

Purnomo AS, Mori T, Kamei I, Td N et al (2010) Application of mushroom waste medium from *Pleurotus ostreatus* for bioremediation of DDT-contaminated soil. Int Biodeter Biodegr 64:397–402

Ramos JL, Marques S, Dillewijn PV, Espinosa-Urgel M et al (2011) Laboratory research aimed at closing the gaps in microbial bioremediation. Trends Biotechnol 29(12):641–647

Ramsay MA, Swannells RP, Shipton J, Duke WA (2000) Effect of bioremediation on the microbial community in oiled mangrove sediments. Mar Pollut Bull 41:413–419

Rayu S, Karpouzas DG, Singh BK (2012) Emerging technologies in bioremediation: constraints and opportunities. Biodegradation 23:917–926

Robinson C, Bromssen MV, Bhattacharya P, Haller S et al (2011) Dynamics of arsenic adsorption in the targeted arsenic-safe aquifers in Matlab, south-eastern Bangladesh: insight from experimental studies. Appl Geochem 26:624–635

Salem HM, Eweida EA, Farag A (2000) Heavy metals in drinking water and their environmental impact on human health. ICEHM 2000. Cairo University, Egypt, pp 542–556

Sasaki Y, Minakawa T, Miyazaki A, Silver S et al (2005) Functional dissection of a mercuric ion transporter MerC from Acidithiobacillus ferrooxidans. Biosci Biochem Biotechnol 69:1394–1402

Sinha RK, Valani D, Sinha S, Singh S et al (2009) Bioremediation of contaminated sites: a low-cost nature's biotechnology for environmental clean up by versatile microbes, plants & earthworms. In: Solid waste management and environmental remediation, pp 1–72

Sukhdhane K, Pandey P, Vennila A, Purushothaman C et al (2015) Sources, distribution and risk assessment of polycyclic aromatic hydrocarbons in the mangrove sediments of Thane Creek, Maharashtra, India. Environ Monit Assess 187:1–14

Sukhdhane KS, Pandey PK, Ajima MN, Jayakumar T et al (2019) Isolation and characterization of phenanthrene degrading bacteria from PAHs contaminated mangrove sediment of Thane Creek in Mumbai, India. Polycycl Aromat Compd 39:73–83

Tang L, Zeng GM, Shen GL, Li YP (2008) Rapid detection of picloram in agricultural field samples using a disposable immune membrane based electrochemical sensor. Environ Sci Technol 42:1207–1212

Tian Y, Liu HJ, Zheng TL, Kwon KK et al (2008) PAHs contamination and bacterial communities in mangrove surface sediments of the Jiulong River estuary, China. Mar Pollut Bull 57:707–715

Tripathi RD, Srivastava S, Mishra S, Singh N et al (2007) Arsenic hazards: strategies for tolerance and remediation by plants. Trends Biotechnol 25(4):158–165

Tuzen M, Saygi KO, Usta C, Soylak M (2008) Pseudomonas aeruginosa immobilized multiwalled carbon nanotubes as biosorbent for heavy metal ions. Bioresour Technol 99(6):1563–1570

USEPA (2004) Cleaning up the nation's waste sites: markets and technology trends. US Environmental Protection Agency, Washington, DC

Veltman K, Huijbregts MAJ, Rye H, Hertwich EG (2012) Including impacts of particulate emissions on marine ecosystems in life cycle assessment: the case of offshore oil and gas production. Integr Environ Assess Manag 7:678–686

Verma JP, Jaiswal DK (2016) Book review: advances in biodegradation and bioremediation of industrial waste. Front Microbiol 6:1–2

Whelan MJ, Coulon F, Hince G, Rayner J et al (2015) Fate and transport of petroleum hydrocarbons in engineered biopiles in polar regions. Chemosphere 131:232–240

Wiatrowski HA, Ward PM, Barkay T (2006) Novel reduction of mercury(II) by mercury-sensitive dissimilatory metal reducing bacteria. Environ Sci Technol 40(21):6690–6696

Wild SR, Jones KC (1986) Biological and abiotic losses of polynuclear hydrocarbons from soils freshly amended with sewage sludge. Environ Toxicol Chem 12:5–12

Wilson CS, Jones KC (1993) Bioremediation of soil contaminated with polynuclear aromatic hydrocarbons (PAHs): a review. Environ Pollut 81:229–249

Wuana RA, Okieimen FE (2011) Heavy metals in contaminated soils: a review of sources, chemistry, risks and best available strategies for remediation. ISRN Ecol:1–20

Xiao J, Guo L, Wang S, Lu Y (2010) Comparative impact of cadmium on two phenanthrene-degrading bacteria isolated from cadmium and phenanthrene co-contaminated soil in China. J Hazard Mater 174:818–823

Yuan SY, Wei SH, Chang BV (2000) Biodegradation of polycyclic aromatic hydrocarbons by a mixed culture. Chemosphere 41:1463–1468

Yuan CG, Lu XF, Qin J, Rosen BP et al (2008) Volatile arsenic species released from Escherichia coli expressing the AsIII S-adenosyl methionine methyl transferase gene. Environ Sci Technol 42:3201–3206

Zeng GM, Chen M, Zeng Z (2013a) Risks of neonicotinoid pesticides. Science 340:1403

Zeng GM, Chen M, Zeng Z (2013b) Shale gas: surface water also at risk. Nature 499:154

Zhang H, Kallimanis A, Koukkou AI, Drainas C (2004) Isolation and characterization of novel bacteria degrading polycyclic aromatic hydrocarbons from polluted Greek soils. Appl Microbiol Biotechnol 65:124–131

Zhang C, Zeng GM, Huang DL, Lai C et al (2014) Combined removal of di(2-ethylhexyl)phthalate (DEHP) and Pb(II) by using a cutinase loaded nanoporous gold polyethyleneimine adsorbent. RSC Adv 4:55511–55518

Zhang R, Wilson VL, Hou A, Meng G (2015) Source of lead pollution, its influence on public health and the countermeasures. Int J Health Anim Sci Food Saf 2:18–31

Zhao XW, Zhou MH, Li QBLYH, He N et al (2005) Simultaneous mercury bioaccumulation and cell propagation by genetically engineered Escherichia coli. Process Biochem 40:1611–1616

Zhou Y, Tang L, Zeng G, Zhang C et al (2016) Current progress in biosensors for heavy metal ions based on DNAzymes/DNA molecules functionalized nanostructures: a review. Sens Actuators B 223:280–294

Effects of Pharmaceutical Waste in Aquatic Life

Malachy N. O. Ajima and Pramod K. Pandey

Abstract

The waste from pharmaceutical products may cause harm to the aquatic life when it get washed into the environment through erosion or when it is discharged untreated into drinking water, underground water and into the aquatic environment which usually cause adverse effects to targeted and nontargeted aquatic organisms and indirectly affect the public health. The sources of pharmaceutical wastewater effluent which are mainly from point and nonpoints are noted to emanate from hospitals, municipal waste, domestics and industrials sources. Unfortunately, most of the pharmaceutical active ingredients are not properly removed, using most wastewater treatment methods before they get into aquatic environment where they cause toxic effects to living organisms. Different types of pharmaceutical wastes from groups of pharmaceuticals including antibiotics, analgesic, hormones, antidepressant, antihypertensive, contraceptive and steroids have been identified and reported to cause physiological dysfunctions, ranging from reproductive, haematological, behavioural, mutagenic, carcinogenic to physiological and enzymological effects. However, there is a need for proper monitoring and documentation of the potential health hazards/or implications, caused by these pharmaceutical wastes to the environment as well as aquatic life for possible mitigation.

M. N. O. Ajima (✉)
Department of Fisheries and Aquaculture Technology, Federal University of Technology, Owerri, Nigeria
e-mail: malachy.ajima@futo.edu.ng

P. K. Pandey
ICAR-Directorate of Coldwater Fisheries Research, Bhimtal, Uttarakhand, India

© The Author(s), under exclusive license to Springer Nature Singapore Pte Ltd. 2021
P. K. Pandey, J. Parhi (eds.), *Advances in Fisheries Biotechnology*,
https://doi.org/10.1007/978-981-16-3215-0_25

Keywords

Pharmaceuticals · Aquatic life · Hazards · Toxic effects

25.1   Introduction

Different types of pharmaceuticals including antibiotics, antihypertensive, contraceptive hormones, anti-inflammatory drugs, analgesic, and antiepileptic drugs are used to prevent, cure and in the treatment of different illness or diseases both in humans and animals. Continuous increase in the usage of pharmaceuticals to treat illness, for disease prevention, influx of wastewaters which are not treated enter into the environment, may increase the availability of drugs in the water body (Fekadu et al. 2019; Williams et al. 2019). Discharges from pharmaceutical industries, hospital effluents, animal and human excretion, improper discarding of unused or expired medicines, contaminated pharmaceutical products, waste released from research foundations and drugs development establishments could also increase the availability of pharmaceuticals and their metabolites into the environment. The residues and the metabolites from the pharmaceuticals usually find their ways into the aquatic ecosystem untreated, especially through effluents from treatment plant where they directly cause ecological impact on the aquatic organisms and indirectly on human health. In the process, most of the pharmaceutical wastes are not adequately broken down during the treatment and consequently, an appreciable part is discharged into the aquatic environment through wastewater effluent. The study shows that the traditional method of wastewater treatment does not naturally remove the targeted pharmaceuticals. The presence and persistence of pharmaceutical wastes in aquatic ecosystem could be deleterious owing to their reaction with other potential contaminants as well as newly emerging xenobiotics.

Further, the continuous influx of the pharmaceuticals and their metabolites may bioaccumulate into aquatic organisms and result in long term chronic effects as well as bio-magnify through food chain, thereby, militating against the survival of the organisms that inhabit such an environment. The effect of pharmaceuticals residues are usually not adequately monitored within aquatic wildlife, indicating the knowledge gap in determining the extent of exposure as well as the route of the chemicals in the organism (Miller et al. 2018). Usually, pharmaceutical wastes are designed to cross biological membranes. However, the rate of assimilation and internal concentrations of the compound are paramount. The availability and concentration of pharmaceutical waste entering or present in aquatic environment depend largely on the quantity of drugs, used by patients, the extent at which the sewage is being treated as well as the efficiency of the wastewater treatment plants. Understanding the potentials for pharmaceutical waste to cause adverse effects to the aquatic ecosystem, requires the ability to evaluate the occurrence or composition of these wastes to entire group of organisms that constitute the aquatic biota such as fish, invertebrates, plants and algae. However, there is a positive relationship that exists among the most regularly consumed groups of pharmaceutical products and their occurrence in the aquatic ecosystem.

In general, concentrations of pharmaceuticals and their metabolites, found in waste water treatment work (WWTW) effluent, range from ng L^{-1} to µg L range, in contrast, pharmaceuticals hardly supersedes 100 ng L in surface waters (Trudeau et al. 2005). Level of transformation of pharmaceuticals that develop in the body of a patient who takes drugs, the rate of degradation in the waste water system that receives water and the manner the compound partitions into column body, to a large extent determines the concentration of pharmaceuticals that can be found in an aquatic environment (Winker et al. 2008).

Metabolism of drugs, that occurs, can vary appreciably between compounds in such a way that some compounds are excreted completely either through urine or faeces from the parent compound without undergoing metabolism whereas other undergo complete metabolism.

In general, pharmaceutical wastes can be hazardous, readily inflammable, abrasive, or extremely unstable and could cause irritation to body cells. Some pharmaceuticals are genotoxic or mutagenic as well as harmful when discharged into the aquatic environment and can cause tumours and reproductive abnormalities to inhabitant aquatic organisms. The potential effects of pharmaceuticals waste in aquatic environment, especially in fish and other aquatic organisms, are explored in this paper.

25.2 Sources of Pharmaceutical Wastes

The main pathway of pharmaceutical wastes or products to aquatic ecosystem is through sewage treatment plant (STP) effluent or urban waste water treatment plants (WWTP) after voiding as a consequence of patient consumption (Corcoran et al. 2010). Drugs and their metabolites can be detected in domestic waste, effluents from hospitals and farming waste due to the fact that they are used by human beings. However, point and diffuse sources are the major pathways in which pharmaceuticals are discharged into the aquatic environment.

Therefore, sources of pharmaceuticals can be classified into two major groups, namely point and nonpoint or diffused sources.

25.2.1 Point Source Pharmaceutical Wastes

Point source is said to be a singular, noticeable origin of contaminant such as a canal or a drainage system which is usually emptied into an aquatic ecosystem. Usually, pharmaceutical and industrial wastes and hospital effluents are major point source of pollution which is normally released into water bodies.

Majority of the waste water is released into the sewage treatment plants through the sewage system. Improper human transformation of drug and excretion into waste drainage system are also some of source of discharging pharmaceuticals wastes into the aquatic ecosystem. Hospital and industrial waste water as well as domestic pathway through sewage treatment plants has been the major channel of pharmaceutical waste that usually enters into the aquatic ecosystem (Ternes 1998).

Some studies have investigated the occurrence of pharmaceuticals in the sewage treatment plants and found that the removal of the pharmaceuticals usually undergo incomplete processes. Besides water bodies, soil zone could be adversely affected by the discharges from waste water. The application of the waste water in farming activities can give room to the availability of some pharmaceutical products and other pollutant in the soils through irrigation processes (Chen et al. 2011; Fenet 2012; Durán-Álvarez et al. 2012).

The septic tank is one of the most important sources of the pharmaceuticals such as ibuprofen, paracetamol, salicylic acid, and triclosan that enter into the aquatic environment (Conn et al. 2010). Seepage of septic system could lead to contamination of water reserve by discharging some pharmaceutical wastes into the aquatic ecosystem. Also concentrations which ranged between 0.058 and $0.9\mu g\ L^{-1}$ of verapamil has been detected in aquatic ecosystem (Al-Rifai et al. 2007; Khan and Ongerth 2004).

25.2.2 Nonpoint Source or Diffused Pharmaceutical Waste Pollution

It is difficult to identify the exact location or the source where the nonpoint source pollution originates. Seepage that emanate from waste treatment apparatus, domestic and municipal runoff, wastes from agricultural runoff and individual point source discharges are the examples of nonpoint source pollution. Usually, expired and unused drugs as well as adulterated pharmaceutical materials such as injections and biological products applied for treatments which are deposited by domestic households and health care treatment industries to the environment are also nonpoint source of pharmaceutical waste. According to European Environmental Agency, diffused pollution can be the result of an array of operations or activities that have no precise point of discharge. Farming waste is a major source of diffuse pollution. However, deposits from atmospheric as well as rural habitations can also be a noble source. Organic contaminants include pharmaceutical wastes entering into the soil and water reserve through different pathways, with the important one being sewage sludge. Sewage sludge application onto the land surface is an important diffused source of the pharmaceutical waste entering into the soils as well as the freshwater ecosystem (Lapworth et al. 2012) (Fig. 25.1). High pharmaceuticals concentration can be observed in the biosolids, including thiabendazole and other numerous pharmaceuticals such as caffeine and carbamazepine (Kinney 2008). Owing to high solubility of pharmaceuticals with appreciable concentration in addition to some organic compound in biosolid, groundwater pollution from discharge of biosolid to soil and runoff surface could occur (Lapworth et al. 2012).

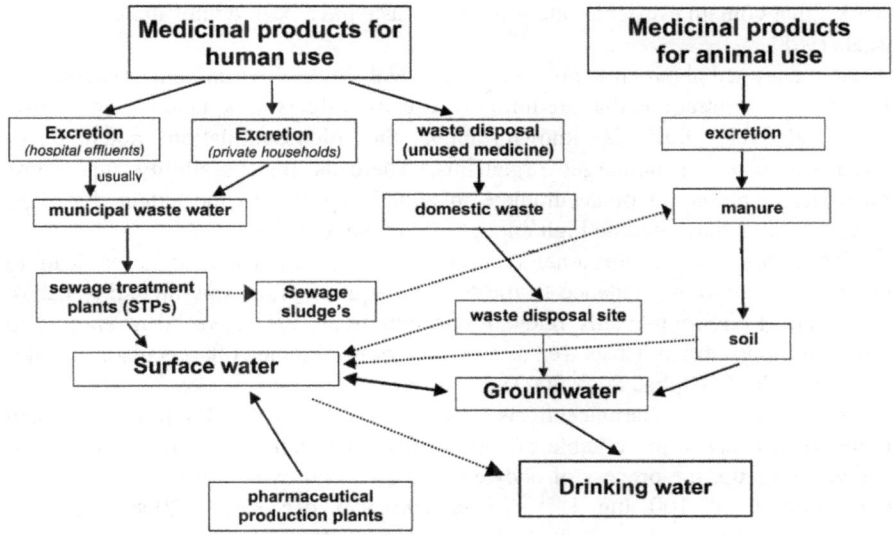

Fig. 25.1 Sources and pathways of pharmaceutical wastes to the aquatic environment as modified from Heberer (2002)

25.3   Effects of Pharmaceutical to Fish

Fish are the most important vertebrate organisms that are vulnerable to pharmaceuticals in aquatic environment, owing to environmental function and similarity of their physiological processes in comparison to mammals. Adverse effects in the populations are usually predicted from laboratory acute and chronic toxicity data in aquatic organisms including algae, crustaceans, and fish (Versteeg et al. 2005). Due to the fact that the primary mode of entry into the aquatic environment for most pharmaceutical waste is through STPs, the environmental risk assessment for the aquatic compartment is paramount. However, in order to understand the concentration that can cause potential effects to aquatic organisms, predicting the concentration is very vital. It is critical to know the effects on organisms across a wide range of concentrations, which will allow us to identify the level of concentrations at which certain compounds can have adverse effects on fish. In general, concentrations of pharmaceutical wastes or metabolites available in the environment are usually too low to illicit any observable direct adverse effects to aquatic organisms, except during chronic and acute laboratory experiments. Some studies reported the effects of certain pharmaceutical waste on natural water bodies that caused certain degrees of abnormalities in fish. Ethinyl estradiol (EE2) which was found to be the main estrogenic contaminant of STP effluents in UK Rivers caused endocrine disruption in fish (Desbrow et al. 1998). High incidences of intersexuality in roach, *Rutilus rutilus*, associated with exposure to sewage treatment works, that contain estrogens and estrogens waste have been demonstrated (Jobling et al. 1998).

As mentioned above, the most pharmaceuticals in the environment are present at very low concentrations that are unlikely to cause effects in aquatic species. However, relatively little is known about the bioaccumulation potential of pharmaceuticals in nontarget organisms. There is the possibility that some pharmaceuticals could bioaccumulate in aquatic species to the extent that their concentrations may become high enough to initiate an effect.

The occurrence of hormonal compounds in the environment could lead to endocrine-disrupting incidence to most of the aquatic organisms including fishes. However, at environmentally relevant concentrations, EE2 have been reported to induce feminisation in fish as well as induction of the female yolk precursor in males (Örn et al. 2006; Chikae et al. 2003).

Other classes of pharmaceuticals waste such as the NSAIDs have also been reported to cause some notable effects on fish. As such, indomethacin has been shown to disrupt the process of oocyte maturation and ovulation in zebra fish at a concentration of 100 mg L^{-1} (Lister and van der Kraak 2008) while at concentrations of μg L^{-1} of ibuprofen, pattern of spawning in Japanese medaka was altered (Flippin et al. 2007). Diclofenac hinders the stimulation of prostaglandin synthesis in the kidney in brown trout at environmentally relevant concentrations of 0.5–50μg L^{-1} (Hoeger et al. 2005) and caused DNA damage and physiological dysfunction in *Oreochromis niloticus* at sub lethal concentrations (Pandey et al. 2017). Also in rainbow trout, diclofenac induced glomeruloneophritis, necrosis of endothelial cells, and hyaline droplet degeneration in the kidney as well as hyperplasia and hypertrophy in gills at exposure level between 1 and 5μg L^{-1} concentrations (Schwaiger et al. 2004).

Ketoconazole, clotrimazole, and fadrozole pharmaceuticals at 11.1 nM concentration have been found to affect steroidogenesis and reproductive success of fish through the inhibition of steroidogenic enzymes including CYP11a, CYP17, and aromatase (Monteiro et al. 2000; Hinfray et al. 2004). Also a decrease in protein level was observed in *Clarias gariepinus* upon exposure to clotrimazole (Melefa et al. 2020). The potential effects of verapamil, a group of calcium channel blockers pharmaceuticals, have caused DNA damages, neurotic effects, molecular responses as well as oxidative imbalance in various tissues of *O. niloticus* after chronic exposure at sublethal concentrations between 0.14 and 0.57 mg L^{-1} (Ajima et al. 2017, 2020). Further, sublethal concentrations of verapamil significantly altered behavioural potentials, antioxidant biomarkers, haematological parameters and morphological responses in *Oncorhynchus mykiss* (Li et al. 2011). Also the behavioural ability of male Siamese fighting fish was impacted negatively by verapamil on exposure at relatively low concentration (Kania et al. 2015).

25.4 Effects of Pharmaceuticals on Other Aquatic Organisms

Besides fishes, pharmaceutical wastes can cause certain degrees of effects to other aquatic life including algae, mussels and other aquatic organisms. In the aquatic ecosystem, algae are often used as sentinel species in environmental monitoring, as an indicator species, to assess environmental conditions and from geological samples as indicators of past environmental conditions, suggesting that algae can be sensitive to environmental changes, including those caused by pharmaceutical wastes. Adverse effects on algae in aquatic environment might lead to serious ecological results, especially in the food chain. Pharmaceutical wastes can probably affect the algae survival usually by reducing the rate of photosynthesis through impairing the functions of chloroplasts. Availability of pharmaceutical waste in an ecosystem can lead to eutrophication and usually lead to algal mortality in such environment and indirectly affect ecological niche.

Increase in concentration of carbamazepine and diclofenac pharmaceuticals lead to the adverse effects on the chloroplasts in algae (Vannini et al. 2011). Relatively low concentration of sulfamethoxazole pharmaceuticals affected the photosynthetic apparatus of algae (Liu et al. 2011). Pharmaceutical waste products in water can form a mixture of various heavy metals and organic pathogens that are likely to have potential mutagenic and genotoxic effect when living organisms are exposed to it.

Adverse effects of some pharmaceutical drugs on certain aquatic organisms have been explored. In some of the studies, environmental concentration of diclofenac altered stability of lysosomal membrane and cyclooxygenase (COX) reactions as well as DNA damage in *Perna perna* (Fontes et al. 2018), and caused genotoxicity in *Mytilus galloprovincialis* (Mezzelani et al. 2018). Survival and reproduction capacity of *Folsomia candida* were impaired by diclofenac toxicity (Chen et al. 2015) while definitive and toxic effects potentials of diclofenac on developmental stages of Xenopus embryos was reported (Chae et al. 2015). Studies have shown that verapamil induced toxic effects on *Daphnia magna* after the treatment at low concentrations (Villegas-Navarro et al. 2003).

Four drugs, erythromycin, fluoxetine, naproxen and gemfibrozil belonging to different therapeutic classes, were selected by El-Bassat et al. (2012) to examine their toxicity to selected plankton organisms from different trophic levels: algae (*Chlorella vulgaris* and *Ankistrodesmus falcatus*), protozoa (*Paramecium caudatum*), rotifera (*Brachionus calyciflorus*) and cladocera (*Daphnia longispina*). LC_{50} values for three of the drugs were between 12 and 82 mg L^{-1}, with algae and protozoans being most sensitive. Fluoxetine showed LC_{50} values between 40 and 830µg L^{-1}, algae again being most sensitive. The low concentrations of fluoxetine showed enhanced growth rates of *B. calyciflorus* and *D. longispina*. Even at low test concentrations, erythromycin decreased the growth rates of all the test organisms. *Paramecium caudatum* was the species most sensitive to naproxen exposure. After 24 h, gemfibrozil had the least effect on all tested organisms. All the surviving tested organisms underwent oxidative stress to different degrees as a result of drug exposure. Although the test concentrations were above those commonly found in the environment, the occurrence of sublethal effects at all test concentrations was observed (El-Bassat et al. 2012).

Grabicova et al. (2015) found azithromycin and sertraline as bioaccumulative pharmaceuticals in Hydro-psyche. Even pharmaceuticals present at low levels in water were found in benthic organisms at relatively high concentrations (up to 85 ng g^{-1} w.w. for azithromycin). Consequently, the uptake of pharmaceuticals via the food web could be an important exposure pathway for the wild fish population.

25.5 Management of Pharmaceutical Wastes

Pharmaceutical waste can emanate from many activities and locations in a health care system. A compounding pharmacy on site could generate drug waste. Waste of pharmaceuticals poses an appreciable treatment and management challenges. Small quantities of pharmaceutical waste at households can be thrown away in the municipal waste stream. Large quantities kept at pharmacies, distribution centres, hospitals, etc. must be managed to minimize the risk of release or to exposure to workers and the possible effect it may cause to aquatic biota. Usually, the category of such waste includes expired, unused, and contaminated pharmaceutical products including vaccines and biological products used for therapy. Prescription and over-the-counter drugs end up as pharmaceutical waste including paraphernalia used in pharmacies such as gloves, masks, bottles, etc. Normally, solid pharmaceutical waste is generally easy to handle and package, but liquid waste poses more challenges in confining the waste and minimizing risk of release.

25.6 Treatment of Pharmaceutical Wastes

In the US, the EPA's Land Disposal Restriction requires treatment of pharmaceuticals before disposal. Treatment is aimed at changing the chemical structure of the medicines. The treated medicine should be acceptable for disposal with no worries of it getting into the aquatic ecosystem where it can cause harm to aquatic organism or even harm people. Some methods of treating pharmaceutical wastes are highlighted and discussed as below:

25.6.1 Incineration

Incineration is a high-temperature oxidation process that involves combustion of the organic portion of biomedical waste (BMW) components, producing gaseous emissions including steam, carbon dioxide, nitrogen oxides, particulate matter, and other toxic substances and inorganic solid residues such as ash. Incineration is an appealing option for the waste management with a heterogeneous waste stream, as many streams with pharmaceuticals tend to be included. In addition, under suboptimal combustion, carbon monoxide and hazardous pollutants such as dioxins and furans may be emitted. Incineration significantly reduces waste volumes, and eliminates pathogens from BMWs. However, it still remains the top used method of medical waste treatment in less developed countries. During the process of incineration, ash from these incinerators must be disposed off in a secured landfill. Other studies have shown that the major and common method of infectious medical waste treatment method in the developed countries is by incineration in which waste is burned to high temperatures (i.e. 1200 °C) and in the process the volume size is reduced and what remains is residual ash. The remaining ash is then dumped at landfill sites and then buried. This process ensures that infectious waste is sterilized and reduced in volume size to ash which in turn reduces cost of transportation to landfill sites.

The challenging issues regarding incineration are the disposal of ash and the treatment of gaseous pollutants, containing furans, dioxins and mercury. In addition, incineration has the advantage of reducing the volume size by 90% of the treated products. Other thermal technologies that have hardly been used for medical waste treatment include gasification, pyrolysis and plasma treatment method.

However, the main disadvantage of incineration of medical waste is the emissions and toxic pollutants dioxins, furans, and mercury arising from burning of the waste. Due to different compositions, burning of infectious waste produces toxic gases into the environment; hence, this method is highly controlled in developed countries as the emitted harmful gases, released into the atmosphere, may affect human health.

25.6.2 Autoclaving

The current known alternative for incineration is autoclaving. This method involves treatment of infectious waste by adding dry heat or steam to elevate the temperature of infectious waste to values sufficient enough to get rid of any microbial contamination. Autoclaving uses saturated steam in direct contact with the BMW in a pressure vessel at time lengths and temperatures sufficient to kill the pathogens. The Biomedical Waste Rules specify the minimum temperature, pressure, and residence time for autoclaves for safe disinfection. The operation requires qualified technicians, and medium investment and operating costs. The BMW is continuously tumbled in the chamber during the process. The advantageous part of the autoclave treatment process is that after waste treatment, the remaining waste can be disposed at the municipal solid waste (MSW) landfill site in the same way as noninfectious waste. Another advantage of autoclave treatment method of infectious medical waste over incineration is that it does not produce pollutants generated from PVC and other products such as mercury, furan and dioxin that are emitted into the environment during incineration.

There are also disadvantage in the use of autoclaving as an infectious waste treatment technology. Autoclave process has heat waste through steam to eliminate the pathogens without direct burning the waste and keeping its appearance like before, the resultant waste after treatment does not distinguish itself from untreated infectious waste, hence, giving the perception that untreated infectious waste is being dumped on landfill sites.

25.6.3 Microwaving

The application of an electromagnetic field over the BMW trigger the liquid in the waste to vibrate and heat up, eradicating the contagious components by transmission. The equipment is capable only if the ultraviolet radiation touches the waste products. The process of microwaving requires smashing to a satisfactory size and humidification. Microwaving provides debris that can be land loaded with municipal waste. The advantages of this treatment technology are its small electrical energy needs no steam requirement. The disadvantages include the need for qualified technicians and frequent breakdown of shredders.

References

Ajima MNO, Pandey PK, Kumar K, Poojary N (2017) Neurotoxic effects, molecular responses and oxidative stress biomarkers in Nile tilapia, *Oreochromis niloticus* (Linnaeus, 1758) exposed to verapamil. Comp Biochem Physiol Pt C 196:44–52

Ajima MNO, Pandey PK, Kumar K, Poojary N, Gora AH (2020) Verapamil caused biochemical alteration, DNA damage and expression of hepatic stress-related gene biomarkers in Nile tilapia, *Oreochromis niloticus*. Comp Clin Pathol 29(1):135–144

Al-Rifai JH, Gabelish CL, Schäfer AI (2007) Occurrence of pharmaceutically active and non-steroidal estrogenic compounds in three different wastewater recycling schemes in Australia. Chemosphere 69:803–815

Chae JP, Park MS, Hwang YS, Min BH, Kim SH, Lee HS, Park MJ (2015) Evaluation of developmental toxicity and teratogenicity of diclofenac using Xenopus embryos. Chemosphere 120:52–58

Chen F, Ying GG, Kong LX, Wang L, Zhao JL, Zhou LJ, Zhang LJ (2011) Distribution and accumulation of endocrine-disrupting chemicals and pharmaceuticals in wastewater irrigated soils in Hebei, China. Environ Pollut 159:1490–1498

Chen G, den Braver MW, van Gestel CA, van Straalen NM, Roelofs D (2015) Ecotoxicogenomic assessment of diclofenac toxicity in soil. Environ Pollut 199:253–260

Chikae M, Ikeda R, Hasan Q, Morita Y, Tamiya E (2003) Effects of alkylphenols on adult male medaka: plasma vitellogenin goes up to the level of estrous female. Environ Toxicol Pharmacol 15:33–36

Conn KE, Lowe KS, Drewes JE, Hoppe-Jones C, Tucholke MB (2010) Occurrence of pharmaceuticals and consumer product chemical in raw wastewater and septic tank effluent from single-family homes. Environ Eng Sci 27:347–356

Corcoran J, Winter MJ, Tyler CR (2010) Pharmaceuticals in the aquatic environment: a critical review of the evidence for health effects in fish. Crit Rev Toxicol 40:287–304

Desbrow C, Routledge EJ, Brighty GC, Sumpter JP, Waldock M (1998) Identification of estrogenic chemicals in STW effluent: 1. Chemical fractionation and in vitro biological screening. Environ Sci Technol 32:1549–1558

Durán-Álvarez JC, Prado-Pano B, Jiménez-Cisneros B (2012) Sorption and desorption of carbamazepine, naproxen and triclosan in a soil irrigated with raw wastewater: estimation of the sorption parameters by considering the initial mass of the compounds in the soil. Chemosphere 88:84–90

El-Bassat RA, Touliabah HE, Harisa GI (2012) Toxicity of four pharmaceuticals from different classes to isolated plankton species. Afr J Aquat Sci 37(1):71–80

Fekadu S, Alemayehu E, Dewil R, Bruggen BV (2019) Pharmaceuticals in freshwater aquatic environments: a comparison of the African and European challenge. Sci Total Environ 654:324–337

Fenet H (2012) Carbamazepine, carbamazepine epoxide and dihydroxycarbamazepine sorption to soil and occurrence in a wastewater reuse site in Tunisia. Chemosphere 88:49–54

Flippin JL, Huggett D, Foran CM (2007) Changes in the timing of reproduction following chronic exposure to ibuprofen in Japanese medaka *(Oryzias latipes)*. Aquat Toxicol 81:73–83

Fontes MK, Gusso-Choueri PK, Maranho LA, de Souza Abessa DM, Mazur WA, de Campos BG, Guimaraes LL, de Toledo MS, Lebre D, Marques JR, Felicio AA, Cesar A, Almeida EA, Pereira CDS (2018) A tiered approach to assess effects of diclofenac on the brown mussel *Perna perna*: a contribution to characterize the hazard. Water Res 132:361–370

Grabicova K, Grabic R, Blaha M, Kumar V, Cerveny D, Fedorova G, Randak T (2015) Presence of pharmaceuticals in benthic fauna living in a small stream affected by effluent from a municipal sewage treatment plant. Water Res 72(1):145–153

Heberer T (2002) Occurrence, fate, and removal of pharmaceutical residues in the aquatic environment: a review of recent research data. Toxicol Lett 131(1–2):5–17

Hinfray N, Porcher JM, Brion F (2004) Inhibition of rainbow trout (Oncorhynchus mykiss) P450 aromatase activities in brain and ovarian microsomes by various environmental substances. Comp Biochem Physiol C 144:252–262

Hoeger BA, Köllner B, Dietrich DR, Hitzfeld B (2005) Water-borne diclofenac affects kidney and gill integrity and selected immune parameters in brown trout *(Salmo trutta* f. *fario)*. Aquat Toxicol 75:53–64

Jobling S, Nolan M, Tyler CR, Brighty G, Sumpter JP (1998) Widespread sexual disruption in wild fish. Environ Sci Technol 32:2498–2506

Kania BF, Debski B, Wrońska D, Zawadzka E (2015) Verapamil–L type voltage gated calcium channel inhibitor diminishes aggressive behaviour in male Siamese fighting fish. Pol J Vet Sci 18:401–406

Khan SJ, Ongerth JE (2004) Modelling of pharmaceutical residues in Australian sewage by quantities of use and fugacity calculations. Chemosphere 54:355–367

Kinney CA (2008) Bioaccumulation of pharmaceuticals and other anthropogenic waste indicators in earthworms from agricultural soil amended with biosolid or swine manure. Environ Sci Technol 42:1863–1870

Lapworth DJ, Baran N, Stuart ME, Ward RS (2012) Emerging organic contaminants in groundwater: a review of sources, fate and occurrence. Environ Pollut 163:287–303

Li ZH, Velisek J, Zlabek V, Grabic R, Machova J, Kolarova J, Li P, Randak T (2011) Chronic toxicity of verapamil on juvenile rainbow trout (*Oncorhynchus mykiss*) effects on morphological indices, haematological parameters and antioxidant responses. J Hazard Mater 185:870–880

Lister A, Van der Kraak G (2008) An investigation into the role of prostaglandins in zebrafish oocyte maturation and ovulation. Gen Comp Endocrinol 159:46–57

Liu BY, Nie XP, Liu WQ, Snoeijs P, Guan C, Tsui MT (2011) Toxic effects of erythromycin, ciprofloxacin and sulfamethoxazole on photosynthetic apparatus in *Selenastrum capricornutum*. Ecotoxicol Environ Saf 74:1027–1035

Melefa TD, Mgbenka BO, Aguzie IO, Andong FA, Nwakor U, Nwani CD (2020) Morphological, haematological and biochemical changes in African catfish *Clarias gariepinus* (Burchell 1822) juveniles exposed to clotrimazole. Comp Biochem Physiol Pt C 236:108815. https://doi.org/10. 1016/j.cbpc.2020.108815

Mezzelani M, Gorbi S, Fattorini D, d'Errico G, Consolandi G, Milan M, Bargelloni L, Regoli F (2018) Long-term exposure of *Mytilus galloprovincialis* to diclofenac, ibuprofen and Ketoprofen: insights into bioavailability, biomarkers and transcriptomic changes. Chemosphere 198:238–248

Miller TH, Bury NR, Owen SF, MacRae JI, Barron LP (2018) A review of the pharmaceutical exposome in aquatic fauna. Environ Pollut 239:129–146

Monteiro PRR, Reis-Henriques MA, Coimbra J (2000) Polycyclic aromatic hydrocarbons inhibit in vitro ovarian steroidogenesis in the flounder (Platichthyus flesus L.). Aquat Toxicol 48:549–559

Örn S, Yamani S, Norrgren L (2006) Comparison of vitellogenin induction sex ratio and gonad morphology between zebrafish and Japanese medaka after exposure to 17a ethinylestradiol and 17ß-trenbolone. Arch Environ Contam Toxicol 51:237–243

Pandey PK, Ajima MNO, Kumar K, Poojary N, Kumar S (2017) Evaluation of DNA damage and physiological responses in Nile tilapia, *Oreochromis niloticus* (Linnaeus, 1758) exposed to sub-lethal diclofenac (DCF). Aquat Toxicol 186:205–214

Schwaiger J, Ferling H, Mallow U, Wintermayr H, Negele RD (2004) Toxic effects of the non-steroidal anti-inflammatory drug diclofenac: part I. histopathological alterations and bioaccumulation in rainbow trout. Aquat Toxicol 68:141–150

Ternes T (1998) Occurrence of drugs in German sewage treatment plants and rivers. Water Res 32:3245–3260

Trudeau VL, Metcalfe CD, Mimeault C, Moon TW (2005) Pharmaceuticals in the environment: drugged fish? In: Mommsen TP, Moon TW (eds) Biochemistry and molecular biology of fishes, vol 6, pp 475–493

Vannini C, Domingo G, Marsoni M, De Mattia F, Labra M, Castiglioni S, Bracale M (2011) Effects of a complex mixture of therapeutic drugs on unicellular algae Pseudokirchneriella subcapitata. Aquat Toxicol 101:459–465

Versteeg DJ, Adler AC, Cunningham VL (2005) Environmental exposure modeling and monitoring of human pharmaceutical concentrations in the environment. In: Willams RT (ed) Human pharmaceuticals: assessing the impacts on aquatic ecosystems. SETAC Press, Pensacola, FL, pp 303–311

Villegas-Navarro A, Rosas LE, Reyes JL (2003) The heart of *Daphnia magna* effects of four cardioactive drugs. Comp Biochem Physiol C 136:127–134

Williams M, Kookana RS, Mehta A, Yadav SK, Tailor BL, Maheshwari B (2019) Emerging contaminants in a river receiving untreated wastewater from an Indian urban Centre. Sci Total Environ 647:1256–1265

Winker M, Tettenborn F, Faika D, Gulyas H, Otterpohl R (2008) Comparison of analytical and theoretical pharmaceutical concentrations in human urine in Germany. Water Res 42:3633–3640

Biosafety and Bio-Security for Sustainable Aquaculture Development

26

Saurav Kumar and Tapas Paul

Abstract

Intensification of aquaculture practices has induced occurrence of several stressors in fish production system and export of fisheries products. These include emerging and trans-boundary aquatic animal diseases, antimicrobial resistance, invasive species, risk with wild fish used for breeding program and outbreak of epidemic diseases. In this prelude, bio-security and biosafety are two major components to minimize the risk of occurrence of these stressors and promoting sustainable aquaculture development. The major objectives of bio-security are cultured animal health management, public and human health management, and incorporating best management practices in aquaculture system. Specific-pathogen-free (SPF) brood stock can be used as tool of bio-security for eliminating the spread of the virus from brooders to post-larvae during hatchery operation and nursery rearing. In Indian scenario, lack of scientific knowledge and interest to follow standard operating procedures (SOP) are the driven forces for poor implementation of bio-security measures in culture system. Marine Products Export Development Authority (MPEDA) has started several new initiatives to enhance the bio-security measures and creating awareness among small and marginal farmers. Biosafety is the prevention of large-scale loss of biological integrity including humans and the ecosystem. However, current biosafety protocols in aquaculture focus mainly on transgenic and DNA vaccine; so, it widens broad area of future research. The present chapter highlighted the role of bio-security and biosafety in the sustainable development of aquaculture for quality food production.

S. Kumar (✉) · T. Paul
ICAR-Central Institute of Fisheries Education, Mumbai, Maharashtra, India
e-mail: saurav@cife.edu.in

© The Author(s), under exclusive license to Springer Nature Singapore Pte
Ltd. 2021
P. K. Pandey, J. Parhi (eds.), *Advances in Fisheries Biotechnology*,
https://doi.org/10.1007/978-981-16-3215-0_26

Keywords

Biosafety · Bio-security · Specific-pathogen-free · Disease · Sustainable aquaculture

26.1   Introduction

World aquaculture production grew faster as compared to those of other major food and dietary produce sectors, registering a yearly growth rate of 5.5% in 2018 and global production of 114.5 million tonnes (FAO 2020). In 2017, the intakes of fish and seafood accounted for 17% of protein altogether obtained from animals and this fraction has been growing continuously (FAO 2020). India stands second in global aquaculture, next to china, with a production of 5.7 million tonnes (FAO 2020). In the coming years, every country in the world is going to face the perilous challenge of meeting the desired quantity and quality of nutritious foodstuff to feed the growing population that is expected to reach 9.6 billion by 2050 (United Nations 2013). Consequently, the demand for aquaculture produces both in terms of quantity and diversity will also increase in future; it is estimated that by 2030 aquaculture will contribute about 62% of fish for direct human consumption (Kobayashi et al. 2015). However, due to the intensification of aquaculture sector for meeting global food demands, disease outbreaks in all diverse form of aquaculture (freshwater, brackish water and mariculture) systems has become a serious constraint to fish production and trading of fish products (Kaoud Hussein 2015; Mishra et al. 2017; Behringer et al. 2020). Diseases, caused by several pathogens such as parasites, bacteria, viruses and fungus, pose a considerable threat to aquaculture productivity leading to significant economic losses by causing mortality up to 100% in farmed fishes (Leung and Bates 2013). On the other hand, an increase in the incidence of emerging pathogens, having zoonotic potential in aquaculture, has led to the emergence of serious concerns in terms of human and cultured animals' health. Therefore, bio-safety is becoming one of the important aspects in developing sustainable aquaculture. The utilization of the existing unexplored aquatic resources in India, the adoption of farming practices in the cage and pen culture systems and the introduction of the exotic fish make the system vulnerable with regard to the aspects of biosafety. Moreover, culture of animals of exotic origin possesses a risk to biological resources of the aquatic environments, especially when they escape. Under such conditions, aquaculture practices should mandatorily consider the introduction of an inclusive biosafety program. The biosafety program has several important components such as risk assessment, hazard identification and risk analysis. Risk analysis takes into account the impact of a hazard it is likely to cause and its consequences and severity. Four hazardous components are identified in the open water aquaculture practices i.e., (1) escape of stocked alien species, (2) alien species that can hybridize with native species, (3) escape of cultured stocks into populations of the same species and (4) escape of genetically engineered organisms. These hazards may have adverse effects at multiple levels from genes to populations to the community in existing aquatic ecosystem. Therefore, the use of adequate preventive measures and mapping of these measures through an active monitoring program is likely to result in reduction of risk in aquaculture. Monitoring aquaculture of operations makes it possible to determine the ecological impacts due to such escape to (1) detect the occurrence of a hazard and initiate remedial measures to reduce its persistence and minimize adverse consequences; and (2) assess the cultured stock's health due to the occurrence which can be applied to analyze the possible risk to other aquaculture operations. Moreover, biosafety programs should lead to a systematic evaluation and solid evidence of the degree of ecological safety versus the risk of an aquaculture operation. It should have a mechanism for mid-course corrections using biosafety measures and adaptive learning that will improve future biosafety decisions and measures.

In brief, the sustainable growth and development of aquaculture need prevention from infectious diseases, monitoring and management of risk associated with the invasive and exotic fish (Samayanpaulraj et al. 2020). The tools like bio-security and biosafety in aquaculture system, which include the best management practices, enable to minimize the risk of occurrence of infectious diseases and impact of alien species on host and ecosystem, respectively. The present chapter emphasizes the role of bio-security and biosafety in the sustainable development of aquaculture for quality food production.

26.2 Bio-Security in Aquaculture

Bio-security is defined as a strategic and integrated approach that encompasses both policy and regulatory frameworks, aimed at analyzing and managing risks relevant to the environment including human, animal and plant health (FAO 2007). In aquaculture, bio-security refers to the protection of farmed fish species from infectious diseases as well as conserving aquatic biodiversity and minimizing the risk to human health, associated with cultivation and consumption of aquatic food products (Subasinghe and Bondad-Reantaso 2006). It regulates the introduction of genetically modified organisms (GMOs) or invasive alien species in aquaculture system by putting appropriate measures and controlling the spread of infectious organisms into individuals, populations or ecosystems. Bio-security is intricately related to the selection of suitable and candidate fish species, managing fish health and environment, sustainable aquaculture production, genetics and biotechnological interventions for improving food quality and safety (Subasinghe et al. 2001).

26.3 Need of Bio-Security in Aquaculture

26.3.1 Trans-Boundary Aquatic Animal Diseases

In recent times, there is an increase in the incidence of trans-boundary aquatic animal diseases (TAAD), causing serious socioeconomic losses in the countries worldwide. The major key factor for the rise of TAAD is the increase in domestic and international trade through which pathogens are spreading to new areas along with fish and its products. TAAD has a serious negative impact on aquaculture systems, causing loss of fish production, decreasing the profit and livelihood of people, altering aquatic biodiversity, and acting as a barrier to inclusive growth of aquaculture industry. Researchers have estimated that the occurrence of TAAD has resulted in economic losses ranging from USD17.5 to 650 million in shrimp industry (Bondad-Reantaso et al. 2005; Flegel et al. 2008). Virus and bacteria constitute 80% of the causative agent of TAAD and some major examples are as follows:

- White spot syndrome virus (WSSV)
- Yellow head virus
- Epizootic ulcerative syndrome (EUS)
- Koi herpes virus (KHV)
- Infectious salmon anemia (ISA) and sea lice
- Viral nervous necrosis (VNN)

WSSV and EUS listed in OIE diseases for aquatic animals are the most serious pathogens globally which spread over many countries in Asia, Africa, and America. KHV had significant impact on ornamental fishes such as koi carp and wild carp species. European countries have incurred losses amounting to millions of dollars and mortality due to occurrence of ISA, sea lice and protozoan parasite *Bonamia ostreae* due, as it spread with the culture of Pacific cupped oysters (*Crassostea gigas*). Transboundary movements of viral pathogens are a major threat to the shrimp industry because these pathogens infect bloodstock and are capable of vertical transmission to infect larvae and finally enter into grow-out ponds through the stocking of infected post-larvae. Thus, screening of pathogens from brood stock and seed is a major key point in the success of shrimp aquaculture system and the main component of bio-security.

26.3.2 Antimicrobial Resistance in Aquaculture

Over the years, many antimicrobial agents have been used for controlling and preventing infectious diseases in fish and shrimp (OIE 2011). Among them, antibiotics are the most common for treating bacterial infections in aquaculture. Currently, the most frequently used antibiotics include oxytetracycline, florfenicol, and sarafloxacin along with other antibiotics such as quinolones, oxolinic acid, ciprofloxacin, enrofloxacin, sulfamethazine, and gentamicin. However, the usage of antibiotics for controlling bacterial diseases in aquaculture has been proven unsustainable and barren due to the development of antibiotic resistance in pathogens, leading to one of the greatest human health challenges of the twenty-first century (Cabello et al. 2016). Continuous long-term use of antibiotics in aquaculture has promoted the evolution and spread of antibiotic resistance genes which can pose a threat to human health, and animals through various pathways. This type of resistance can be established either by mutation in a gene, modifying the structure of the ribosomal target, or by acquiring plasmids, transposons. Thus, it can be transferred through horizontal gene transfer (Martínez 2008; Yang et al. 2018).

In aquaculture, large quantities of antiparasitic drugs and chemicals are used indiscriminately, resulting in the development of resistance in the parasites. Even though the other modes of control measures such as vaccinations (Dadar et al. 2017; Xu et al. 2020) and phytotherapeutics (Kumar et al. 2012; Kumari et al. 2019; Zhou et al. 2020) have been employed, but most of them have exhibit limited success in controlling the incidence of parasitic infections. Although, several medicines have been used to control infectious agents, the withdrawal period of those medicines is yet to be evaluated for most of the commercially important fish species. So, keeping all these in view, there is urgent need to strengthen the bio-security measures in aquaculture system.

26.3.3 Risk Associated with Wild Fish Used for Breeding Program

It is observed that brood stock, caught from the wild, carry several pathogenic strains of microbes including virus which don't show any symptoms of infection in adult and transmitted to post-larvae. When these post-larvae are released into grow-out ponds they develop infection and cause huge mortality and in turn an economic loss to farmers. Import of any brood stock for seed production and subsequent grow-out culture must be regulated by performing risk analysis and quarantine procedures, followed by a diagnostic test for identifying an unknown virus which may pose threat to indigenous species.

26.3.4 Risk Associated with Live Feed

Live feeds such as algae, *Artemia*, polychaetes and squid meat are widely used as feed during larval rearing for economically important fish and shrimp and are potent vector that carry varied pathogens into the system. Thus, the sensitive and adequate screening of pathogens from live feed is an essential component of specific pathogen free animal production. Adequate attention has not been paid to this aspect.

26.3.5 Crustacean Culture

Crustacean can carry the pathogens for a longer time than other vertebrates and the pathogenic strains cause chronic infection without showing any gross sign of disease (Flegel 2006) due to their immune tolerance behavior. Therefore, crustacean culture system should follow strict bio-security measures. There is a need to develop efficient bio-security measures which deals exclusively with different aspects of shrimp culture.

26.4 Major Goals of Bio-Security

The following are the major goals of any aquaculture program:

1. Cultured animal health management—obtaining healthy stocks and optimizing their health and immunity through good husbandry.
2. Pathogen management—preventing, reducing or eliminating pathogens.
3. Public health management—educating and managing staff and visitors.

26.5 Developing and Using Specific-Pathogen-Free (SPF) Brood Stock as a Tool of Bio-Security in Aquaculture

Specific-pathogen-free (SPF) brood stock can be used in shrimp aquaculture farms to eliminate the spread of virus from brooders to post-larvae which are further stocked in rearing ponds. Biosecure hatcheries are free from entry of virus or their host during hatchery operation. Major components in biosecure hatcheries are as follows:

- SPF brood stocks which are reared in captive condition for the production of post-larvae results in higher sustainable production with no risk of pathogen transmission.
- Local breeding centers can be established maintaining proper bio-security measures which could potentiate the supply of brood stock and post-larvae for grow-out culture.
- Grow-out culture ponds should be covered with nets (bio-fence) properly for not allowing any insects carriers to enter into the water body.
- Direct use of brood stock from any foreign SPF labs must not be permitted without proper disease diagnostic and quarantine procedures.
- SPF shrimp are free of one or more specific pathogens which can be reliably diagnosed and pose a significant threat to the industry.
- Mixed culture of shrimps should not be carried out to avoid transmission of the virus from one species to others. For example, rearing of captured *Penaeus monodon* and exotic SPF *L. vannamei* in an Asian shrimp hatchery may result in a transfer of PstDNV from *P. monodon* to *L. vannamei* at the larval stage. Further, it is reported that co-cultivation of *P. monodon* and *Fenneropenaeus*

indicus larvae may cause white muscle disease and high mortality in both the species due to *Macrobrachium rosenbergii* nodavirus (MrNV) (Ravi et al. 2009).

- For producing SPF stocks, zero water exchange system can be encouraged in culture ponds as an exchange of untreated water is responsible for the introduction of disease-causing pathogens in the aquatic system. Zero water exchange system provides an eco-friendly and cost-effective alternative to conventional shrimp culture (Pruder 2004). Further, if water exchange is done, the input water must be treated with the water disinfection process to eliminate the transmission of pathogens.

26.6   Bio-Security at International Levels

Bio-security in aquaculture can be defined as a component of overall aquatic animal health management. In Australia, a primary attempt has been adequately placed under a national aquatic animal health management strategy (AQUAPLAN6) to bring together and coordinate the various elements of aquatic animal health management viz. (1) International linkages (2) Quarantine (3) Surveillance, monitoring and reporting (4) Awareness (5) Public and environmental health and aquaculture (6) Research and development and (7) Legislation, policies and jurisdiction. In addition to national level bio-security, there are international aquatic animal health programs and initiatives as well that try to promote and harmonize what is being done at the national level. Concerning aquatic animals, these include the OIE, the FAO and the Network of Aquaculture Centers in Asia-Pacific (NACA). For example, the OIE sets standards (recognized by the World Trade Organization) for translocation of aquatic animal products, risk analysis and evaluation/recognition of the competent authorities, as well as identifying globally significant (notifiable) animal diseases, coordinating the collation and reporting the health status of member countries concerning these notifiable diseases.

26.7   Status of Bio-Security in Aquaculture Farms of India

Due to large-scale production of Pacific White Shrimp *Litopenaeus vannamei* over the last few decades, demand for SPF brood stock has increased exponentially in different parts of the country. Rajiv Gandhi Centre for Aquaculture (RGCA) has established brood stock multiplication center for *L. vannamei* with a view to provide consistent high-quality SPF brood stock, having high growth rate, resistance to disease, and high survivability. RGCA has collaborated with M/s. Oceanic Institute, Hawaii for successful implementation of this project and supply of SPF brood stock. M/s. Oceanic Institute, Hawaii is providing the germplasm (post-larvae) and post-larvae (PL) are reared at RGCA to form brood stock which is then screened and supplied to shrimp hatcheries all over the country. This facility is capacitated to supply around 45,000 brood stock to the shrimp industry each year. India is one of the leading countries in shrimp production with a total production of *P. vannamei* as

618,678 MT, *P. monodon* as 54,901 MT and *M. rosenbergii* as 7222 MT (MPEDA 2019). However, many small and marginal farmers in India don't follow the standard operating procedures (SOP) or bio-security measures during culture period. A study on measuring the bio-security implementation rate in shrimp hatcheries showed that 50% of small and large and 25% of medium hatcheries carry out screening of pathogens (Raja et al. 2012). Lack of scientific knowledge and interest to follow SOP are the major factors for the poor implementation of bio-security measures. Keeping in mind these aspects, Marine Products Export Development Authority (MPEDA) has launched several new guidelines and schemes such as farm enrollment and block chain technology for the proper tracing of shrimp farms in different states of the country. Several task forces are employed for inspection of shrimp hatcheries and aqua shops to examine the implementation rate of bio-security measures and in 2018–19, 479 hatcheries and 1029 aqua shops were inspected (MPEDA 2019). Several new initiatives are taken in a direction toward bio-security certification which includes:

- Adoption of Best Management Practices (BMPs) and Standard Operating Procedures (SOP) approved by MPEDA in all shrimp farms.
- Antibiotic-free shrimp seed production pilot project in Andhra Pradesh using bacteriophage and probiotics.
- Antibiotic-free shrimp seed certification program.
- Testing the Produce—Pre Harvest Test (ELISA) for antibiotics.
- Using farm inputs with proper labeling.
- Quality declaration of produce by the farmer attested by MPEDA.

26.8 Biosafety in Aquaculture

In aquaculture, the main focus is on good health of the cultured aquatic animals. However, the implications of aquaculture activities have a direct impact on the wellbeing of both human and environmental health. Moreover, continued growth of the sector has bearing on public health. A large number of unmanaged and poorly managed fish ponds are likely to act as breeding grounds for many vectors of waterborne diseases. Many of the infectious pathogens having zoonotic potential such as schistosoma, black grub diaginians and bacteria namely *Aeromonas* spp. and *Edwardsiella tarda* are a serious concern during handling of cultured fish. Though the zoonotic aspects of aquaculture diseases are indisputably vital, fish diseases that can be transmitted to humans are few, and limited to certain pathogenic helminths for which fish act as the intermediate host. Hence it poses a risk to human health and biosafety concept, applied to the industry and laboratory, is also applicable to aquaculture. Biosafety is the prevention of large-scale loss of biological integrity including humans and the ecosystem. The basic components in biosafety are to assess the biological risk and monitoring it to minimize the loss. However, the set guidelines and regulations, decontamination and disinfection, disposal and shipment of biological materials, wastes, facility and equipment design, medical surveillance, practices and risk assessments are integral factors to underline the biosafety in aquaculture.

26.9   Approach of Biosafety in Aquaculture

Although biosafety, in comparison to bio-security, is an ancient term and practiced in many firms, yet relatively poorly defined exists in the aquaculture sector. In limited form, biosafety discussed in the Convention on Biological Diversity (CBD), mainly refers to environmental and human health safeguards regarding living modified organisms (LMOs), produced by biotechnology associated to gene transfer or transgenic. Remarkably, gene transfer is still not economical and commercially viable in the aquaculture sector. Altogether there are several pilot projects and research programs in many parts of the world that are developing commercially important transgenic fish, such as salmon, catfish, carps, and tilapia. The conventional approaches and biotechnological techniques, such as chromosome manipulation, sex-reversal, hybridization, and conventional selective breeding are more common in aquaculture. Therefore, narrow scope of the current biosafety protocols exists in the field of aquaculture i.e., focusing primarily on transgenic, and DNA vaccine or such modified fish should be carefully considered. All the wild relatives of domesticated aquatic species are still found in nature; biosafety protocols, or similar regulations, should eventually strive to protect these resources while allowing for the development of aquaculture and international trade.

26.10   Conclusion

Biosafety and bio-security measures will reduce the risk of catastrophic losses from infectious disease, alien species and low-level losses due to disease or imbalance of ecological diversity that, over time, can also greatly affect the entire aquatic resources, germplasm and aquaculture industry. Based on continuous monitoring of biological risk in the aquaculture system, need-based development of guidelines and regulatory measures are an integral component. Further, adoption of standard operating procedures for aquaculture is the central theme of biosafety for public and environmental health, making aquaculture a sustainable industry to produce food and facilitate nutritional security to the increasing population.

References

Behringer DC, Wood CL, Krkošek M, Bushek D (2020) Disease in fisheries and aquaculture. Marine Disease Ecology, p 183
Bondad-Reantaso MG, Subasinghe RP, Arthur JR, Ogawa K, Chinabut S et al (2005) Disease and health management in Asian aquaculture. Vet Parasitol 132(3–4):249–272

Cabello FC, Godfrey HP, Buschmann AH, Dölz HJ (2016) Aquaculture as yet another environmental gateway to the development and globalisation of antimicrobial resistance. Lancet Infect Dis 16(7):e127–e133

Dadar M, Dhama K, Vakharia VN, Hoseinifar SH, Karthik K, Tiwari R, Khandia R, Munjal A, Salgado-Miranda C, Joshi SK (2017) Advances in aquaculture vaccines against fish pathogens: global status and current trends. Rev Fish Sci Aquacult 25(3):184–217

FAO (2007) The state of world fisheries and aquaculture. http://www.fao.org/3/a-a0699e.pdf

FAO (2020) The state of world fisheries and aquaculture 2020. In: Sustainability in action. Rome https://doi.org/10.4060/ca9229en

Flegel TW (2006) Disease testing and treatment. In: Boyd CE, Jory D, Chamberlain GW (eds) Operating procedures for shrimp farming. Global shrimp OP survey results and recommendations. Global Aquaculture Alliance, St. Louis, pp 98–103

Flegel TW, Lightner DV, Lo CF, Owens L (2008) Shrimp disease control: past, present and future. In: Diseases in Asian Aquaculture VI, vol 505. Fish Health Section, Asian Fisheries Society, Manila, pp 355–378

Kaoud Hussein A (2015) Article review: heavy metals and pesticides in aquaculture: health problems. Eur J Acad Essays 2(9):15–22

Kobayashi M, Msangi S, Batka M, Vannuccini S, Dey MM, Anderson JL (2015) Fish to 2030: the role and opportunity for aquaculture. Aquacult Econ Manag 19(3):282–300

Kumar S, Raman RP, Kumar K, Pandey PK, Kumar N, Mohanty S, Kumar A (2012) In vitro and in vivo antiparasitic activity of Azadirachtin against *Argulus* spp. in *Carassius auratus* (Linn. 1758). Parasitol Res 110(5):1795–1800

Kumari P, Kumar S, Ramesh M, Shameena S, Deo AD, Rajendran KV, Raman RP (2019) Antiparasitic effect of aqueous and organic solvent extracts of Azadirachta indica leaf against Argulus japonicus in Carassius auratus. Aquaculture 511:634175

Leung TL, Bates AE (2013) More rapid and severe disease outbreaks for aquaculture at the tropics: implications for food security. J Appl Ecol 50:215–222

Martínez JL (2008) Antibiotics and antibiotic resistance genes in natural environments. Science 321:365–367

Mishra SS, Rakesh D, Dhiman M, Choudhary P, Debbarma J, Sahoo SN, Barua A, Giri BS, Ramesh R, Ananda K, Mishra CK (2017) Present status of fish disease management in freshwater aquaculture in India: state-of-the-art-review. J Aquacult Fish 1:003

MPEDA (2019) Marine products export development authority. Govt. of India. www.mpeda.gov.in

OIE (2011) Aquatic animal health code, 14th edn. World Organisation for Animal Health, Paris. www.oie.int/en/international-standard-setting/aquatic-code/access-online/

Pruder GD (2004) Bio-security: application in aquaculture. Aquacult Eng 32(1):3–10

Raja S, Kumaran M, Ravichandran P (2012) Implementation of bio-security measures in commercial shrimp hatcheries in India, pp 1–6

Ravi M, Nazeer Basha A, Sarathi M, Rosa Idalia HH, Sri Widada J, Bonami JR, Sahul Hameed AS (2009) Studies on the occurrence of White Tail Disease (WTD) caused by MrNV and XSV in hatchery-reared post-larvae of *Penaeus indicus* and *P. monodon*. Aquaculture 292:117–120

Samayanpaulraj V, Sivaramapillai M, Palani SN, Govindaraj K, Velu V, Ramesh U (2020) Identification and characterization of virulent *Aeromonas hydrophila* Ah17 from infected *Channastriata* in river Cauvery and in vitro evaluation of shrimp chitosan. Food Sci Nutr 8 (2):1272–1283

Subasinghe RP, Bondad-Reantaso MG (2006) Bio-security in aquaculture: international agreements and instruments, their compliance, prospects and challenges for developing countries. In: Scarfe AD, Lee C-S, O'Bryen P (eds) Aquaculture bio-security: prevention, control and eradication of aquatic animal disease. Blackwell, Ames, pp 9–16

Subasinghe RP, Bondad-Reantaso MB, McGladdery SE (2001) Aquaculture development, health and wealth. In: Subasinghe RP, Bueno P, Phillips MJ, Hough C, Mcgladdery SE, Arthur JR (eds) Aquaculture in the Third Millennium. Technical Proceedings of the Conference on

Aquaculture in the Third Millennium, Bangkok, Thailand, 20–25 February 2000. FAO, Bangkok, NACA and Rome, pp 167–191

United Nations, Department of Economic and Social Affairs, Population Division (2013) World Population Prospects: The 2012 Revision, Key Findings and Advance tables (Working Paper No. ESA/P/WP.227). Retrieved from the United Nations website: http://esa.un.org/wpp/Documentation/pdf/WPP2012_%20KEY%20FINDINGS.pdf

Xu H, Xing J, Tang X, Sheng X, Zhan W (2020) The effects of CCL3, CCL4, CCL19 and CCL21 as molecular adjuvants on the immune response to VAA DNA vaccine in flounder (*Paralichthys olivaceus*). Dev Comp Immunol 103:103492

Yang Y, Song W, Lin H, Wang W, Du L, Xing W (2018) Antibiotics and antibiotic resistance genes in global lakes: a review and meta-analysis. Environ Int 116:60–73

Zhou S, Dong J, Liu Y, Yang Q, Xu N, Yang Y, Gu Z, Ai X (2020) Anthelmintic efficacy of herbal medicines against a monogenean parasite, *Gyrodactylus kobayashii*, infecting goldfish (*Carassius auratus*). Aquaculture 521:734992

Specific Pathogen Free (SPF) Shrimps in Aquaculture

27

Suresh Eswaran

Abstract

Shrimp is one of the most traded commodities in the world aquaculture. The farmed shrimp production has surpassed the production of wild caught shrimps. The main reason attributed to this improvement is the domestication and development of Specific pathogen free (SPF) shrimps especially SPF Pacific white shrimp (*Penaeus vannamei*). SPF stocks are produced with stringent surveillance system including accurate diagnostic methods. SPF shrimp has many advantages over the non–SPF stocks. SPF brood stock development program was started in Hawaii and many other countries have also adopted to produce SPF stocks. The chapter discusses in detail about SPF shrimp in aquaculture.

Keywords

SPF · Shrimp · Quarantine · Aquaculture

27.1 Introduction

Shrimp is one of the most important crustacean species farmed in aquaculture throughout the world (Bondad-Reantaso et al. 2012). The commercially cultivated shrimp species are Pacific White Shrimp (*P. vannamei*), Western blue shrimp (*P. stylirostris*), and Green tiger shrimp (*P. monodon*) (Salunke et al. 2020). Among all these species, *P. vannamei* emerged as the most important and dominant species in the world. Occurrence of disease has become one of the biggest threats to shrimp farming industry for the past two decades (Karunasagar and Ababouch

S. Eswaran (✉)
Institute of Fisheries Postgraduate Studies, Tamil Nadu Dr. J. Jayalalithaa Fisheries University, Chennai, Tamil Nadu, India

© The Author(s), under exclusive license to Springer Nature Singapore Pte Ltd. 2021
P. K. Pandey, J. Parhi (eds.), *Advances in Fisheries Biotechnology*, https://doi.org/10.1007/978-981-16-3215-0_27

2012). Despite the disease problem faced by the industry, currently, the farmed shrimp production has surpassed the wild shrimp production. This has become possible mostly by the development of specific pathogen free (SPF) stocks. The production of SPF stocks was possible due to the domestication, advance disease diagnostic methods and bio-secure facilities (Lightner and Redman 2012). This chapter is dealing about the concept of SPF and the methodologies to develop SPF shrimp stocks.

27.2   Concept of SPF

SPF stocks can be defined as group of animals that have been tested negative for specific pathogens with most appropriate and accurate diagnostic methods for a period of at least 2 years, raised in strict biosecure facilities and fed with biosecure feeds (Wyban 2019). SPF status can be claimed only when those animals are under routine surveillance program for the specific pathogens including advanced diagnostic methods. Hence, SPF status indicates its sanitary status only and does not give any indication about its genetic background (Alday-Sanz 2018).

It is imperative to distinguish the differences among Specific Pathogen Free (SPF), Specific Pathogen Resistant (SPR) and Specific Pathogen Resistant (SPT). As mentioned earlier, SPF indicates its sanitary status only. SPF stocks meaning that they are free from specific pathogens but not necessarily free of all pathogens. Specific pathogen resistant (SPR) animals can be resistant to specific pathogens when they are exposed to. Specific pathogen tolerant (SPT) animals are tolerant to a specific disease. Still, they can be infected with pathogens but may not develop the disease (Alday-Sanz 2018).

27.3   Methodologies to Produce SPF Shrimp Stock

SPF stocks are produced with adequate bio-secure facilities, accurate diagnostic methods including histology, microbiology and molecular methods. There are no single accepted guidelines to produce SPF stocks throughout the world. The steps that are described here are general concepts to produce SPF shrimp stocks (Lightner 2011).

27.3.1 Identification of Candidate SPF Stocks

Identification of suitable candidate stock is the first step to produce SPF stock. The candidate stock can be either from wild or from farms. To begin the process of developing an SPF stock, a candidate stock of interest is identified (Lightner 2011).

27.3.2 Testing for Pathogens

Once identified, candidate SPF stocks have to be tested for specific pathogens (Table 27.1). Therefore, samples are taken from the candidate stocks and tested for specific pathogens with utmost appropriate and diagnostic methods. If none of the samples are found to be positive, the stocks/population is designated as founder population (F_0). If the samples are found to be positive, the stock is not selected to produce F_0 populations (Lightner 2011).

27.3.3 Primary Quarantine

The population that is designated as F_0 is collected and kept in primary quarantine for rearing. The rearing time of F_0 stock can be from 30 days to 1 year. The duration time in primary quarantine varies from species to species. While maintaining F_0 stocks in the primary quarantine, the stocks are periodically monitored for any sign of disease. In addition to that they are sampled and tested regularly to detect specific pathogens. If any pathogens of concern are detected during the test, the entire stock is destroyed. Those stocks that are found to be free from specific pathogens alone are transferred to separate secondary quarantine facility (Lightner 2011).

27.3.4 Secondary Quarantine

Specific pathogens free F_0 stocks are acquired from the primary quarantine and transferred to the secondary quarantine facility. They are reared in this facility to mature and bred to produce F_1 generation. The F_1 stocks are maintained in secondary quarantine for further testing for specific pathogens of concern. Then, F_1 stocks are tested for pathogens with the most appropriate and accurate pathogen detection

Table 27.1 OIE listed crustacean diseases (OIE 2020)

Sl. no.	Disease name	Pathogens name
1.	Taura syndrome	Acute hepatopancreatic necrosis disease
2.	Crayfish plague	*Aphanomyces astaci*
3.	Necrotising hepatopancreatitis	Hepatobacter penaei (necrotizing hepatopancreatitis)
4.	Infectious hypodermal and hematopoietic necrosis	Infectious hypodermal and haematopoietic necrosis virus (IHHNV)
5.	Infectious myonecrosis	Infectious myonecrosis virus
6.	White tail disease	*Macrobrachium rosenbergii* nodavirus (white tail disease)
7.	Taura syndrome	Taura syndrome virus (TSV)
8.	White spot disease	White spot syndrome virus (WSSV)
9.	Yellow head disease	Infection with yellow head virus genotype 1

methods. If F_1 stocks are confirmed that these are free from pathogens, they are designated as Specific Pathogen Free (SPF) stocks. Then, these are used to produce domesticated lines of SPF shrimp (Lightner 2011).

Once the stock is declared SPF of specified diseases/pathogens, it is of paramount importance to maintain the SPF status. Hence, the domesticated SPF stocks are subjected to routine surveillance program. The surveillance program should be scheduled and targeted. In addition to classical methods such as histopathology and microbiology, molecular tools have become indispensable to any surveillance program. If any stock to be claimed SPF status, molecular diagnostic methods are necessary to declare its status.

27.4   Advantages and Limitations of SPF

SPF offer many advantages over non-SPF stocks. SPF brood stocks are available throughout the year to produce seeds. They can also be utilized to conduct selection program to improve the important economic traits such as fast growth rate, disease resistance, feed conversion ratio, survival rate etc. SPF stocks are free from specific pathogens; therefore use of chemicals to control diseases can be reduced considerably. Since, SPF stocks are domesticated; they are able to mate for better reproductive success in captive conditions. The other advantage of the SPF is to find the origin of shrimp stocks through traceability (Barman et al. 2012).

Even though SPF stocks have become savior of the shrimp farming industry throughout the world, they also have some drawbacks. SPF stocks are free of specific pathogens for which they have been tested but they may have other viral pathogens despite the rigorous screening and may escape detection. For example some reports said that shrimps in Brazil and Colombia were contaminated with Taura Syndrome Virus (TSV) due to the shipment of SPF shrimp from Hawaii (Brock 1997). This was because, at the time, TS was not known to have a viral cause and therefore went unchecked in SPF protocols. In addition to that novel diseases may arise from mutations of previously non-pathogenic organisms, i.e., the highly mutable RNA viruses. Hence, there remains a possibility that importation of SPF shrimp may not rule out simultaneous importation of pathogens. Another complaint from the importing countries is that SPF stocks are bred with entirely consisting of sibling and produce inbred SPF stocks intentionally (Bierne et al. 2000). This means that future generations of animals will show poor performance with regard to many important traits compared to the control populations.

27.5   Status of SPF Around the World

Initially, the SPF brood stocks were produced and supplied by one country (Hawaii). Currently, other countries such as Florida, Thailand, Mexico, Singapore and Indonesia also joined to produce SPF brood stocks. USA and Madagascar produce

SPF *P. monodon* brood stocks. But, commercial farming of SPF *P. monodon* is yet to take place (Barman et al. 2012).

27.6 Conclusion

Shrimp farming is going to continue its momentum in aquaculture industry. Shrimp farming reached its current stage by leap and bound despite facing many problems. Disease has been and is going to be the biggest threat to the industry. Without any doubt, SPF has helped to produce the shrimp where it has surpassed the wild shrimp production. But SPF alone cannot help industry for the sustainable production. Since, new diseases are emerging often in the shrimp farming industry; a long-term approach is required to combat the diseases. Farmed shrimp has huge potential for genetic improvement (Nguyen 2015). Therefore, development of improved shrimp line should be given top priority. Genetic improvement can be attained with classical selective breeding and advanced technologies like genome-based genetic selection and more the advanced genome editing tools like CRISPR/Cas9 (Clustered Regularly Inter-spaced Short Palindromic Repeats) and TALEN (Transcription activator-like effector nucleases) (Elaswad and Dunham 2018). These technologies will ultimately lead to improved growth rate, disease resistance and other economic traits in aquaculture species.

References

Alday-Sanz V (2018) Specific pathogen free (SPF), specific pathogen resistant (SPR) and specific pathogen tolerant (SPT) as part of the biosecurity strategy for Whiteleg shrimp (*Penaeusvannamei* Boone 1931). Asian Fish Sci 31S:112–120

Barman D, Vikash K, Suvra R (2012) Specific pathogen free shrimps: their scope in aquaculture. World Aquaculture: 67–68

Bierne N, Beuzart I, Vonau V, Bonhomme F, Bedier E (2000) Inbreeding depression in a Penaeid shrimp. Aquaculture 184:203–219

Bondad-Reantaso MG, Subasinghe RP, Josupeit H, Cai J, Zhou X (2012) The role of crustacean fisheries and aquaculture in global food security: past, present and future. J Invertebr Pathol 110:158–165

Brock JA (1997) Special topic review: Taura syndrome, a disease important to shrimp farms in the Americas. World J Microbiol Biotechnol 13:415–418

Elaswad A, Dunham R (2018) Disease reduction in aquaculture with genetic and genomic technology: current and future approaches. Rev Aquacult 10(4):876–898

Karunasagar I, Ababouch L (2012) Shrimp viral diseases, import risk assessment and international trade. Indian J Virol 23(2):141–148

Lightner DV (2011) Status of shrimp diseases and advances inshrimp health management. In: Diseases in Asian Aquaculture VII, pp 121–134

Lightner DV, Redman RM (2012) Development of specific pathogen-free (SPF) shrimp stocks and their application to sustainable shrimp farming. In: Infectious disease in aquaculture. Woodhead Publishing, Sawston, pp 277–317

Nguyen N (2015) Genetic improvement for important farmed aquaculture species with a reference to carp, tilapia and prawns in Asia: achievements, lessons and challenges. Fish Fish 17 (2):483–506

OIE (World Animal Health Organization) (2020) Aquatic animal health code, 21st edn https://www.oie.int/animal-health-in-the-world/oie-listed-diseases-2020/

Salunke M, Kalyankar A, Khedkar CD, Shingare M, Khedkar GD (2020) A review on shrimp aquaculture in India: historical perspective, constraints, status and future implications for impacts on aquatic ecosystem and biodiversity. Rev Fish Sci Aquac. https://doi.org/10.1080/23308249.2020.1723058

Wyban J (2019) Selective breeding of *Penaeusvannamei*: impact on world aquaculture and lessons for future. J Coast Res 86:1–5

Enzymes from Aquatic Resources and Their Application in Food and Cosmetic Industry

28

R. Jeyashakila, B. Sivaraman, and G. Jeyasekaran

Abstract

Aquatic organisms are one of the important sources of enzymes. Marine animals, plants, invertebrates and microorganims are the major contributors. The rapid increase in the commercial utilization of aquatic resources for processing also generates huge quantities of solid wastes. Fish wastes are also valuable sources of many functional proteins and enzymes. Various seafood enzymes such as protease, lipase, chitinase, transglutaminase, isolated and purified from fish wastes find application in food and cosmetic industries. Other marine enzymes derived from aquatic organisms find application in cosmetic products are also described in this chapter.

Keywords

Seafood enzymes · Food industry · Cosmetic industry

28.1 Introduction

Enzymes are unique protein molecules produced by living organisms. They act as a cellular catalyst to bring out specific reaction very quickly in the cell. Enzymes find application in various industries such as paper, leather, textile, pharmaceuticals, chemicals, biofuels, detergents, animal feed, food and cosmetics (Adrio and Demain 2014).

Enzymes are discovered way back in 1833 as very special active proteins. Diastase is the first enzyme extracted as a thermo labile substance from

R. Jeyashakila (✉) · B. Sivaraman · G. Jeyasekaran
Department of Fish Quality Assurance and Management, Fisheries College and Research Institute, Tamil Nadu Dr. J. Jayalalithaa Fisheries University, Thoothukudi, Tamil Nadu, India

© The Author(s), under exclusive license to Springer Nature Singapore Pte Ltd. 2021
P. K. Pandey, J. Parhi (eds.), *Advances in Fisheries Biotechnology*,
https://doi.org/10.1007/978-981-16-3215-0_28

Table 28.1 Classification of enzymes and their mode of action with examples

Enzyme classes	Action	Examples
EC 1—Oxidoreductases	Catalyze oxidation reduction reactions	Dehydrogenases, oxygenases, peroxidases, oxidases, hydroxylases, reductases, catalases
EC 2—Transferases	Transfer chemical group from one molecule to another or within a single molecule (e.g. a methyl or phosphate group)	Amino-transferase, phosphoryl-transferase, glycosyl-transferase, methyl-transferase, acyl-transferase, kinases
EC 3—Hydrolases	Use water to cleave a single molecule into two molecules	Esterases, phospholipases, glycosidases, amidases, peptidases, ribonucleases
EC 4—Lyases	Cleave molecules using a non-hydrolytic process, leaving double bonds or split molecules by adding groups to double bonds	Decarboxylases, hydratases, synthases, aldolases, dehydratases
EC 5—Isomerases	Change isometric structures by intramolecular rearrangements	Racemases, epimerases, cis-isomerases, trans-isomerases, intramolecular transferases
EC 6—Ligases	Create a covalent chemical bond to join two molecules using energy from ATP	ATP synthetases from C-O, C-S, C-N, and C-C bonds

malt that converts starch into sugar. In 1926, the protein nature of the enzyme is finally confirmed and urease was successfully extracted from Jack beans (Summer 1926). However, the new era began after the use of cell-free enzyme renin in cheese making.

Enzymes are made of 100 to 1000 amino acids arranged in a very specific and unique sequential order. The amino acids fold into a unique shape allowing the enzymes to carry out specific enzyme reaction. Enzymes work based on shape recognition by forming a complex with the substrate. The reaction proceeds when the reactive site of the enzyme locks into the substrate. For example, maltose is made up of two glucose molecules. The enzyme, maltase acts on maltose and releases the two glucose molecules. Maltase is very specific to maltose and breaks about 1000 maltose bonds per second.

Some enzymes consist only of proteins and no other chemical groups, while some enzymes require additional chemical groups known a cofactor. Cofactor can be either inorganic or an organic molecule (coenzyme). Specific cofactors are mainly metals such as zinc, iron, magnesium, copper, etc. Coenzymes are predominantly derivatives of vitamins.

The International Union of Biochemistry and Molecular Biology developed a nomenclature for enzymes and given the EC numbers. Each enzyme is described by a sequence preceded by "EC". There are six classes of enzymes and their mode of action with examples are given in Table 28.1.

28.2 Seafood Enzymes

Aquatic organisms survive in extreme environments like high pressure, wide temperatures, salinity, dissolved oxygen, etc. and hence form a unique source of enzymes. The rapid increase in the commercial fish processing activities generates huge quantities of solid wastes. Solid wastes usually represent 20–60% of the initial raw material and constitute whole waste fish, fish heads, viscera, skin, bones, scales, swimbladder, fins, blood, liver, gonads and some muscle tissue (Awarenet 2004). Fish wastes are valuable resources of proteins, enzymes, lipids and minerals having high biological value (Kim et al. 2006). Various seafood enzymes such as protease, lipase, chitinase, and transglutaminase have been isolated and purified from fish wastes. The important enzymes extracted and purified from seafood are briefly explained below:

28.2.1 Nucleotide Degrading Enzymes

ATP nucleotide present in fish tissue initially undergoes autolytic degradation to form hypoxanthine (Hx) and later degraded by bacterial enzymes to uric acid. Enzymes involved in ATP degradation pathway are ATPase, AMP deaminase, 5-nucletotidase, nucleotide phosphorylase, inosine nucleosidase, and xanthine oxidase (Gill 2000). AMP deaminase (EC 3.5.4.6) that converts AMP to IMP isolated from the skeletal muscle of cod (Dingle and Hines 1967, 1971) and thorn back ray (Makarewicz 1969). The enzyme, 5-nucletotidase (EC 3.1.3.5) responsible for the conversion of IMP to inosine extracted from haddock, halibut, Atlantic cod, Alaska pollack, American plaice and winter flounder (Dingle and Hines 1971). It is a membrane bound enzyme present in some fish species, while in other fish, it is found in cytosol. Nucleotide phosphorylase (EC 2.4.2.1) and inosine nucleosidase (EC 3.2.2.2) that catalyze the conversion of inosine to hypoxanthine are found in Pacific lingcod (Tarr 1958) and Atlantic cod (LeBlanc 1987). Xanthine oxidase (EC 1.1.3.22) that converts hypoxanthine to xanthine found in a variety of fish species.

28.2.2 Myosin ATPases

Myosin ATPase activity occurs in pre-rigor and rigor mortis stages in fish, during the onset of spoilage. Muscle myosin ATPase isolated from several fish species such as cod, pollack, tunas, tilapia, carp, trout, mackerel, yellowtail, milkfish and shark (Ochiai and Chow 2000).

28.2.3 Phospholipases

Lipases that degrade phospholipids are grouped into two categories—acylhydrolases and phosphodiesterases. Phospholipase (PLA1) splits sn-1 fatty acyl ester bond, Phospholipase 2 splits sn-2 fatty acyl ester bond, Phospholipase B splits both or remaining ester in lysophospholipid and Phospholipase C attacks diglyceride-phosphate link and Phospholipase D splits choline-phosphate link. PLA1 isolated from the muscle of bonito (*Euthynnus pelamis*) (Hirano et al. 1997) catalyzes hydrolysis and transacylation. PLA2 isolated muscle of pollack *Pollachius virens* (Audley et al. 1978), pyloric caeca and hepatopancreas of red sea bream *Pagrus major* (Ono and Iijima 1998), liver of trout Salmon Gairdner (Neas and Hazel 1985) and muscle of cod *Gadus morhua* (Aaen et al. 1995).

28.2.4 Lipases

Members of lipases are lipoprotein lipase (LPL), hepatic lipase (HL) and pancreatic lipase (PL). In addition, pancreas synthesizes another lipolytic enzyme that catalyzes the hydrolysis of acylglycerols, phospholipids, cholesterol esters and vitamin esters known as bile salt activated lipases (BAL). Fish possess mammalian type PLs that are isolated from pancreas of rainbow trout *Oncorhynchus mykiss* (Leger et al. 1977) and from hepatopancreas of sardine, *Sardinella longiceps* (Mukundan et al. 1986). A homologue of BAL reported in fish species such as anchovy, striped bass, pink salmon, leopard shark, rainbow trout and dog fish (López-Amaya and Marangoni 2000). Fish lipases are also purified from pyloric caeca of cod (Gjellesvik et al. 1992), hepatopancreas of red sea bream (Iijima et al. 1998) and pancreas of Atlantic salmon (*Salmo salar*) (Gjellesvik et al. 1994).

28.2.5 Transglutaminases

Transglutaminase (TGase), otherwise known as glutamine γ glutamyl transferase (EC 2.3.2.13) is a transferase that catalyzes the acyl transfer reaction between γ carboxyamide groups of glutamine residues in proteins, peptides and primary amines. TGase is found in tissues of mammals, birds, fish, shell fish, microorganism, and plants. TGase purified from muscle of carp (*Cyprinus carpio*) and liver of red sea bream (Kishi et al. 1991; Yasueda et al. 1994). In surimi-based seafood products, the network formation of myosin due to crosslinking by endogenous TGase is responsible for "setting" and "suwari" formation (Kamath et al. 1992). "Setting" strengthens the texture of surimi-based products.

28.2.6 TMAO Degrading Enzymes

Trimethylamine oxide (TMAO) is an osmoregulatory non-protein nitrogenous compound found mostly in marine fish. TMAO is either synthesized or degraded by the enzymatic reaction in fish. TMA monooxygenase that catalyzes the synthesis of TMAO from TMA is present in methylotropic bacteria in calanoid copepods and in some marine fish (Sotelo and Rehbein 2000). TMA monooxygenase found in liver of herring, rainbow smelt and two Antartic fish species (Raymond and De Vries 1998; Raymond 1998). TMA dehydrogenase (EC 1.5.99.7) that catalyzes the oxidative demethylation of TMA to DMA and FA present in methylotropic bacteria. TMAO reductase that converts TMAO into TMA (a compound responsible for the off odour in fish) present in spoilage bacteria such as *Shewanella, Alteromonas, Photobacterium* and *Vibrio* (Regenstein et al. 1982; Huss 1988). TMAO demethylase (EC. 4.1.2.32) that degrades TMAO into DMA and formaldehyde (FA), in the absence of oxygen is abundant in fish kidney and viscera. TMAO demethylase identified in several species of marine fish and marine invertebrates, mainly mollusks such as cod, saithe, haddock, ling, blue ling, blue whiting, Alaska pollack, cusk, whiting, red hake, herring, cucumber fish, blue crab, bivalve, and squid (Sotelo and Rehbein 2000).

28.2.7 Proteinases

Proteinases catalyze the cleavage of peptide bonds with the participation of water molecules as reactants. Digestive proteinases from marine animals are classified on the basis of their similarity to proteinases as trypsin-like, chymotrypsin-like, chymosin-like and cathepsin-like; and on the basis of pH sensitivity as acid, neutral or alkaline proteinases (Simpson 2000). As per International Union of Biochemists (IUB), marine proteinases are classified as acid or aspartate proteinases, serine proteinases, thiol or cysteine proteinases or metalloproteinases.

Acid or aspartyl proteinases have their catalytic sites composed of carboxyl groups of two aspartic acid residues. Three common aspartyl proteinases isolated from the gut of marine organisms are pepsin (EC 3.4.23.1), chymosin (EC 3.4.23.4), and gastricin (EC 3.4.23.3). Pepsin and pepsin-like enzymes extracted from the digestive glands of marine animals such as Atlantic cod (Martinez and Olsen 1989), capelin (Gildberg and Raa 1983), Greenland cod (Squires et al. 1986), Polar cod (Arunchalam and Haard 1985), sardine (Noda and Murakami 1981) and dogfish (Merrett et al. 1969). Chymosin-like enzymes isolated from gastric mucosa of yong and adult harp seals (36). Gastricins obtained from gastric juices of salmon (Fruton and Bergmann 1940) and hake (Sanchez-Chiang and Ponce 1981).

Serine proteinases have a serine residue, together with an imidazole and an aspartyl carboxyl group in their catalytic site. Serine proteinases are present in the pancreatic juices, pyloric ceca and intestines of animals. Three major serine proteinases purified from the digestive glands of marine animals are trypsin (EC 3.4.21.4), chymotrypsin (EC. 3.4.21.1), and elastase (EC 3.4.21.11). Trypsin isolated from several marine animals such as anchovy (Martinez et al. 1988), Greenland cod (Simpson and Haard 1984a, b), Atlantic cod (Asgeirsson and Bjarnasson 1989), capelin (Sanchez-Chiang and Ponce 1981), mullet (Guizani et al. 1991), sardine (Noda and Murakami 1981), catfish (Yoshinaka et al. 1984), crayfish (Pfleiderer et al. 1967) and cunner (Simpson and Haard 1985). Chymotrypsins isolated from several marine fish species like anchovy, Altantic cod (Heu et al. 1995), capelin and herring (Kalac 1978), rainbow trout (Kristjansson and Nielson 1992) and spiny dogfish (Racicot 1984). Elastase characterized from Altantic cod (Raa and Walther 1989), carp (Cohen et al. 1981), catfish, eel, bluefin tuna, seabass, stingray (Yoshinaka et al. 1985), Dover sole (Clark et al. 1985), monkfish (Lansing et al. 1953), tuna (Zendzian and Bernard 1967), and yellowtail (Yoshinaka et al. 1985). Collagenases purified from the digestive glands of crabs, lobsters and prawn (Eisen et al. 1973; Nip et al. 1985).

Thiol proteinases have cysteine and histidine residues in their catalytic sites. Cathepsin B (3.4.22.1) isolated from the digestive glands of marine animals. Thiol proteinases found in hepatopancreas of marine animals and more than 90% of activity is characterized in short-finned squids. Cathepsin B isolated from horse clam (Reid and Rauchert 1976), mussel (Zeef and Dennison 1988), and surf clam (Chen and Zall 1986). Metalloproteinases (EC 3.4.24) contain one tyrosine residue and one imidazole residue in their catalytic sites. They are found in the muscle tissue of rockfish (Warrier et al. 1985), carp (Makinodan et al. 1979) and squid (Okamoto et al. 1993).

28.2.8 Polyphenoloxidase

Polyphenoloxidase, PPO (EC 1.10.3.1) is responsible for enzymatic browning in fruits, vegetables and seafood, especially crustaceans. PPO causes a type of discoloration called melanosis in lobsters, shrimps and carbs. PPO is a copper containing enzyme having two Cu binding sites, CuA and CuB. PPOs are localized in hemolymph in an inactive form, which is activated by proteases, lipids, or polysaccharides (Kim et al. 2000). PPO is abundant in the gills of crustaceans, which contain hemolymph or blood. PPOs purified from blood cells of crayfish *Pacifastacus leniusculus* (Aspán and Söderhäll 1991), American horseshoe crab (Aspán et al. 1995), and spiny lobster (Gaykema et al. 1984).

28.2.9 Lipoxygenases (LOXs)

Lipoxygenase (EC 1.13.11.12) catalyzes the incorporation of oxygen at 5, 12, or 15 positions in polyunsaturated fatty acids such as arachidonic acid (ARA) to form hydroperoxide derivatives. LOXs are present in platelet, leucocyte, reticulocyte, basophil and neutrophil in animal tissues (Pan and Kuo 2000). In marine animals, LOX activity found in skin and gill of trout (German and Kinsella 1985), ayu (Zhang et al. 1992), and sardine (Mohri et al. 1990) and in eggs of starfish (Meijer et al.

1984), sea urchin (Hawkins and Brash 1987) and grey mullet. LOX activity in fish skin and gill is responsible for the flavour in fresh and processed fish. LOX activity is found in hemolymph, but not in muscle, gut and hepatopancreas of fish. Three types of LOX such as 5-LOX, 12-LOX, and 15-LOX are found in the tissue of shrimp and grey mullet. 12-LOX has the highest activity than 5 or 15-LOX. 12-LOX found in various organs of rainbow trout (Knight et al. 1995). LOXs are present in freshwater green algae, *Oscillatoria* sp. (Beneytout et al. 1989), marine microgreen algae, *Enteromorpha intestinalis* (Kuo et al. 1996), *Ulva lactuca* (Kuo et al. 1997), and *Ulva conglobata* (Hu and Pan 1999), marine red algae (Moghaddan and Gerwick 1990) and micro algae, *Chlorella* (Zimmerman and Vick 1973). LOXs impart sea weedy fresh fish flavour and produce antimicrobial substances.

28.3 Application of Enzymes in Food Industry

Food and feed remain the largest industrial sectors for the use of enzymes. In food industries, enzymes are first used in making cheese and fruit juices, since 1930. The major breakthrough started around 1960s, when enzymes are manufactured on a large scale for use in starch industry. The traditional acid hydrolysis was completely replaced by alpha amylases and glucoamylases to convert starch into glucose.

The use of seafood enzymes in various applications emerged as a new subject for food scientists, particularly food biochemists in recent decades. Seafood enzymes find application in improvement of food flavour/texture and as an indicator of quality.

28.3.1 Enzymes as Indicators of Quality

Adenosine triphosphate (EC 3.6.1.3), ATPase activity is associated with pre-rigor and onset of rigor mortis. ATP content in pre-rigor fish varies from 0.69 to 5.19 mg% P (average 2.4 mg%P), while in post-rigor fish it occurs in small amounts 0.7 to 0.8 mg%P. Measurement of ATPase activity serve as an index for assessing the freshness of specific fish and fishery products such as red fish, tilapia, and Nova Scotia mackerel. Mg^{2+} ATPase activity is an accurate index of change in quality of chill and frozen stored fish (Yamanka and Mackie 1971; Seki and Narita 1980). Ca^{2+} ATPase activity is also used to evaluate the quality of frozen surimi from Alaska Pollack (Katoh et al. 1979; Kawashima et al. 1973).

Lactate dehydrogenase, LDH (EC 1.1.127), a key enzyme in glycolytic pathway is found in many species of fish, which accumulates lactic acid. LDH isoforms found in skeletal muscle of tuna and flounder, flatfish (Cahn et al. 1962), tilapia (Curran et al. 1986), mrigal, tilapia, pearl spot, milk fish and mullet (Nambudiri and Gopakumar 1990). The protein denaturation in frozen fish caused by the accumulation of free fatty acids later leads to disappearance of LDH activity. So, the measurement of LDH activity is an indication of protein denaturation in frozen fish. Several commercial LDH enzyme assay kits are also available.

Lysosomal and mitochondrial enzymatic activities can differentiate fresh (unfrozen) fish from frozen thawed fish. Enzymes that play major role in fish muscle damage are alpha glucosidases (EC 3.2.1.20), beta N-glucosaminidases (EC 3.2.1.30) and acid phosphatases (EC 3.1.3.2). Glutamate oxaloacetate transaminase (GOT) activity present in mitochondria of carp fillets can differentiate fresh fish fillets (Chhatbar and Velankar 1977). Malate dehydrogenase is an excellent index to determine fresh and frozen cod and haddock (Hamm and Masic 1971). Cytochrome oxidase activity and lysosomal glycose oxidase increased several folds in frozen and thawed trout (Barbagli and Crescenzi 1981; Yoshioka 1985). Lipoamide reductase and 5'AMP deaminase activity increased in frozen and thawed farmed fish and shellfish (Janicke and Knopf 1968). The measurement of lysosomal enzyme activities is however not adopted for routine testing of fish quality.

28.3.2 Enzymes as Flavour Biogenesis

Lipoxygenases are intentionally added to foods for the purpose of bleaching of pigments and/or for the generation of aroma compounds. LOXs are used along with other enzymes. LOXs isolated from shrimp hemolymph and grey mullet produce seafood aroma due to the presence of marine PUFA (Pan and Kuo 1994). The use of fish LOX as an industrial biocatalyst is limited due to its instability. Addition of imidazole compounds had increased the stability of fish gill LOX (Hsu and Pan 1996). The use of fish-derived LOXs for indirect degradation of carotenoid pigments to produce cooked salmon-like aroma compounds is very promising (Josephson and Lindsay 1986).

28.3.3 Enzymes in Improving Seafood Texture

Cathepsins and calpains are involved in meat tenderization process. Cathepsins B and L of salmon cause muscle softening during spawning migration (Yamashita and Konagaya 1990). Heat-stable alkaline proteinases (HAP) are responsible for "modori-inducing" or "gel-softening" in some fish (Makinodan and Ikeda 1971). Modori-inducing proteinases are present in sarcoplasmic fraction of fish (Toyohara et al. 1990). It is interesting that 80% of calpains, cathepsins B, L and L-like activities remained in frozen mackerel surimi or surimi gel (Jiang et al. 1994). So, the presence of residual proteinases affects the texture of surimi-based products. The soft texture of Pacific whiting, arrow tooth flounder, menhaden white croaker and mackerel are also associated with proteolytic activity.

Transglutaminase (TGase) improves the functionality of casein, beta lactoglobulin and soybean proteins. TGase incorporates essential amino acids and fortifies food proteins. The formation of cross links between different food proteins is very promising in the development of new protein foods, including surimi-based products and edible films. TGase present in fish muscle and surimi wash water influences the texture of cooked surimi. TGase from Alaska Pollack induces gelation of minced fish (Jiang et al. 1994).

28.3.4 Enzymes in Krill Processing

Krill is subjected to partial auto proteolysis to obtain high yield and quality food grade product. Fresh krill is mixed with water (3:1 w/v) for 15–90 min at 12–55 °C to decrease the viscosity, decanted into shell free solution, hydrolysate and chitin shell. Thermal coagulation of protein proceeds at 92–93 °C for 3–5 min, which is then chilled, strained, homogenized and frozen into blocks. Catechol oxidase (EC 1.10.3.1) present in crustaceans is responsible for the darkening process (Kolakowski and Sikorski 2000). Lipases/phospholipases present in alimentary tract of krill causes extensive hydrolysis of phospholipids than triacylglycerols (Ellingsen and Mohr 1981). Krill lipids are resistant to autoxidation despite having high PUFA content, due to the interaction of phospholipids with hydroperoxides (Kolakowski 1988).

28.3.5 Enzymes in Speciality Products

Seafood enzymes are used in preparation of speciality products such as surimi. The quality of surimi products enhanced by the activity of endogenous heat-stable proteinases due to the gel-softening effect. Another important step to produce high quality gel from surimi is the formation of ε-γ-glutamyl lysine cross links in fish proteins by using TGase. TGase modifies the functional properties of surimi, as they can form crosslink with other proteins such as whey protein, meat protein, soybean protein, and gluten. Modification of food proteins by TGase results in many textured products like hamburger, meatballs, canned meat, frozen meat, moulded meat, surimi fish paste, krill paste, baked foods, protein powders, etc. to protect lysine during food production (Díaz-López and García-Carreño 2000).

28.3.6 Enzymes in Enriching PUFA of Fish Oils

Lipases (EC 3.1.1.3) are used to modify marine oils to produce PUFA enriched fish oil (Haraldsson 1990). Lipases isolated from Atlantic cod hydrolyse PUFAs over short-chain acids, which is a unique property. Currently lipases from microorganisms are used for preparing concentrates of n-3 PUFA, and not the ones from fish.

28.3.7 Enzymes in Caviar Production

Caviar is one of the major delicatessen type salted fish products. Caviar refers to riddled and cured roe of sturgeon (*Acipenser* spp. and *Huso huso*). Black caviar is prepared from the eggs of sturgeon and red caviar from salmon. Riddling process is done manually or mechanically. The release of the roe from the supportive connective tissue is done using enzymes. Proteinases release salmon roe from the connective tissue (Sugihara et al. 1973). Cold-adaptive Atlantic cod pepsin gave good results. Cold-active fish enzymes remove mucoprotein layer on newly spawned eggs from walleye and catfish (Gildberg 1993).

28.3.8 Enzymes in Cured Fish Products

Proteolytic enzymes are used in preparation of ripened salted fish, mainly cured herring and cod, and fermented squid. Traditional cured fish products are prepared by partial proteolysis catalyzed by endogenous muscle proteases that aid in tissue degradation during tenderization (Shenderyuk and Bykowski 1990). Trypsin-like enzymes from pyloric caecum of herring and cod contribute for the endopeptidase activity in the ripening process (Ruiter 1972; Simpson and Haard 1984a, b). Squid liver accelerates autolytic fermentation and tenderization of brined Atlantic short-finned squid (Lee et al. 1982; Kolodziejska et al. 1992). Proteases of squid liver extract contains collagenolytic enzymes, which make the cooked squid mantle tender (Kolodziejska et al. 1992).

28.3.9 Enzymes in Protein Hydrolysates

Protein hydrolysate is a heterogeneous mixture of polypeptides and free amino acids made by digestion of whole fish or by-products using proteolytic enzymes, of both endogenous and exogenous origin. Liquid fish hydrolysates prepared from small fish are available in ancient Rome and parts of Southeast Asia. Fish protein hydrolysates (FPH) are used as a source of peptone in the growth media for microorganisms (Raa and Gildberg 1976) in Canada, Norway and Japan. Substrate for FPH production include by-products from fish filleting, by-catch fish from trawlers and small underutilized lean fish species or krill (*Euphausia superba*). The first digestion step is by autolysis of acidified fish with endogenous pepsin and peptidases and the second step is by the addition of exogenous proteases under alkaline condition. Commercial proteases used are pepsin, chymotrypsin, and trypsin of animal origin; bromelain, papain and ficin of plant origin; and alcalase and flavorzyme of microbial origin. FPH produced from Caspian sprat by partial hydrolysis using endogenous proteases (Sikorski and Naczk 1981) and from krill by proteolytic enzymes containing a mixture of endopeptidases and exopeptidases (Karlstam et al. 1991). The main use of FPH is in dietetic foods such as macaroni or bread and in fish soup, fish paste and shellfish analogues, as flavouring compounds. The major feed application is as milk replacers for calves and pigs; and as protein source and attractant in fish feed. FPH stimulates the immune system of animals fed with FPH.

Bitterness is a common problem with FPH, which is due to the formation of peptides containing bulky hydrophobic groups towards their C-terminal. The intensity of bitterness depends on the degree of hydrolysis and specificity of enzymes (Bárzana and García-Garibay 1994). Glutamic acid, polyphosphates, gelatin and glycine are used to mask the bitterness (Roy 1992). Enzymatic treatment using exopeptidases leads to re-synthesis of "plastein" from peptides and amino acids by transpeptidation (Sikorski and Naczk 1981). Plastein formation is facilitated by proteolytic enzymes such as trypsin, chymotrypsin, and pepsin at 37–50 °C (Arai and Fujimaki 1978) and similar cold-adaptive enzymes at low temperature (Haard and Simpson 1994). Cold-adaptive enzymes have narrow specificity for peptide bonds with relatively high activity and hence suggested to improve the pleasant tasting amino acids in FPH (de Vecchi and Coppes 1996).

28.3.10 Enzymes as Seafood Flavorings

Flavour potentiators emphasize meaty flavour in soup, sauces, gravies and other savoury products. They occur naturally in sea tangle (*Laminaria japonica*), and dried bonito tuna. Fresh fish rich in $5'$-AMP is converted into $5'$-IMP by the action of AMP deaminase released during processing. Seafood flavours are used as additives in kamaboko, artificial crab, fish sausage and extrusion fish products (Kawai 1996). Flavours from fish, shellfish (shrimps), and mollusks (oysters, clams) are recovered by the use of endogenous enzymes (Díaz-López and García-Carreño 2000).

28.3.11 Enzymes in Fish Sauces

Fish sauce is enzymatically solubilized and digested fish protein used as a food ingredient and condiment on vegetable dishes. Fish sauce is made from small pelagic fishes such as anchovies and sardines. Fish fermentation involves salting, enzyme hydrolysis and bacterial fermentation. Trypsin and chymotrypsin are involved in tissue solubilization and protein digestion in sauce (Orejana and Liston 1981). Collagenases and elastases are important in the initial digestion of connective tissues (Gildberg and Øverbø 1990; Kristjansson et al. 1995). Cathepsin D is not involved in tissue solubilization during fish sauce fermentation (Reddi et al. 1972). Proteases from squid hepatopancreas used in fish sauce produces taste influencing amino acids and peptides with more flavour in less fermentation time (Raksakulthai and Haard 1992).

28.3.12 Enzymes as Fish Processing Aids

28.3.12.1 Deskinning

Mechanical removal of fish skin is difficult and often damages the fish fillets. A combination of chemical and enzymatic treatment helps in solubilisation of collagenous skin tissue. Deskinning is done by exposing the fish in acetic acid (5% for 5 min) first to loosen the scales and then denaturing the skin collagen at 15–20 °C. The skin of skate wings removed by the mixture of proteases and carbohydrases of fish viscera and the skin of squid by endogenous enzymes containing cathepsin D and B like proteases (Sakai et al. 1981) and also by collagenases of crab hepatopancreas (Sikorski et al. 1995).

28.3.12.2 Descaling

Descaling of certain fish is a problem as the process damages the skin and gives low fillet yield. Fish like haddock, hake, ocean perch, redfish and salmon fetch more price, if scales are thoroughly removed. As mechanical scale removal takes long time, enzymatic descaling is preferred. Descaling is done in a special mixture of fish digestive enzymes (Haard and Simpson 1994; Sikorski et al. 1995).

28.3.12.3 Membrane Removal

Salted cod swim bladder is a product of Iceland having a major market in Italy. Cod swim bladder is generally covered in a thin black membrane, but a white salted product is more preferred. Pepsin from cod and collagenases from crab hepatopancreas used for the removal of black membrane (Haard 1997). A thin collagenous membrane surrounding the cod liver can also be removed by enzymes prior to canning (Stefánsson and Steingrímsdóttir 1990).

28.4   Application in Other Food Sectors

The main use of seafood enzymes is in dairy technology, meat tenderization and juice clarification. Trypsin from stomachless fish has a special ability to digest native proteins than those derived from functional stomach.

28.4.1 Dairy Technology

The major use of enzymes is in the coagulation of milk to make cheese. The term "rennet" is used to describe all milk clotting enzyme preparations used in cheese making. The rennet is an aqueous extract made from the fourth stomach of bovine, sheep, and goats or microorganisms. Milk coagulation process has two stages; enzymatic and non-enzymatic. Milk clotting begins with enzymatic cleavage of chymosin bond of k casein at low pH to form para k casein and a macropeptide. The hydrolysis of k casein renders the casein micelles to coagulate in the presence of calcium at 10 °C. Non-enzymatic aggregation of altered casein micelles forms a gel structure (curd) and whey separates out at 30–39 °C. Cold-adaptive gastric proteases of Atlantic cod pepsin (Brewer et al. 1984) and tuna gastric enzyme (Tavares et al. 1997) are milk coagulating enzymes. An aminopeptidase from squid hepatopancreas used in Cheddar cheese ripening as it reduced bitterness and enhanced the cheese flavour (García-Carreño et al. 1997). Cold-adaptive pepsin is an excellent rennet substitute in the preparation of Cheddar cheese. Chymosin-like protease from the gastric mucosa of harp seal found successful in the preparation of Cheddar cheese (Shamsuzzaman and Haard 1983). The milk clotting enzymes from fish stomach mucosa or shellfish hepatopancreas are inexpensive alternative to rennet substitutes in cheese manufacture.

28.4.2 Prevention of Oxidized Flavour in Milk

Oxidized flavour is a common defect in dairy products. Peroxidase (EC 1.11.1.7) catalyzes oxidative reactions in milk and dairy products. Trypsin from cold-adapted fish has more thermal stability than bovine trypsin and it prevents the oxidized flavours in milk (Simpson and Haard 1984a, b).

28.4.3 Preparation of Infant Milk

Cow milk is humanized by the addition of lysozyme to make it an infant milk. Lysozyme acts as a preservative in reducing bacterial counts in milk, as an alternative to antibiotics Lysozymes from Arctic scallop viscera possess activity 300% higher than that of hen egg white lysozyme (Myrnes and Johansen 1994).

28.5   Seafood Enzymes in Cosmetic Industry

The global cosmetics industry is growing at a compounded annual growth rate (CAGR) of 5.5% and expected to reach USD 758.4 billion by 2025 (Ridder 2020) due to the demand for both body health and beauty care products. Asia Pacific region accounts for the largest market share of nearly 32%, due to large prospective population. The Indian cosmetics market currently stands at USD 6.5 billion and is expected to reach USD 20 billion by 2025 by registering an annual growth rate of 25%.

Enzymes are becoming popular in a variety of application in cosmetic products, including hair dye, skin care, oral care and sun care products. Cosmetics sector is interested in enzymes to enhance the beauty of the skin through improvement of skin appearance and prevention of skin problems. The use of coenzymes and cofactors in cosmetics promote efficient functions of the enzymes, as they are stable low molecular weight compounds that can easily penetrate through the stratum corneum and activate the enzymes. They are relatively easy to formulate into cosmetics as they offer a good degree of safety when applied topically.

Table 28.2 Enzymes used in cosmetic products and their use

Enzyme	Use
Proteases	Anti-aging/anti-wrinkle/peeling agent
Superoxide dismutase	Free radical scavengers, stability agent
Peroxidases	Free radical scavengers
Hyaluronidase	Moisturizing agent
Alkaline phosphatase	Anti-wrinkle agent
Catalase	Free radical scavengers
Lipases	Conditioning agent, skin cleaning agent
Glutathione transferase	Detoxifying agent

The first commercial enzyme used is trypsin isolated from animals to degrade proteins for use in detergents. Microbial protease derived from *Bacillus* spp. was earlier used in washing powders and later onwards large-scale commercialization began. Novozyme of Denmark started a huge business in 1959 by manufacturing detergents using microbial enzymes.

Recently enzymes are used by cosmetic scientists in developing personal care products (PCPs) because of good consumer appeal and improved performance. Proteolytic enzymes such as bromelain and papain are used in PCPs for skin peeling, and smoothening. Table 28.2 gives the list of enzymes used in cosmetic products and their use.

The enzymes used for skin protection have abilities to capture free radicals generated by environmental pollution, microorganisms, sunlight, radiations, etc. Enzyme formulations suitable for topical application are relatively easy to handle. Selection of base and surface-active agent is important to formulate a stable product. Another approach is to increase the efficiency of existing ingredients to improve skin functioning. Many new topical ingredients ranging from salmon caviar to sea urchin spines find application in anti-aging formulations (Draelos 2012). The current use of enzymes in different applications in cosmetics and PCPs are below:

28.5.1 Superoxide Dismutase

Superoxide dismutase (SOD, EC.1.15.1.1) is an oxidoreductase enzyme that catalyzes oxidation and reduction reactions. SOD is ubiquitous in nature and classified on the type of metal bound at the active site and protein folding into four types, namely Cu/Zn-SOD, Fe-SOD, Mn-SOD, and Ni-SOD. Cu/Zn-SOD and Mn-SOD present in marine prokaryotes and eukaryotes; Fe-SOD in prokaryotes, protozoans, algae and plants, and Ni-SOD purified from several soil bacteria of *Streptomyces*. The isoforms are distinguished based on their sensitivity to chemicals. Cu/Zn-SOD is very sensitive to H_2O_2 and cyanide, Mn-SOD is insensitive to H_2O_2 and cyanide, and Fe-SOD is very sensitive to H_2O_2 but not inhibited by cyanide. More precisely, Fe-SOD and Mn-SOD are present in the mitochondria, peroxisomes, and chloroplasts of prokaryotes and protists (Zeinali et al. 2015).

Table 28.3 Superoxide dismutase identified from marine sources

Kingdom	Source
Marine bacteria	*Photobacterium leiognathid and Photobacterium sepia, Synechococcus, Nodularia, Aphanizomenon, Anabaena, Geobacillus*
Marine Chromista	*Lingulodinium polyedrum, Minutocellus polymorphus, Thalassiosira weissflogii, Nitzschia closterium*
Marine Plantae	*Porphyrium cruentum, Tetraselmis gracilis, Bruguiera gymnorrhiza, Avicennia marina, Enteromorpha linza, Sonneratia alba*
Marine Fungi	*Debaryomyces hansenii, Cryptococcus* sp.
Marine Animalia	*Lampanyctus crocodilus, Xiphias gladius, Gadus morhua, Prionace glauca, Mytilus edulis, Crassostrea gigas, Litopenaeus vannamei, Haliotis discus, Alvinella pompejana, Apostichopus japonicus*

SOD is the most effective enzyme used in skin care products. Mann and Leilin discovered it as a green protein in 1938, while McCord and Fridovich discovered its enzymatic activity in 1969. Loreal is the first cosmeceutical company to obtain an EU patent for the use of SOD in cosmetics (EU patent No. 2/287 in 1973).

SOD acts as an antioxidant and protects the cellular components from oxidation by reactive oxygen species (ROS). ROS are produced by cells during normal aerobic respiration during oxidative phosphorylation, and also as a result of injury, ozone, nutrient deficiencies, photo-inhibition, temperature abuse, toxic metals, and UV/gamma rays. Molecular oxygen is reduced to O_2^- (superoxide, a ROS), when it absorbs an excited electron released from compounds of the electron transport chain. The excessive amounts of superoxide are extremely toxic to cells and can damage the macromolecules in cells by denaturing enzymes and proteins, oxidizing unsaturated fatty acids of membrane lipids, fragmenting DNA, and damaging cell membranes. SOD catalyzes dismutation of superoxide (O^{2-}) radical into ordinary molecular oxygen (O_2) and hydrogen peroxide (H_2O_2).

$$2O_2^- + 2H^+ \rightarrow O_2 + H_2O_2$$

The enzymatic activity depends on the body temperature, pH and presence of co-enzymes. The hydrogen peroxide formed in dismutation process is degraded by enzymes, catalase and peroxidases into water. So, the combined action of SOD and catalase keep the levels of superoxide anion and H_2O_2 low. Solar radiation, mainly UV exposure and oxidative stress on the skin cause age-related loss of skin elasticity (Naylor et al. 2011) and defective cellular signalling and photoaging (Lee et al. 2012). SOD binds directly to collagen and protects the breakdown of type I collagen due to oxidative stress.

Marine SODs are derived mainly from Bacteria, Chromista, Plantas, Fungi and Animalia (Zeinali et al. 2015). Table 28.3 provides the different sources of SODs from marine organisms. The SODs are identified in a number of marine bacteria and they mainly belong to Fe-SODs, with few exceptions. Marine dinoflagellate, *Lingulodinium polyedrum*, diatom, *Thalassiosira weissflogii*, a unicellular red alga, *Porphyridium cruentum* possess Mn-SOD activity. A cold-adaptive SOD purified from *Enteromorpha linza*. Cytosolic Cn/Zn-SOD isolated from marine yeast, *Debaryomyces hansenii*. Cu/Zn-SOD isolated from bathophilic teleost (*Lampanyctus crocodilus*), liver of swordfish (*Xiphias gladius*), and an elasmobranch, shark (*Prionace glauca*). Mn-SOD found in rainbow trout (*Salmo irideus*), skin of plaice (*Paralichthys olivaceus*) and hepatopancreas of Japanese flounder (*P. olivaceous*). Mn-SOD activity of carp liver is comparatively higher than in other tissues, while in saithe SOD activity is higher in heart, spleen and roe. The swimbladder of fish species also exhibited high level of SOD activity. All marine crustaceans such as blue crab (*Callinectes sapidus*) and shrimp (*Litopenaeus vannamei*) have cytosolic Mn-SOD, but lack Cu/Zn-SOD. Blue mussel, *Mytilus edulis* contains Cu/Zn-SOD in digestive glands and gills.

28.5.2 Proteases

Proteases are hydrolases that break peptide bonds in proteins. They are used as skin exfoliating agents, mainly to remove old and dead cells and protein fragments from the skin. Cysteine type proteases used in cosmetics are bromelain from pineapple, papain from papaya, actinidin from kiwi and ficin from ficus latex. Serine type proteases are also derived from hatching fluid of salmon known as Aquabeautine XL. Proteases extracted from microalgae, *Nanochloropsis, Phaeodactylum* and *Tetraselmis* are also used in cosmetics.

28.5.3 Lipases

Lipases (triacylglycerol acylhydrolases EC 3.1.1.3) are serine hydrolases that catalyze hydrolysis of triacylglycerols (TAGs) into glycerol and fatty acids. They act as skin cleansers, as it degrades residual oil and makes the skin clean. It acts as a conditioning agent in facial masks, skin cleansers, skin creams and make up removers. Marine fungi, *Aspergillus awamori* and *Rhodotorula mucilaginosa* are promising sources for lipase production. Several marine bacteria produce lipases. They are a thermophilic bacterium *Aeromonas* sp., psychrophilic bacteria *viz. Okhotskensis* sp., *Psychrobacter* sp., *Colwellia* sp., mesophilic bacterium, *Vibrio fischeri* and halophilic bacterium, *Halobacillus*. Lipases purified from hepatopancreas of oil sardine, intestine of tilapia, internal organ of grey mullet, liver on notothenioid fish, viscera of chinook salmon and hoki. Lipases also recovered from marine microalgae, *Nannochloropsis oculata and T-Isochrysis lutea, Phaeodactylum tricornutum, Tetraselmis* sp. and *Burkholderia* sp. (Navvabia et al. 2018).

28.5.4 Glutathione Transferase

Glutathione transferase is a detoxifying enzyme obtained from algae. It catalyzes conjugation of glutathione (GST) through the SH group to electrophilic centre on a wide variety of toxic endogenous or xenobiotic compounds. The conjugation helps to increase their solubility and removal of glutathione with limited toxicity. GST possesses glutathione-dependent peroxidase activity and helps to reduce organic hydroperoxides to alcohols, thereby preventing the formation of reactive carbonyl compounds. Hydroperoxides and reactive carbonyls play important role in ageing progression and other human diseases. GST isolated from photosynthetic algae, *Chlamydomonas reinhardtii* by Chatzikonstantinou et al. (2012).

28.6 Application of Seafood Enzymes in Cosmetic Industry

28.6.1 Enzymes as Anti-Aging Molecules

The superoxide dismutase in combination with catalase is responsible for the protection of proteins from aging due to oxidation. Proteolytic enzymes are frequently used in cosmetics to break down proteins and enable the skin to absorb other components for promoting cell growth and renewal. Papain from papaya and bromelain from pineapple are used as ingredients in cosmetic products for re-surfacing and skin smoothing. These enzymes are useful tools in treating skin aging, acne, congestion and pigmentation. There is a range of other enzyme-based beauty products available to promote the formation of fat and collagen. Diacylglycerol acyl-transferase (DGAT-1) enhances the action of retinoic acid, which accelerates epidermis and hair renewal. Lysyl and prolylhydroxylases help in the synthesis of collagen necessary for the structure of skin along with vitamin C.

28.6.2 Enzymes as Exfoliant

Proteases are good chemical exfoliants, as they dissolve and remove dead cells from the surface of the skin. They are used in combination with other chemical peels, including certain vitamins. To treat oily skin prone for acne, they are used in combination with salicylic acid (e.g. butylated hydroxy anisole). To treat sun-damaged and un-even toned skin, they are used in combination with alpha hydroxy acids (e.g. retinoic acid). SOD is also used with peroxidase as free radical scavengers in cosmetic products. It has an ability to reduce UV-induced erythema. SOD is extracted from yeast and peroxidase is present in aqueous extracts of fennel. Peroxidase prevents bacterial attack, when used in cosmetic formulations, as it consumes oxygen.

28.6.3 Enzymes as Anti-Free Radicals

SODs have the ability to capture free radicals and to prevent damage to skin caused by environmental pollution, bacteria, smoke, sunlight and other harmful factors. They work on the surface of the skin and will not penetrate down to the living cells. SOD is an enzyme that provides functionality, safety, and stability in cosmetic products.

28.7   Conclusion

Aquatic organisms produce unique enzymes having exciting applications in food and cosmetic sectors. The sources of enzymes include marine animals, plants, invertebrates and microorganisms. Large quantities of fish processing discards also form a major source of enzymes. Now-a-days, very few enzymes find application directly in these sectors and many more applications need to find thier way. Isolation, identification and characterization of new enzymes from marine environment therefore provide tremendous opportunities for newer applications in food and cosmetic industries in future.

References

Aaen B, Flemming J, Benny J (1995) Partial purification and characterization of a cellular acidic phospholipase A_2 from cod (*Gadus morhua*) muscle. Comp Biochem Physiol 110B:547–554

Adrio JL, Demain AL (2014) Microbial enzymes: tools for biotechnological processes. Biomol Ther 4:117–139

Arai S, Fujimaki M (1978) The plastein reaction. Theoretical basis. Ann Nutr Aliment 32:701–707

Arunchalam K, Haard NF (1985) Isolation and characterization of pepsin from polar cod, *Boreogadus saida*. Comp Biochem Physiol 80B:467–473

Asgeirsson B, Bjarnasson JB (1989) Purification and characterization of trypsin from the poikilotherm *Gadus morhua*. Eur J Biochem 180:85–94

Aspán A, Söderhäll K (1991) Purification of prophenoloxidase from crayfish blood cells, and its activation by an endogenous serine proteinase. Insect Biochem 21:363–373

Aspán A, Huang TS, Cerenius L, Söderhäll K (1995) cDNA cloning of prophenoloxidase from the freshwater crayfish *Pacifastacus leniusculus* and its activation. Proc Natl Acad Sci U S A 92:939–943

Audley MA, Shetty KJ, Kinsella JE (1978) Isolation and properties of phospholipase a from Pollock muscle. J Food Sci 43:1771–1775

Awarenet (2004) Handbook for the prevention and minimization of waste and valorization of by-products in European agro food industries. Agro-food waste minimization and reduction network (AWARENET). Grow Programme, European Commission, pp 1–7

Barbagli C, Crescenzi GS (1981) Influence of freezing and thawing on the release of cy- tochrome oxidase from chicken liver and from beef and trout muscle. J Food Sci 46:491–493

Bárzana E, García-Garibay M (1994) Production of fish protein concentrates. In: Martin AM (ed) Fisheries processing: biotechnological applications. Chapman & Hall, London, pp 206–222

Beneytout L, Andrianarison RH, Rakotoarisoa Z, Tixier M (1989) Properties of a lipoxygenase in green algae (*Oscillatoria sp*). Plant Physiol 91:367–372

Brewer P, Helbig N, Haard NF (1984) Atlantic cod pepsin—characterization and use as a rennet substitute. Can Inst Food Sci Technol J 17:38–43

Cahn RD, Kaplan NO, Levine L, Zwilling E (1962) Nature and development of lactic dehydrogenase. Science 136:962

Chatzikonstantinou M, Vlachakis D, Chronopoulou E, Papageorgiou L, Papageorgiou A, Labrou NE (2012) The glutathione transferase family of *Chlamydomonas reinhardtii*: identification and characterization of novel sigma class-like enzymes. Algal Res 24:237–250

Chen HC, Zall RR (1986) Partial purification and characterization of cathepsin D-like and B-like acid proteases from surf clam viscera. J Food Sci 51:71–78

Chhatbar SK, Velankar NK (1977) A biochemical test for the distinction of fresh fish from frozen and thawed fish. Fish Technol 14:131–136

Clark J, Macdonald NL, Stark JR (1985) Metabolism in marine flatfish. III. Measurement of elastase activity in the digestive tract of Dover sole (*Solea solea* L.). Comp Biochem Physiol 81B:695–700

Cohen T, Gertler A, Birk Y (1981) Pancreatic proteolytic enzymes from carp (*Cyprinus carpio*). I. Purification and physical properties of trypsin, chymotrypsin, elastase and carboxypeptidase B. Comp Biochem Physiol 69B:639–646

Curran CA, Poulter RE, Brueton A, Jones NR, Jones NSD (1986) Effect of handling treatment on fillet yields and quality of tropical fish. J Food Technol 21:301–310

de Vecchi S, Coppes Z (1996) Marine fish digestive proteases—relevance to food industry and the south-West Atlantic region—a review. J Food Biochem 20:193–214

Díaz-López M, García-Carreño FL (2000) Applications of fish and shellfish enzymes in food and feed products. Chapter 21. In: Haard NF, Simpson BK (eds) Seafood enzymes: utilization and influence on postharvest seafood quality. Marcel Dekker, New York, pp 571–618

Dingle JR, Hines JA (1967) Extraction and some properties of adenosine 5'-monophosphate aminohydrolase from prerigor and postrigor muscle of cod. J Fish Res Bd Can 24(8):1717–1730

Dingle JR, Hines JA (1971) Degradation of inosine 5'-monophosphate in the skeletal muscle of several North Atlantic fishes. J Fish Res Bd Can 28(8):1125–1131

Draelos ZD (2012) Cosmetics, diet, and the future. Dermatol Ther 25:267–272

Eisen AZ, Henderson KD, Jeffrey JJ, Bradshaw RA (1973) A collagenolytic protease from the hepatopancreas of the fiddler crab, Uca pugilator. Purification and properties. Biochemistry 12:1814–1822

Ellingsen T, Mohr V (1981) Antarktisk krill biokjemi og teknologi. Rapport II. Lipider. Seksjon for fiskerikjemi. Institutt for teknisk biokjemi. Norges tekniske hØgskole, Trontheim, pp 1–109

Fruton S, Bergmann M (1940) The specificity of salmon pepsin. J Biol Chem 204:559–560

García-Carreño FL, Raksakulthai R, Haard NF (1997) Processing wastes. Exopeptidases from shellfish. In: Bremmer A, Davis C, Austin B (eds) Making the most of the catch. Hamilton, AUSEAS, pp 37–43

Gaykema WPJ, Hol WGJ, Vereijken JM, Soeter NM, Bak HJ, Beintema JJ (1984) 3.2Å structure of the copper-containing, oxygen-carrying protein *Panulirus interruptus* hemocyanin. Nature 309:23–29

German JB, Kinsella JE (1985) Lipid oxidation in fish tissue. Enzymatic initiation via lipoxygenase. J Agric Food Chem 33:680–683

Gildberg A (1993) Enzymic processing of marine raw materials. Process Biochem 28:1–15

Gildberg A, Øverbø K (1990) Purification and characterization of pancreatic elastase from Atlantic cod (*Gadus morhua*). Comp Biochem Physiol 97B:775–782

Gildberg A, Raa J (1983) Purification and characterization of pepsins from Arctic fish capelin (*Mallotus villosus*). Comp Biochem Physiol 75A:337–342

Gill T (2000) Nucleotide-degrading enzyme. Chapter 2. In: Haard NF, Simpson BK (eds) Seafood enzymes: utilization and influence on postharvest seafood quality. Marcel Dekker, New York, pp 37–68

Gjellesvik DR, Lombardo D, Walther BT (1992) Pancreatic bile salt dependent lipase from cod (*Gadus morhua*): purification and properties. Biochim Biophys Acta 1124:123–134

Gjellesvik DR, Lorens JB, Male R (1994) Pancreatic carboxylester lipase from Atlantic salmon (*Salmo salar*) cDNA sequence and computer-assisted modelling of tertiary structure. Eur J Biochem 226:603–612

Guizani RRS, Marshall MR, Wei CI (1991) Isolation and characterization of a trypsin from the pyloric ceca of mullet (*Mugil cephalus*). Comp Biochem Physiol 98:517–521

Haard NF (1997) Specialty enzymes from marine organisms. Biotecnologia 2:78–85

Haard NF, Simpson BK (1994) Proteases from aquatic organisms and their uses in the seafood industry. In: Martin AM (ed) Fisheries processing: biotechnological applications. Chapman & Hall, London, pp 132–154

Hamm R, Masic D (1971) Influence of freezing and thawing of carp on the sub-cellular distribution of asparate aminotransferase in the skeletal muscle. Arch Fisch 22:121–124

Haraldsson GG (1990) The applications of lipases for modification of facts and oils, including marine oils. In: Voigt MN, Botta JR (eds) Advances in fisheries technology and biotechnology for increased profitability. Technomic Publishing, Lancaster, PA, pp 337–357

Hawkins DJ, Brash AR (1987) Eggs of the sea urchin, *Strongylcentrotus purpuratus*, contain a prominent (11R) and (12R) lipoxygenase activity. J Biol Chem 262(16):7629–7634

Heu MS, Kim HR, Pyeum JH (1995) Comparison of trypsin and chymotrypsin from the viscera of anchovy, *Eugraulis japonica*. Comp Biochem Physiol 112B:557–568

Hirano K, Tanaka A, Yoshizumi K, Tanaka T, Satouchi K (1997) Properties of phospholipase A_1/ transacylase in the white muscle of bonito *Euthynnus pelamis* (Linnaeus). J Biochem 122:1160–1166

Hsu HH, Pan BS (1996) Effects of protector and hydroxyapatite partial purification on stability of lipoxygenase from gray mullet gill. J Agric Food Chem 44:741–745

Hu SP, Pan BS (1999) Aroma modification of fish oil using macro algae lipoxygenase. J Am Oil Chem Soc

Huss HH (1988) Fresh fish—quality and quality changes. FAO fish series No 29. FAO, Rome

Iijima N, Tanaka S, Ota Y (1998) Purification and characterization of bile salt-activated lipase from the hepatopancreas of red sea bream, *Pagrus major*. Fish Physiol Biochem 18:59–69

Janicke R, Knopf S (1968) Molecular weight and quaternary structure of lactic dehydrogenase. Eur J Biochem 4:157

Jiang ST, Lee JJ, Chen HC (1994) Purification and characterization of cathepsin B from ordinary muscle of mackerel (*Scomber australasicus*). J Agric Food Chem 42(5):1073–1079

Josephson DB, Lindsay RC (1986) Enzymic generation of volatile aroma compounds from fresh fish. In: Parliment TH, Croteau R (eds) Biogeneration of aromas. ACS symposium series, vol 317. American Chemical Society, Washington, DC, pp 201–219

Kalac J (1978) Studies on herring (*Clupea harengus L.*) and capelin (*Mallotus villosus*) pyloric ceca protease. III. Characterization of anionic fractions of chymotrypsins. Biologia (Bratslava) 33:939–945

Kamath GG, Lanier TC, Foegeding EA, Hamann DD (1992) Nondisulfide covalent cross- linking of myosin heavy chain in setting of Alaska Pollock and Atlantic croaker surimi. J Food Biochem 16:151–172

Karlstam B, Vincent J, Johansson B, Brynö C (1991) A simple purification method of squeezed krill for obtaining high levels of hydrolytic enzymes. Prep Biochem 21:237–256

Katoh N, Nozaki N, Komatsu K, Arai K (1979) A new method for evaluation of quality of frozen surimi from Alaska pollack. Relationship between myofibrillar ATPase activity and Kamaboko forming ability of frozen surimi. Bull Jpn Soc Sci Fish 48(8):1027–1032

Kawai T (1996) Fish flavor. Crit Rev Food Sci Nutr 36:257–298

Kawashima T, Arai K, Saito T (1973) Studies on muscular proteins of fish-X. the amount of actomyosin in frozen surimi from Alaska-pollack. Bull Jpn Soc Sci Fish 39(5):525–532

Kim J, Marshall MR, Wei C (2000) Polyphenoloxidase. Chapter 10. In: Haard NF, Simpson BK (eds) Seafood enzymes: utilization and influence on postharvest seafood quality. Marcel Dekker, New York, pp 271–315

Kim YJ, Kim HJ, No JK, Chung HY, Fernandes G (2006) Anti inflammatory action of dietary fish oil and calorie restriction. Life Sci 78:2523–2532

Kishi H, Nozawa H, Seki N (1991) Reactivity of muscle transglutaminase on carp myofibrils and myosin B. Nippon Suisan Gakk 57:1203–1210

Knight J, Holland JW, Bowden LA, Hallidary K, Rowley AF (1995) Eicosanoid generating capacities of different tissues from rainbow trout, *Oncorhynchus mykiss*. Lipids 30:451–458

Kolakowski A (1988) Changes in lipids during the storage of krill (*Euphausia superba* Dana) at 3°C. Z Lebensm Unters Forsch 186:519–523

Kolakowski E, Sikorski ZE (2000) Endogenous enzymes in Antarctic krill: control of activity during storage and utilization chapter 18. In: Haard NF, Simpson BK (eds) Seafood enzymes: utilization and influence on postharvest seafood quality. Marcel Dekker, New York, pp 505–529

Kolodziejska I, Pacana J, Sikorski ZE (1992) Effect of squid liver extract on proteins and on the texture of cooked squid mantle. J Food Biochem 16:141–150

Kristjansson MM, Nielson HH (1992) Purification and characterization of two chymotrypsin-like proteases from the pyloric ceca of rainbow trout (*Oncorhynchus mykiss*). Comp Biochem Physiol 101B:247–253

Kristjansson MM, Gudmundsdottir S, Fox JW, Bjarnason JB (1995) Characterization of collagenolytic serine proteinase from Atlantic cod (*Gadus morhua*). Comp Biochem Physiol 110B:707–717

Kuo JM, Hwang A, Hsu HH, Pan BS (1996) Preliminary identification of lipoxygenase in algae (*Enteromorpha intestinalis*) for aroma formation. J Agric Food Chem 44(8):2073–2077

Kuo JM, Hwang A, Yeh DB (1997) Purification, substrate specificity and products of a Ca^{2+}-stimulating lipoxygenase from sea algae (*Ulva lactuca*). J Agric Food Chem 45(6):2055–2060

Lansing AI, Rosenthal TB, Alex M (1953) Presence of elastase in teleost islet tissue. Proc Soc Exp Biol Med 84:689–691

LeBlanc PJ (1987) Approaches to the study of nucleotide catabolism for fish freshness evaluation. MSc thesis, Technical University of Nova Scotia

Lee YZ, Simpson BK, Haard NF (1982) Supplementation of squid fermentation with proteolytic enzymes. J Food Biochem 6:127–134

Lee CW, Park NH, Kim JW, Um BH, Shpatov AV (2012) Study of skin anti-ageing and antiinflammatory effects of dihydroquercetin, natural triterpenoids, and their synthetic derivatives. Bioor ganicheskaia Khimiia 38:374–381

Leger C, Bauchart D, Flanzy J (1977) Some properties of pancreatic lipase in *Salmo gairdnerii* rich: effects of bile salts and Ca+2, gel filtration. Comp Biochem Physiol 57:359–363

López-Amaya C, Marangoni AG (2000) Lipases. Chapter 5. In: Haard NF, Simpson BK (eds) Seafood enzymes: utilization and influence on postharvest seafood quality. Marcel Dekker, New York, pp 121–146

Makarewicz W (1969) AMP-aminohydrolase in muscle of elasmobranch fish. Purification procedure and properties of the purified enzyme. Comp Biochem Physiol 29:1–26

Makinodan Y, Ikeda S (1971) Studies on fish muscle protease. IV. Relation between hi-modori of kamaboko and muscle proteinase. Bull Jpn Soc Sci Fish 37:518–523

Makinodan Y, Hirotsuka M, Ikeda S (1979) Neutral proteinase of carp muscle. J Food Sci 44 (1110):1979

Martinez A, Olsen RL (1989) Characterization of pepsins from cod. US Biochem Corp 161:22–23

Martinez A, Olsen RL, Serra J (1988) Purification and characterization of two trypsin-like enzymes from the digestive tract of anchovy, *Engraulis encrasicholus*. Comp Biochem Physiol 91B:677–684

Meijer L, Guerrier P, Maclouf J (1984) Arachidonic acid, 12- and 15-hydroxye-icosatraenoic acid, eicosapentaenoic acid, and phospholipase A2 induce starfish oocyte maturation. Dev Biol 106:368–378

Merrett TG, Bar-Eli E, van Vunakis H (1969) Pepsinogens A, C, and D from the smooth dogfish. Biochemistry 8:3696–3702

Moghaddan MF, Gerwick WH (1990) 12-lipoxygenase activity in the red marine alga *Gracilariopsis lemaneiformis*. Phytochemistry 29(8):2457–2458

Mohri S, Cho SY, Endo Y, Fujimoto K (1990) Lipoxygenase activity in sardine skin. Agric Biol Chem 54:1889–1991

Mukundan MK, Gopakumar K, Nair MR (1986) Purification of a lipase from the he- patopancreas of oil sardine (*Sardinella longiceps* Linnaceus) and its characteristics and properties. J Sci Food Agric 36:191–203

Myrnes B, Johansen A (1994) Recovery of lysozyme from scallop waste. Prep Biochem 24:69–80

Nambudiri DO, Gopakumar K (1990) Effect of freezing and thawing on press juice and enzyme activity in the muscle of farmed fish and shellfish I.I.F. - I.I.R. UK Commission, Aberdeen, pp 183–187

Navvabia A, Razzaghia M, Fernandes P, Karamid L, Homaei A (2018) Novel lipases discovery specifically from marine organisms for industrial production and practical applications. Process Biochem 70:61–70

Naylor EC, Watson RE, Sherratt MJ (2011) Molecular aspects of skin ageing. Maturitas 69:249–256

Neas NP, Hazel JR (1985) Partial purification and kinetic characterization of the microsomal phospholipase A2 from thermally acclimated rainbow trout (*Salmo gairdneri*). J Comp Physiol B 155:461–469

Nip WK, Moy JH, Tzang YY (1985) Effect of purging on quality changes of ice-chilled freshwater prawn, *Machrobranchium rosenbergii*. J Food Technol 20:9–15

Noda M, Murakami K (1981) Studies on proteinases from the digestive organs of sardine— purification and characterization of three alkaline proteinases from the pyloric ceca. Biochim Biophys Acta 658:17–26

Ochiai Y, Chow C (2000) Myosin ATPase. Chapter 3. In: Haard NF, Simpson BK (eds) Seafood enzymes: utilization and influence on postharvest seafood quality. Marcel Dekker, New York, pp 69–89

Okamoto Y, Otsuka-Fuchino H, Horiuchi S, Tamiyu T, Matsumoto JJ, Tsuchiya T (1993) Purifica- tion and characterization of two metalloproteinases from squid mantle muscle, myosinase I and myosinase II. Biochim Biophys Acta 1161:97–104

Ono H, Iijima N (1998) Purification and characterization of phospholipase A_2 isoforms from the hepatopancreas of red sea bream, *Pagrus major*. Fish Physiol Biochem 18:135–147

Orejana FM, Liston J (1981) Agents of proteolysis and its inhibition in Patis (fish sauce) fermenta- tion. J Food Sci 47:198–203

Pan BS, Kuo JM (1994) Flavor of shellfish and kamaboko flavourants. In: Shahidi F, Botta JR (eds) Seafoods: chemistry, processing technology and quality. Blackie Academic & Professional (Chapman & Hall), Glasgow, pp 85–114

Pan BS, Kuo JM (2000) Lipoxygenases. Chapter 11. In: Haard NF, Simpson BK (eds) Seafood enzymes: utilization and influence on postharvest seafood quality. Marcel Dekker, New York, pp 317–336

Pfleiderer VG, Zwilling R, Sonneborn H-H (1967) Eine protease vom molekulargewicht 11,000 and eine trypsinahinche frackion aus *Astacus fluviatilis*. Z Physiol Chem 348:1319–1331

Raa J, Gildberg A (1976) Autolysis and proteolytic activity of cod viscera. J Food Technol 11:619–628

Raa J, Walther BT (1989) Purification and characterization of chymotrypsin, trypsin and elastase like proteinases from cod (*Gadus morhua* L.). Comp Biochem Physiol 93B:317–324

Racicot WF (1984) A kinetic and thermodynamic comparison of bovine and dogfish chymotrypsins. Ph.D. thesis, University of Massachusetts, Amherst, MA

Raksakulthai N, Haard NF (1992) Correlation between the concentration of peptides and amino acids and the flavour of fish sauce. ASEAN Food J 7:86–90

Raymond JA (1998) Trimethylamine oxide and urea synthesis in rainbow smelt and some other northern fishes. Physiol Zool 71:515–523

Raymond JA, De Vries AL (1998) Elevated concentrations and synthetic pathways of trimethylamine oxide and urea in some teleost fishes of McMurdo Sound, Antartica. Fish Physiol Biochem 18:387–398

Reddi PK, Constantinides SM, Dymsza HA (1972) Catheptic activity in fish muscle. J Food Sci 37:643–648

Regenstein JM, Schlosser MA, Samson A, Fey M (1982) Chemical changes of trimethylamine oxide during fresh and frozen storage of fish. In: Martin RE, Flick GJ, Hebard CE, Ward DR (eds) Chemistry and biochemistry of marine food products. AVI Publ Comp, Westport, CT, pp 137–148

Reid RGB, Rauchert K (1976) Catheptic endopeptidases and protein digestion in the horse clam *Tresus capax* (Gould). Comp Biochem Physiol 54B:467–472

Ridder (2020). https://www.statista.com/statistics/585522/global-value-cosmetics-market/

Roy G (1992) Bitterness: reduction and inhibition. Trends Food Sci Technol 3:85–91

Ruiter A (1972) Substitution of proteases in the enzymatic ripening of herring. Ann Technol Agric 21:597–605

Sakai J, Sakaguchi Y, Matsumoto JJ (1981) Acid proteinase activity of squid mantle muscle: some properties and subcellular distribution. Comp Biochem Physiol 70B:791–794

Sanchez-Chiang L, Ponce O (1981) Gastricsinogens and Gastricsins from *Merluccius gayi*—purification and properties. Comp Biochem Physiol 68B:251–257

Seki N, Narita N (1980) Changes in ATPase activities and other properties of carp myofibrillar proteins during ice storage. Bull Jpn Soc Sci Fish 46(2):207–213

Shamsuzzaman K, Haard NF (1983) Evaluation of harp seal gastric protease as a rennet substitute for Cheddar cheese. J Food Sci 48:179–182

Shenderyuk VI, Bykowski PJ (1990) Salting and marinating of fish. In: Sikorski ZE (ed) Seafood: resources, nutritional composition, and preservation. CRC, Boca Raton, FL, pp 147–162

Sikorski ZE, Naczk M (1981) Modification of technological properties of fish protein concentrates. CRC Crit Rev Food Sci Nutr 14:201–230

Sikorski ZE, Gildberg A, Ruiter A (1995) Fish products. In: Ruiter A (ed) Fish and fishery products. CAB International, Wallingford, pp 315–346

Simpson B (2000) Digestive proteinases from marine animals. Chapter 8. In: Haard NF, Simpson BK (eds) Seafood enzymes: utilization and influence on postharvest seafood quality. Marcel Dekker, New York, pp 191–213

Simpson BK, Haard NF (1984a) Purification and characterization of trypsin from greenland cod (*Gadus ogac*). 1. Kinetic and thermodynamic characteristics. Can J Biochem Cell Biol 62:894–900

Simpson BK, Haard NF (1984b) Trypsin from Greenland cod as a food-processing aid. J Appl Biochem 6:135–143

Simpson BK, Haard NF (1985) Characterization of the trypsin fraction from cunner, *Tautogolabrus adspersus*. Comp Biochem Physiol 80B:475–480

Sotelo CG, Rehbein H (2000) TMAO degrading enzymes. In: Haard NF, Simpson BK (eds) Seafood enzymes: utilization and influence on postharvest seafood quality. Marcel Dekker, New York, p 167

Squires EJ, Haard NF, Feltham LAW (1986) Pepsin isozymes from Greenland cod, *Gadus ogac*. I. Purification and physical properties. Can J Biochem Cell Biol 64:205–214

Stefánsson G, Steingrímsdóttir U (1990) Application of enzymes for fish processing in Iceland—present and future aspects. In: Voigt MN, Botta JR (eds) Advances in fisheries technology and biotechnology for increased profitability. Technomic Publishing, Lancaster, PA, pp 237–250

Sugihara T, Yashima C, Tamura H, Kawasaki M, Shimizu S (1973) Process for preparation of ikura (salmon egg). US Patent No 3759718

Summer JB (1926) The isolation and crystallization of the enzyme urease: preliminary paper. J Biochem 69:435–441

Tarr HLA (1958) Lingcod muscle purine nucleotide phosphorylase. Can J Biochem Physiol 36:517–530

Tavares FP, Baptista JAB, Marcone MF (1997) Milk-coagulating enzymes of tuna fish waste as a rennet substitute. Int J Food Sci Nutr 48:169–176

Toyohara H, Kinoshita M, Shimizu Y (1990) Proteolytic degradation of threadfin bream meat gel. J Food Sci 55:259–260

Warrier SBK, Ninjoor V, Nadkarni GB (1985) Involvement of hydrolytic enzymes in the spoilage of Bombay duck (*Harpodon nehereus*). In: Technical report harvest and postharvest technology of fish. Society of Fisheries Technology (India), pp 470–472

Yamanka A, Mackie IM (1971) Changes in the activity of a sarcoplasmic adenosine triphosphatase during iced-storage and frozen storage of cod. Bull Jpn Soc Sci Fish 37(11):1105–1109

Yamashita M, Konagaya S (1990) High activities of cathepsins B, D, H and L in the muscle of chum salmon in spawning migration. Comp Biochem Physiol 95B:149–152

Yasueda H, Kumazawa H, Motoki M (1994) Purification and characterization of a tissue- type transglutaminase from red sea bream (*Pagrus major*). Biosci Biotechnol Biochem 58:2041–2045

Yoshinaka R, Sato M, Suzuki T, Ikeda S (1984) Enzymatic characterization of anionic trypsin of the catfish *Parasilurus asotus*. Comp Biochem Physiol 77B:1–6

Yoshinaka R, Sato M, Tanaka M, Ikeda S (1985) Distribution of pancreatic elastase and metalloproteins in several species of fish. Comp Biochem Physiol 80B:227–233

Yoshioka K (1985) Differentiation of freeze thawed fish from fresh fish by the examination of blood. Bull Jpn Soc Sci Fish 51:1331–1336

Zeef AH, Dennison C (1988) A novel cathepsin from the mussel (*Perna perna* Linne). Comp Biochem Physiol 90B:204–210

Zeinali F, Homaei A, Kamrani E (2015) Sources of marine superoxide dismutases: characteristics and applications. Int J Biol Macromol 79:627–637

Zendzian EN, Bernard EA (1967) Distribution of pancreatic ribonuclease, chymotrypsin and trypsin in vertebrates. Arch Biochem Biophys 122:699–713

Zhang CH, Hirano T, Suzuki T, Shirai T (1992) Enzymatically generated specific volatile compounds in ayu tissues. Nippon Suisan Gakk 58:559–565

Zimmerman DC, Vick BA (1973) Lipoxygenase in *Chlorella pyrenoidosa*. Lipids 8:264–266

Biotechnological Approaches to Valorization of Fish Biowastes and Their Potential Applications

29

Jerusha Stephen, Manjusha Lekshmi, Binaya Bhusan Nayak, and Sanath Kumar

Abstract

Fish production and utilization activities generate significant quantities of biological wastes, the disposal of which is not only an economic burden but also an environmental concern. The conventional fish waste utilization methods, such as their conversion into fish meal and bio fertilizer, are economically not lucrative owing to their limited applications. With the increasing aquaculture production of fish and shellfish, the industry needs novel ways of utilizing fish biowaste generated from processing industry for better economic returns. Fish waste is a known source of useful proteins such as collagen and is rich in extractable lipids, vitamins, and minerals. These components of fish waste have broad applications in the medical, pharmaceutical, and food industries. Some of the key extractable components of biomedical importance from fish and shellfish wastes are collagen/gelatin, amino acids, peptides, chitin, astaxanthin, lipids, and calcium. Biotechnological means of fish waste utilization involve enzyme- and bacteria-mediated hydrolysis to derive biologically active compounds and production of microbial enzymes using fish wastes as substrates. Prospective studies on fish waste utilization should establish methods for utilizing fish waste on an industrial scale, so that the management of waste will be a sustainable and economically viable activity.

Keywords

Biotechnology · Fish Waste · Hydrolysate · Bioactivity · Peptide · Biomedical · Valorization

J. Stephen · M. Lekshmi · B. B. Nayak · S. Kumar (✉)
Post-Harvest Technology, ICAR-Central Institute of Fisheries Education, Mumbai, Maharashtra, India
e-mail: sanathkumar@cife.edu.in

© The Author(s), under exclusive license to Springer Nature Singapore Pte Ltd. 2021
P. K. Pandey, J. Parhi (eds.), *Advances in Fisheries Biotechnology*,
https://doi.org/10.1007/978-981-16-3215-0_29

29.1   Introduction

The global demand for food fish has witnessed a sharp rise due to increasing population. About 88% of 179 million tonnes of fish produced in 2018 was consumed directly, while the remaining 12% of fish produced was used for non-food purposes such as fish meal and oil (FAO 2018). However, one of the aspects between fish production and consumption, that has missed critical attention, is the loss of fish due to wastage. An estimated 27% of the fish produced globally is lost through wastage between landing and consumption (FAO 2018). While this represents the loss of food fish, fish waste can have a negative environmental effect if not managed scientifically. Fish waste comprises of head, fin, bones, scales, and the intestine. Shrimp aquaculture generates shells and heads as waste. Some of the practical ways of utilizing these could be in the form of fish hydrolysates, which can be used in animal feeds. Fish wastes can also be used as fertilizers. The utilization of fish wastes can earn additional revenue for the industry (Arvanitoyannis and Kassaveti 2008). Fish oils and fish meal are the two common products derived from fish biowaste, whereas chitin is derived from shrimp shell waste. The increasing cost of disposing aquaculture wastes has necessitated the utilization of multiple fish wastes for the production of valuable products. Enzymatic and bacterial interventions can yield biologically active products from fish wastes, potentially useful in the pharmaceutical and food industries. Fish waste utilization can reduce the environmental impact of wastes generated from fish processing plants and the domestic use.

29.2   Fish Biowaste and Its Utilization

Processing of fish and shellfish generates significant quantities of wastes in the form of heads, bones, scales, fins, skins, and entrails (Anh et al. 2011). Depending on the fish species and the processing method, the wastage can be as high as 75% (Alves et al. 2017). A study involving 14 European countries estimated that the waste generated from fish could range from 40–70%. Disposal of fish wastes in landfills incurs considerable costs to the processing plants and has environmental impacts. Waste disposal is an enormous economic burden to the processing plants unless by-products that match the cost of waste management in terms of value are generated from the wastes (Ghosh et al. 2016). However, an example from the U.K. suggests that although 50% of the cod goes as waste, the by-products produced from the wastes can generate only <10% of the value of fish lost as waste (Ghosh et al. 2016). In other words, there is a dire need to discover and develop potential ways of utilizing fish wastes for economic gains. One of the preferred conventional ways of fish waste utilization is its conversion into fish meal, a preferred protein-rich ingredient in aquaculture feeds. About 60% of the fish meal produced globally is utilized by aquaculture. Although small in terms of global production, other by-products derived from fish wastes include fish oil, organic fertilizers, silage, fish protein concentrates, etc. The fish by-product industry is gaining prominence, and several countries have industry-scale utilization of fish wastes for by-product preparation. Some of the important products of value from fish wastes include fish sauces, leather from fish skin, astaxanthin, chitosan, gelatin, calcium from fish bones, enzymes, and pharmaceutical ingredients from some fish by-products (FAO 2018). Ready-to-eat snacks or products such as biscuits, crispies, nuggets, etc. are produced in some countries.

Although the production of fish hydrolysates from fish waste is not new, the health beneficial and biomedical properties of hydrolysates or the compounds derived from them have become the focus of research only recently. The bioactive peptides derived from the sciaenid fish *Argirosomus regius* and the gilthead sea bream *Sparus aurata* showed growth inhibitory activities against the human colon and breast cancer cells, similar to the established anticancer molecule etoposide (Kandyliari et al. 2020). The consumption of fish protein hydrolysates may not have an acute and immediate effect on health, but could serve as a better nutraceutical supplement and prophylactic measure in the long run (Alvares et al. 2018). The fish waste hydrolysate-derived bioactive peptides with angiotensin I converting enzyme (ACE) inhibitory activity have been reported from threadfin breams (*Nemipterus japonicus*) (Gajanan et al. 2016), mrigal (*Cirrhinus mrigala*) (Elavarasan et al. (2016), Atlantic salmon (*Salmo salar*) (Neves et al. 2017a), Nile tilapia (*Oreochromis niloticus*) (Borges-Contreras et al. 2019) and yellow fin sole (*Limanda aspera*) (Jung et al. 2006). By far, the potential health benefits of fish hydrolysates have been predicted based on the analysis their biochemical composition and in vitro bioactivities, although limited studies have used animal models (Wergedahl et al. 2004; Bjørndal et al. 2013; Dale et al. 2019). Prospective studies, using animal models, are necessary to establish the nutraceutical properties of fish protein hydrolysates and promote them as dietary supplements.

29.2.1 Fish Scale

Fish scale is generated in large quantities in processing facilities and in the domestic fish markets. Depending on the fish species, scales constitute 1–2% of the body weight. Due to their rigid architecture and chemical nature, fish scales are generally recalcitrant to enzymatic degradation and pose a significant disposal challenge. On the other hand, fish scales are good sources of type 1 collagen, calcium, and hydroxyapatite. These properties of fish scales demand biotechnological interventions to utilize and derive useful compounds from them and reduce the environmental impacts due to their discard.

29.2.1.1 Chemical Composition of Fish Scales

Fish scales are composed of both organic (40–50%) and inorganic materials (7–25%)(Chinh et al. 2019). The organic components include collagen, scleroproteins, fat, lecithin, and vitamins, while the inorganic components are mainly the hydroxyapatite and calcium phosphate (Chinh et al. 2019). Scales are mesodermal in origin, composed of cross-plies of collagen fibers arranged in three layers that contain hydroxyapatite between and within the fibers (Brown and

Wellings 1968; Gil-Duran et al. 2016). With 15–69% minerals by weight, fish scale structurally resembles bones and teeth (Zhu et al. 2012). Ultrastructural studies have revealed that fish scale has three distinct regions, namely the limiting layer (LL), external elasmodine (EE), and internal elasmodine (IE) (Arola et al. 2018). LL is the most external and highly mineralized layer of the scale, containing randomly arranged thin collagen fibers of 3–50 nm diameters. EE layer consists of unidirectionally aligned collagen fibers of 100–160 nm, forming a laminate-like structure (Zylberberg and Nicolas 1982; Garrano et al. 2012). IE layer is the same as EE but has a lower mineral content.

29.2.1.2 Fish Scale as a Source of Type-1 Collagen

Although the biomedical applications of type-1 collagen were known, there has been a renewed research interest for its use in various contexts. Currently, collagen is sourced mainly from bovine and porcine skin. Fish skin, bone, and scales contain type-1 collagen, and the studies have demonstrated that type-1 collagen can be extracted from them (Zhang et al. 2009; Gu et al. 2011). Sankar et al. (2008) extracted collagen from *Lates calcarifer* scales and prepared them in the form of sheets. The fish collagen sheet was porous and exhibited fibrous nature and had sufficient tensile strength to serve as a wound dressing material.

Collagen extraction from scales employs acids, pepsin, microorganisms, or their enzymes. The method for pepsin-solubilized collagen is well established and involves deproteinization using alkali, followed by acid demineralization. Collagen is repeatedly extracted, using acetic acid (acid-soluble collagen) and subsequently digested with pepsin (pepsin-soluble collagen) (Duan et al. 2009). Studies have reported the extraction of type-1 collagen from the scales of Egyptian Nile tilapia (Bhagwat and Dandge 2016), red drum fish (Chen et al. 2016), tilapia (Sujithra et al. 2013), and sea bass (Sankar et al. 2008). The yield of type-1 acid-soluble collagen from the carp scales was 1.35% on a dry weight basis and consisted of two $\alpha1$ and one $\alpha2$ chain (Duan et al. 2009).

The physical properties of fish collagen, amino acid composition, etc. can vary depending on the fish species and its geographical origin, so also the yield. Vietnamese carp (*Cyprinus carpio*) scales yielded 13.6% of type-1 collagen, composed of $\alpha1$ and $\alpha2$ chains with molecular masses of 139 and 129 kDa, respectively, and a β chain (Chinh et al. 2019). Amino acids such as threonine, arginine, proline, serine, alanine, glutamic acid, and glycine were abundant, while methionine, tryptophan, and tyrosine were in meager amounts (Chinh et al. 2019). The rare amino acid tryptophan was also present in the fish collagen. The yield of fish collagen from scales can vary from 1.35 to 50% depending on the species and its origin. The lowest reported yield of fish scale collagen was from carps, whereas the highest yield was from sardine scales (Nagai et al. 2004; Duan et al. 2009). Others report have shown the yield of 37.50% from red sea bream and 41.5% from sea bass (Nagai et al. 2004).

29.2.1.3 Bioactivities of Fish Scale Collagen Hydrolysates

The whole, type-1 collagen derived from different sources has medical, pharmaceutical, and cosmetic applications. On the other hand, collagen peptides, generated by enzyme hydrolysis, are increasingly finding applications in biomedical and pharmaceutical fields owing to their bioactivities. The increased hydroxyproline levels in human plasma after collagen hydrolysate ingestion has been proposed to benefit health (Shigemura et al. 2014). In a study, collagen hydrolysate derived by using ginger protease was found to contain X-Hyp-Gly-type tripeptides which elevated the alkaline phosphatase activity of pre-osteoblasts, suggesting that when used as a functional food, this hydrolysate can help to manage and treat bone diseases such as the osteoporosis (Taga et al. 2018). Similar studies with animal collagen suggest alleviation of UAV-induced damage in human dermoblasts by hen collagen hydrolysate (Wang et al. 2019), anti-inflammatory activities of chicken collagen hydrolysate (Offengenden et al. 2018), inhibition of fibrinogen/thrombin clotting (Maruyama et al. 1993) and immunomodulatory activities of yak bone collagen hydrolysate (Gao et al. 2019). In the light of these findings, there is an immense scope to establish bioactivities of fish collagen hydrolysates and promote their production and utilization in pharmaceutical, cosmetic, livestock and food industries.

A few studies on fish collagen hydrolysate have reported bioactivities of collagen peptides. Zamorano-Apodaca et al. (2020) reported that all peptide fractions of fish collagen hydrolysates exhibited intense antioxidant activities. The acid-soluble collagen used in their study was derived from a mixed by-product of different fish species. In another study, salmon collagen was hydrolyzed using marine bacterial collagenase, and the resultant hydrolysate exhibited excellent anti-freezing and antioxidant activities (Wu et al. 2018). An alcalase-derived collagen hydrolysate from *Thunnus orientalis* bone could promote osteoblast proliferation and differentiation, suggesting that this product can be potentially useful in bone regeneration (Ding et al. 2019). Small collagen peptides, prepared from silver carp collagen hydrolysate, improved water holding capacity of a silver carp protein isolate-based gel (Abdollahi et al. 2018). A novel approach (Pal and Suresh 2017) derived various bioactive peptide sequences from carp swim bladder collagen by in silico proteolysis and suggested that the carp collagen hydrolysates can have potential applications in functional foods and nutraceuticals. A study has shown that fish scales can be used as low-cost materials to absorb azo dye (Ooi et al. 2017). Fish scales can be used to adsorb astaxanthin from processing plant effluents (Stepnowski et al. 2004). Fish scales are also used as substrates for *Bacillus altitudinis* for the production of alkaline protease and amino acid-rich aqua hydrolysate (Harikrishna et al. 2017).

Fahmi et al. (2004) hydrolyzed seabream scales using an alkaline protease treatment and evaluated the inhibitory effects of collagen peptides on angiotensin I converting enzyme (ACE I). These investigators found inhibition of ACE I by the scale hydrolysate, and in naturally hypertensive experimental rats, oral administration of collagen peptides significantly decreased the blood pressure. Four peptides were isolated from scale hydrolysate, and the ACE inhibitory activities of peptides were five-folds higher than that of the crude hydrolysate (Fahmi et al. 2004). The

ACE inhibitory activity and antioxidant properties of milkfish (*Chanos chanos*) scale gelatin were reported by Huang et al. (2018). Collagen, derived from fish scale, has low mechanical and degradation stabilities. Hence, methylation and 1, 4-butanediol diglycidyl ether (BDE) cross linking steps are necessary to improve its physicochemical properties. Such chemically modified collagen promoted the growth of blood and lymphatic vessels in a murine model, indicative of its wound healing properties (Wang et al. 2017).

29.2.1.4 Bacterial and Enzymatic Degradation of Fish Scale Biowaste

Biodegradation of fish scales, using bacteria, is a straightforward way of remediating fish scales. Several bacteria produce potent collagenase enzymes that can degrade fish collagen, producing collagen peptides and amino acids (Fahmi et al. 2004; Nagai et al. 2004; Liu et al. 2009; Huang et al. 2015). Alternatively, direct hydrolysis of fish scales with trypsin or pepsin enzymes has shown to produce antioxidant peptides in the case of croaker and seabream scales (Fahmi et al. 2004; Wang et al. 2013). Enzymes such as papain and flavourzyme have been used for the hydrolysis of scales and the production of bioactive collagen peptides (Huang et al. 2015).

29.2.1.5 Bacterial Collagenases and Their Application in Fish Scale Biodegradation

Collagenase enzymes are produced by diverse bacterial species, both pathogenic and non-pathogenic. Collagen has a triple helical structure made up of three polypeptide chains. Hydrogen bonds stabilize the triple helix, and the collagen structure's stability is disturbed when hydrogen bonds are disrupted, resulting in the formation of gelatin (Harrington 1996). Collagens are classified into at least 19 distinct types based on their length, degree of hydroxylysine glycosylation, the ratio of hydroxylated and non-hydroxylated residues, and these are associated with different tissues. The susceptibility of collagenases to enzymatic hydrolysis varies depending on the collagen type and its molecular structure. True collagenases can cleave the helical regions of the collagen. In contrast, others such as pepsin, trypsin, chymotrypsin, papain, and other proteolytic enzymes can degrade the non-helical structures of collagen and the gelatin (Harrington 1996).

The production of collagenase enzymes is very critical for the pathogenicity of some bacteria. Clostridia such as *Clostridium perfringens*, *C. histolyticum*, and *Vibrio* spp. produce collagenases that help them to invade soft tissues and cause infections. These bacterial collagenolytic enzymes include some serine proteases and metalloproteases (Zhang et al. 2015). Although considered as virulence factors, bacterial collagenases can be used for producing bioactive collagen peptides in vitro. Studies have shown that marine bacteria are good sources of collagenolytic bacteria (Yang et al. 2017), and these supposedly have higher catalytic activities against fish bone and skin collagen. A few molds are known to produce collagenase enzymes that can be employed in fish scale hydrolysis.

29.2.1.6 Application of Bacterial Collagenases in Fish Scale Hydrolysis

Studies on the application of collagenase-producing bacteria or molds for remediation of fish scale biowaste are sparse. Nandy et al. (2014) reported the isolation of fungi that degraded fish scale powder from the mangroves of Sundarbans. A strain of *Aspergillus niger* actively degraded fish scales, using glucose as the extraneous carbon source (Basu et al. 2005). There are no published studies that have employed bacteria directly for the degradation of fish scales. The structural integrity of the fish scale requires that the collagenase enzyme penetrates into deeper layers to degrade the collagen. This involves the destabilization of hydroxyapatite fine structure. There is tremendous scope for direct hydrolysis of fish scales to obtain bioactive collagen peptides. This approach will serve twin goals of fish scale waste remediation and the production of useful bioactive compounds. If bacteria can degrade scales into soluble compounds, this will address the formidable challenge of managing fish scale wastes.

29.2.2 Fish Skin and Its Derivatives

During fish processing, the collagen-containing waste materials can account for 30% of the total by-products (Blanco et al. 2017). Collagen is found in fish skin, fin, and bones (Nagai and Suzuki 2000). Fish skin consists of type 1 collagen and type 5 collagen as major and minor components (Yata et al. 2001). Investigations into skin-derived collagen from Albacore tuna (*Thunnus alalunga*), Dog shark (*Scoliodon sorrakowah*), and Rohu (*Labeo rohita*) have revealed the presence of α1, α2, and β chains similar to that of calf-skin type 1 collagen (Hema et al. 2013). The presence of α3 chain in many teleost fishes make certain fish skin collagen unique and different from the collagen of vertebrates (Kimura et al. 1987). In the context of bovine spongiform encephalopathy related issues associated with collagen of bovine origin, fish scale has no known zoonotic attributes and is likely to be preferred by the consumers. From ethical and religious point of view also, fish collagen has better acceptability in comparison with its bovine and porcine counterparts (Alves et al. 2017).

Fish skin collagen is perhaps one of the earliest to be explored for potential biomedical applications. Based on scientific evidence, some of the useful properties of fish skin collagen include wound healing, tissue regeneration, enhanced angiogenesis (Zhang et al. 2011), cell proliferation of facial fibroblast cells (Chai et al. 2010), bone repair, and bone augmentation (Pallela et al. 2012). One of the major drawbacks of fish-derived collagen is its low melting point. However, this property may differ with the fish species. A study reports the isolation of skin collagen with thermal stability, resembling mammalian collagen from the marine eel fish (*Evenchelys macrura*) (Veeruraj et al. 2013). Biotechnological interventions are necessary to improve this property of marine collagen so that it can find wide applications. Enzyme hydrolysis-derived fractions of collagen have demonstrated versatile bioactive properties. Among the various enzymes employed, alcalase has been found to be more suitable for fish protein hydrolysate preparation

(Wisuthiphaet et al. 2016; Rasli and Sarbon 2019). The active peptide fractions derived from the alcalase enzyme hydrolysis of fish skin gelatin have antioxidant activities (Hosseini et al. 2017). The tryptic hydrolysate of *Johnius belengerii* skin gelatin yielded peptides that showed superoxide, carbon-centered 1, 1-diphenyl-2-picrylhydrazyl (DPPH) radical scavenging activities (Mendis et al. 2005). The freshwater catfish skin and tilapia skin collagen peptides possess angiotensin-converting enzyme (ACE) inhibitory properties (Sungperm et al. 2020). Using molecular docking approach, the tilapia skin gelatin hydrolysates were found to be non-competitively inhibiting the ACE (Ling et al. 2018). The cod skin collagen polypeptides have good moisture absorption and retention properties and could find application in alleviating UV-induced damage (Hou et al. 2012). The glycyl endo-peptidase mediated hydrolysis of unicorn leatherjacket fish skin gelatin yielded bioactive peptides (Karnjanapratum et al. 2016). These peptides protected DNA from oxidative damage, reduced oxidative damage by H_2O_2 in HepG2 cells, and lowered the pro-inflammatory cytokine and nitric oxide production. Also, the human colon cancer cells (Caco-2) proliferation was reduced significantly in vitro (Karnjanapratum et al. 2016). The seabass *(Lates calcarifer)* skin gelatin hydrolysate also showed similar DNA protective, immunomodulatory, antioxidative, and antiproliferative properties (Sae-leaw et al. 2016).

29.2.3 Fish Head and Bone

The fish head is an important by-product of fish processing, which is rich in protein and calcium. Like other protein-rich fish wastes, fish head forms a suitable substrate for hydrolysate preparation. The fish head wastes could be fermented, using lactic acid bacteria to obtain protein and amino acid-rich fish sauce with clean taste (Gao et al. 2020). The Nile Tilapia processing waste hydrolysate showed inhibitory activity against *Listeria monocytogenes* and *Salmonella enterica* serovar Typhimurium (Trang and Pasuwan 2018). Extensive work has been done on the antioxidant potentials of enzyme-mediated hydrolysates of *Labeo rohita* (Bruno et al. 2019), skipjack tuna (*Katsuwonus pelamis*) (Kharyeki et al. 2018; Zhang et al. 2019), and common carp head wastes (*Cyprinus carpio)* (Elavarasan et al. 2014; Reyhani and Jafarpour 2017). The hydrolysate from Atlantic salmon trimmings showed the dipeptidyl peptidase IV (DPP-IV) enzyme inhibitory activity, suggesting anti-diabetic properties of the fish bioactive peptides in the preparation (Neves et al. 2017a). *Rachycentron canadum* (cobia) head protein hydrolysate, prepared using papain, was tested for its ACE inhibitory activities in spontaneously hypersensitive rats (Yang et al. 2013). The study identified the protein fractions of size between 173 and 494 Da, which improved ACE inhibitory effects after diges-tion with gastrointestinal proteases. This bioactive protein hydrolysate decreased the systolic pressure significantly in spontaneously hypertensive rats (Yang et al. 2013). The bioactive peptides derived from gill, bone, skin, and head protein hydrolysates of sciaenid fish *Argirosomus regius* and gilthead sea bream *Sparus aurata* inhibited the growth of human colon and breast cancer cells (Kandyliari et al. 2020). Several independent studies have demonstrated bioactivities such as the angiotensin I converting enzyme (ACE) inhibitory activity of hydrolysate-derived peptides from threadfin breams (*Nemipterus japonicus*) (Gajanan et al. 2016), *Cirrhinus mrigala* (Elavarasan et al. 2016), Atlantic salmon (*Salmo salar*) (Neves et al. (2017a), Nile tilapia (*Oreochromis niloticus*) (Borges-Contreras et al. 2019) and yellow fin sole (*Limanda aspera*) (Jung et al. 2006).

The consumption of fish protein hydrolysates might not have an immediate effect on health, but could serve better as a nutraceutical supplement and prophylactic measure (Alvares et al. 2018). The isolation of bioactive peptides and their incorporation in food can have significant human health benefits.

29.2.4 Viscera

For preparing protein hydrolysate from *Pangasianodon hypophthalmus* viscera, the enzymatic treatments using pepsin and papain enzymes were found to be superior to the acid and alkali treatments. The protein hydrolysate of *P. hypophthalmus* viscera using pepsin enzyme showed higher antioxidant activity and better quality than the chemically extracted hydrolysate (Hassan et al. 2019). The red scorpionfish (*Scorpaena notata*) viscera protein hydrolysate, prepared using the serine protease, exhibited remarkable in vitro antibacterial activity (Aissaoui et al. 2017). The fish viscera can also serve as a rich source of endogenous proteases (Vannabun et al. 2014; Murthy et al. 2018). The alkaline protease extracted from *Sardinella aurita* was used to produce protein hydrolysate from the same fish waste (Bougatef et al. 2008). The protein hydrolysate prepared with this endogenous protease showed higher ACE inhibitory activity than those extracted with other proteases like crude proteases from *Bacillus licheniformis* NH1 and *Aspergillus clavatus* ES, alcalase, and chymotrypsin (Bougatef et al. 2008).

29.3   Shellfish Biowastes and Their Utilization

The shellfish processing yields carapace and shell wastes. The average head and shell waste yield from the shrimp is around 60% by weight of the whole shrimp. The shell contains 15–40% chitin, and its amount in the whole marine environment has been estimated at about 1560 million tons (Kurita 2006). In India, its availability is estimated to be one lakh tons annually (Gopakumar 2002).

29.3.1 Chitin

Chitin, originally known as *"fongine,"* was first discovered by Braconnot in 1811 from fungi. Following this, Odier in 1823 extracted chitin from the exoskeleton of a mayfly named it as *"chitine"* meaning "covering" (Crini 2019). Chemically chitin is a biopolymer made up of β-(1$\rightarrow$4)-N-acetyl-D-glucosamine. Chitin, produced by various organisms, is the second most abundant natural biopolymer, next to cellulose. Chitin is a stable compound and occurs as ordered crystalline microfibrils, forming the structural components in the arthropods exoskeleton and the fungal cell walls. Commercial chitin is produced from the shell wastes of crab and shrimp. The global market for chitin, chitosan and its derivatives is expected to grow at 6.3% in the next 5 years and reach 53 million USD in 2024.

29.3.1.1 Isomorphs of Chitin

In nature, chitin exists in three isomorphic structures, namely α, β and γ chitin. The α-chitin is found in the cell wall of fungi and in the exoskeleton of arthropods like krill, shrimp, crab, and insects. The β-chitin occurs in the squid pens and the tubes of polychaete annelid worms like vestimetiferan and pogonophoran tubeworms (Younes and Rinaudo 2015). The γ-chitin is the least studied and is found in insect cocoon. The most commonly occurring form is the α-chitin, followed by β-chitin. These different forms of chitins arise from the difference in molecular scale packaging and its polar properties. The α-chitin sheets are tightly packed with hydrogen bonds between the sheets. The presence of hydrogen bonds makes the α-chitin more water-resistant. The absence of such hydrogen bonds in the β-chitin makes them swell in polar solvents (Nisha et al. 2016). The γ-chitin is very rare and is more similar to the α-chitin (Kaya et al. 2017; Moussian 2019)

29.3.1.2 Process of Chitin Extraction

The key raw material for chitin production is the cuticle of aquatic crustaceans like shrimps and crabs. The chitin, along with complex protein networks and calcium carbonate deposition, produces the hard shell in shrimp and crab. Hence, extraction of chitin from shrimp shell follows two primary processes, namely demineralization, and deproteinization. Additionally, a decolorization can be performed to remove pigments. In the chemical method, the demineralization is carried out using acids followed by alkali to remove the proteins and some pigments (deproteinization). The bulk chemical, used for chitin production, has a toll on the environment. Hence, newer and greener methods for chitin extraction from shell waste are of immense interest. The application of enzymes and/or microbial cultures that produce acids and enzymes are the two main approaches in the bio-extraction of chitin.

29.3.2 Biological Extraction of Chitin

The biotechnological processes can be employed for chitin extraction from shell wastes, using proteolytic enzymes or a fermentation process, resulting in deacetylation of chitin yielding chitosan, and oligosaccharides. Biologically extracted chitin and its derivatives possess similar or improved characteristics in comparison to chemically processed chitin (Beaney et al. 2005). Table 29.1 compares the pros and cons of chemical and biological extraction of chitin.

Table 29.1 Comparison of chemical and biological methods of chitin extraction

S. no	Chemical method	Biological method
1	Demineralization is by the acid treatment	Demineralization is by the organic acids like lactic acid, produced by the bacteria through conversion of an added carbon source
2	Deproteinization is by alkali treatment	Deproteinization is by proteases, secreted into the fermentation medium or by the addition of proteases or proteolytic bacteria
3	Effluent treatment after acid and alkaline extraction of chitin may cause an increase in the cost of chitin	Bio-extraction cost can be optimized by reducing the cost of carbon source. Solubilized proteins, pigments, and minerals may be used as human and animal nutrients, which can bring additional revenue (Gadgey and Bahekar 2017)
4	Considerably harmful to physicochemical properties of chitin and leads to low molecular weight and deacetylation, thus reducing the quality of chitin	Chitin is of low degree of deacetylation and depolymerization. The high molecular weights chitin is homogeneous and of superior quality
5	The effluent contains chemicals that are deleterious to the environment unless treated	It is eco-friendly and is gaining attraction due to the green chemistry
6	The order of the steps (deproteinization and demineralization) does not have a significant effect on the quality and the yield of chitin	During enzymatic deproteinization, the minerals existing in the cuticles may decrease the accessibility of the proteases, affecting deproteinization efficiency. Thus, the demineralization should be done first
7	Shorter processing time compared to biological methods	Longer processing time compared to chemical methods
8	Huge chemical consumption	Lower energy input
9	No further treatment is required	To obtain highly purified chitin, a biological process must be completed by further mild chemical treatment to remove the residual protein and minerals
10	Widely adapted	Still limited to laboratory-scale studies

29.3.2.1 Enzymatic Deproteinization

In the enzymatic deproteinization process, the shell is usually pre-treated with organic acids to achieve demineralization. The deproteinization efficiency depends on various factors such as enzyme/substrate (E/S) ratio, incubation time, the surface area of the substrate, temperature, pH, and pretreatment of the shell (demineralization). In addition to chitin, the enzymatic deproteinization yields recoverable protein hydrolysates and pigments, which are highly valuable (Antunes-Valcareggi et al. 2017). The chitosan derived from enzymatically deproteinized chitins has higher molecular weight and similar acetylation degree compared to that obtained through a chemical process (Younes et al. 2014). Table 29.2 gives a brief account of different

Table 29.2 Enzymatic production of chitin

Raw material	Pre-treatment	Enzyme	Deproteinization	Deproteinization efficiency (%)	Reference
Litopenaeus vannamei	Lactic or acetic acid treatment for 20 min at 25 °C	*Streptomyces griseus* protease	At E/S ratio of 55 U g^{-1}; particle size (50–25μm); pH 7 and incubation time 3 h	91.1	Hongkulsup et al. (2016)
Shrimp shell		*Bacillus mojavensis* A21 and *Balistes capriscus*	Hydrolysis at 45 °C; E/S ratio of 20 U mg^{-1} for 3 h	77 $\pm$ 3 and 78 $\pm$ 2	Younes et al. (2014)
Metapenaeus monoceros	Cooked at 90 °C for 20 min	*Bacillus cereus* SV1	3 h at 40 °C at pH 8; E/S ratio 20	88.8 $\pm$ 0.4	Manni et al. (2010)
Shrimp shell		Aspartic proteases		91.4	Deng et al. (2020)
L. vannamei	Lactic acid	Trypsin and Ficin	E/S ratio 0.1–0.05% w/w followed by mild-alkali treatment (0–2% NaOH; 60 °C and 30 min)	92	Marzieh et al. (2019)
Crab shell	Glutamic acid at 1:10 m/v	Alkaline protease	E/S ratio 1500 Ug^{-1}; pH 9.0, at 55 °C for 6 h	–	Ding et al. (2020)
Shrimp shell		*Micromonospora chaiyaphumensis* S103 alkaline protease	Hydrolysis at 45 °C for 3 h; pH 8.0 and with an E/S ratio 20 U mg^{-1}	93	Mhamdi et al. (2017)
		Bacillus invictae alkaline proteases	Hydrolysis at 50 °C with an E/S ratio of 10 U mg^{-1} after 3 h	76	Hammami et al. (2017)
			Fermentation with *Bacillus invictae*	82	
Portunus segnis		Serine alkaline protease subtilisin, from *Melghiribacillus thermohalophilus* Nari2AT	pH 10 and 75 °C; E/S ratio of 20 U mg^{-1}	65	Mechri et al. (2019)
Metapenaeus monoceros				82	

Portunus segnis shells		Crude digestive alkaline proteases from the viscera of *P. segnis*	E/S ratio of 5 U mg^{-1} after 3 h incubation at 50 °C	84.69 ± 0.65	Hamdi et al. (2017)
Penaeus keralhurus shells				91.06 ± 1.4	
Crayfish shell waste powder	*Bacillus coagulans* LA204 10% as innoclum	Proteinase K	48 h at 50 °C with 5% (w/v) glucose, proteinase k 1000 U g^{-1}, and	93.00	Dun et al. (2019)
Acetes spp.		Papain	pH −7.0, E/S (%)−2.1 at 52 °C	96.50	Dhanabalan et al. (2020)
		Pepsin	pH −3.1, E/S (%)−1.5 at 37 °C	89.50	

enzymatic chitin production with particular reference to its deproteinization efficiency.

29.3.2.2 Fermentation

The addition of exogenous proteases for large-scale deproteinization is not economical, unless cheap sources of such enzymes are identified, and also the deproteinized shells require an additional demineralization treatment. The fermentation method addresses these issues. In the fermentation process, the microorganisms capable of producing acid for demineralization and protease for deproteinization steps are added to the shell waste. A cheap carbon source is added to aid in acid production, such as the fruit wastes, which yields a superior quality chitin (Arbia et al. 2019; Tan et al. 2020b). The co-utilization of plant waste as a carbon source for demineralization and deproteinization in the bacteria-mediated fermentation process is a rational approach toward circumventing the use of harmful chemicals.

One-Step Fermentation

Choosing the right bacterial strain is crucial for the fermentation of shell waste and subsequent chitin production. In one-step fermentation, a single type of bacterium carries out both demineralization and deproteinization simultaneously. Fermentation of *Penaeus merguiensis* shell waste, using *Pseudomonas aeruginosa,* resulted in 82% demineralization and 92% deproteinization, and 6 days of fermentation yielded 41% chitin (Sedaghat et al. 2017). A different study reported 96% demineralization and 89% deproteinization of *P. monodon* shell waste, using *P. aeruginosa* strain A2 (Ghorbel-Bellaaj et al. 2011). The porcelain crab *Allopetrolisthes punctatus* shell waste fermented with *Lactobacillus plantarum sp. strain* 47 at 10% inoculum and 15% sucrose for 60 h yielded 99.6% demineralized and 95.30% deproteinized chitin (Castro et al. 2018). From crab shell waste, chitin was extracted using *Bacillus pumilus* culture with soybean curd residue *okara,* serving as a cheap source of nutrients for the bacterium (Taokaew et al. 2020). The fermentation efficiency can be improvised through simple interventions such as pH adjustment and re-inoculation (Neves et al. 2017b). Chitin production from freshwater shrimp shell waste through fermentation was improved with pH adjustment, and re-inoculation of the *Lactobacillus plantarum* culture resulted in 99% deproteinization and 87% demineralization after 72 h (Neves et al. 2017b).

Two-Step Fermentation

Bacteria that produce the required levels of organic acids might not produce sufficient amounts of proteases. In similar lines, the preferred candidate for deproteinization of shell waste might not perform demineralization with the same efficiency. To overcome this, two different bacteria can be utilized for optimal chitin production. This is called two-step fermentation or successive fermentation (Jung et al. 2007). Production of chitin from red crab shell waste through two-step fermentation employed *Lactobacillus paracasei subsp. tolerans* KCTC-3074 for demineralization, and *Serratia marcescens* FS-3 for deproteinization (Jung et al. 2007). Fermentation of the shell waste with proteolytic *S. marcescens* FS-3 for

5 days (deproteinization), followed by fermentation with *L. paracasei* subsp. *tolerans* KCTC-3074 (demineralization) was found to give better results. In this case, at the end of deproteinization on day 5, the pH value dropped from 6.90 to 5.89. On successive addition of *L. paracasei* subsp. *tolerans* KCTC-3074, the pH dropped to 3.62 on day 7. This two-step fermentation yielded 94.30% demineralization and 68.90% deproteinization (Jung et al. 2007). In another study by Ploydee and Chaiyanan (2014), demineralization was carried out first with *Lactobacillus pentosus* L7, followed by deproteinization with *Bacillus thuringiensis* SA to extract chitin and chitosan from *L. vannamei* shell waste. Glucose was added for lactic acid production. The biologically extracted chitin yielded superior quality chitosan on further processing. The viscosity of chitosan derived from biologically extracted chitin was found to be superior to the chemically produced chitosan (Ploydee and Chaiyanan 2014). Table 29.3 gives an account of various combinations of microbes reportedly used in two-step fermentation.

Three-Step Fermentation
In a study by Zhang et al. (2017a), chitin and chitosan were produced through a three-step fermentation. The first step involved deproteinization, using *Serratia marcescens* B742, followed by demineralization using *Lactobacillus plantarum* ATCC 8014, and chitin deacetylation into chitosan using *Rhizopus japonicus* M193. The degree of deacetylation and molecular mass of produced chitosan was 81.23% and 512.06 kDa, respectively (Zhang et al. 2017a). These three-step fermentation parameters have been optimized for large scale industrial production by the same researchers (Zhang et al. 2017b), and this process is highly promising.

29.3.3 Chitin Hydrolyzing Enzymes

The utilization of chitin is hampered due to its poor solubility. On the other hand, chitin derivatives like chitosan and oligosaccharides find diverse applications in biomedical fields. Chitooligosaccharides are depolymerized chitosan products with improved solubility, low molecular weights, and low viscosity (Xu et al. 2020). The conversion of chitin into its derivatives can also be carried out through enzymes such as chitinases (Navarrete-Bolaños et al. 2020), chitosanases, chitin deacetylases (Zhang et al. 2014), N-acetyl glucosaminyl transferase (Zhang et al. 2007) and lytic polysaccharide monooxygenases (Kaczmarek et al. 2019). *Aeromonas caviae,* isolated from the marine environment, produced chitinases, which yielded 93% N-acetyl-glucosamine from chitin (Cardozo et al. 2017). These microbial enzymes can be used to produce certain value-added nitrogen-containing amino acids like tyrosine and L-DOPA, which cannot be synthesized chemically from chitin (Ma et al. 2020).

The chitin modifying enzymes are gaining research interest due to their vast applications. The bottleneck lies with the chitin's native structure. The crystalline structure of chitin hinders enzymatic binding with the reactive polymer sites. Mild chemical treatments and the application of specifically engineered enzyme cocktails

Table 29.3 Bacterial combinations employed in two-step fermentation

Raw material	Acid-producing bacteria	Demineralization efficiency (%)	Proteolytic bacteria	Deproteinization efficiency (%)	Reference
L. vannamei shell waste	*Lactobacillus rhamnoides*	97.50	*Bacillus amyloliquefaciens*	96.8	Liu et al. (2020)
Shrimp shell powder	*Lactobacillus plantarum* ATCC 8014	93.00	*Serratia marcescens* B742	94.5	Zhang et al. (2012)
Red crab shell waste	*Lactobacillus paracasei subsp. tolerans* KCTC-3074	94.30	*S. marcescens* FS-3	68.9	Jung et al. (2007)
L. vannamei shell waste	*Lactobacillus pentosus* L7	NA	*Bacillus thuringiensis* SA	NA	Ploydee and Chaiyanan (2014)
Shrimp waste	*Lactobacillus brevis*	66.45	*Rhizopus oligosporus*	96.00	Aranday-García et al. (2017)
Krill shell waste	*L. plantarum*	84.60	*B. subtilis*	88.90	Chen et al. (2017)
Lobster shell waste	*L. plantarum*	89.59	*S. marcescens* db11	87.90	Chakravarty et al. (2018)

can overcome this issue (Kaczmarek et al. 2019). One-step chemical treatment of chitin with hydrochloric acid or ionic liquid (1-ethyl-3-methylimidazolium acetate) yields colloidal chitin with higher enzyme accessibility. The bacterium *Chitinolytic bactermeiyuanensis* SYBC-H1 was shown to reduce the crystalline structure of chitin and yield N-acetyl-D-glucosamine (Zhang et al. 2018). The chitinase extracted from *Streptomyces albolongus* ATCC 27414 converted colloidal chitin into N-acetylglucosamine and N, N′-diacetylchitobiose with a yield of 61.49% at 48 h. The addition of a short homogenization step increased the yield to 76.11% (Li et al. 2019).

29.3.4 Applications of Biologically Extracted Chitin and Its Derivatives

Chitin and its derivatives are known for their hypocholesterolemic effects. The chitin and its derivatives from the lobster *Homarus americanus* tail shell showed significant bile binding capacity. The bile binding ability of chitin and chitosan could be due to the electrostatic and hydrophobic interactions. Chitin and its oligosaccharides showed a higher binding affinity with sodium chenodeoxycholate and sodium deoxycholate than sodium cholate and sodium taurocholate (Xu et al. 2020). This could be attributed to the higher hydrophobicity exhibited by chitin. The higher molecular weight of chitin leads to aggregation of anions on its surface, which binds with the bile acids. Since the chitooligosaccharide is soluble, its interaction with bile acid is akin to the water-soluble fibers with a positive correlation between bile acid-binding capacity and viscosity (Zacherl et al. 2011; Xu et al. 2020). Chitin oligosaccharides were derived directly from the shell waste using *Bacillus cereus* strain SSW1 which exhibited antioxidant properties and could also serve as a prebiotic (Ismail and Emran 2020). The chitinolytic system of the fungus *Trichoderma harzianum* was harnessed to prepare chitooligosaccharides that showed antibacterial and anticancerous activities (Rafael et al. 2019). The red pigment prodigiosin produced by *Serratia marcescens* TNU01 is known for its antioxidant, anticancerous, and acetylcholinesterase inhibitory activities. The shell wastes rich in protein were found as a suitable nutrient source for prodigiosin production by *S. marcescens* TNU01 (Nguyen et al. 2020). N-acetylneuraminic acid, an important sialic acid found in human cells, can be produced directly from chitin through microbial degradation, using chitinolytic *S. marcescens* (Yan and Fong 2018). The bioconverted chitosan using *Asperigellus flavus* exhibits antimicrobial properties against human pathogens such as *Streptococcus* spp., *Staphylococcus aureus, Escherichia coli, Pseudomonas aeruginosa,* and *Candida albicans* (Yonis et al. 2019).

29.3.4.1 Environmental Application

The shell dust of the commercially cultivated snail *Bellamya bengalensis* was found to be a cheap and useful material for biosorption and removal of cadmium from aqueous solutions (Aditya and Hossain 2018). The biologically extracted chitosan was found to be a suitable antibacterial agent in bioethanol production (Tanganini et al. 2020). The chitin from shrimp shell waste could be used in a microbial fuel cell for energy recovery using the electroactive and chitinolytic *Aeromonas hydrophila* (Li et al. 2017).

29.3.5 Protein Hydrolysate

Protein hydrolysate from shrimp head is obtained using endogenous enzymes (autolysis) or bacteria that produce hydrolytic enzymes (Guo et al. 2019). Shrimp head fermented with *B. licheniformis* OPL-007 yielded supernatant rich in phenols, amino acids, and other polysaccharides, exhibiting strong antioxidant properties (Mao et al. 2013). The enzymatic protein hydrolysate derived from blue crab waste (*Callinectes sapidus*) was observed to be rich in astaxanthin, with antioxidant activities comparable to that of commercially available astaxanthin and butylated hydroxytoluene (BHT) (Antunes-Valcareggi et al. 2017). The production of chitin through fermentation with *Lactobacillus acidophilus* FTDC3871 yielded protein hydrolysate as a by-product rich in carotenoid (Tan et al. 2020a). The antioxidant properties of carotenoid pigments are well established. These pigments can be extracted from crustacean shell wastes along with the protein hydrolysate. The papain-mediated hydrolysis of *Parapenaeopsis stylifera* shell waste yielded carotenoprotein containing $114 \pm 0.02 \mu g \ g^{-1}$ carotenoid (Pattanaik et al. 2020). Bordenave et al. (2002) reported that 50μg of shrimp protein hydrolysate showed 57% ACE inhibition. The speckled shrimp *Metapenaeus monoceros* waste, when fermented with the proteolytic bacterium *Anoxybacillus kamchatkensis* M1V yielded protein hydrolysate with interesting bioactive properties. The protein hydrolysate showed inhibition against four enzymes, namely acetylcholinesterase (AChE), tyrosinase, amylase, and angiotensin I convertase (ACE), along with a strong antioxidant activity (Mechri et al. 2020). Since AchE enzyme inhibition is considered as a treatment option for Alzheimer's disease, this shrimp waste hydrolysate a high potential (Kabir et al. 2019; Saxena and Dubey 2019). The ACE inhibitors are used for treating hypertension, especially in diabetic patients, myocardial infarction, chronic kidney disease, glomerular dysfunction etc. (Herman et al. 2020). The ACE inhibitory activity of the shrimp hydrolysate (IC50 $71.52 \pm 1.48 \mu g \ mL^{-1}$) was higher than the clinically used ACE inhibitor captopril (IC50 $85.33 \pm 1.26 \mu g \ mL^{-1}$) (Mechri et al. 2020).

29.4 Conclusion

Unutilized fish and shellfish wastes represent the loss of valuable proteins, peptides, fat, minerals, and other micronutrients. Together with the environmental concerns and the high costs associated with fish waste disposal, the need to explore novel ways of utilizing fish wastes for human and animal use is envisaged. The existing knowledge on the potential biomedical applications of fish biowastes needs to be translated into the production of high economic value products. The establishment of commercial-scale extraction methods for useful compounds from fish waste and their conversion into beneficial by-products is necessary to attract fisheries entrepreneurs' interests to consider this area as an investment opportunity. Continued research efforts to diversify waste utilization and its valorization into high commercial value products are critically needed to realize the future goal of sustainable fisheries.

References

Abdollahi M, Rezaei M, Jafarpour A, Undeland I (2018) Sequential extraction of gel-forming proteins, collagen and collagen hydrolysate from gutted silver carp (*Hypophthalmichthys molitrix*), a biorefinery approach. Food Chem 242:568–578

Aditya G, Hossain A (2018) Valorization of aquaculture waste in removal of cadmium from aqueous solution: optimization by kinetics and ANN analysis. Appl Water Sci 8:68. https://doi.org/10.1007/s13201-018-0712-z

Aissaoui N, Chobert J-M, Haertlé T et al (2017) Purification and biochemical characterization of a neutral serine protease from *Trichoderma harzianum*. Use in antibacterial peptide production from a fish by-product hydrolysate. Appl Biochem Biotechnol 182:831–845. https://doi.org/10.1007/s12010-016-2365-4

Alvares TS, Conte-Junior CA, Pierucci AP et al (2018) Acute effect of fish protein hydrolysate supplementation on vascular function in healthy individuals. J Funct Foods 46:250–255. https://doi.org/10.1016/j.jff.2018.04.066

Alves AL, Marques ALP, Martins E et al (2017) Cosmetic potential of marine fish skin collagen. Cosmetics 4:39. https://doi.org/10.3390/cosmetics4040039

Anh PT, Dieu TTM, Mol AP et al (2011) Towards eco-agro industrial clusters in aquatic production: the case of shrimp processing industry in Vietnam. J Clean Prod 19:2107–2118

Antunes-Valcareggi SA, Ferreira SRS, Hense H (2017) Enzymatic hydrolysis of blue crab (*Callinectes sapidus*) waste processing to obtain chitin, protein, and astaxanthin-enriched extract. Int J Environ Agric Res 3:81–92

Aranday-García R, Román Guerrero A, Ifuku S, Shirai K (2017) Successive inoculation of *Lactobacillus brevis* and *Rhizopus oligosporus* on shrimp wastes for recovery of chitin and added-value products. Process Biochem 58:17–24. https://doi.org/10.1016/j.procbio.2017.04.036

Arbia W, Arbia L, Adour L et al (2019) Characterization by spectrometric methods of chitin produced from white shrimp shells of *Parapenaeus longirostris* by *Lactobacillus helveticus* cultivated on glucose or date waste. Algerian J Environ Sci Technol 5:933–955

Arola D, Murcia S, Stossel M et al (2018) The limiting layer of fish scales: structure and properties. Acta Biomater 67:319–330

Arvanitoyannis IS, Kassaveti A (2008) Fish industry waste: treatments, environmental impacts, current and potential uses. Int J Food Sci Technol 43:726–745

Basu B, Das M, Banik AK (2005) Biodegradation of fish scale by *Aspergillus niger*: an enzymatic and scanning electron microscopic study. J Food Sci Technol 42:387–391

Beaney P, Lizardi-Mendoza J, Healy M (2005) Comparison of chitins produced by chemical and bioprocessing methods. J Chem Technol Biotechnol 80:145–150. https://doi.org/10.1002/jctb.1164

Bhagwat PK, Dandge PB (2016) Isolation, characterization and valorizable applications of fish scale collagen in food and agriculture industries. Biocatal Agric Biotechnol 7:234–240

Bjørndal B, Berge C, Ramsvik MS, Svardal A, Bohov P, Skorve J, Berge RK (2013) A fish protein hydrolysate alters fatty acid composition in liver and adipose tissue and increases plasma carnitine levels in a mouse model of chronic inflammation. Lipids Health Dis 12:143

Blanco M, Vázquez JA, Pérez-Martín RI, Sotelo CG (2017) Hydrolysates of fish skin collagen: an opportunity for valorizing fish industry by-products. Mar Drugs 15:131. https://doi.org/10.3390/md15050131

Bordenave S, Fruitier I, Ballandier I et al (2002) HPLC preparation of fish waste hydrolysate fractions. Effect on guinea pig ileum and ACE activity. Prep Biochem Biotechnol 32:65–77. https://doi.org/10.1081/PB-120013162

Borges-Contreras B, Martínez-Sánchez CE, Herman-Lara E et al (2019) Angiotensin-converting enzyme inhibition in vitro by protein hydrolysates and peptide fractions from mojarra of nile tilapia (*Oreochromis niloticus*) skeleton. J Med Food 22:286–293. https://doi.org/10.1089/jmf.2018.0163

Bougatef A, Nedjar-Arroume N, Ravallec-Plé R et al (2008) Angiotensin I-converting enzyme (ACE) inhibitory activities of sardinelle (*Sardinella aurita*) by-products protein hydrolysates obtained by treatment with microbial and visceral fish serine proteases. Food Chem 111:350–356. https://doi.org/10.1016/j.foodchem.2008.03.074

Brown GA, Wellings SR (1968) Collagen formation and calcification in teleost scales. Z Für Zellforsch Mikrosk Anat 93:571–582

Bruno SF, Kudre TG, Bhaskar N (2019) Effects of different pretreatments and proteases on recovery, umami taste compound contents and antioxidant potentials of *Labeo rohita* head protein hydrolysates. J Food Sci Technol 56:1966–1977. https://doi.org/10.1007/s13197-019-03663-3

Cardozo FA, Gonzalez JM, Feitosa VA et al (2017) Bioconversion of α-chitin into N-acetyl-glucosamine using chitinases produced by marine-derived *Aeromonas caviae* isolates. World J Microbiol Biotechnol 33:201. https://doi.org/10.1007/s11274-017-2373-8

Castro R, Guerrero-Legarreta I, Bórquez R (2018) Chitin extraction from *Allopetrolisthes punctatus* crab using lactic fermentation. Biotechnol Rep 20:e00287. https://doi.org/10.1016/j.btre.2018.e00287

Chai HJ, Li JH, Huang HN, Li TL, Chan YL, Shiau CY, Wu CJ (2010) Effects of sizes and conformations of fish-scale collagen peptides on facial skin qualities and transdermal penetration efficiency. J Biomed Biotechnol 2010:757301

Chakravarty J, Yang C-L, Palmer J, Brigham CJ (2018) Chitin extraction from lobster shell waste using microbial culture-based methods. Appl Food Biotechnol 5:141–154. https://doi.org/10.22037/afb.v5i3.20787

Chen S, Chen H, Xie Q et al (2016) Rapid isolation of high purity pepsin-soluble type I collagen from scales of red drum fish (*Sciaenops ocellatus*). Food Hydrocoll 52:468–477

Chen X, Jiang Q, Xu Y, Xia W (2017) Recovery of chitin from antarctic krill (*Euphausia superba*) shell waste by microbial deproteinization and demineralization. J Aquat Food Prod Technol 26:1210–1220. https://doi.org/10.1080/10498850.2015.1094686

Chinh NT, Manh VQ, Trung VQ et al (2019) Characterization of collagen derived from tropical freshwater carp fish scale wastes and its amino acid sequence. Nat Prod Commun 14:1934578X19866288. https://doi.org/10.1177/1934578X19866288

Crini G (2019) Historical review on chitin and chitosan biopolymers. Environ Chem Lett 17:1623–1643. https://doi.org/10.1007/s10311-019-00901-0

Dale HF, Madsen L, Lied GA (2019) Fish-derived proteins and their potential to improve human health. Nutr Rev 77(8):572–583. https://doi.org/10.1093/nutrit/nuz016

Deng J-J, Mao H-H, Fang W et al (2020) Enzymatic conversion and recovery of protein, chitin, and astaxanthin from shrimp shell waste. J Clean Prod 271:122655. https://doi.org/10.1016/j.jclepro.2020.122655

Dhanabalan V, Balange AK, Nayak BB et al (2020) Effect of proteolytic enzymes on the extent of deproteinisation of *Acetes* shell residues. J Exp Zool India 23:947–951

Ding D, Yu T, Du B, Huang Y (2019) Collagen hydrolysate from *Thunnus orientalis* bone induces osteoblast proliferation and differentiation. Chem Eng Sci 205:143–150

Ding H, Lv L, Wang Z, Liu L (2020) Study on the "glutamic acid-enzymolysis" process for extracting chitin from crab shell waste and its by-product recovery. Appl Biochem Biotechnol 190:1074–1091. https://doi.org/10.1007/s12010-019-03139-2

Duan R, Zhang J, Du X et al (2009) Properties of collagen from skin, scale and bone of carp (*Cyprinus carpio*). Food Chem 112:702–706

Dun Y, Li Y, Xu J et al (2019) Simultaneous fermentation and hydrolysis to extract chitin from crayfish shell waste. Int J Biol Macromol 123:420–426. https://doi.org/10.1016/j.ijbiomac.2018.11.088

Elavarasan K, Kumar VN, Shamasundar BA (2014) Antioxidant and functional properties of fish protein hydrolysates from fresh water carp(*Catla catla*) as influenced by the nature of enzyme. J Food Process Preserv 38:1207–1214. https://doi.org/10.1111/jfpp.12081

Elavarasan K, Shamasundar BA, Badii F, Howell N (2016) Angiotensin I-converting enzyme (ACE) inhibitory activity and structural properties of oven- and freeze-dried protein hydrolysate from fresh water fish (*Cirrhinus mrigala*). Food Chem 206:210–216. https://doi.org/10.1016/j.foodchem.2016.03.047

Fahmi A, Morimura S, Guo H-C et al (2004) Production of angiotensin I converting enzyme inhibitory peptides from sea bream scales. Process Biochem 39:1195–1200

FAO (2018) The State of World Fisheries and Aquaculture 2018: meeting the sustainable development goals. FAO, Rome

Gadgey KK, Bahekar A (2017) Studies on extraction methods of chitin from crab shell and investigation of its mechanical properties. Int J Mech EngTechnol 8:220–231

Gajanan PG, Elavarasan K, Shamasundar BA (2016) Bioactive and functional properties of protein hydrolysates from fish frame processing waste using plant proteases. Environ Sci Pollut Res 23:24901–24911. https://doi.org/10.1007/s11356-016-7618-9

Gao S, Hong H, Zhang C et al (2019) Immunomodulatory effects of collagen hydrolysates from yak (*Bos grunniens*) bone on cyclophosphamide-induced immunosuppression in BALB/c mice. J Funct Foods 60:103420

Gao P, Li L, Xia W et al (2020) Valorization of Nile tilapia (*Oreochromis niloticus*) fish head for a novel fish sauce by fermentation with selected lactic acid bacteria. LWT 129:109539. https://doi.org/10.1016/j.lwt.2020.109539

Garrano AMC, La Rosa G, Zhang D et al (2012) On the mechanical behavior of scales from *Cyprinus carpio*. J Mech Behav Biomed Mater 7:17–29

Ghorbel-Bellaaj O, Hmidet N, Jellouli K et al (2011) Shrimp waste fermentation with *Pseudomonas aeruginosa* A2: optimization of chitin extraction conditions through Plackett–Burman and response surface methodology approaches. Int J Biol Macromol 48:596–602. https://doi.org/10.1016/j.ijbiomac.2011.01.024

Ghosh PR, Fawcett D, Sharma SB, Poinern GEJ (2016) Progress towards sustainable utilisation and management of food wastes in the global economy. Int J Food Sci 2016:3563478

Gil-Duran S, Arola D, Ossa EA (2016) Effect of chemical composition and microstructure on the mechanical behavior of fish scales from *Megalops Atlanticus*. J Mech Behav Biomed Mater 56:134–145

Gopakumar K (2002) Textbook of fish processing technology. Indian Council of Agricultural Research, New Delhi

Gu R-Z, Li C-Y, Liu W-Y et al (2011) Angiotensin I-converting enzyme inhibitory activity of low-molecular-weight peptides from Atlantic salmon (*Salmo salar* L.) skin. Food Res Int 44:1536–1540

Guo N, Sun J, Zhang Z, Mao X (2019) Recovery of chitin and protein from shrimp head waste by endogenous enzyme autolysis and fermentation. J Ocean Univ China 18:719–726. https://doi.org/10.1007/s11802-019-3867-9

Hamdi M, Hammami A, Hajji S et al (2017) Chitin extraction from blue crab (*Portunus segnis*) and shrimp (*Penaeus kerathurus*) shells using digestive alkaline proteases from *P. segnis* viscera. Int J Biol Macromol 101:455–463. https://doi.org/10.1016/j.ijbiomac.2017.02.103

Hammami A, Hamdi M, Abdelhedi O et al (2017) Surfactant- and oxidant-stable alkaline proteases from *Bacillus invictae*: characterization and potential applications in chitin extraction and as a detergent additive. Int J Biol Macromol 96:272–281. https://doi.org/10.1016/j.ijbiomac.2016.12.035

Harikrishna N, Mahalakshmi S, Kumar KK, Reddy G (2017) Fish Scales as potential substrate for production of alkaline protease and amino acid rich aqua hydrolyzate by *Bacillus altitudinis* GVC11. Indian J Microbiol 57:339–343

Harrington DJ (1996) Bacterial collagenases and collagen-degrading enzymes and their potential role in human disease. Infect Immun 64(6):1885–1891

Hassan MA, Xavier KAM, Gupta S et al (2019) Antioxidant properties and instrumental quality characteristics of spray dried *Pangasius* visceral protein hydrolysate prepared by chemical and enzymatic methods. Environ Sci Pollut Res 26:8875–8884. https://doi.org/10.1007/s11356-019-04144-y

Hema GS, Shyni K, Mathew S et al (2013) A simple method for isolation of fish skin collagen—biochemical characterization of skin collagen extracted from albacore tuna (*Thunnus alalunga*), dog shark (*Scoliodon sorrakowah*) and rohu (*Labeo rohita*). Ann Biol Res 4:271–278.

Herman LL, Padala SA, Annamaraju P, Bashir K (2020) Angiotensin converting enzyme inhibitors (ACEI). In: StatPearls [Internet]. StatPearls Publishing, Treasure Island, FL

Hongkulsup C, Khutoryanskiy VV, Niranjan K (2016) Enzyme assisted extraction of chitin from shrimp shells (*Litopenaeus vannamei*). J Chem Technol Biotechnol 91:1250–1256. https://doi.org/10.1002/jctb.4714

Hosseini SF, Ramezanzade L, Nikkhah M (2017) Nano-liposomal entrapment of bioactive peptidic fraction from fish gelatin hydrolysate. Int J Biol Macromol 105:1455–1463. https://doi.org/10.1016/j.ijbiomac.2017.05.141

Hou H, Li B, Zhang Z et al (2012) Moisture absorption and retention properties, and activity in alleviating skin photo damage of collagen polypeptide from marine fish skin. Food Chem 135:1432–1439. https://doi.org/10.1016/j.foodchem.2012.06.009

Huang C-Y, Wu C-H, Yang J-I et al (2015) Evaluation of iron-binding activity of collagen peptides prepared from the scales of four cultivated fishes in Taiwan. J Food Drug Anal 23:671–678

Huang C-Y, Tsai Y-H, Hong Y-H et al (2018) Characterization and antioxidant and angiotensin i-converting enzyme (ACE)-inhibitory activities of gelatin hydrolysates prepared from extrusion-pretreated milkfish (*Chanos chanos*) Scale. Mar Drugs 16:346. https://doi.org/10.3390/md16100346

Ismail SA, Emran MA (2020) Direct microbial production of prebiotic and antioxidant chitin-oligosaccharides from shrimp by-products. Egypt J Aquat Biol Fish 24:181–195. https://doi.org/10.21608/ejabf.2020.98021

Jung W-K, Mendis E, Je J-Y et al (2006) Angiotensin I-converting enzyme inhibitory peptide from yellow fin sole (*Limanda aspera*) frame protein and its antihypertensive effect in spontaneously hypertensive rats. Food Chem 94:26–32. https://doi.org/10.1016/j.foodchem.2004.09.048

Jung WJ, Jo GH, Kuk JH et al (2007) Production of chitin from red crab shell waste by successive fermentation with *Lactobacillus paracasei* KCTC-3074 and *Serratia marcescens* FS-3. Carbohydr Polym 68:746–750. https://doi.org/10.1016/j.carbpol.2006.08.011

Kabir MT, Uddin M, Begum M et al (2019) Cholinesterase inhibitors for Alzheimer's disease: multitargeting strategy based on anti-Alzheimer's drugs repositioning. Curr Pharm Des 25:3519–3535

Kaczmarek MB, Struszczyk-Swita K, Li X et al (2019) Enzymatic modifications of chitin, chitosan, and chito oligosaccharides. Front Bioeng Biotechnol 7:243. https://doi.org/10.3389/fbioe.2019.00243

Kandyliari A, Golla JP, Chen Y et al (2020) Antiproliferative activity of protein hydrolysates derived from fish by-products on human colon and breast cancer cells. Proc Nutr Soc 79(OCE2): E282. https://doi.org/10.1017/S002966512000230X

Karnjanapratum S, O'Callaghan YC, Benjakul S, O'Brien N (2016) Antioxidant, immunomodulatory and antiproliferative effects of gelatin hydrolysate from unicorn leatherjacket skin. J Sci Food Agric 96:3220–3226. https://doi.org/10.1002/jsfa.7504

Kaya M, Mujtaba M, Ehrlich H et al (2017) On chemistry of γ-chitin. Carbohydr Polym 176:177–186. https://doi.org/10.1016/j.carbpol.2017.08.076

Kharyeki EM, Rezaei M, Khodabandeh S, Motamedzadegan A (2018) Antioxidant activity of protein hydrolysate in skipjack tuna head. Fish Sci Technol 7:57–64

Kimura S, Ohno Y, Miyauchi Y, Uchida N (1987) Fish skin type I collagen: wide distribution of an α3 subunit in teleosts. Comp Biochem Physiol Pt B Comp Biochem 88:27–34. https://doi.org/10.1016/0305-0491(87)90074-5

Kurita K (2006) Chitin and chitosan: functional biopolymers from marine crustaceans. Marine Biotechnol 8:203. https://doi.org/10.1007/s10126-005-0097-5

Li S-W, He H, Zeng RJ, Sheng G-P (2017) Chitin degradation and electricity generation by *Aeromonas hydrophila* in microbial fuel cells. Chemosphere 168:293–299. https://doi.org/10.1016/j.chemosphere.2016.10.080

Li J, Huang W-C, Gao L et al (2019) Efficient enzymatic hydrolysis of ionic liquid pretreated chitin and its dissolution mechanism. Carbohydr Polym 211:329–335. https://doi.org/10.1016/j.carbpol.2019.02.027

Ling Y, Liping S, Yongliang Z (2018) Preparation and identification of novel inhibitory angiotensin-I-converting enzyme peptides from tilapia skin gelatin hydrolysates: inhibition kinetics and molecular docking. Food Funct 9:5251–5259. https://doi.org/10.1039/C8FO00569A

Liu W, Li G, Miao Y, Wu X (2009) Preparation and characterization of pepsin-solubilized type I collagen from the scales of snakehead (*Ophiocephalus argus*). J Food Biochem 33:20–37

Liu Y, Xing R, Yang H et al (2020) Chitin extraction from shrimp (*Litopenaeus vannamei*) shells by successive two-step fermentation with *Lactobacillus rhamnoides* and *Bacillus amyloliquefaciens*. Int J Biol Macromol 148:424–433. https://doi.org/10.1016/j.ijbiomac.2020.01.124

Ma X, Gözaydın G, Yang H et al (2020) Upcycling chitin-containing waste into organonitrogen chemicals *via* an integrated process. Proc Natl Acad Sci 117:7719–7728. https://doi.org/10.1073/pnas.1919862117

Manni L, Ghorbel-Bellaaj O, Jellouli K et al (2010) Extraction and characterization of chitin, chitosan, and protein hydrolysates prepared from shrimp waste by treatment with crude protease from *Bacillus cereus* SV1. Appl Biochem Biotechnol 162:345–357. https://doi.org/10.1007/s12010-009-8846-y

Mao X, Liu P, He S et al (2013) Antioxidant properties of bio-active substances from shrimp head fermented by *Bacillus licheniformis* OPL-007. Appl Biochem Biotechnol 171:1240–1252. https://doi.org/10.1007/s12010-013-0217-z

Maruyama S, Nonaka I, Tanaka H (1993) Inhibitory effects of enzymatic hydrolysates of collagen and collagen-related synthetic peptides on fibrinogen/thrombin clotting. Biochim Biophys Acta Protein Struct Mol Enzymol 1164:215–218

Marzieh M-N, Zahra F, Tahereh E, Sara K-N (2019) Comparison of the physicochemical and structural characteristics of enzymatic produced chitin and commercial chitin. Int J Biol Macromol 139:270–276. https://doi.org/10.1016/j.ijbiomac.2019.07.217

Mechri S, Bouacem K, Jabeur F et al (2019) Purification and biochemical characterization of a novel thermostable and halotolerant subtilisin SAPN, a serine protease from *Melghiribacillus thermohalophilus* Nari2AT for chitin extraction from crab and shrimp shell by-products. Extremophiles 23:529–547. https://doi.org/10.1007/s00792-019-01105-8

Mechri S, Sellem I, Bouacem K et al (2020) Valorization of shrimp *Metapenaeus monoceros* bio-waste as a source of bioactive protein hydrolysate. In: MOL2NET, International Conference

Series on Multidisciplinary Sciences. UEA, Puyo, Ecuador. http://sciforum.net/conference/mol2net-06

Mendis E, Rajapakse N, Kim S-K (2005) Antioxidant properties of a radical-scavenging peptide purified from enzymatically prepared fish skin gelatin hydrolysate. J Agric Food Chem 53:581–587. https://doi.org/10.1021/jf048877v

Mhamdi S, Ktari N, Hajji S et al (2017) Alkaline proteases from a newly isolated *Micromonospora chaiyaphumensis* S103: characterization and application as a detergent additive and for chitin extraction from shrimp shell waste. Int J Biol Macromol 94:415–422. https://doi.org/10.1016/j.ijbiomac.2016.10.036

Moussian B (2019) Chitin: structure, chemistry and biology. In: Yang Q, Fukamizo T (eds) Targeting chitin-containing organisms. Springer, Singapore, pp 5–18

Murthy LN, Phadke GG, Unnikrishnan P et al (2018) Valorization of fish viscera for crude proteases production and its use in bioactive protein hydrolysate preparation. Waste Biomass Valor 9:1735–1746. https://doi.org/10.1007/s12649-017-9962-5

Nagai T, Suzuki N (2000) Isolation of collagen from fish waste material—skin, bone and fins. Food Chem 68:277–281. https://doi.org/10.1016/S0308-8146(99)00188-0

Nagai T, Izumi M, Ishii M (2004) Fish scale collagen. Preparation and partial characterization. Int J Food Sci Technol 39:239–244

Nandy V, Bakshi M, Ghosh S et al (2014) Potentiality assessment of fish scale biodegradation using mangrove fungi isolated from Indian Sundarban. Int Lett Nat Sci 14:68–76

Navarrete-Bolaños JL, González-Torres I, Vargas-Bermúdez VH, Jiménez-Islas H (2020) A biotechnological insight to recycle waste: analyzing the spontaneous fermentation of shrimp waste to design the hydrolysis process of chitin into n-acetylglucosamine. Rev Mex Ing Quím 19:263–274. https://doi.org/10.24275/rmiq/Bio544

Neves AC, Harnedy PA, O'Keeffe MB, FitzGerald RJ (2017a) Bioactive peptides from Atlantic salmon (*Salmo salar*) with angiotensin converting enzyme and dipeptidyl peptidase IV inhibitory, and antioxidant activities. Food Chem 218:396–405. https://doi.org/10.1016/j.foodchem.2016.09.053

Neves AC, Zanette C, Grade ST et al (2017b) Optimization of lactic fermentation for extraction of chitin from freshwater shrimp waste. Acta Sci Technol 39:125–133

Nguyen VB, Nguyen DN, Wang S-L (2020) Microbial reclamation of chitin and protein-containing marine by-products for the production of prodigiosin and the evaluation of its bioactivities. Polymers 12:1328. https://doi.org/10.3390/polym12061328

Nisha S, Seenivasan A, Vasanth D (2016) Chitin and its derivatives: structure, production, and their applications. In: International conference on signal processing, communication, power and embedded system (SCOPES), Paralakhemundi, Odisha, India, pp 127–130

Offengenden M, Chakrabarti S, Wu J (2018) Chicken collagen hydrolysates differentially mediate anti-inflammatory activity and type I collagen synthesis on human dermal fibroblasts. Food Sci Human Wellness 7:138–147

Ooi J, Lee LY, Hiew BYZ et al (2017) Assessment of fish scales waste as a low cost and eco-friendly adsorbent for removal of an azo dye: equilibrium, kinetic and thermodynamic studies. Bioresour Technol 245:656–664

Pal GK, Suresh PV (2017) Physico-chemical characteristics and fibril-forming capacity of carp swim bladder collagens and exploration of their potential bioactive peptides by in silico approaches. Int J Biol Macromol 101:304–313

Pallela R, Venkatesan J, Janapala VR, Kim SK (2012) Biophysicochemical evaluation of chitosan-hydroxyapatite-marine sponge collagen composite for bone tissue engineering. J Biomed Mater Res A 100:486–495

Pattanaik SS, Sawant PB, Xavier KAM et al (2020) Characterization of carotenoprotein from different shrimp shell waste for possible use as supplementary nutritive feed ingredient in animal diets. Aquaculture 515:734594. https://doi.org/10.1016/j.aquaculture.2019.734594

Ploydee E, Chaiyanan S (2014) Production of high viscosity chitosan from biologically purified chitin isolated by microbial fermentation and deproteinization. Int J Polym Sci 2014:162173. https://www.hindawi.com/journals/ijps/2014/162173/. Accessed 6 Sep 2020

Rafael O-HD, Fernándo Z-GL, Abraham P-T et al (2019) Production of chitosan-oligosaccharides by the chitin-hydrolytic system of *Trichoderma harzianum* and their antimicrobial and anticancer effects. Carbohydr Res 486:107836. https://doi.org/10.1016/j.carres.2019.107836

Rasli HI, Sarbon NM (2019) Preparation and physicochemical characterization of fish skin gelatin hydrolysate from shortfin scad (*Decapterus macrosoma*). Int Food Res J 26:287–294

Reyhani PS, Jafarpour A (2017) Effects of degree of hydrolysis on functional properties and antioxidants activity of hydrolysate from head and frame of common carp (*Cyprinus carpio*) fish. Iran J Food Sci Technol 14:113–124

Sae-leaw T, O'Callaghan YC, Benjakul S, O'Brien NM (2016) Antioxidant, immunomodulatory and antiproliferative effects of gelatin hydrolysates from seabass (*Lates calcarifer*) skins. Int J Food Sci Technol 51:1545–1551. https://doi.org/10.1111/ijfs.13123

Sankar S, Sekar S, Mohan R et al (2008) Preparation and partial characterization of collagen sheet from fish (*Lates calcarifer*) scales. Int J Biol Macromol 42:6–9. https://doi.org/10.1016/j.ijbiomac.2007.08.003

Saxena M, Dubey R (2019) Target enzyme in Alzheimer's disease: Acetylcholinesterase inhibitors. Curr Top Med Chem 19:264–275

Sedaghat F, Yousefzadi M, Toiserkani H, Najafipour S (2017) Bioconversion of shrimp waste *Penaeus merguiensis* using lactic acid fermentation: an alternative procedure for chemical extraction of chitin and chitosan. Int J Biol Macromol 104:883–888. https://doi.org/10.1016/j.ijbiomac.2017.06.099

Shigemura Y, Kubomura D, Sato Y, Sato K (2014) Dose-dependent changes in the levels of free and peptide forms of hydroxyproline in human plasma after collagen hydrolysate ingestion. Food Chem 159:328–332. https://doi.org/10.1016/j.foodchem.2014.02.091

Stepnowski P, Ólafsson G, Helgason H, Jastorff B (2004) Recovery of astaxanthin from seafood wastewater utilizing fish scales waste. Chemosphere 54:413–417. https://doi.org/10.1016/S0045-6535(03)00718-5

Sujithra S, Kiruthiga N, Prabhu MJ, Kumeresan R (2013) Isolation and determination of type I collagen from tilapia (*Oreochromis niloticus*) waste. Int J Eng Technol 5:5

Sungperm P, Khongla C, Yongsawatdigul J (2020) Physicochemical properties and angiotensin i converting enzyme inhibitory peptides of freshwater fish skin collagens. J Aquat Food Prod Technol 29:650–660. https://doi.org/10.1080/10498850.2020.1788683

Taga Y, Kusubata M, Ogawa-Goto K et al (2018) Collagen-derived X-Hyp-Gly-type tripeptides promote differentiation of MC3T3-E1 pre-osteoblasts. J Funct Foods 46:456–462. https://doi.org/10.1016/j.jff.2018.05.017

Tan JS, Sahar A, Lee CK, Phapugrangkul P (2020a) Chitin extraction from shrimp wastes by single step fermentation with *Lactobacillus acidophilus* FTDC3871 using response surface methodology. J Food Process Preserv 44:e14895. https://doi.org/10.1111/jfpp.14895

Tan YN, Lee PP, Chen WN (2020b) Microbial extraction of chitin from seafood waste using sugars derived from fruit waste-stream. AMB Express 10:17. https://doi.org/10.1186/s13568-020-0954-7

Tanganini IC, Shirahigue LD, Altenhofen da Silva M et al (2020) Bioprocessing of shrimp wastes to obtain chitosan and its antimicrobial potential in the context of ethanolic fermentation against bacterial contamination. 3 Biotech 10:135. https://doi.org/10.1007/s13205-020-2128-3

Taokaew S, Zhang X, Chuenkaek T, Kobayashi T (2020) Chitin from fermentative extraction of crab shells using okara as a nutrient source and comparative analysis of structural differences from chemically extracted chitin. Biochem Eng J 159:107588. https://doi.org/10.1016/j.bej.2020.107588

Trang HTH, Pasuwan P (2018) Screening antimicrobial activity against pathogens from protein hydrolysate of rice bran and Nile Tilapia by-products. Int Food Res J 25(5):2157–2163

Vannabun A, Ketnawa S, Phongthai S et al (2014) Characterization of acid and alkaline proteases from viscera of farmed giant catfish. Food Biosci 6:9–16. https://doi.org/10.1016/j.fbio.2014.01.001

Veeruraj A, Arumugam M, Balasubramanian T (2013) Isolation and characterization of thermostable collagen from the marine eel-fish (*Evenchelys macrura*). Process Biochem 48:1592–1602. https://doi.org/10.1016/j.procbio.2013.07.011

Wang B, Wang Y-M, Chi C-F et al (2013) Isolation and characterization of collagen and antioxidant collagen peptides from scales of croceine croaker (*Pseudosciaena crocea*). Mar Drugs 11:4641–4661. https://doi.org/10.3390/md11114641

Wang JK, Yeo KP, Chun YY et al (2017) Fish scale-derived collagen patch promotes growth of blood and lymphatic vessels in vivo. Acta Biomater 63:246–260. https://doi.org/10.1016/j.actbio.2017.09.001

Wang X, Hong H, Wu J (2019) Hen collagen hydrolysate alleviates UVA-induced damage in human dermal fibroblasts. J Funct Foods 63:103574. https://doi.org/10.1016/j.jff.2019.103574

Wergedahl H, Liaset B, Gudbrandsen OA, Lied E, Espe M, Muna Z, Mørk S, Berge RK (2004) Fish protein hydrolysate reduces plasma total cholesterol, increases the proportion of HDL cholesterol, and lowers acyl-CoA:cholesterol acyltransferase activity in liver of Zucker rats. J Nutr 134:1320–1327

Wisuthiphaet N, Klinchan S, Kongruang S (2016) Fish protein hydrolysate production by acid and enzymatic hydrolysis. Appl Sci Eng Prog 9:261–270

Wu R, Wu C, Liu D et al (2018) Antioxidant and anti-freezing peptides from salmon collagen hydrolysate prepared by bacterial extracellular protease. Food Chem 248:346–352. https://doi.org/10.1016/j.foodchem.2017.12.035

Xu W, Mohan A, Pitts NL et al (2020) Bile acid-binding capacity of lobster shell-derived chitin, chitosan and chitooligosaccharides. Food Biosci 33:100476. https://doi.org/10.1016/j.fbio.2019.100476

Yan Q, Fong SS (2018) Design and modularized optimization of one-step production of N-acetylneuraminic acid from chitin in *Serratia marcescens*. Biotechnol Bioeng 115:2255–2267. https://doi.org/10.1002/bit.26782

Yang P, Jiang Y, Hong P, Cao W (2013) Angiotensin I converting enzyme inhibitory activity and antihypertensive effect in spontaneously hypertensive rats of cobia (*Rachycentron canadum*) head papain hydrolysate. Food Sci Technol Int 19:209–215. https://doi.org/10.1177/1082013212442196

Yang X, Xiao X, Liu D et al (2017) Optimization of collagenase production by *Pseudoalteromonas* sp. SJN2 and application of collagenases in the preparation of antioxidative hydrolysates. Mar Drugs 15:377. https://doi.org/10.3390/md15120377

Yata M, Yoshida C, Fujisawa S et al (2001) Identification and characterization of molecular species of collagen in fish skin. J Food Sci 66:247–251. https://doi.org/10.1111/j.1365-2621.2001.tb11325.x

Yonis RW, Luti KJ, Aziz GM (2019) Statistical optimization of chitin bioconversion to produce an effective chitosan in solid state fermentation by *Asperigellus flavus*. Iraqi J Agric Sci 50 (3):916–927

Younes I, Rinaudo M (2015) Chitin and chitosan preparation from marine sources. structure, properties and applications. Mar Drugs 13:1133–1174. https://doi.org/10.3390/md13031133

Younes I, Hajji S, Frachet V et al (2014) Chitin extraction from shrimp shell using enzymatic treatment. Antitumor, antioxidant and antimicrobial activities of chitosan. Int J Biol Macromol 69:489–498. https://doi.org/10.1016/j.ijbiomac.2014.06.013

Zacherl C, Eisner P, Engel K-H (2011) In vitro model to correlate viscosity and bile acid-binding capacity of digested water-soluble and insoluble dietary fibres. Food Chem 126:423–428. https://doi.org/10.1016/j.foodchem.2010.10.113

Zamorano-Apodaca JC, García-Sifuentes CO, Carvajal-Millán E et al (2020) Biological and functional properties of peptide fractions obtained from collagen hydrolysate derived from mixed by-products of different fish species. Food Chem 331:127350. https://doi.org/10.1016/j.foodchem.2020.127350

Zhang D, Wang PG, Qi Q (2007) A two-step fermentation process for efficient production of penta-N-acetyl-chitopentaose in recombinant *Escherichia coli*. Biotechnol Lett 29:1729–1733. https://doi.org/10.1007/s10529-007-9462-y

Zhang J, Duan R, Tian Y, Konno K (2009) Characterisation of acid-soluble collagen from skin of silver carp (*Hypophthalmichthys molitrix*). Food Chem 116:318–322. https://doi.org/10.1016/j.foodchem.2009.02.053

Zhang Z, Wang J, Ding Y, Dai X, Li Y (2011) Oral administration of marine collagen peptides from Chum Salmon skin enhances cutaneous wound healing and angiogenesis in rats. J Sci Food Agric 91:2173–2179

Zhang H, Jin Y, Deng Y et al (2012) Production of chitin from shrimp shell powders using *Serratia marcescens* B742 and *Lactobacillus plantarum* ATCC 8014 successive two-step fermentation. Carbohydr Res 362:13–20. https://doi.org/10.1016/j.carres.2012.09.011

Zhang H, Yang S, Fang J et al (2014) Optimization of the fermentation conditions of *Rhizopus japonicus* M193 for the production of chitin deacetylase and chitosan. Carbohydr Polym 101:57–67. https://doi.org/10.1016/j.carbpol.2013.09.015

Zhang Y-Z, Ran L-Y, Li C-Y, Chen X-L (2015) Diversity, structures, and collagen-degrading mechanisms of bacterial collagenolytic proteases. Appl Environ Microbiol 81:6098–6107. https://doi.org/10.1128/AEM.00883-15

Zhang H, Yu H, Qian Y, Chen S (2017a) Production of chitin and chitosan using successive three-step microbial fermentation. J Polym Mater 34(1):123–127

Zhang H, Yun S, Song L et al (2017b) The preparation and characterization of chitin and chitosan under large-scale submerged fermentation level using shrimp by-products as substrate. Int J Biol Macromol 96:334–339

Zhang A, Wei G, Mo X et al (2018) Enzymatic hydrolysis of chitin pretreated by bacterial fermentation to obtain pure N-acetyl-D-glucosamine. Green Chem 20:2320–2327. https://doi.org/10.1039/C8GC00265G

Zhang L, Zhao G-X, Zhao Y-Q et al (2019) Identification and active evaluation of antioxidant peptides from protein hydrolysates of skipjack tuna (*Katsuwonus pelamis*) Head. Antioxidants 8:318. https://doi.org/10.3390/antiox8080318

Zhu D, Ortega CF, Motamedi R et al (2012) Structure and mechanical performance of a "modern" fish scale. Adv Eng Mater 14:B185–B194. https://doi.org/10.1002/adem.201180057

Zylberberg L, Nicolas G (1982) Ultrastructure of scales in a teleost (*Carassius auratus* L.) after use of rapid freeze-fixation and freeze-substitution. Cell Tissue Res 223:349–367. https://doi.org/10.1007/BF01258495

Correction to: Advances in Fisheries Biotechnology

Pramod Kumar Pandey and Janmejay Parhi

Correction to:
P. K. Pandey, J. Parhi (eds.), *Advances in Fisheries Biotechnology*,
https://doi.org/10.1007/978-981-16-3215-0

The original version of this book was inadvertently published with the incorrect affiliation (Organizational Division) as "Reproduction" for authors Dr. Janmejay Parhi Dr. Ananya Khatei and Dr. Dibyajyoti Sahoo in the Front Matter and Chapters 2, 5 and 11. It has been updated as "Fish Genetics and Reproduction".

The updated version of these chapters can be found at
https://doi.org/10.1007/978-981-16-3215-0
https://doi.org/10.1007/978-981-16-3215-0_2
https://doi.org/10.1007/978-981-16-3215-0_5
https://doi.org/10.1007/978-981-16-3215-0_11

© The Author(s), under exclusive license to Springer Nature Singapore Pte Ltd. 2022
P. K. Pandey, J. Parhi (eds.), *Advances in Fisheries Biotechnology*,
https://doi.org/10.1007/978-981-16-3215-0_30

Lightning Source UK Ltd.
Milton Keynes UK
UKHW022105160223
417164UK00012B/111

9 789811 632174